DOUBLE STARS, PHYSICAL PROPERTIES AND GENERIC RELATIONS

Double Stars, Physical Properties and Generic Relations

Proceedings of IAU Colloquium No. 80 held at Lembang, Java,
3–7 June 1983

Edited by

BAMBANG HIDAYAT
Lembang, Java

ZDENĚK KOPAL
Manchester, U.K.

and

JÜRGEN RAHE
Bamberg, F.R.G.

Reprinted from

Astrophysics and Space Science, Vol. 99, Nos. 1/2

D. Reidel Publishing Company

Dordrecht / Boston

ISBN-13: 978-94-009-6374-0   e-ISBN-13: 978-94-009-6372-6
DOI: 10.1007/ 978-94-009-6372-6

TABLE OF CONTENTS

(Double Stars, Physical Properties and Generic Relations)

v

INTRODUCTION

The Bosscha Observatory in Lembang, Java, Indonesia, celebrated in 1983 its 60th anniversary. Since its foundation, the physical properties of binary systems have formed a major research topic of this observatory. Until 1970, the study of visual binaries and the determination of orbits received most emphasis. Since then, also the evolution of close binary systems, such as X-ray binaries, Wolf–Rayet binaries and binary pulsars, has been researched with priority in Lembang. It seemed thus appropriate that a Colloquium devoted to the study of binary systems be held in Lembang at the time of the Observatory's anniversary.

In the Colloquium, the role of wide double (and multiple) systems received special emphasis – not only because of the long tradition of visual binary research at Lembang; but also because their role in documenting stellar evolution has been largely overlooked in recent decades, and needs to be brought into focus with the information forthcoming from close binaries.

The Colloquium covered the physical properties of visual as well as close binary systems, and their generic relations, in the broadest possible sense. It was sponsored by the International Astronomical Union as IAU Colloquium No. 80 ('Double Stars, Physical Properties and Generic Relations').

After the official opening ceremony, the meeting started with a discussion on the future of astronomy in Asia. The scientific sessions began with the 'V. Bappu Memorial Lecture on the Evolution of Binary Systems', presented by Z. Kopal. Subsequent sessions were divided into four following parts: (1) Physical properties of double and multiple stars; occurrence of binaries in different stellar systems; new observational methods; (2) Wide systems and their evolution; (3) Close binaries; new observations and their interpretation; close binary evolution; (4) Generic relations between wide and close systems.

The Scientific Organizing Committee consisted of K. D. Abhyankar, A. H. Batten, J. Dommanget, O. G. Franz, E. P. J. van den Heuvel, K. A. van der Hucht, J. Ibrahim, Y. Kondo, Z. Kopal, D. C. Morton, J. Rahe (chairman), D. Sugimoto, W. Sutantyo, A. R. Upgren, and B. Warner. Chairman of the Local Organizing Committee was B. Hidayat. The meeting took place in Bandung, Java, from June 3 to 7, 1983 and was attended by about 60 astronomers from 16 different countries.

Shortly after the Colloquium, a solar eclipse of several minutes' length occurred on Java, and astronomers attending the meeting had the chance to experience this 'once in a lifetime' astronomical event.

Most memorable of IAU Colloquium No. 80 are the many interesting presentations and stimulating – often spirited – discussions, the generous hospitality of the Bosscha Observatory, and – last but not least – the observation of the solar eclipse in the fascinating Javanese surroundings.

BAMBANG HIDAYAT, ZDENĚK KOPAL, and JÜRGEN RAHE

Astrophysics and Space Science **99** (1984) 1.

EVOLUTION OF BINARY SYSTEMS AND THEIR GENERIC RELATIONS

Vainu Bappu Memorial Lecture*

ZDENĚK KOPAL

Department of Astronomy, University of Manchester, England

(Received 9 August, 1983)

Prologue: I rise to address you today with sadness in my heart – sadness which I trust is shared by all those here present; for as is underlined by its subtitle, our beloved colleague Vainu Bappu – past President of the International Astronomical Union, and life-long student of the subject of our Colloquium – is no longer in this world to take his rightful place among us; and, therefore, my remarks which follow can be dedicated only to his memory.

I feel sad all the more as Vainu Bappu was also once (albeit for a rather short time) a student of mine at Harvard University; and the course he took with me on eclipsing binaries in the spring term of 1948 may have been the beginning of his life-long interest in binary systems, and in problems arising from their existence. Fate separated us by thousands of miles for most part of our subsequent lives; but brought us together in the end again: namely, in the spring of 1982 when Vainu Bappu invited me to give a course of lectures at the Indian Institute for Astrophysics founded by him at Bangalore. Their main topic was to be the Fourier analysis of the light changes of eclipsing variables, and to last several weeks; but the course soon overflowed its originally-intended scope into more general problems connected with the evolution of double and multiple systems of stars.

It was a real pleasure to see Vainu in my audience once more after one-third of a century – looking not a day older than I remembered him from our Harvard days – with his keen interest in the subject undiminished by time; and his brilliant remarks were enjoyed by the audience as much as by myself. But all good things have got to have an end; and this end came when Vainu and other friends took me to the Bangalore airport on the day of my departure. The connecting flight to Bombay was late; but astronomers never waste their time: we immediately started to discuss the problems of evolution of the binary systems, and continued until my flight was called for boarding about an hour later. Of course we did not solve any; but our minds were full of them as we shook hands at the gate, and made another date to discuss them in August at Patras, Greece. Alas, it was not to be; for not many days after the XVIIIth General Assembly of the I.A.U. (over which Vainu was to preside) started, grievous news came from Munich of his untimely passing. Requiescat in pace!

When, therefore, the Organizing Committee of this Colloquium honoured me with an invitation to deliver the Bappu Memorial Lecture at Bandung, which more fitting subject could I choose for it than that we discussed at Bangalore during our last hour together in this world? And so it will be – in the hope that especially the astronomers of the younger generation – both those here present as well as those who may read what I have to say – will be able to advance our subject beyond the limits to which its milestones were carried by astronomers of my age. Not that they will all solve them; this is too much to hope; but step-by-step advances must continue relentlessly, as time goes on, to make us understand better the phenomena which we observe in the sky.

Some 30–40 yr ago – when the world was recuperating from its most recent holocaust, and the problems of the evolution of double and multiple stars began to claim serious attention – the situation facing us could well be described by the words of the Bible: "the harvest is large, but workers are few". Now – thirty years later – this is no longer the case. The number of astronomers interested in these problems has increased

---

* Delivered on 3 June, 1983 at the Lembang-Bamberg IAU Colloquium No. 80 on 'Double Stars: Physical Properties and Generic Relations', held at Bandung, Indonesia.

Astrophysics and Space Science **99** (1984) 3–21. 0004–640X/84/0991–0003$02.85.

by at least one order of magnitude; and brought in its wake such a flood of 'scenarios' aiming to account for what we observe that a casual perusal of contemporary literature on the subject is more likely to be-wilder, rather than enlighten, the student entering this field at the present time. This fact alone reveals – if anything – that many of the problems at issue do not admit as yet of unique solutions; and that the outcome of our attempts at their solutions depends, do not so much on given observational constraints, as on *ad hoc* assumptions introduced to render the problem tractable, or determinate at all. What may, therefore, be a better service to Bappu's memory than to stress what we do *not* know as yet, or what remains still uncertain, rather than to indulge in further proliferation of hypothetical scenarios and pass them off as gospel truth. We must, above all, allow ourselves to be guided by the observations; and this is what I shall endeavour to do in this lecture.

1. Introduction

Before we come to grips with individual aspects of our problem, it befits to define first what we mean by double or multiple systems of the stars. In what follows, we shall consider as such the associations of stars which mutual attraction compels to revolve around the common centre of gravity for a time span that is long in comparison with orbital periods of such configurations. For most systems which we shall have an opportunity to recall in this lecture, this disparity will amount to many orders of magnitude – a fact which will make their gravitational liaison indissoluble – till 'death does them part' in (say) the holocaust of a supernova explosion or by other means of comparable violence.

Our current knowledge of the frequency-distribution of double stars in mutual separation is so far but very incomplete – largely due to observational selection which hampers discovery of pairs within certain ranges of separation, and favours others. The upper limit is set by properties of the fluctuating gravity field in which a binary pair happens to be situated, and which tends to dissolve it (cf. Chandrasekhar, 1944); while the lower limit is given by the dimensions of the constituent stars. In that part of the galactic spiral arm which happens to be our celestial home, this upper limit comes close to half a parsec (cf. Kopal, 1978; p. 10), corresponding to orbital periods of the order of 10^8 yr; while, at the lower limit, the orbital dimensions may amount to 10^4 km (corresponding to the dimensions of white dwarfs), and periods of revolution to only minutes of our time. Moreover, a discovery of pairs still smaller in size – and revolving in seconds rather than minutes – can be expected with confidence in the future.

All such objects constitute a huge reservoir of binary configurations in their own right; and represent probably the major part of stellar population of our Galaxy. If their separations are large, we refer to them as 'wide' (which, in our proximity in space, can manifest themselves as 'visual') binaries; while if this separation becomes comparable with the dimensions of the constituent stars (or does not exceed them by more than one order of magnitude), we speak of 'close' binaries – requiring completely different (spectroscopic, photometric) methods of discovery.

In more recent times, the custom has begun to take root to refer to the latter as 'interacting' binaries; but to me this term does not seem to offer any advantage. For – by definition – the components of *all* binaries are bound to interact; gravitationally alone if they are wide; and gravitationally as well as hydrodynamically (or hydromagnetically) if they are close. To refer only to the latter as 'interacting' is, therefore, illogical – merely calling the same thing by a new and longer name – and as long as no better nomenclature cna be proposed, in what follows we shall continue to refer to the two groups of binaries as 'wide' and 'close', respectively – with the understanding that the components of wide systems interact only gravitationally, whereas in close systems they can interact also hydrodynamically, or even in a more complicated manner. Both belong to the topic of our discussion, and their generic relations should be of equal interest to us.

The *evolution* of binary or multiple systems – be these close or wide – commenced to emerge as one of the central problems of contemporary astrophysics since the latter 1940's, when the general framework of stellar evolution was being placed on a sound physical basis. Much of it was, to be sure, foreshadowed by the earlier work of Eddington and his contemporaries in the first half of this century; but it was not till the work of Bethe and others that the evolution of matter under conditions prevalent in stellar interiors could be described in terms of exothermic nuclear reactions. In particular, it became possible then for the first time to relate the rate of energy production of the stars with their mass (and chemical composition) in a quantitative manner.

As long as a star is single, its mass, chemical composition, and age remain independent parameters which cannot be uniquely deduced from the observations. However, the double (and multiple) star systems – be these close or wide – constitute an extreme type of stellar associations, which remain inseparable for time-intervals exceeding the age of our Galaxy (cf. Chandrasekhar, 1944), and must have originated from pre-existing gaseous substrate so well-mixed (by turbulence) that the chemical composition of the material constituting their components may initially have been indistinguishable. Morever, their formation must have occurred at (very approximately) the same time; so that their present aages must likewise be essentially identical*. Such stars could, therefore, have differed only in their initial mass; and if so, the evolutionary tracks of the components could subsequently begin to differentiate only on account of this fact: the larger the mass, the faster should be the rate of nuclear evolution – with all consequences which this may entail.

When we turn to confront this simple consequence of the theory of nuclear evolution with what we actually observe, we find that the observations verify these theoretical expectations (within the limits of observational errors) *as long as both components remain on the Main Sequence*. In such a case, the more massive component invariably turns out to be the larger and the hotter of the two; and if their mass-ratio is very close to one, the components remain virtual twins.

This, however, continues to be the case only as along as both stars happen to be on the Main Sequence and derive their energy output from a conversion of internal

---

* For massive stars (with, say, $m \gg 3 \odot$) the rate of Kelvin contraction towards the Main Sequence is such that the individual components should have ignited their hydrogen within less than 10^6 yr of each other. In systems with one component very much less massive (say, of the order of $1 \odot$) than its mate, the time interval between their respective births could amount to 10^7 yr or more – but still very short in comparison with the total span of their subsequent evolution.

hydrogen into helium. Once, however, at least one component of the binary pair has departed from the Main Sequence, the theoretical amulet appears suddenly to have lost its charm; for it is no longer the more massive component which continues to lead on the evolutionary track of the pair, but, instead, the lead has passed on to the less massive one – in flagrant contradiction to theoretical expectations. This, moreover, appears to be true of all types of binary systems *regardless of proximity of their components* – visual systems like Sirius or Procyon (in which a typical Main-Sequence star is attended by less massive white dwarfs) are even more pronounced examples of such a situation than (say) Algol or other similar close binaries, in which this phenomenon first happened to attract attention. This perplexing fact earned for itself the name of an *evolutionary paradox*, which began to stare us in the face since about the middle of this century, and has continued to do so ever since. In what follows we wish to describe the present state of this problem, and attempt to foresee the way in which its solution should be sought.

2. Evolutionary Paradox

The first step towards an identification of the cause of the paradox outlined in the preceding paragraphs may appear to be simple, and ascribable to a breakdown of our tacit assumption that the stars evolve along the tracks of constant mass. Indeed, a study of the physical properties of double stars discloses these to be compatible with the observations only as long as their components remain on the Main Sequence; but not necessarily beyond the hydrogen-burning stage. Certainly the existence of such close pairs as Algol – on which a subgiant of gK0 spectrum is 4.7 times less massive than the principal Main-Sequence component of spectrum B8 – or a wide binary like Sirius – in which a Main-Sequence A0-star is attended by a white dwarf 2.3 times less massive – forces us to recognize the fact that this could not be true unless the present secondary (i.e., less massive) components of such systems were once more massive of the two; and attained their present state only after *losing* a large part of their initial mass some time after an incipient shortage of hydrogen forced them to abandon the Main Sequence. Thus far a general agreement exists that (short of abandoning the basic tenets of nuclear evolution of the stars – and this should be considered only as our last resort) the components of binary systems must lose a large part of their initial masses – to enable them eventually to satisfy the Chandrasekhar limit and become white dwarfs.

The question is only: when does it happen and why? Is, moreover, the binary nature pre-requisite for such an act, or is this bound to happen to every star which has reached the necessary stage – even though the phenomenon may become observable only in binary systems? It is on these questions that opinions still differ, and the final answer is not yet in sight; for this answer is intimately connected with the *physics* of the processes causing the loss of mass of the stars, and not – strictly speaking – with the astronomy of binary systems.

Let us attempt to explain why this is the case – and this part of my talk could almost be given a Shakespearian title of 'Comedy full of Errors'. It is as though Nature – that greatest of teasers – has dangled before our eyes many misleading clues to test our intelligence, and to watch with interest how long it may take us to disentangle them.

This story commenced with the work of Schoenberg and Chandrasekhar (1942), demonstrating that when hydrogen exhaustion in the central parts of a star drops below a certain limit (about 12% by mass), its core begins to shrink; and a conservation of the potential energy of the configuration as a whole then requires that this shrinkage of the core of a star must be compensated by an expansion of its outer parts (the 'mirror effect') causing the star to grow in size.

As far as single stars are concerned (and it is these alone that Schoenberg and Chandrasekhar had in mind) this argument remains unanswerable. But not necessarily so in close binary systems; for there it is the total energy of the system as a whole which should be conserved; and dissipative processes (like dynamical tides) allow for an exchange of potential energy between components and the kinetic energy of the system. Both of these processes operate on the Kelvin time-scale; and the efficiency of exchange depends on the viscosity (plasma, or turbulent) of stellar matter. In the absence of its more accurate knowledge, it is impossible to estimate the extent to which the Schoenberg–Chandrasekhar 'mirror effect' should be applicable to the components of binary systems with any assurance; but, for the sake of subsequent discussion, let us assume that this is indeed the case.

If the stars were single (or the component of a wide binary – like Sirius or Procyon), the post-Main Sequence expansion caused by the 'mirror effect' could continue unchecked as long as a shrinkage of the core keeps providing the surplus potential energy for expansion of the outer layers. However, in close binaries, the proximity of the companion will surround the expanding star with an invisible barrier – in the form of the *Roche limit*, defined as the largest *closed* equipotential volume capable of containing the star's mass. If and when a given star has reached this limit (due to the operation of the 'mirror-effect'), its further growth in size may get *arrested*; and a continuing tendency to expand could bring about an actual loss of mass.

That this may indeed be the case was supported by an independent discovery by Crawford (1955) and the present speaker (Kopal, 1954, 1955) that, in a whole group of close binaries (including Algol), the secondary (less massive) component just about fills in its Roche limit while the primary's mass remains well inside this limit. I bestowed on these the name of 'semi-detached' systems (Kopal, 1955); and their existence has become since one of the cornerstones of the modern double-star astronomy, whose physical implications will be discussed below. For the moment, I should like to stress a frequently-overlooked fact that it is virtually impossible to prove whether or not any one particular system is actually semi-detached. In principle, this can be done by a comparison of the fractional dimensions of the 'contact' component (obtainable, for eclipsing variables, from an analysis of their light curves) with the fractional dimensions of its corresponding Roche lobe (obtainable from the spectroscopically-determined mass ratio). Both these quantities are, however, known to us only within certain limits of observational errors; and their coincidence renders a contact nature of the respective star only the more probable, the smaller the range of the respective errors. If, however, the fractional dimensions of the components tend – as they do – to coincide with those of their Roche limits for a whole group of systems independently observed, the probability that such coincidences are not accidental becomes so strengthened as to render the existence of such 'contact' stars tantamount to an observed fact. The question is only about its meaning; but once we raise it, we find ourselves at once in deep waters.

In looking back at the interpretation of this fact when its was discovered 30 yr ago, we cannot but confess that these early attempts amounted to but little more than jumping to conclusions in naive belief that Nature should be comprehensible in the light of such knowledge as we possessed at that time. Alas, clouds soon began to gather in the sky over such a presumption; and Nature soon gave us a salutary lesson for our impatience and loose thinking. In what way did we deserve it?

3. Mass Transfer

Shortly after the existence of contact components was discovered in close binary systems, and coupled with the expected effects of developing hydrogen shortage, Hoyle (1955) put forward an attractive hypothesis of 'mass transfer' between components of such systems, which certainly did not lack merit and can be summarized as follows.

When the hydrogen abundance in the deep interior of a Main-Sequence star drops below 12% by weight, its core begins to shrink and outer layers expand towards the Roche limit; having attained it, the latter begins to 'leak' through the Lagrangian point L_1 (at which the effective gravity vanishes and the equilibrium becomes neutral) to 'overflow' on to the secondary (up to that time, less massive) component, and augment its mass to the extent to which the erstwhile primary may become the secondary star of smaller mass, but evolved from the Main Sequence*. By a further extension of the same argument, Hoyle conjectured ... "a possibility that the predatory star will be forced to make amends for its former behaviour by returning material to the (at present) fainter star, which it robbed of mass so unfeelingly in the past. In the interest of cosmic justice it is to be hoped that this happens; but whether it does or not is unsure" (cf. Hoyle, *op cit.*; p. 200). And – we may add – the doubts on whether or not this entire process is actually operative (or, at least, responsible for a complete explanation of our evolutionary paradox) still continue to be with us almost 30 yr later; for the following reasons.

First, as had been pointed out at the very beginning of our subject (Kopal, 1954, 1955; Crawford, 1955), in every single system known at that time, it was *the less massive component which appeared to fill its Roche limit*; while the fractional dimensions of its more massive mate remained well interior to this limit. The question why we do not observe systems at the immediately preceding stage, in which the (originally) more massive star begins to expand towards its Roche limit and start disgorging mass which

---

* Can – to leave no stone unturned – the secondary components in semi-detached systems still be in the pre-Main-Sequence stage of Kelvin contraction? Scarcely so; for (quite apart from difficulties with the time-scale), no cause is known (cf. Kopal, 1954; p. 685) why their contraction should be arrested at the Roche limit.

would reverse the role of the two components, may have been at least partly answered by Morton (1960) and Smak (1962), who pointed out that the stellar evolution, which could take the system through that stage, unrolls on the Kelvin time-scale – with sufficient rapidity for few if any such stars to be 'caught in the act' of exchanging their roles at any particular time. A fuller discussion of our problem at this stage can be found, e.g., in Plavec (1968) or Paczynski (1971), and need not be repeated in this place.

Such views could have been seriously considered at the time when they were first put forward; for the sample of data then available was limited. In the 24 yr which have elapsed since, the sample of known semi-detached systems has at least trebled – and still no case of transitional stage was caught in the net of our observations – a fact which would require this act to occur all the more rapidly to escape detection. It is this fact which led a predominant majority of theoretical investigators of the respective stage of stellar evolution to choose only *massive* binary systems for their studies – in which the evolution (on the Kelvin as well as nuclear time-scale) can proceed indeed at a sufficiently fast rate. However, such a strategy ignores the fact that systems so massive are *very rare* per unit volume of galactic space (though not so rare in our catalogues of bright stars, as observational selection favours their discovery). As is well known, a large majority of stars in our Galaxy – in fact, some nine-tenths of them – possess masses smaller than that of the Sun; and (as far as we know) a high percentage of such systems form likewise binary systems. Your present speaker pointed out a number of them which appear to have reached semi-detached state (cf. Kopal, 1971) in spite of the smallness of their mass – systems in which the Kelvin time-scale may be longer than that of nuclear time-scale of more massive stars – but their existence has been greeted by most protagonists of the 'mass-exchange' scheme only with an embarrassed silence.

The main weakness of all schemes postulating a mere exchange of mass between the components – such that the total mass of the system remains conserved – is, however, an inadequate physical basis of the processes by which this exchange is to take place. The conventional view that the reason of the mass loss from contact configurations is low gravity prevalent there – so that any hypothetical transfer requires but a minimum amount of energy to make it operative – lands us on the horns of a dilemma. For a low-velocity mass transfer is to be accomplished fast enough to escape detection, the density of mass being so transferred must be very high to accomplish the purpose; and a flux so dense would have to absorb light effectively enough to deform the observed light curves of close eclipsing binaries to an extent which has not been verified by the observations. The same amount of mass can, of course, be transferred by a star at lower densities if the material moves faster; but then not all of it is likely to be acquired by its mate, and some can escape from the system – thus violating the assumption that the total mass of the system should remain constant.

The actual means of mass transfer by gas streams from one star to another will be considered in more detail in the next section, in the light of the constraints imposed upon it by the observations. In doing so we shall find that the principal weakness of most schemes of this type proposed so far was the *assumption* that *mass is only exchanged between components, but none is lost to the system*. From the physical point of view, an initial assumption of this kind is certainly unwarranted; for, if true, it should rather follow as a consequence if the investigation leads to such a conclusion; but to assume it in advance represents, in effect, an undue interference with the physical basis of the problem. It does possess, however, one merit; and this is why it had been so widely adopted in the past: namely, it simplifies computations performed on so restricted a basis.

The question can, of course, be then asked: what meaning can the results of such computations possess, and how legitimate is it to compare their outcome with observations? The reason for such doubts will transpire more fully in subsequent parts of my address; and may explain why we conjectured that William Shakespeare could have been tempted to entitle the contents of this section as a 'Comedy full of Errors'. Indeed, he may have gone further and called it 'Much Ado About Nothing' if the observed facts at the basis of our discussion would not demand explanation. And if, perchance, some readers may have found our comments on the contemporary scene too frivolous, they can only turn some pages more in Shapespeare's *Collected Works* to re-name it: 'As You Like It'.

4. Gas Streams

To paraphrase a witty remark of Lippmann to Poincaré (which concerned the exponential law of error distribution; cf. Poincaré, 1896), "everybody believes in the gas-streams in close binary systems: the observers, because they think that the existence of such streams can be proved by mathematicians; and the mathematicians, because they believe that such streams have been established by the observations". It is certainly true that the ideas exposed in the preceding section would not have been received with such a ready ear by so many investigators in the past, had it not been for the fact that they appeared to derive support from certain observed facts – mainly spectroscopic – which antedated the emergence of our 'evolutionary paradox' and seemed to offer an easy way out of our difficulties: namely, that several close binaries exhibited lines in their spectra whose Doppler shifts did not correspond to the orbital motion of their components, but were seriously at variance with them. The first example of such lines – its so-called B5 lines – in the spectrum of β Lyrae were discovered by Bĕlopolsky before the end of the last century (cf. Bĕlopolsky, 1893, 1897); and, somewhat later, recalcitrant metallic lines (mainly of Fe and Mg) whose Doppler shifts did not follow the orbital motion of either component were discovered also in Algol (cf. Barney, 1923). More recently, Carpenter (1930) found that the hydrogen lines in the spectrum of U Cephei – a well-known eclipsing system of virtually circular orbit – exhibited Doppler shifts indicative of highly asymmetric radial-velocity curve of its A0 (later reclassified to B8) component – simulating, in fact, a spurious eccentricity e as large as 0.47! This was certainly a anomalous phenomenon, to which astronomers of that time (led by Henry Norris Russell) preferred to adopt an ostrich attitude – until such a posture was made untenable by the pioneer work of Otto Struve and his school between 1940–1950. For Struve not only fully confirmed the genuine nature of Carpenter's earlier results for U Cephei (cf. Struve, 1944), but detected at least another pair – namely, RZ Scuti (cf. Neubauer and Struve, 1945) where a similar effect was even more conspicuous.

As is well known, Struve and his followers (for a summary of their views, cf., e.g., Struve, 1950) ascribed the origin of the anomalous observed Doppler shifts to *gas streams* within the respective systems; and their views – graphically illustrated by the familiar 'elephant trunks', stretching from the conical point of the contact component to its more massive mate – became for many years almost a trademark for this line of thought. And when, in the mid-1950's we came to face our 'evolutionary paradox', it was only too tempting to identify such streams with the mechanism of mass-transfer between the components described in the preceding section. Yet – as it happens so often in the history of science – a little more patience in the interpretation of the observational evidence could have held us back from premature jumping to conclusions, and made us think whether or not such a view can be justified also on other physical grounds.

In an attempt to answer this question, let us return to the gist of the argument advanced in the preceding section. As the star expands at a sufficiently slow rate, the value of the potential of all forces acting upon the surface remains constant, but its gradient (i.e., the gravitational acceleration) does not. When the star has eventually reached its Roche limit, the surface potential attains a minimum value it can possess for any closed configuration; while the gravitational acceleration – varying over the surface and diminishing in the direction of its mate – actually vanishes at the conical point (identical with the Lagrangian point L_1). This fact, by itself, need not cause any mass to escape; for its equilibrium there is merely neutral, and the Roche limit represents only a *static* configuration. A smallness of gravity in the neighbourhood of L_1) should only make it *easier* for *small* perturbations to remove mass from there than from any other part of the star's surface.

Such considerations prompted in the past several investigators (Kopal, 1956, 1957, 1959; Gould, 1957, 1959; Plavec and Kříž, 1965; and others) to consider a hypothetical outflow of mass from contact configurations as a problem of *particle mechanics*, within the framework of the restricted problem of three bodies. In retrospect, however, all this mechanical approach was doomed to constitute scarcely more than a numerical exercise, with little or no relevance to the physics of our underlying problem (except, perhaps, that the periodic orbits of this type may represent limiting cases of steady-state hydrodynamic flow, obtaining when its density is allowed to approach zero). The reason why this should be so is the fact that an appeal to the restricted problem of three bodies could be physically justified only *if the mean free path of the particles ejected by the expanding star are long in comparison with the scale of their motions;* so that the mutual *collisions* of moving particles (i.e., a *pressure* generated by them) can be *disregarded.*

Can this, however, be true of gas particles (atoms, or ions) which can leave a trace in the observed spectra of binary systems? The equivalent widths of anomalous line profiles in the spectra of U Cep (or RZ Sct) can, in principle, disclose the number of atoms (or ions) along the line of sight capable of producing the observed effects in the spectra. For U Cephei, the deformed lines are essentially those of hydrogen; and absorption properties of hydrogen are well known. In making use of them, Batten (1974) found that the number of hydrogen particles necessary to account for the observed spectroscopic anomalies should be between 10^{12}–10^{13} atoms per cc. However, at such

densities *the mean free path of the respective particles would be many orders of magnitude
smaller than the dimensions of the respective flow* – i.e., in the 'elephant trunk' of mass
transfer – a fact which discloses that pressure within it cannot be disregarded; and that,
in the absence of any (unknown) force to contain the material in transit to its gravitational
pipelines, the respective gas should *disperse* into space before reaching its desired
destination.

Byt worse is yet in store for those who may wish to save the simple interpretations
at all cost. For spectroscopic anomalies observed by Struve and others in the Balmer
lines of hydrogen can, of course, be caused only by *neutral* hydrogen present along the
line of sight; but how much of the *total* hydrogen abundance can remain there in neutral
state? The answer must be sought in Strömgren's theory of H II regions, developed and
applied extensively to gaseous nebulae and interstellar matter (for their latest presentations, cf., e.g., Osterbrock, 1974; or Spitzer, 1978). A direct transfer of the results
presented in these books, and obtained for (say) planetary nebulae to binary systems
is not possible, because of a great difference in certain parameters involved in such
problems: in planetary nebulae the dimensions are large but densities low (and also the
exciting stars hotter); while for circumstellar gas in close binary systems the opposite
is the case; and these differences may amount to many orders of magnitude.

Only the first steps to investigate the physical properties prevalent in Strömgren zones
of close binaries have so far been made (cf. Kopal, 1981). The results indicate, however,
that the bulk of available hydrogen in (say) U Cephei can remain in neutral state only
in the atmosphere of its primary component of B8 spectrum, where ample supply of free
electrons (provided by ionization of light metals) makes hydrogen recombination
practically instantaneous. However, between the two components, a diminishing density
(including that of the electrons) will keep an increasing fraction of hydrogen ionized –
so that the total amount of hydrogen present in between the stars (let alone outside the
system) should be very much *larger* than that of neutral hydrogen alone. The mean free
path of its constituents (different as it may be for neutral or charged particles) amounts,
in turn, to so tiny a fraction of the flow scale that the absurdity of treating dynamical
phenomena in such a medium in terms of particle mechanics – in which collisions are
ignored and mean free path considered infinite – is glaringly evident. If so, it follows
that the gas envelopes capable of impressing observable features in the composite
spectra of such systems are no mere 'exospheres' whose particles can move in ballistic
trajectories, but genuine 'extended atmospheres'. Therefore it is hydrodynamics – rather
than particle mechanics – to which we should appeal in our efforts to place a study of
gas motions in close binary systems on an adequate physical basis.

That this must be so was realized already several years ago; and many investigations
attempted to carry out such a programme (cf. Biermann, 1971; Prendergast and Taam,
1974; Sorensen *et al.*, 1975; Lubow and Shu, 1975; Budding, 1981; and others). If, in
spite of protracted efforts, the progress in this field has been slow, this is due to inherent
difficulties – yet to be overcome before any meaningful comparisons between theory and
observations can be attempted.

Let us mention at least a few – if alone as tasks which we must eventually face and accomplish. First, spectroscopic observations indicate that (at least, the neutral component of) gas streams in close binary systems move with velocities of the order of 100 km s^{-1}; while the velocity of sound in hydrogen gas at a temperature of 10^4 deg (a typical ionization temperature, indicated by the spectra) should be close to 12 km s^{-1}. If so, it follows that the gas motions in question should be *hypersonic*, and be characterized by Mach numbers of the order of 10 (or more). This fact rules out, however, any possibility of *linearization* of hydrodynamic motions (inherent in most investigations quoted above); for if we did so, we would rule out possible formation of *shock waves*, which may indeed arise and play an important role in the spectra – especially the bow-shocks!

Secondly, we mentioned already that a considerable fraction of circumstellar gas in close binary systems may be *ionized*; and if so, the *viscosity* μ of the plasma (mainly hydrogen) component should be by several orders of magnitude higher than if the same gas were neutral (cf. Chapman, 1954; or Oster, 1957). As, moreover, the density ρ of the medium in question is probably quite low (10^{13} protons per cc would correspond to a plasma density of only 10^{-11} g cm^{-3}) its *kinematic viscosity* μ/ρ may become enormous; and the terms factored by it may dominate the respective equations of motion.

Add to this the fact that the Reynolds numbers of the flow prove to be so large that such motions are probably *turbulent* as well (and characterized by high turbulent viscosity into the bargain); and we find ourselves confronted with hypersonic flow in viscous turbulent media – about the worst accumulation of attributes one can ascribe to any flow! This flow should now be theoretically treated in three-dimensional space; with time constituting the fourth independent variable (which can be dispensed with only if the flow is steady). No wonder that not much headway has been made with the solution of the equations of motion (subject to appropriate boundary conditions) governing such flows!

Fortunately, this may not be really necessary for the main objective of our inquiry, which is to ascertain the cause of the 'evolutionary paradox' described in the preceding sections; and account for the ways in which evolving components of close binaries can dispense with their mass. The causes of such processes are probably *internal* to each star; and the mechanism of ejection, the 'stellar winds'. To this aspect of our problem we propose to turn in the next section; and to conclude the present we shall return to spectroscopic anomalies concerning asymmetric radial-velocity curves, mentioned already earlier in this section, as observed, e.g., by Carpenter or Struve in the system of U Cephei. For these investigators, the source of such anomalies was to be sought in gas streams between the stars. But whatever their motions, hydrogen in them should be largely ionized and incapable of absorbing in Balmer lines. To localize their origin, we should seek to identify regions where hydrogen can remain predominantly in neutral state; and this is, of course, the atmosphere of the star itself.

As is well known, the observed radial velocities of the stars are measured from the Doppler shifts of spectral lines formed in the atmospheres of the stars in question; and the use of such shifts as indicators of orbital motions entails a tacit assumption that *the atmospheric layers in which the measured spectral lines originate are at rest with respect to the star's centre of mass.* This assumption underlies indeed all work on the radial velocities of the stars carried out in the past; but regardless of what may be true if the star is single, can this continue to be the case in binary systems in which their components *irradiate* each other from close proximity?

As is well known, such an irradiation produces *heating* by many hundreds of degrees, which is bound to stir *thermal convection* in the respective atmosphere – steady-state gas motion if axial rotation of the irradiated star is synchronized with orbital motion of the illuminating source, but a non-steady one if this source rises and sets over the illuminated stars on account of asynchronism between rotation and revolution. It is interesting to recall, in this connection, that the principal components of both U Cep and RZ Sct – exhibiting strongly skew-symmetric radial-velocity curves in spite of their circular orbits – rotate much *faster* than they revolve (cf. Struve, 1949); and, therefore, experience conspicuous 'day-and-night' variations of temperature over their surfaces.

The effects of such an irradiation on the buoyancy of atmospheric gas, and the motions arising therefrom, cannot – we repeat – be ignored; the question remains yet only as to their efficiency. Can these represent the main, or contributory, cause of anomalous Doppler shifts exhibited by U Cephei, RZ Scuti and many other similar systems? In spite of some recent work by Kirbiyik and Smith (1976) or Kopal (1980), the quantitative answer is not yet known; but at least a possibility exists that this may be indeed the case.

5. Mass Loss

The principal objections to the mass-exchange scheme as outlined earlier in Section 3 are twofold: namely, the fact that the 'evolutionary paradox' raised in Section 2 is encountered, not only in close binary systems of sufficient mass for the effects of differential evolution to become noticeable in 10^7–10^8 yr; but in all types of binary system regardless of their mass or proximity (i.e., whether these are close or wide). Algol, as well as Sirius, appear to be equally prone to become the victims of it – possessing, as they do, evolved components of mass distinctly smaller than that of their Main-Sequence mates. A good part of the paradoxical nature of such a situation goes back, to be sure, to the restriction that *the systems in question evolve on constant mass;* and that the role of this evolution is limited to a mere *exchange* of mass overflowing from one star to another. We stressed, however, already that such a restriction is wholly arbitrary, and not required by the observations: in fact, the opposite is the case; and the aim of the present section will be to explain why in more detail.

In order to do so, let us first turn our attention to the observed predominance of contact components of low mass, and a virtual absence of systems in which the more massive is being caught in the process of expansion towards its Roche limits. This is equally true of massive systems (whose rate of evolution is sufficiently high to make such a scheme at least a theoretical possibility) as well as of binaries whose total mass is of the order of only one solar mass – in which the Kelvin time-scale of evolution may become much longer than that of nuclear evolution of more massive stars, and thus increase the probability of catching the low-mass systems in the elusive act of mass exchange; and observing in 'slow motion' what happens rapidly in those of greater mass. None were caught in it so far – regardless of whether their mass is large or small; and this is only bound to strenghten our previous doubts on whether such a metamorphosis of the components as envisaged by the mass-exchange scheme takes place at all!

Such doubts are further aggravated by the fact that – as is well known – a star of one solar mass requires not less than 10^{10} yr to exhaust enough hydrogen in its interior to evolve from the Main Sequence. And yet we know eclipsing systems – like V471 Tauri in the Hyades cluster – which (like its parent cluster) is no more than 600 million years old; and within this time the less massive component of this system of total mass of 1.4 ⊙ has already become a white dwarf!

This case (and other similar ones which could be adduced) represent a veritable *reductio ad absurdum* of the process of mass exchange between components which should keep the total mass of the system constant; and the final verdict on such a possibility cannot be too long delayed. But if further arguments are needed, let us recall that the same 'evolutionary paradox' (for the solution of which the whole process of mass-exchange was originally invented) exhibited by Algol-like close binary systems is likewise encountered – and in even more extreme form – in wide systems of the Sirius or Procyon type, in which a typical Main-Sequence star (of luminosity class V) is attended by a white dwarf. And these are no isolated examples; according to a recent catalogue by Agayev *et al.* (1982), over 60 visual binaries in our proximity are known to possess white-dwarfs as less massive components; and no doubt many more will be discovered in the future.

Consider, in particular, once more the system of Sirius whose physical properties are well known. At the present time its white dwarf component represents only 30% of the total mass of this system; 70% being stored in the Main-Sequence A0 star of luminosity class V. These two stars must have formed a dynamical partnership throughout all their past; and their ages be closely the same. If, however, the secondary component could have attained a white-dwarf stage while the primary still lingers on the Main Sequence, the conclusion is inevitable that the present Sirius B must have once been the more massive component of the two – which embarked on its evolution with a good deal more than the present mass of 2.3 ⊙ possessed by Sirius A – but *lost* most part of it at some stage of its subsequent evolution while the present A0 star still continued to linger on the Main Sequence. Indeed, this must have happened if Sirius B managed to pass the Chandrasekhar limit to become a white dwarf.

What had happened to this missing mass? In the case of Sirius (and many other such systems) the velocity of escape from the gravitational field of an evolving star is but marginally different from one which would enable this mass to escape altogether from the system; and the range of velocities for which mass particles could escape from the star but are gravitationally constrained to remain a part of the system (and thus provide material for potential accretion by its mate) is so narrow as to make any transfer of mass in this manner extremely inefficient. In other words, most part of the mass lost by the future white dwarf must have been lost to the system as a whole, and expelled into interstellar space.

What kind of a physical process could have led to such an expulsion? By 1955 – when the 'evolutionary paradox' began to be considered a significant scientific problem – little was known about the ways in which stars can divest themselves of mass: accretion, rather than loss, was then the main object of attention. Since the advent of the space-age in 1957, and a gradual opening of ultraviolet spectra of the stars, the situation began to change drastically. In the years past we learned much about 'stellar winds' and the sources of energy which produce and sustain them. The observations in recent years with the International Ultraviolet Explorer telescope led to a discovery that even seemingly 'normal' stars can expel matter through their coronae with a flux entailing a mass loss equivalent to 10^{-6} to $10^{-5} \odot$ per annum; in the form of a plasma heated to several million degrees, and escaping with velocities between 10^3–10^4 km s^{-1} (cf., e.g., McCluskey and Kondo, 1981). Moreover, the outflow of this mass appears to be isotropic (cf. Kondo *et al.*, 1982) and barely affected by the proximity of any stellar companion.

Indeed, it begins to appear to us now that the phenomena, known previously (from the optical parts of the spectra) to be characteristic of the stars of the Wolf-Rayet type, may be much more common than had been thought hitherto; and need not be necessarily caused by binary nature of such objects. At least we do know of many single stars exhibiting similar characteristics as the subgiants in semi-detached binary systems, but unhampered by any Roche limits.

Should this mean that an approach to such limits in close binary systems has nothing to do with mass loss by such stars? Not necessarily so; for it is at least possible that an approach to such a limit may give rise to (or stimulate) sub-surface convection which, in turn, feeds energy into the maintenance of powerful stellar winds; the latter representing a sufficient mechanism for mass removal to meet all our needs.

The high velocity of such winds (generally in excess of 1000 km s^{-1}) can account naturally for the escape of the requisite amount of mass from the system per unit time; while the high temperature of the respective plasma (which may attain several million degrees) renders it well-nigh transparent at optical frequencies; and thus does not affect too much the light curves of close eclipsing systems observed through the optical window of our atmosphere or beyond – perhaps as far as the Lyman limit. The difficulties connected with a low-velocity, high-flux mass removal – set forth in the previous sections – completely disappear when hot stellar wind is invoked to replace neutral-gas streams; and the role of the Roche limit may appear to us in a completely different light – as stimulating convection rather than lowering the gravity in the first place (the two may, of course, be connected).

Have we, therefore, arrived at a stage at which we can at last say... 'and so it all ends happily'? Not by a long shot; for our present scheme too contains steps which are hypothetical, and likewise unproven by a more exact analysis. A conclusion that the mass lost by individual components in the course of their post-Main Sequence evolution escapes mostly from the system (and only very little of it can be transferred – let alone exchanged between them) is, in my opinion, so well-founded by existing observational data as to be virtually unanswerable; and the probability that this mass-removal occurs mainly through the medium of high-velocity stellar winds at high temperatures certainly to deserves serious consideration.

The existence in close binary systems of 'contact components' – i.e., of stars which fill the largest closed equipotentials capable of containing their mass – must, however, be likewise accepted as an established fact; but its physical significance remains still not clear. It may be that a sudden onset of convection and consequent stellar wind is indeed the cause of mass removal at the Roche limit; but this has not yet been proven and remains so far a hypothesis (albeit, to my mind, more likely than any other that had been proposed for the purpose). And we should also not lose sight of the fact that a large-scale mass loss must affect, at some stage, also the components of wide binary systems (like Sirius or Procyon) – pairs so wide that none of their components could approach (let alone attain) their respective Roche limits, even if the initial mass of the companion had been several times as large as it is at present*.

And this is, ladies and gentlemen, where we stand today. In grappling with problems mentioned as still open in this lecture, we should keep in mind that the principal reasons of this state is not only the complexity of the underlying equations of our problem (described in the preceding two sections of this paper), but mainly the fact that *the boundary conditions constraining their solutions cannot be specified from observations at our disposal sufficiently to render the solutions of such equations unique.* In order to obtain theoretical models which can be compared with the observations, the deficiency in observed facts must, therefore, be supplemented by intuition of the investigator to fill the gaps; and this where we can find ourselves on very slippery ground.

To give an example of such a situation, consider the case of a 'mass-exchange' mentioned earlier in Section 3 of this address. Even in its simplest physical form (in which the respective gas flow is replaced by a stream of particles which do not interact with each other), the underlying physical problem can be mathematically expressed in terms of three second-order differential equations of the restricted problem of three bodies, governing the positions $x(t)$, $y(t)$, $z(t)$ of such particles as a function of the time t. A unique specification of the motions of particles in three-dimensional space calls, therefore, for a knowledge of three initial positions x, y, z as well as of their velocities $\dot{x}$, $\dot{y}$, $\dot{z}$. If we identify the xy-plane with one tangent to the celestial sphere, and the z-direction with that of the line of sight, the photometric as well as spectroscopic observations can provide some empirical knowledge of the positions x and y, and of the velocity component $\dot{z}$, but cannot furnish any information on the remaining initial values of z, $\dot{x}$, and $\dot{y}$. In order to integrate actual trajectories, these remaining initial

---

* This follows from a well-known integral of the two-body problem with variable mass (cf. Hadjidemetriou, 1963), asserting that if the loss of mass is isotropic (which it will be if it occurs at high speed), a product of the semi-latus rectum of relative orbit and the total mass of the system must be conserved. The present semi-major axis of the relative orbit of Sirius is known to be 20.1 astr. units. If, therefore, the mass of the present white dwarf ($0.98\,\odot$) was initially (say) 5 times as large – rendering the initial total mass of the system to have been close to $7\,\odot$ rather than the present $3\,\odot$, this would have reduced the present size of the orbit only to 8.6 AU = 1850 solar radii – still rendering the Roche limits of the (then) more massive star very much larger than a star of mass $5\,\odot$ could ever hope to attain in the course of its evolution.

conditions must, therefore, be only guessed at; and the outcome then depends, of course, wholly on the correctness of such a guess. In other words, by choosing these in a suitable manner, we can transfer a mass particle from anywhere to anywhere within the system, and with an arbitrary velocity.

But need the results so obtained have anything to do with reality? Astronomers working on this part of our vineyard in the past would, perhaps, have been less than human if they did not at least try to explore the consequences of such arbitrary assumptions; the trouble only arises if they mistake their assumptions for established facts. To do so may actually become counterproductive, and inhibit further advances rather than contribute to them.

Worse comes, moreover, when such assumptions begin to resonate; and are accepted as truth because somebody else said so before (for 'they cannot all be wrong'!). 'A definite study of the herd instinct of astronomers is yet to be written' remarked recently Fernie (1969) on a similar situation elsewhere in astronomy, "but there are times when we resemble nothing so much as a herd of antelope, heads down in tight parallel formation, thundering with firm determination in a particular direction across the plain. At a given signal from the leader, we whirl about and, with equally firm determination, thunder off in quite a different direction, still in a tight parallel formation." A not very complimentary picture, perhaps, but containing more than a grain of truth. Did we indeed spend too much time in the past running with our heads down, and not thinking enough en route as we thunder across the plains?

Or – to quote another caustic comment which the late Oliver Heaviside (creator of the operational calculus) once made on a similar occasion – "almost everyone agrees with this hypothesis; and, therefore, the hypothesis is almost certainly wrong"; by which he meant that Nature but very seldom discloses her secrets at a first try; and that meticulous attention to every detail of the problem is necessary to force their eventual surrender.

More specific comments on such a situation have already been made (cf., e.g., Kopal, 1978, p. 472; or 1981, pp. 555–557), and need not be repeated at this time. However, the foregoing general remarks should make it perhaps abundantly clear that, in grappling with our current problems, we are still far from being out of the woods. But in the course of this time we have learned to get better acquainted with what remains yet to be done, and to learn from our past mistakes to avoid new pitfalls. Above all, let us strive not to be caught by posterity in practising a 'Procrustean science' – in which by emphasis and omission, disregard of unpleasant facts or accumulation of superfluous hypotheses, we not only may be trying to fit known facts on to the Bed of Procrustes of our preconceived opinions, but also offend the wisdom of 'Occam's razor' that 'entia non sunt multiplicanda praeter necessitatem'. And – above all – whenever our tentative conclusions may (as they did so many times in the past) turn out to be false, let us keep in mind that a quest for truth in science is seldom a monotonous process; and in the case of any disappointment, remember the words which William Shakespeare had Cassius say in his tragedy on Julius Caesar: 'The fault, dear Brutus, lies not in our stars, but in ourselves.'

6. Conclusions

Having progressed so far in our narrative, let us attempt to summarize the present – and not necessarily final – state of our subject in the following terms:

(1) In order to account for the phenomena exhibited by binary stars in terms of presently accepted theories of nuclear evolution, it is necessary to postulate a large-scale loss of mass by their components in post-Main-Sequence stage. Moreover, such a loss must be suffered by all types of binaries – close as well as wide – by systems like Algol as well as Sirius; and no doubt its cause affects all stars – be these single or multiple; in binaries it becomes merely more evident and impossible to ignore.

(2) The mass lost is very probably carried away at high speed – far above the velocity of escape from the gravitational field of the system – essentially isotropically; and only that part of it may be transferred from one component to another as is intercepted by the target surface.

(3) Most of this mass is probably expelled in the form of 'stellar winds' of hydrogen-helium plasma, at velocities between 10^3–10^4 km s^{-1}, and temperatures in the range of 10^6–10^7 degrees. It is this temperature that renders this plasma essentially transparent in the optical domain of the spectrum. The velocity with which the star as a whole may expand towards the Roche limit need not have anything in common with the velocity of ejection beyond this limit by stellar winds – the two may differ by many orders of magnitude.

(4) The energy necessary to raise the flux of corpuscular radiation to the level sufficient for the primary (more massive) and secondary (less massive) components to reverse their roles stems – like for single stars – probably from sub-surface convection. Whether or not an approach of the star's surface to its Roche limit in close binary systems stimulates convection to this extent is as yet uncertain, but remains a distinct possibility. It must, however, be also kept in mind that a mass loss of the same order of magnitude is likely to occur also in wide binaries – like Sirius or Procyon – whose components could scarcely have attained their Roche limits even at the maximum state of expansion permitted by the mirror-effect.

(5) Spectroscopic anomalies which have been observed at optical wavelengths are more likely produced by gas-streams in stellar atmospheres – arising from mutual irradiation of the two stars – rather than by circumstellar gas; for hydrogen in the latter medium should be largely ionized, and incapable of absorption in Balmer lines.

The total amount of gas in between the stars – ionized as well as neutral – may become observable only if its density attains the values at which the mean free path of the constituent particles becomes smaller by many orders of magnitude than the dimensions of the respective system. But if so, its motions must be governed by the laws of hydrodynamics rather than particle mechanics; and the kinematic viscosity of its ionized component (larger by several orders of magnitude than that which the same gas would possess in neutral state) may become so large as to necessitate retention of the Navier–Stokes terms in the respective equations of motion for an adequate description of the reality.

Moreover, since the motion of gas relative to the stars which are embedded in it is likely to be hypersonic (characterized by Mach numbers of the order of 10 or more), shock-wave phenomena – in particular, the bow-shocks in front of the leading hemisphere of less massive components – should give rise to phenomena which may be spectroscopically observable.

(6) If the hydrodynamical phenomena likely to occur in close binary systems remain still but incompletely investigated from the theoretical point of view, the principal reason is not only the complexity of fundamental equations which should control such phenomena, but – above all – the fact that the boundary conditions which constrain the solutions of such equations are *not* specified by the observations sufficiently to render such solutions *unique*. And as long as this is the case, it must be kept in mind that the scenarios postulating various 'shells', 'rings', 'accretion discs', etc. invoked by many less patient confrères to account for the few observed facts at our disposal are wholly dependent on the assumptions by which our intuition (or imagination) must supplement these facts to make any comparison between theory and observations possible at all.

That we often do this is, of course, inevitable; for this is how science advances in the long run. But, in doing so, we should not proclaim such contraptions as gospel truth. Indeed, we must constantly keep in mind that the scenarios so conjured up remain constructions of our own making until the underlying assumptions can be verified by independent evidence, and shown to be consistent with basic physics. And always keep in the back of your mind the unpleasant thought that while any (finite) number of observed facts which may be in agreement with a given hypothesis does not yet establish its validity (they can only increase its probability), a single well-established fact which is definitely at variance with such a hypothesis is sufficient to disprove it!

References

Agayev, A. G., Guseinov, O. H., and Novruzova, H. I.: 1982, *Astrophys. Space Sci.* **81**, 5.
Barney, I. W.: 1923, *Astron. J.* **35**, 95.
Batten, A. H.: 1974, *Publ. Dominion Astrophys. Obs.* **14**, 191.
Bělopolsky, A. A.: 1893, *Mem. Soc. Spettr. Ital.* **22**, 101.
Bělopolsky, A. A.: 1897, *Mem. Soc. Spettr. Ital.* **26**, 135.
Biermann, P.: 1971, *Astron. Astrophys.* **10**, 205.
Budding, E.: 1981, in F. D. Kahn (ed.), *Investigating the Universe*, D. Reidel Publ. Co., Dordrecht, Holland, p. 271.
Carpenter, F. M.: 1930, *Astrophys. J.* **72**, 205.
Chandrasekhar, S.: 1944, *Astrophys. J.* **99**, 54.
Chapman, S.: 1954, *Astrophys. J.* **120**, 151.
Crawford, J. A.: 1955, *Astrophys. J.* **121**, 71.
Fernie, J. D.: 1969, *Publ. Astron. Soc. Pacific* **81**, 707.
Gould, N. L.: 1957, *Publ. Astron. Soc. Pacific* **69**, 541.
Gould, N. L.: 1959, *Astron. J.* **64**, 136.
Hadjidemetriou, J.: 1963, *Icarus* **2**, 440.
Hoyle, F.: 1955, *Frontiers of Astronomy*, Heinemann, London, p. 195.
Kirbiyik, H. and Smith, R. C.: 1976, *Monthly Notices Roy. Astron. Soc.* **176**, 103.
Kondo, Y., Feibelman, W. A., and West, D. K.: 1982, *Astrophys. J.* **252**, 208.
Kopal, Z.: 1954, *Mém. Roy. Soc. Liège* **15**(4), 684.
Kopal, Z.: 1955, *Ann. Astrophys.* **18**, 375.

Kopal, Z.: 1956, *Ann. Astrophys.* **19**, 298.
Kopal, Z.: 1957, in G. H. Herbig (ed.), 'Non-Stable Stars', *IAU Symp.* **3**, 123ff.
Kopal, Z.: 1959, *Close Binary Systems*, Chapman-Hall and John Wiley, London and New York, Chapter VII.
Kopal, Z.: 1971, *Publ. Astron. Soc. Pacific* **83**, 521.
Kopal, Z.: 1978, *Dynamics of Close Binary Systems*, D. Reidel Publ. Co., Dordrecht, Holland, p. 10.
Kopal, Z.: 1980, *Astrophys. Space Sci.* **70**, 329; **71**, 65.
Kopal, Z.: 1981, in E. B. Carling and Z. Kopal (eds.), *Photometric and Spectroscopic Binary Systems*, D. Reidel Publ. Co., Dordrecht, Holland, p. 535.
Lubow, S. H. and Shu, F. H.: 1975, *Astrophys. J.* **198**, 383.
McCluskey, G. E. and Kondo, Y.: 1981, *Astrophys. J.* **246**, 464.
Morton, D. C.: 1960, *Astrophys. J.* **132**, 146.
Neubauer, F. J. and Struve, O.: 1945, *Astrophys. J.* **101**, 240.
Oster, L.: 1957, *Z. Astrophys.* **42**, 228.
Osterbrock, D. E.: 1974, *Astrophysics of Gaseous Nebulae*, Freeman and Co., San Francisco.
Paczynski, B.: 1971, *Ann. Rev. Astron. Astrophys.* **9**, 183.
Plavec, M.: 1968, *Adv. Astron. Astrophys.* **6**, 202.
Plavec, M. and Kříž, S.: 1965, *Bull. Astron. Inst. Czech.* **16**, 297.
Poincaré, H.: 1896, *Calcul des Probabilités*, Paris, p. 149.
Prendergast, K. H. and Taam, R. E.: 1974, *Astrophys. J.* **189**, 125.
Schoenberg, M. and Chandrasekhar, S.: 1942, *Astrophys. J.* **96**, 161.
Smak, J.: 1962, *Acta Astron.* **12**, 28.
Sorensen, S. A., Matsuda, T., and Sakurai, T.: 1975, *Astrophys. Space Sci.* **33**, 465.
Spitzer, L.: 1978, *Physical Processes in Interstellar Medium*, John Wiley, New York.
Struve, O.: 1944, *Astrophys. J.* **99**, 222.
Struve, O.: 1949, *Monthly Notices Roy. Astron. Soc.* **109**, 487.
Struve, O.: 1950, *Stellar Evolution*, Princeton Univ. Press, London.
Prendergast, K. H. and Taam, R. E.: 1974, *Astrophys. J.* **189**, 125.
Schoenberg, M. and Chandrasekhar, S.: 1942, *Astrophys. J.* **96**, 161.
Smak, J.: 1962, *Acta Astron.* **12**, 28.
Sorensen, S. A., Matsuda, T., and Sakurai, T.: 1975, *Astrophys. Space Sci.* **33**, 465.
Spitzer, L.: 1978, *Physical Processes in Interstellar Medium*, John Wiley, New York.
Struve, O.: 1944, *Astrophys. J.* **99**, 222.
Struve, O.: 1949, *Monthly Notices Roy. Astron. Soc.* **109**, 487.
Struve, O.: 1950, *Stellar Evolution*, Princeton Univ. Press, London.

THE IMPORTANCE OF WIDE-SYSTEM STUDIES FOR STELLAR EVOLUTION AND GALACTIC DYNAMICS*

J. DOMMANGET

Royal Observatory, Belgium

(Received 30 June, 1983)

Abstract. Wide pairs and wide multiple systems have been too much neglected during many years by visual double star astronomers with the argument that only close visual pairs (short periods) may lead to mass-determinations in a relatively short time interval. But mass-determination should not be considered as the only interest of double star astronomy, even if it is of a fundamental nature.

Today, it appears that researches on the origin and the evolution of the wide systems are urgently wanted, not only for the understanding of the evolution of the stellar medium, but also for a better knowledge of galactic dynamics. Some examples are given.

Presently, the main task for double star specialists will be an important improvement in double and multiple star census for all kind of systems: close, medium, and wide. The Hipparcos satellite will probably add some important informations in that respect but ground-based observations also remain of the highest importance (radial velocities, photometry, astrometry, etc.).

1. Introduction

The discovery of the first double stars showing orbital motion has immediately led to the idea of the possible determination of the stellar masses with high accuracy because it was the first time that a perfect example of the two-body theory in its full pureness was discovered in the Universe.

In his report to M. D'Ouvaroff, President of the Imperial Academy of Sciences in St. Petersbourg (1837), F. G. G. Struve wrote:

"... Si deux soleils sont liés par l'attraction, il doit y avoir des mouvements en lignes courbes continues. L'astronomie, dans les siècles passés, ne connaissait de pareils mouvements que dans le système solaire; les étoiles doubles nous les offrent dans l'immense éloignement des étoiles fixes. C'est en observant ces mouvements que nous finirons par en découvrir les lois. Si les lois de la gravitation universelle de Newton sont la plus sublime découverte qu'ait faite l'esprit humain dans le cours de plusieurs milliers d'années, nous sommes bien près d'être à même de déterminer si ces lois n'appartiennent qu'au système solaire, ou si elles sont communes à l'univers entier. L'astronomie marche donc vers une nouvelle époque qui datera du moment où l'on fera voir que la mécanique céleste ne se borne pas aux phénomènes du système solaire, mais peut s'appliquer aux mouvements des étoiles fixes...."

The fact that the gravitational laws are applicable to all bodies in the Universe and that as a consequence the stellar masses can be determined through the orbital motion of the binaries, has dominate all double star astronomy for a long time and remains even today, for many astronomers, the basic reason for observing visual double stars.

However the regular discovery of new systems of all types by the visual method as well as by other techniques, such as spectroscopic, astrometric, speckle, and occulta-

* Paper presented at the Lembang-Bamberg IAU Colloquium No. 80 on 'Double Stars: Physical Properties and Generic Relations', held at Bandung, Indonesia, 3–7 June, 1983.

Astrophysics and Space Science **99** (1984) 23–28. 0004–640X/84/0991–0023$00.90.

tions, has led to the occurence of an exceptionally high frequency of binaries – at least in the surrounding of the Sun – in comparison with all other stellar objects: single stars, triple systems, etc.

Unfortunately, this seems not to have brought any change in the attitude of astronomers regarding medium wide and wide pairs. Very close pairs, because of the astrophysical problems they revealed, have created a very interesting field of research for many astrophysicists, while close visual pairs remained the only interesting objects for astrometrists in view of orbit computations and mass determinations.

Of course, astrometric observations as well as astrophysical observations of components of medium wide and wide pairs do not favour a particularly great enthusiasm. But we have to decide whether we want to know more about binaries or whether we just want to conduct our activities in small areas ignoring the characteristics of the whole domain which we are interested in.

I would like to take the present opportunity to recall how important may be any interest for these particular but very numerous presently neglected binaries.

2. The Fundamental Problem of Double Star Astronomy

In addition to their exceptional frequency, the binaries show some important orbital features that should be clearly and continuously kept in mind.

Their periods have values regularly distributed between a few days and some thousand years without any particular gaps. Similarly, the semi-axis major (or the distance between the components when no orbits are known) are also spread over a large interval of values between some hundredths and a few thousands of AU, also without any particular gap.

Now, about the orbital eccentricity, a regular distribution of its values is observed between 0 and 1 with a sensible decrease of the frequency from the smallest values to the highest ones. Here also, no particular gaps are observed.

Knowing this, the most puzzling of all questions that arise in the field of double star astronomy is: "How have the medium wide and the wide binaries been formed?"

Of course, for the very close pairs, nobody will imagine any other formation process than those admitting a common origin of their components, starting from a unique protostellar cloud. At the contrary, for systems having sufficiently separated components, the situation is less evident and different opinions are present.

It seems that only two different ways may be accepted for the genesis of these binaries:

(a) a formation mechanism leading immediately to the presently observed orbital characteristics;

(b) a formation as close pairs, followed by an orbital evolution changing the original orbit into the presently observed one.

A third way – the formation by capture – is no longer considered because even if the change of an hyperbolic orbit into an elliptic one is perfectly possible, it cannot statistically lead to the high frequency of binaries presently observed.

In case of item (a), one should admit that there is no difference between the formation of components of wide pairs and the formation of single stars. That means that, in both cases, the space distribution of the condensation centers leading to protostars was at random. However, the observed distribution diagram of the distance between any star or a component of a binary and all others in its surrounding shows a typical discontinuity around $A = 2 \times 10^3$ AU, as may be seen on Figure 1. As a consequence, the diagram for the binary components cannot be considered as part of that of the single stars. This fact seems to be sufficient for rejecting item (a).

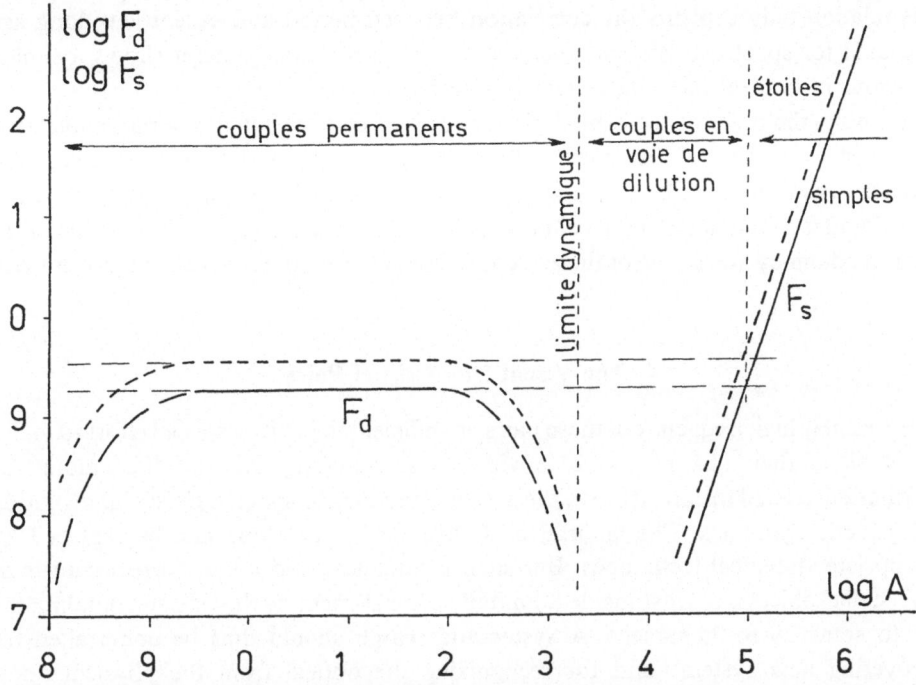

Fig. 1.   Schematic distribution diagram of the true distances between the components of binaries and between single stars, following a statistical research of the author.
(*Comm.. Obs. Roy. de Belg.*, Série B, No. 70, 1971.)

In case of item (b), wide pairs would appear as the result of an orbital evolution of close pairs. The first mechanism that should be considered seems to be a secular mass-loss that increases the period and the semi-axis major. About the eccentricity the situation is somewhat more complicated but it does not need to be considered here for the present discussion. The most crucial point for adopting this process lies in the need of admitting that *the great majority of stars are loosing mass* secularly and quite substantially.

It is thus clear that if we want to know more about the genesis of binaries, we should increase our interest for the *non-close pairs*, that may be considered in two different categories: the visual orbital pairs and the visual non-orbital pairs.

3. The Visual Orbital Pairs

They are the only visual systems of which the orbital characteristics are known. From their study (some 900), some important results should be mentioned.

First, the assumption that there exists a permanent secular mass-loss for all components of any luminosity classes, has led to the discovery of the following important relation between mass and eccentricity in the case of pure binary systems*:

$$e^{2.8} \mathfrak{M}_{AB} < 3.60 .$$

This relation fully explains the correlation between period and eccentricity long ago suspected for spectroscopic and visual systems. At the same time, it shows a similar behaviour of both of these categories of binaries.

Secondly, the poles of the orbital planes seem to be distributed in a particular way: inside space elements of the order of 20 to 30 parsecs, the orbital planes would show a certain organisation**. If this is really so, stars inside such elements would have been formed under a common dynamical process and one could imagine the formation of the stellar medium by steps, one of these steps being the formation of such space elements.

4. The Visual Non-Orbital Pairs

The extremely high frequency of these pairs, is sufficient to justify a special effort to make use of all of their characteristics in view to substantially increase the amount of information needed to contribute to the above mentioned researches on stellar evolution and galactic dynamics. The ignorance of their orbital elements can be replaced by appropriate statistical techniques. But such techniques need a first correct census of these stellar objects, at least inside a limited space element as the one surrounding the Sun to some 20 to 25 parsecs. A systematic search should thus be undertaken for discovering new systems and for recognizing the optical from the physical ones. Parallaxes, proper motions as well as radial velocities of their components are urgently needed.

The main object of such a systematic survey should be to correctly establish the distribution diagram illustrated by Figure 1. This means that the survey should concern systems where the components may show separation reaching 0.01 parsec. Using Hertzsprung's formula[†]:

$$\log \rho < 3.0 + 1.50I - 0.20m ,$$

where I = color index, m = magnitude of brightest component, one finds, for stars of

---

* IAU Colloquium No. 59, Trieste, 1980, Proceedings, 1981, p. 507.

** *Bull. Astron. de l'Obs. Roy. de Belg.* **VI**, 6, 1968, p. 246.

[†] *Comm. Obs. Roy. de Belg., série B*, No. 17, 1967, p. 36.

class V, the following apparent separations:

Spectral type of A \ m_A	8	10	12	14	16
B	8″	3″	1″3	0″6	0″2
A	25	10	4	2	0.6
F	79	32	13	5	2
G	180	71	28	11	5
K	370	150	59	23	9
M	3700	1500	590	234	93

This table shows that for instance a very 'classical' star of spectral type FV or GV with an apparent magnitude of the order of 10 to 12, may have a companion at a distance that may reach one arc-min! Some such systems are well known but how many have not been discovered and who will be interested in starting a systematic survey of such systems?

5. Multiple Stars

The genesis of the stellar medium does not concern only single stars and simple binaries but also multiple systems. Statistics on multiple systems is far from being complete, and their genesis remains a puzzle.

Here, also a correct census is urgently needed. A particularly important effort seems to be made for discovering close companions to components of already known visual binaries. But, once again, no special survey seems to have been undertaken to discover additional wide components to known spectroscopic or visual pairs.

Without such a correct census of the multiple systems, no useful statistics may be conducted to recognize any law or correlations concerning mass distribution, dynamical and astrophysical characteristics of their components, or to define a limit between multiple systems and open clusters.

6. Conclusion

From this rough survey of the problems related to double star astronomy it appears that medium wide and wide pairs are of a very great importance but that their study is extraordinarily neglected. Their high frequency and the difficulties of organizing in the most efficient way, a survey of discoveries, are fundamentally responsible of this situation. But this does not change the urgent need of undertaking this tremendous amount of work to complete the present statistical material.

Ground-based equipment will be needed in the field of photometry and spectroscopy (spectral classification and radial velocities) but expected new space techniques will be of the greatest efficiency. In particular the HIPPARCOS astrometry satellite may not only improve the census of double and multiple systems by new discoveries, but may also, by accurate parallaxes and proper motion determinations, make usable dynamical criteria to recognise optical from physical systems.

We must keep in mind that as long as a sufficiently correct census of all double and multiple systems will not have been realized, no valuable statistical research will be possible in this important field of astronomy for stellar evolution and galactic dynamics.

DUPLICITY ON THE MAIN SEQUENCE*

T. HERCZEG

Department of Physics and Astronomy, University of Oklahoma, Norman, Oklahoma, U.S.A.

(Received 3 October, 1983)

Abstract. A review paper or a lecture like the following one, will best serve 'its' conference by giving an overview of the basic facts, and an impartial review of current debates, also by trying to point out some apparently crucial questions whose solutions, we hope, will determine the line of future research.

Because these stars are essentially unevolved, beyond the topic of multiplicity on the Main Sequence looms the fundamental problem of the formation of binary star systems. Thus we are going to concentrate on the following questions: the fraction of stars that are formed in binary and multiple systems, the distribution of mass ratios for unevolved systems, the role of very wide pairs and the smallest known stellar or substellar masses. We will pay special attention to nearby binary stars. On the other hand, we do not have the space to discuss in any detail the binaries in extragalactic systems, in the upper regions of the HR-diagram; they are practically all evolved systems.

1. The Percentage of Multiplicity

It is a commonplace in double star astronomy that about half of the stars in galaxy are in double or multiple systems, that is, if we study three stars closer, we may expect that one of them turns out to be double. The true figure of duplicity might be even higher, but more accurate statements about the percentages of multiplicity are much harder to come by. The problem is the great difficulty of carrying out reliable statistical surveys based upon reasonably complete and homogeneous samples. Setting, for instance, a suitable magnitude limit for a survey means completeness up to a certain distance only if we restrict the survey to a narrow range of absolute magnitudes. This is the case with Abt's and Levy's well known study of nearby solar type stars but not with Aitken's important surveys of visual duplicity. Even if the survey is more or less complete to a given space limit, there is an inhomogeneity introduced by a 'magnitude equation': spectroscopic data are usually less accurate for fainter objects and visual detection will miss, with increasing distance, closer companions which remain unresolved.

Studies about the distribution of mass ratios, eccentricities, angular momenta and similar data are equally or probably more exposed to observational bias than the question of mere duplicity of multiplicity. As to this latter point, we can say that several independent studies suggest, quite concordantly, a fraction of duplicity around 60–65%, if one makes an allowance for 'missed objects'. Heintz even arrives at a multiplicity of 80–85% (1969). In this context, the multiplicity ratio is taken as

$$\frac{\text{no. of components in double or multiple systems}}{\text{no. of all stars considered}}$$

---

* Paper presented at the Lembang–Bamberg IAU Colloquium No. 80 on 'Double Stars: Physical Properties and Generic Relations', held at Bandung, Indonesia, 3–7 June, 1983.

Astrophysics and Space Science **99** (1984) 29–39. 0004–640X/84/0991–0029$01.65.

This means asking the question: how many stars, in general, are formed as members of a multiple system? Sometimes a different question is asked: given a certain class of stars, such as WR-stars, Am-stars, novae, what is the percentage of these objects in doubles or multiples: is perhaps duplicity an important, possibly characteristic feature of the class? In such cases the duplicity ratio is modified to

$$\frac{\text{no. of systems with at least one component of the specified type}}{\text{no. of all stars of this type}}$$

In the following table, percentages based on this ratio are indicated by an asterisk.

A selected list of recent (more or less so) statistical studies of stellar duplicity present itself as follows:

nearby stars:	$r \leq 5$ pc	multiplicity: 59%	
	$r \leq 10$ pc	55%	Woolley *et al.* (1970) and
	$r \leq 20$ pc	45%	Gliese's Catalogue (1980)
Main Sequence,	B-M	50%	Jaschek and Gomez (1970)
O-type stars		36%*	Garmany *et al.* (1980)
early B-type stars	(B2–B5, IV–V)	50%	Abt and Levy (1978)
late B-type stars	(B7–B9)	45%	Wolff (1978)
A–F star around N galactic pole		40–45%	Hill *et al.* (1976)
solar type stars	(F3–G2, IV–V)	53%	Abt and Levy (1976)
M-type dwarfs		39%	Worley (1969)
B2–F5 (Fehrenbach prism study)		'well above 50%'	Gieseking (1980)
Among evolved stars:			
Giants (K III)		30%*	Jaschek and Gomez (1970)
WR stars		53%*	Lamontagne and Moffat (1982)

We may add two counts of multiplicity among the brightest and the brighter stars in the sky. These samples are very far from being homogeneous but refer to objects which, on the whole, should be the best studied ones:

| brightest stars ($V \leq 1.65$) | multiplicity 60% | |
| *Catalogue of Bright Stars* | 55% | (Hoffleit and Jaschek, 1982) |

In the same catalogue we also find a tabulation of the frequencies of increasingly large multiple systems, starting with

$$N = 2, n = 1715 \text{ systems, then}$$
$$N = 3, n = 675,$$
$$N = 4, n = 237, \text{ etc.}$$

The highest multiplicities are: $N = 15, n = 1$ and $N = 17, n = 1$. It is quite obvious that these unusually high multiplicities still need a specific study of confirmation.

Gieseking's investigation, mentioned in the tables of statistical studies above, is remarkable as it represents the first large scale application, with the aim of binary star statistics, of a promising modern technique: the basis of his statistical analysis was essentially improved by the addition of 900 stars observed with the Fehrenbach prism. In this way large bodies of homogeneous observational material can be gained although the accuracy is somewhat inferior to that of slit spectroscopy and the detection of double-lined binaries is not favorable. A potentially even more powerful method is Fellgett's photoelectric radial-velocity spectrometer, capable of an accuracy of ± 0.5 km s^{-1} and thus comparable with the best high dispersion determinations. The gain in observation and reduction time is enormous; there is a limitation to later spectral types but in this respect the Fehrenbach prism complements the spectrometer very well. On the 'other end' of the distribution, among the visual binaries, separations $\leq 0''.1$ are in the range of interferometric methods and even the lunar occultations could be used for statistical purposes although this needs a sustained and well organized effort: during the 18.6 year period of the revolution of the nodal line, some 9% of the stars in the BD or SAO catalogues could be checked by this method.

Returning to various attempts to distill a definitive ratio of stellar multiplicity out of these statistics, we may summarize the outcome by the qualitative statement: it is generally assumed that a single ratio exists for all unevolved, Main-Sequence stars (see, for instance, Jaschek and Gomez, 1970) and this percentage may be rather high, substantially above the often quoted 50%. Even the reversed question seems not to be far-fetched: are there, after all, single stars? – although the answer to this question is almost certainly affirmative. A further question can be added: are planets around many, perhaps most, of these single stars? There is very little objective ground for an answer to this question as yet, although occasionally the attempt has been made to extrapolate the mass distribution of the secondary components to values as small as 1/1000 solar mass. All this is, however, very uncertain since the planetary system is a markedly different structure from binary and multiple stars, also star clusters, and may have been formed in its own very specific way. It is by no means certain that the maxim: "A planetary system can be considered to be a binary (or multiple) system in which the mass ratio is very large" alone will help us much in understanding the cosmogony of our solar system.

2. Mass-Ratios and Double Star Formation

Statistics of the mass ratios in binary systems as well as the distribution according to angular momentum (usually expressed by linear separations or periods) are at least as important as the multiplcity percentages. We do hope to obtain information concerning the formation mechanisms of binary stars. Thus we are going to consider here, however briefly, four studies representing this area of current research. Their results are not in very good agreement, to say the least, illustrating how far we are from the answer to the problem of binary formation.

In his already cited 1980 paper, Gieseking also discusses the mass ratios in binaries of the spectral range B2–F5. This was a review article and a detailed publication of the data is expected to follow. Yet it is obvious that the material, augmented by a large number of objective prism observations, is far more homogeneous than that of a traditional catalogue study. Gieseking's result, that mass ratios around $q = 1$ are virtually non-existent and the mass ratios peak near 0.25 (see his Figure 7), is unexpected and surprising in the extreme. One's immediate reaction is that the method of observation must very strongly discriminate against double-lined binaries and with it against mass ratios higher than 0.65 or 0.7; a remark in a similar sense has been made in the paper. Nevertheless, Gieseking does interpret this distribution as supporting Lucy's earlier views that binary origins by fission should favor low mass ratios.

Lucy himself found, if not in direct contradiction, certainly in no agreement with these results, that double-lined spectroscopic binaries (SB2's) show a remarkably sharp peak at $q = 0.97$, making nearly identical components quite common (Lucy and Ricco, 1979).

Lucy and Ricco based their work on Batten's 6th catalogue of spectroscopic binary orbits (1967). Unlike Virginia Trimble in an earlier 'more ambitious investigation', they considered only double-lined binaries, about 180 systems. This restriction has the advantage that no assumption is necessary concerning the unknown inclinations and the selection of unevolved systems is more reliable, since we may expect that evolutionary effects lead preferably to lower mass ratios. On the other hand, this restriction is *ab initio* strongly biassed toward high mass ratios, as the secondary spectrum tends to remain invisible if $q < 0.7$. Lucy and Ricco's claim is, essentially, that even in the limited interval $0.6 < q < 1.0$, there exists a conspicuous peak between $q = 0.95$ and $q = 1.0$ (occasional values of $q > 1.0$ are considered observational errors). Since the material in the catalogue offers a statistically rather poor sample, no effort was spared to show that the peak so near $q = 1$ is not a consequence of selection effects nor close binary evolution. (The latter is actually not too surprising.)

But what exactly is meant when the authors say that: "many close binaries ... with intermediate and small total masses ... are formed by a mechanism, that, in its ideal form, would create binaries with identical components"? Many, but not all? What if the form of the mechanism is "not ideal"? Are they in favor of several different processes of binary formation? If we add namely the grey mass of some 800 single-lined binaries (SB1's), the picture would almost certainly change considerably. We then may expect a more bimodal distribution with a second maximum perhaps around $q = 0.3$ or $q = 0.4$, even if one could eliminate all the evolved pairs – not an easy problem if we have a good number of single-lined non-eclipsing systems. Thus we see two careful and extensive studies coming up with conclusions that do not seem well compatible with each other. If we turn to a third investigation we find another, different result yet we see that the final picture still not at hand.

3. The Solar Type Stars and Nearby Binaries

The best available information about mass ratios is still the already mentioned investigation by Abt and Levy (1976) of nearby solar type stars (F3–G2, IV–V). The main advantages of this study are: (1) it is 'objective', to use the authors' word, covering all objects within a certain magnitude limit in a homogeneous treatment; (2) we may hope that for these nearby stars spectroscopic and visual detections complement each other, making the sample virtually complete; (3) evolutionary effects play no significant role due to the relatively late spectral types.

There was some criticism raised about this statistic; the critics seem to have their so-called valid points but I do not think these would bring down the whole study. To mention one example: the homogeneous treatment could be improved by standardizing the exposure times, photographic densities, even waiting for similar seeing conditions, but it does not seem necessary to adhere to such strict rules in order to detect radial velocity variations of the order of 2–3 km s^{-1}. There certainly are weak points: the whole material is relatively small, 123 stars, leaving important 'bins' of the mass ratio and orbital period represented by only 2–4, even 1–2 objects; the time coverage of some systems is insufficient; finally, the estimate of the minimum detectable velocity is perhaps somewhat optimistic – thus influencing the 'completeness corrections' applied by the authors.

The main results of this investigation still deserve our full attention. Concerning the mass ratios, they are twofold:

(1) For the primary mass, $M_1 = 1.2 M_\odot$, the maximum frequency is at $M_2 = 1.2 M_\odot$ ($q = 1$), then the frequency of mass ratios show a slow decline, approximately with $M_2^{0.4}$. This is valid for 'short' periods, $P \leq 100$ years.

(2) For $P \geq 100$ years, the frequency of secondaries increases rapidly as one goes to smaller mass ratios, following the van Rhijn distribution.

The difference between these two groups is interpreted as a consequence of their entirely different history of formation: fission of fast rotating protostars (1) vs separate formation from contracting protostars (2). This part of the discussion illustrated again the importance of these statistics for the theories of binary star formation.

Corrections for 'missed' objects is an essential part of this study. For most mass ratios and periods, visual or spectroscopic detection is, indeed, possible. There are, however, some very difficult combinations. If we consider the smallest mass ratio used by Abt and Levy, $M_1 = 1.2 M_\odot$, $M_2 = 0.0075 M_\odot$, we have to deal with the following possibilities:

Period	Max. angular sep. ($0\rlap{.}''08 < p < 0\rlap{.}''12$)	Max. V for primary
10^{-2} yr		8.8 km s^{-1}
10^{-1} yr	$\sim 0\rlap{.}''02$	4.1 km s^{-1}
1 yr	$0\rlap{.}''09 - 0\rlap{.}''13$	1.9 km s^{-1}
10 yr	$0\rlap{.}''4 - 0\rlap{.}''6$	0.8 km s^{-1}
100 yr	$1\rlap{.}''9 - 2\rlap{.}''8$	0.4 km s^{-1}

Abt and Levy assume that 2 km^{-1} amplitude in the radial component of V_1 can be detected. Considering a magnitude difference of 8 or 9 between components of such a disparity of masses, we may conclude that the cases:

with $P = 10^{-2}$ yr and $P = 100$ yr are certainly ,
with $P = 10^{-1}$ yr and $P = 10$ yr marginally detectable ,

while a system with $P = 1$ year will almost certainly escape detection. Corrections for incompleteness at this mass ratio are a difficult matter and the low mass end of the distribution remains correspondingly uncertain.

A point-by-point discussion of Abt and Levy's distribution of the mass ratios, summarized in their diagram, see Figure 6 of the paper cited, does indeed suggest that cases with $M_1 = M_2 = 1.2 M_\odot$ are most frequent, forming a shallow maximum at $q = 1$. For the rest of the q-values the distribution *could be* considered – for 'short period' systems – as having a constant frequency, independent of M_2. This is, interestingly enough, the distribution shown by nearby visual binaries, as it will be presented immediately. The low-mass end of this distribution remains an open question. It is difficult to dispute the markedly different behavior of 'long period' pairs in Abt and Levy's data; the dividing point seems, however, somewhat arbitrary and it can be placed closer to $P \sim 10$ yr. It should be noted, on the other hand, that there is hardly even a

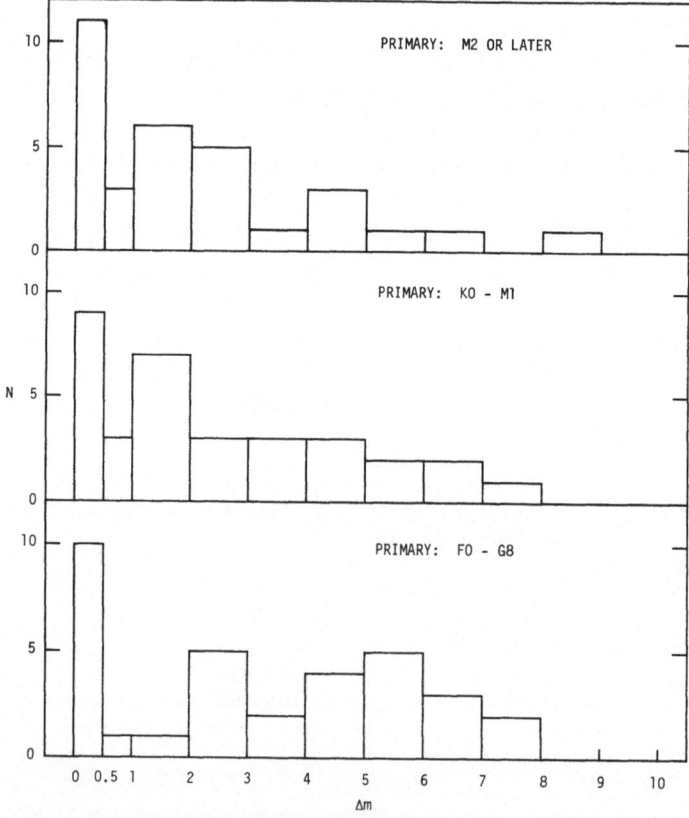

Fig. 1.   Distribution of the magnitude differences among 98 visual pairs, based on a 15 hr A.R. interval of the *Catalogue of Nearby Stars*. For all systems, $r < 20$ pc.

hint of a van Rhijn distribution among the secondaries of nearby visual binaries – the more remarkable since for these long periods Abt and Levy, too, have to rely on visual pairs.

It is worth noting that the distribution of periods (or semi-major axes) for this sample is, as Abt and Levy found, not bimodal. Similar distributions with a single maximum around $P = 5$ to 10 yr was found also by Kuiper, Heintz, and others. How is this distribution compatible with two different mechanisms of the binary star formation?

The question seems justified whether we can find out more about the distribution of mass ratios by looking at the nearby binaries, $r < 20$ pc. The material is condensed in two important catalogues but this does not make a number of follow-up studies and checks in the literature unnecessary. A common investigation of nearby double and multiple stars is underway between the present author and Michael Gainer (St. Vincent College, Latrobe, Pennsylvania); as of now, the A.R. intervals 0^h–3^h and 12^h–24^h are covered and the following remarks are based on this partial study, encompassing about 60% of the available material.

Since these double stars are almost exclusively Main-Sequence pairs, the mass ratios can be translated into Δm-values between the components. The histograms of Figure 1 show a similarity of Abt and Levy's diagrams, with the modifications proposed above. There is a distinct maximum at $\Delta m = 0$, followed by a distribution down to $\Delta m = 10$, which can be best characterized by a constant value. (Both our Figure 1 and the Abt–Levy distributions are plotted with a logarithmic scale along the abscissa.) The maximum at $\Delta m = 0$ to 0.5, corresponding to about $q = 1.0$ to 0.85 is unlikely to be result of observational selection: these are relatively bright pairs of $3''$–$5''$ separation, on the average, and a magnitude difference of 0.5 can hardly affect the chances of a discovery to such a marked extent. One has to bear in mind, of course, that the two samples are not identical: ours corresponds rather to the 'long period' group of Abt and Levy's.

4. Very Wide Binaries (cpm Pairs)

It is certainly not without importance for the theory of binary star formation that the distribution curve of the semi-major axes (actually: the observed separations) has a long, pronounced tail, reaching from a few hundreds or perhaps thousands AU to up to 5×10^4 AU or more, a substantial fraction of a parsec. In many cases the common origin is not immediately visible in the telescope and proper motion studies contribute strongly to the discoveries. It is well known, for instance, that the visual pair α Centauri AB has a companion at a distance of $2°$ – Proxima – which shares the parallax and the space motion of the visual pair and undoubtedly forms a triple system with it. The projected separation is about 9500 AU, the orbital period may be of the order of half a million years (as compared with the 80-year period of the close pair).

The dividing line between 'ordinary' visual binaries and cpm pairs is, of course, a matter of convenience. Abt and Levy, quite consistently, call visual binaries those pairs for which an orbit can already be calculated and all the other physical pairs cpm binaries. (Most double-star astronomers would probably not call a system like, for instance, the spectacular pair β Cygni a cpm binary, in spite of a separation over 30" and a period which might be in the range of ten thousands of years; yet it is the unchanged configuration, the common proper motion, that indicates the binary nature of the pair.)

The number of cpm components in triple and multiple systems is surprisingly high. So for instance among 16 triple stars in the surveyed section of Gliese's *Catalogue*, no less than 7 are discovered by common proper motion. Luyten (1971) estimated that the space density of wide pairs may be as high as 0.003 pc^{-3}.

Among them the widest known pairs are:

36 UMa and BD 57° 1266, and
α PsA and HD 216803;

in both cases the projected separations amount to about 5×10^4 AU. The orbital periods are of the order of 10 million years, the (average) orbital velocities around 200 m s^{-1}. The existence and 'life expectancy' of such wide systems, in view of stellar perturbations, poses interesting problems. Yet there can be little doubt that these wide pairs do exist, although the observed common proper motion should be confirmed by parallax and radial velocity information – not always available. As illustration of a strong case of two components belonging to the same system, in spite of nearly 2° distance between them in the sky, we may quote the data of the Formalhaut system:

	α PsA	HD 216803
Parallax	0".149 ± 0".008	0".128 ± 0".008
μ_α	0".386	0".326
μ_δ	−0".161	−0".158
Radial vel.	+6 km s^{-1} (class A)	+10 km s^{-1} (class C)

This seems to be a definitive case of a very wide double system. It is tempting to link the existence of these wide pairs to capture processes having occurred in the denser part of a star cluster – now completely dispersed – where these stars were originally formed. We know from numerical treatment of the *n*-body problem that this process of double star formation is possible and the pairs formed in this way have characteristic separations of several hundreds or thousands of AUs.

5. Unseen Companions

Among nearby stars we hope to find the lowest luminosities and lowest masses on the Main Sequence. The study of nearby stars can also supply the information about a whole new class of objects of possibly very low mass: black dwarfs or objects with 'substellar masses' and – conceivably – even members of other planetary systems.

The lowest known luminosity is ascribed to Gliese 752 = BD 4° 4084B; this component of a visual binary, classified M5e ±, has M_V = 18.6; for comparison, the A-component, classified M3.5e, has M_V = 10.31. (The parallax is 0".173 and therefore this absolute magnitude is well determined.) This luminosity is still unquestionably 'stellar': Jupiter, at full phase, would appear 7.8 mag. fainter than Gliese 752, at the same distance.

The lowest masses for observed ('seen') stars, as known today, are in the visual systems:

Ross 614: $0.11 M_\odot + 0.007 M_\odot$ (Probst, 1979), and
Wolf 424: $0.067 M_\odot + 0.064 M_\odot$ (Heintz, 1977);

the uncertainty of the masses in Wolf 424 is given $\pm 0.015 M_\odot$. The absolute visual magnitudes are 16.6 (Ross 614B) and 15.1 (Wolf 424). For possibly even smaller masses we have to turn to the so-called unseen companions.

This group of objects is set apart by the very special observing technique they require; they are more properly called, as part of a wider family, astrometric binaries. Our best hopes to find 'substellar' masses and perhaps even planets around other stars, rest with this way of investigation. The planets are a controversial question we do not want to touch upon here: it mainly concerns the nearby Barnard's star and its possible planetary companions. This particular field is something almost a monopoly of a handful of specialists among positional astronomers and a colleague from the wider fields of binary star astronomy, paying a visit to this *hortus conclusus*, can only register with some astonishment the conspicuous divergence of opinion among the specalists. Perhaps we may accept David C. Black's judgement (1980), himself a visitor to this topic that "any perturbations to the motion of Barnard's star are at or below the level of present astrometric observational accuracy".

Substellar masses or black dwarfs, however, may be a different type of evidence. Following Kumar's work, they are generally expected in the range of $0.01 M_\odot$ to $0.06-0.07 M_\odot$, that is, from 10 to about 60–70 Jupiter masses. If frequent enough, they may, indeed represent a new class of galactic objects. They may be quite numerous, as products of fragmentation in the contracting prestellar cloud, near the lower end of the mass spectrum (which we do not know with sufficient accuracy). Their possible discovery is linked to membership in binary or multiple systems and we have to add, that present astrometric evidence is rather meager, hardly supporting the idea of a wide class of new objects. Most of the prospective candidates listed in an earlier paper by Kumar (1966) are not be found in more recent lists, and the evidence for substellar masses narrowed down to four cases, none of them completely unambiguous.

We may consider, as basis of a discussion, van de Kamp's compilation in his review article (1975), some of the data seem, however, far from being definitive. It is somewhat discouraging, for instance, to learn that in the case of Gliese 873 = EV Lac, the period was recently revised from 28.9 yr to 45 yr (Van de Kamp, 1981). There are only four stars which may possibly have a 'black dwarf' companion in the system:

ε Eri, BD 68° 946, BD 43° 4305 and Stein 2051.

In case of ε Eri, even the existence of a perturbation is questioned, see Heintz (1978). The triple system Stein 2051 exhibits the perturbation beyond doubt, even spectaculary, but the range of possible masses turns out to be $0.02 M_\odot$ to $0.17 M_\odot$, depending also on the mass of the C-component, a white dwarf. Thus the unseen companion of the M-type component can be itself a late type M-dwarf. Difficulties of this type of a study are particularly well shown in the case of BD 68° 946 = Ci 18,2354. Here a recent rediscussion of the system by Lippincott (1977) resulted in a marked modification of the important orbital element α, the semi-major axis of the photocentric orbit: α went from $0''.102$ to $0''.033 \pm 0''.002$. Heintz is ready to exclude the case at the time being, before more reliable data can be secured. On the other hand, Miss Lippincott points out that the observational basis was substantially increased before this revision and the entire material was remeasured with a Grant-type machine, that is, it was done more objectively than earlier. The change in the masses was strongly downward; the minimum mass of the unseen companion stands now at $0.009 M_\odot$, although this value corresponds – as usual – to the hypothesis of a magnitude difference $\Delta m \to \infty$ between the components, meaning a dark companion.

We mentioned these cases as illustrations to the point that, owing to the difficulty of the measurements, their interpretation is by no means straightforward. It seems fair to say that not a single case of unquestionably substellar masses have been found yet, but we may have one or two good candidates. The frequent view of popular works and even textbooks that unseen astrometric companions refer to 'black dwarfs' (expressed, for instance, by the unfortunate phrase in German literature: *planetenaehnliche Stern-begleiter*) is simply not correct. They refer in most cases to late Main Sequence companions.

This fact is also expressed by the value of the mean mass, $0.3 M_\odot$, for unseen companions, used by van de Kamp (1981) in an attempt to estimate the mass density of these objects. Revising an earlier figure, he proposed $N = 0.07$ pc^{-3} for the number density, and $0.021 M_\odot$ pc^{-3} for the mass density of the unseen companions. This is a small number statistic and the number density is still comparable with the number density of stars, thus a figure that at first glance appears unexpectedly high.

6. Notes on Particular Objects Around the Lower Main Sequence

Concluding this survey, we are going to add, at least in the form of a few time (and space) restricted remarks, some interesting finds about binaries 'around' the Main Sequence (subgiants and subdwarfs) and about a particular class of binaries on the Main Sequence: the low mass contact systems (W Ursae Majoris type). All these remarks are based on unpublished work.

The study of nearby binaries, mentioned earlier, provides us almost exclusively with unevolved systems on the Main Sequence. Near the spectral types F and early G, however, we find a few evolved binaries of the combination subgiant + Main-Sequence star. Among the nearby *subgiants*, we expect to find some very old field stars and in fact, most of the subgiants in binary systems in Gliese's catalogue follow the M67 evolutionary track. Subgiants can be found, of course, among the single stars as well but membership in binary systems can help us to define the position in the HR diagram more accurately.

There are about ten *subdwarf* binaries within 20 pc, mostly marked by high space velocities, 100 km s^{-1} s to 250 km s^{-1} s. Not surprisingly, in typical cases both components are subdwarfs and pairs with nearly identical components are frequent. Their 'vertical' distance from the Main Sequence can reach 6–7 mag.

On the Main Sequence itself there are a few very close detached binaries among nearby stars; they are 'on the verge' of interaction between the components. Castor C = YY Geminorum is the best known example with $P = 19.5$ hr: it is not interacting in the form of any substantial mass exchange but both components show enhanced stellar activity. On the other hand, the number of W UMa type contact systems is surprisingly low: one in Gliese's catalogue (44i Bootis), none in Abt and Levy's sample. Their space density is certainly not the highest among eclipsing binaries and it may be as low as $1-2 \times 10^{-5}$ pc^{-3}. One of them, the binary i Bootis is member in a triple system and this enables us to obtain a good determination of the mass of the contact pair. This turns out unexpectedly low: not much larger than $1.2 M_\odot$. A somewhat similar case, at a greater distance is VW Cephei. Here the system data are less well determined but they, too, suggest an undermassive contact pair of $1.3-1.5 M_\odot$. The spectral types are G2 + G2 resp. G5 + K0. It would be very interesting to see other binary masses of this group derived from triple and multiple systems.

References

Abt, H. and Levy, S.: 1976, *Astrophys. J. Suppl.* **30**, 241, see also in T. Gehrels (ed.), *Protostars and Planets*, Univ. Arizona Press, 1978.

Abt, H. and Levy S.: 1978, *Astrophys. J. Suppl.* **36**, 241.

Batten, A. H.: 1967, *Publ. DAO* **13**, 119.

Garmany, C. D., Conti, P. S., and Massey, P.: 1980, *Astrophys. J.* **242**, 1063.

Gieseking, F.: 1982, *Mitt. Astron. Ges.* **57**, 143.

Gliese, W.: 1969, *Veröff. Astron. Rechen-Instr. Heidelberg*, No. 22, see also Gliese, W. and Jahreiss, H.: 1979, *Astron. Astrophys. Suppl.* **38**, 423.

Heintz, W. D.: 1969, *J. Roy. Astron. Soc. Canada* **63**, 275.

Heintz, W. D.: 1978, *Astrophys. J.* **220**, 931.

Hill, G., Allison, E. *et al.*: 1976, *Mem. Roy. Astron. Soc.* **82**, 69.

Hoffleit, D. in collab. with Jaschek, C.: 1982, *Catalogue of Bright Stars*, 4th ed., Yale Univ. Obs.

Jaschek, C. and Gomez, A. E.: 1970, *Publ. Astron. Soc. Pacific* **82**, 809.

Kumar, S. S.: 1967, in M. Hack, *Colloq. on Late Type Stars*, Trieste.

Lamontagne, R. and Moffat, A. F. J.: 1982, in C. W. H. de Loore and A. J. Willis (eds.), 'Wolf–Rayet Stars: Observations, Physics, Evolution', *IAU Symp.* **99**, 283.

Lippincott, S. L.: 1977, *Astron. J.* **82**, 925.

Luyten, W. J.: 1971, *Astrophys. Space Sci.* **11**, 49.

Lucy, L. B. and Ricco, E.: 1979, *Astron. J.* **84**, 401.

van de Kamp, P.: 1975, *Ann. Rev. Astron. Astrophys.* **13**, 295.

van de Kamp, P.: 1981, *Stellar Paths*, D. Reidel Publ. Co., Dordrecht, Holland.

Wolff, S. C.: 1978, *Astrophys. J.* **222**, 556.

Woolley, R. v. d. R. *et al.*: 1970, Royal Greenwich Obs. Annals, No. 5.

Worley, C. E.: 1969, in J. Kumar (ed.), *Low-Luminosity Stars*, Gordon and Breach, New York.

BINARY STATISTICS AND STAR FORMATION*

HANS ZINNECKER

Royal Observatory, Edinburgh, Scotland

(Received 2 November, 1983)

Abstract. Binary statistics, in particular the distributions of mass ratios and orbital periods, are reviewed in an attempt to obtain clues to possible star formation and cloud fragmentation processes. Various observational selection effects which hamper the establishment of the true distributions are discussed. Four different theories of binary formation are compared (fission, fragmentation, capture, and the disintegration of small star clusters), none of which can be ruled out. We conclude that there may be many ways to form binary systems. The dominant mode of binary formation could be ring fragmentation or disc fragmentation depending upon whether the distribution of mass ratios is found to decrease or to increase towards small mass ratios. Future speckle interferometric measurements of a sufficiently large sample of close visual binaries are suggested to settle this important observational question. The present paper is special in that it brings together a wealth of useful information, both observational and theoretical, in one place.

Atoms form molecules,
*stars form binaries.***

1. Introduction

1.1. BATTEN'S BINARY STATISTICS

The total frequency of binary and multiple systems as well as the distribution of the mass ratio among the components and the distribution of orbital periods are important data and constraints for the theory of star formation. It is well known that at least half of the stars in the solar neighborhood belong to binary or multiple systems (cf. Heintz, 1969; Abt, 1979, 1983); in the classical book on the subject (Batten, 1973) the statistics of binary and multiple systems read as follows (approximate figures):

$\frac{1}{2}$ of all stars are binaries or belong to systems of higher multiplicity;

$\frac{1}{3}$ of all binary systems belong to triple systems;

$\frac{1}{4}$ of all triple systems belong to quadruple systems, etc., ...

A binary star which deserves to be mentioned in the present context of star formation is the spectroscopic binary Delta Orionis (actually a triple system, since there is a distant companion) the spectrum of which led to the discovery of the interstellar medium by Hartmann in 1904: in contrast to the stellar lines which were shifting periodically in time, with a period of 5.7 days, its spectrum also showed 'stationary' Ca II (H and K) absorption lines which were correctly interpreted as being due to the calcium ions of an intervening interstellar gas.

---

* Paper presented at the Lembang-Bamberg IAU Colloquium No. 80 on 'Double Stars: Physical Properties and Generic Relations', held at Bandung, Indonesia, 3–7 June, 1983.
** We quote Su-Shu Huang in IAU-Colloquium No. 33 = Revista Mexicana, Vol. 3 (1977). Many papers related to the present problem can be found there.

Astrophysics and Space Science **99** (1984) 41–70. 0004–640X/84/0991–0041$04.50.

1.2. BASIC QUESTIONS, DEFINITIONS, AND NUMBERS

Given that a substantial fraction of stars formed in pairs, one may, quite generally, raise the following two fundamental questions, neither of which is satisfactorily answered at present:

(1) Do binary systems preferentially occur with components of roughly equal masses or with components of largely unequal masses?

(2) Do binary systems preferentially form close pairs or wide pairs?

Here we define the binary mass ratio as $q = M_B/M_A$, where M_A is the primary and M_B is the secondary ($M_A > M_B$ or $q < 1$ initially; however mass exchange or mass loss may cause $q > 1$ at a later stage). The distinction between close pairs and wide pairs is adopted as usual, i.e.: close pairs are close enough to be able to exchange mass during stellar evolution while wide pairs are not; the dividing line then is at a separation of a few AU for the binary components.

Observationally, there are two main categories of binary stellar systems: those which can be resolved into two components by a telescope, and those which appear to be one object visually, but can be identified as consisting of two stars due to their periodically variable spectrum. The former are called visual binaries (VB), the latter spectroscopic binaries (SB). The spectroscopic binaries are subdivided into two classes: spectroscopic binaries with single lines (SB1's) and with double lines (SB2's). For SB1's it is sufficient to observe a line shifting with time, whereas for SB2's it is necessary to observe a line splitting. SB2's directly allow the determination of the mass ratio of the binary components; SB1's only yield the mass function of the system (i.e., the quantity $Q = M_B^3 \sin^3 i/(M_A + M_B)^2$], so that the additional information about the primary mass and the inclination angle i is required, if the mass ratio is to be inferred. (The mass function must not be confused with the Initial Mass Function (IMF) which describes the relative proportions with which stars of different masses are born. The existence of unresolved binaries in star counts influences the definition of the IMF: see Appendix B.) The threshold in the projected radial velocity of the orbital motion for the detection of spectroscopic binaries is discussed in Appendix A. In the most recent SB-Catalogue (Batten *et al.*, 1978) with almost 1000 systems listed roughly $\frac{1}{3}$ are SB2's and the remaining $\frac{2}{3}$ are SB1's; 15–20% of the SB1's are eclipsing binaries, many of which have an evolved component. There are only a few dozens eclipsing SB2's. It is to them, however, that we owe almost all our knowledge of stellar masses and the calibration of the stellar mass luminosity relation (see Popper, 1967, 1980: Blaauw, 1981).

1.3. THE SCOPE OF THE PAPER

The main purpose of the present article will be to review some statistical data about binary stellar systems and to analyse them as potential clues to the theory of star formation. Unfortunately, most of the statistical data are severely biased by observational selection effects. Any catalogue of binary stars (e.g. IDS-*Catalogue of Visual Binary Stars* or the 6th and 7th *Catalogue of the Orbital Elements of Spectroscopic Binary Stars*) is bound to represent a very incomplete sample as will become evident in the course of the discussion below. Moreover the catalogue data are not a priori corrected for possible changes in the orbital elements due to interactions between close binary components.

In the past, observations of interacting close binaries have often been used to test the theory of stellar evolution. An evolutionary sequence has emerged (see the review by Trimble, 1983). In the future, observations of non-interacting wide binaries should be used to constrain the theory of star formation. It is hoped that the present article will provide a stimulus for the observers to attack this task.

The paper is divided into two main parts, an observational one on binary statistics (Section 2) and a theoretical one on the origin of binary and multiple stars (Section 3). A final chapter summarises the major conclusions and lists some suggestions for future work. A brief outline of the contents of Sections 2 and 3 is seen from Table I.

TABLE I

2. Binary Statistics (Observations)
   2.1. The Total Binary Frequency
   2.2. The Distributions of the Mass Ratio and the Orbital Period
   2.3. Multiple Systems
   2.4. Young Double Stars (Pop. I)
   2.5. Old Double Stars (Pop. II)
   2.6. Binaries in Other Galaxies
3. On the Origin of Binary and Multiple Systems (Theory)
   3.1. The Angular Momentum Problem in Star Formation
   3.2. Theories of the Formation of Binary Stars
   3.3. Implications from Binary Statistics

Subsections 2.4, 2.5, and 2.6 can be omitted on first reading. Two appendices deal with (A) the detection of binaries and (B) the Initial Mass Function corrected for the existence of binary stars.

Stellar rotation and its relation to star formation will not be discussed in the present paper, since there is an adequate, recent discussion of this by Franco (1983) (see also Woolfson (1978), Wesson (1979), Vogel and Kuhi (1981), Wolff *et al.* (1982), Guthrie (1983), Gray (1982), and Fleck (1982)).

2. Binary Statistics

2.1. THE TOTAL BINARY FREQUENCY

We shall not discuss the total binary star frequency on the Main Sequence. This has been done very carefully in a recent paper by Gieseking (1983). We wish to quote his summary: "Many authors (especially spectroscopists) tend to overestimate the fraction of detected binaries. But paradoxically, many authors at the same time may underestimate the total binary frequency, because overestimating their detection probability and/or overestimating the volume of the binary parameter space covered by their observations, they tend to underestimate the number of binaries which escaped detection. (A meaningful comparison between published binary frequencies is only possible, if they refer to equivalent volumes of binary parameter space.) The details of a possible variation of the binary frequency along the Main Sequence as well as its true value is not yet well established. There is no observational evidence for the binary frequency to vary strongly along the Main Sequence. There is observational evidence which suggests a nearly 100% frequency of binary and multiple systems among Main-Sequence objects."

Judging from the binary frequency of the immediate solar vicinity (taken at face value), we would infer a smaller fraction of the total binary frequency. For instance, van de Kamp's (1971) list of nearby stars (within a sphere of radius 5.2 pc around the Sun) contains 60 visible stars (including the Sun itself) of which 32 are single stars. The other 28 consist of 11 binary systems and 2 triple systems. However, the statistics of both spectroscopic and visual binaries are certainly not complete for small mass ratios, even for van de Kamp's ultralocal sample of stars. In some cases, there is a suspicion that the single stars are not really single but are circled by an invisible companion. Barnard's star is an example of such a system. McCarthy (1983) reports detection of unseen companions to 15 nearby stars due to infrared speckle observations.

Recently, Poveda *et al.* (1982) obtained a true fraction of visual binaries and multiples among field stars as high as 90%, with most of the companions remaining undetected. In order to produce a relatively uniform and homogeneous group of double and multiple stars free of optical and spurious systems and suited for statistical analysis, they applied a 'filter' to the about 70 000 entries of the updated *Index Catalogue of Visual Double Stars* (IDS-Catalogue) resulting in a 'filtered' catalogue with nearly 20 000 entries eliminated. It was on the basis of this catalogue that they found the duplicity frequency quoted above. They also found that this catalogue is largely complete for pairs brighter than 10th magnitude and with brightness differences less than 1 magnitude. Interestingly enough, roughly one out of three field stars turns out to be a visual multiple; thus there are far more multiple systems than indicated in Batten's (1973) book to which we referred in the Introduction.

Lastly, we note that the evaluation of true total binary frequencies is not possible without knowing at least the frequency distributions of the mass ratio and the separation of the binary components.

2.2. FREQUENCY DISTRIBUTION OF THE MASS RATIO AND THE ORBITAL PERIOD

2.2.1. *Mass Ratio*

The frequency distribution of mass ratios $f(q)$ was first investigated by Kuiper (1935) who found $f(q) = 2(1 + q)^{-2}$. The distribution is shown in Figure 1 as a challenge for everybody.

Kuiper (1935) believed this result to be true for all (close and wide) binaries. Heintz (1969) favored $f(q) \sim q^{1/2}(1 + q)^{-3}$ for all (spectroscopic and visual) binaries, while Popov (1970) gave $f(q) \sim q^2$ for spectroscopic binaries. Later, Trimble (1974, 1978) analyzed the *Sixth Catalogue of the Orbital Elements of Spectroscopic Binary Stars* and

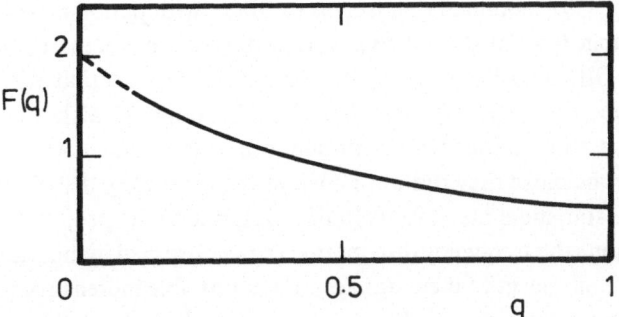

Fig. 1. Possible distribution of initial mass ratios for binary stars (after Kuiper, 1935).

its supplements, and suggested a bimodal frequency distribution of mass ratios with two peaks at $q \sim 0.25$ and $q \sim 1$. Staniucha (1979) analyzed the *Seventh Catalogue of the Orbital Elements of Spectroscopic Binary Stars* (Batten *et al.*, 1978) confirming the bimodal structure of the frequency distribution: the q-distribution of SB1's peaks at $q \sim 0.2$–0.3 that of SB2's at $q \sim 1$. However, as mentioned before, these 'results' are strongly biased by selection effects and evolutionary effects. Evolutionary effects have been discussed by Kruszewski (1967), Paczynski (1971), Thomas (1977), Shu and Lubow (1981), and Giuricin *et al.* (1983). One selection effect which contributes to the difference in the characteristic mass ratio for SB1's and SB2's could be that for a mass ratio near unity (SB2's) the primary and the secondary component of the binary system can move very much relative to the center of mass, while for a moderately small mass ratio (SB1's) only the slow motion of the primary relative to the center of mass can be observed (the fast moving secondary is too faint). The observed q-distribution of SB's is then interpreted in the following way (Gieseking, private communication): The q-distribution is bound to be bimodal, i.e. to have an apparent gap, because of the very low probability of detection of systems with moderate q. From $q = 1$ (for which the velocity amplitude and the probability of detection is largest) towards $q = 0$, the probability of detection starts to decrease due to decreasing double line splitting until there remains only one blended line. For smaller and smaller q the lines of the secondary component begin to disappear so that the lines of the primary component become cleaner and cleaner, eventually allowing the recognition of clear Doppler shifts (second maximum of the SB-detection probability). For $q \to 0$ the probability of detection starts to decrease again, since the velocity amplitude goes to zero. Another selection effect is that observers tend to study SB2's more carefully than SB1's; thus SB2's are overrepresented in Batten *et al.*'s (1978) catalogue.

In a detailed statistical study Lucy and Ricco (1979) claimed that the peak at $q \sim 1$ for SB2's is a real feature. A similar conclusion was reached by Kraicheva *et al.* (1979) who tried to calculate the initial q-distribution from the observed one taking into account models for the evolutionary effects.

Lucy and Ricco (1979)* confined themselves to SB2's with periods less than 25 days and masses in the range $0.5-10\,M_\odot$. Of course, they did realize that there is a paramount selection effect for SB2's to have $q \simeq 1$; however, they argued that the decrease to smaller q is too steep to be caused by undetected SB2's alone. Kraicheva *et al.* (1979)* considered spectroscopic binaries with semi-major axes less than 1 AU. Their conclusions may be shaky, because they did not consider the frequency distribution of $\sin^3 i$; they only invoke the statistical mean value (0.68). But, due to the projection effect, even for random inclinations the frequency distribution of the observed inclinations is proportional to $\sin i$. Since this was not taken into account, a possible increase of the frequency distribution of the mass ratio to small mass ratios is underestimated.

Abt and Levy (1976; see also Abt, 1977, 1978) made a study of 76 systems and derived results which are believed to be not seriously affected by selection effects. Their sample of primary components contained essentially all F3–G2 Main-Sequence stars brighter than $V = 5.5$ mag and north of $-20°$ declination. The narrow spectral range corresponds to a stellar mass of $1.2-1.3\,M_\odot$. The sample included 36 spectroscopic binaries (4 SB2's, 32 SB1's), 19 visual binaries, and 21 common proper motion (CPM) pairs. For binaries with periods less than 100 yr, they found a different distribution of the mass ratio than for binaries with periods greater than 100 yr. They attribute this difference to the operation of two distinct binary formation mechanisms in the two regimes.

Huang (1977) has noted that the discrepancy between the old distribution of mass ratios derived by Kuiper (1935) and the new distribution derived by Abt and Levy (1976) may be due to the condition imposed by Kuiper that the distribution is a function of $M_2/(M_1 + M_2)$ rather than M_2/M_1.

TABLE II

Distribution of mass ratios according to Abt (1978) for the sample of Abt and Levy (1976). Note that the total number of binary systems is $\Sigma_q\,N(q) = 88$ (not 76) (12 systems have been added on the basis of incompleteness calculations)

	$q \in (1, \frac{1}{2})$	$q \in (\frac{1}{2}, \frac{1}{4})$	$q \in (\frac{1}{4}, \frac{1}{8})$	$q \in (\frac{1}{8}, \frac{1}{16})$	$q < \frac{1}{16}$	
$N(q)$	13	16	11	8	2	$(P < 100\text{ yr})$
$N(q)$	8	11	19	–	–	$(P > 100\text{ yr})$

Table II illustrates the distribution of mass ratios (grouped in bins of factors of 2) as inferred from Abt (1978) adding up the numbers for the various period segments.

In our own opinion, the trend in Table III is not clear-cut, and the small numbers involved in the statistics should give rise to caution.

It seems that the number of long-period binaries increases with decreasing mass ratio (much like the distribution proposed by Kuiper, 1935), while the distribution of short-period binaries, if we do not put too much statistical weight on the figure for the lowest mass ratio, stays rather constant (contrary to what is fitted to the data by Abt, 1978, viz. $N \sim q^{1/3}$).

* Both papers are very valuable approaches to the problem in question.

Ducati and Jaschek (1982) have criticized Abt and Levy's procedure to derive mass ratios from their data (it is not the data that has been criticized). They conclude that the existing data permits one only to assert that SB1's with small ratios are more frequent than those with large mass ratios (cf. Jaschek and Ferrer, 1972; Jaschek, 1976).

Taken together, the situation for *spectroscopic* binaries is inconclusive as far as their distribution of mass ratios is concerned.

For the 14 (!) high-quality *visual* binaries listed in the review by Popper (1980) the mean value of $M_2/(M_1 + M_2)$ is 0.4, i.e., the mean value of $q = M_2/M_1$ is $\frac{2}{3}$; cf. also the list of Harris *et al.* (1963).

2.2.2. *Orbital Periods*

Frequency distributions of the orbital periods of binary stars have been given by Kuiper (1935, 1955), Brosche (1964), van Albada and Blaauw (1967), Heintz (1969), Kraicheva *et al.* (1979a, b), Staniucha (1979), and by Abt and Levy (1976; see also Abt, 1977, 1978). In principle, the binary period distribution is far easier to establish than the binary mass ratio distribution. In practice there is considerable difficulty in the range from 1 to 10 AU and the corresponding periods (cf. Blaauw, 1981).

We shall first discuss the distribution of the orbital periods for spectroscopic binaries. This distribution has a rather sharp maximum at a period of a few days (Kraicheva *et al.*, 1979a; Staniucha, 1979).

The corresponding distribution of orbital angular momentum per unit mass peaks at 5×10^{18} cm^2 s^{-1} for SB1's and 9×10^{18} cm^2 s^{-1} for SB2's (Kraicheva *et al.*, 1979a). (For comparison, the specific angular momentum of the solar system is 1.5×10^{17} cm^2 s^{-1}, 98% of which is in the orbital motion of Jupiter.)

The apparent characteristic period of a few days reflects the convolution of two effects. The first is that period distribution increases initially for increasing periods or semi-major axes, the second is that the probability of detection decreases with increasing periods or semi-major axes due to the fact that the velocity amplitudes become smaller and smaller. An additional effect is that observations become more and more tedious once the periods become too long.

We now proceed to discuss the period distribution found by Abt and Levy (1976) and given in Abt (1977, 1978). As previously stated, the sample chosen by Abt and Levy is supposed to represent a largely unbiased sample.

Our Figure 2 reproduces Figure 1 from the paper by Abt (1977) and shows the observed histogram of binary periods for the survey done by Abt and Levy (1976).

The most remarkable feature of the histogram is its broad, unimodal shape. The binary periods span a range of at least 6 orders of magnitude from less than a day to more than a thousand years. We also see that there is enough of an overlap between spectroscopic binaries and visual binaries that a bimodal distribution did not develop. A similar unimodal period distribution was also obtained by Heintz (1969) based on binaries of all spectral types, down to apparent magnitude 9. A paper on wide binaries in the solar neighborhood by Retterer and King (1982) gives a nice comparison of both

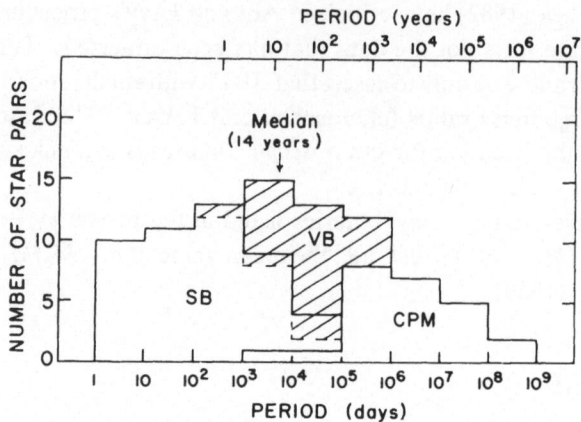

Fig. 2.   Best-bet distribution of orbital periods for binary stars (from Abt, 1977).

distributions. Note that, unlike Table II, Figure 2 is not corrected for incompleteness. Such a correction would probably be only a minor affair. It is very unlikely that it would change the unimodal distribution to a bimodal distribution. Earlier work concentrating on the period distribution of B0 to B5 stars (van Albada and Blaauw, 1967) did, however, seem to indicate a bimodal distribution. Abt and Levy (1976) remark that if they studied early-type stars where rotational broadened lines prevent the discovery of many of the spectroscopic binaries with long periods, and the greater distances of most of the stars prevent detection of most of the short period visual pairs, they would expect a bimodal distribution. Further investigation (Gieseking, 1983) shows hat the development of an apparent bimodal period distribution is generally expected and can be interpreted as being due to the low detection probability of periods intermediate between those characteristic for spectroscopic and visual systems. Notice also that the period distribution of SB's in Figure 2 has a maximum at about 1 yr; comparison with the maximum of the period distribution from analysis of the 6th and 7th Catalogue of SB's, being of the order of a few days, clearly shows that many of the longer period SB's have a low detection probability.

One more interesting aspect, common to the Heintz (1969) and the Abt (1978, 1979) survey, should be pointed out: both found that the period distribution depended only weakly on the spectral class of the binary (see, however, Brosche and Hoffmann, 1979).

Given the large spread of binary orbital periods, it is useful to define a physical classification for all binaries in terms of their orbital period (Table III). The nomenclature is chosen to be symmetrical and simple, and the corresponding figures are easy to remember.

A few comments concerning Table III may be in order:

For the extremely close systems ('contact systems') $P < 10^{-3}$ yr pertains to the low mass pairs, whereas $P \lesssim 5 \times 10^{-3}$ yr would pertain to high mass pairs. Of course, there is a range of periods in between. In the binary classification table the values given are for the low mass pairs, because they are by far the more numerous ones. For the

TABLE III

Physical classification of binary systems
(P = orbital period in years)

Extremely close binaries [a]	$P < 10^{-3}$
Very close binaries	$10^{-3} \lesssim P \lesssim 1$
Close binaries	$1 \lesssim P \lesssim 10$
Wide binaries	$10 \lesssim P \lesssim 10^2$
Very wide binaries	$10^2 \lesssim P \lesssim 10^3$
Extremely wide binaries [b]	$P > 10^3$

[a] Contact systems.
[b] CPM-systems.

extremely wide systems ('Common Proper Motion systems') the theoretical upper limit of the semi-major axis is $\sim 10^4$ AU (Retterer and King, 1982) corresponding to orbital periods $\sim 10^6$ yr. Statistical methods are required to separate physical pairs and merely optical pairs (see Poveda *et al.*, 1982).

Whether or not a single formation process is able to explain the huge dispersion of binary periods, all the way from the extremely close to the extremely wide systems, is a major issue of binary star formation theories.

2.2.3. *Correlation between Mass Ratio and Separation*

Having discussed the distribution of the mass ratio and the distribution of the orbital periods (or equivalently the distribution of the separation) of binary systems, it is natural to ask if there are any correlations between these elements. It turns out that small mass ratios clearly tend to go with small separations for the case of the observed spectroscopic binaries (Staniucha, 1979). It is quite obvious, though, that the correlation in part comes from the observational selection which acts against the discovery of binaries with large separation and very small mass ratio*. On the other hand, evolutionary effects such as mass transfer, will change the separation in a systematic way (see Giuricin *et al.*, 1983, for Algol binaries).

On the assumption of mass conservation and conservation of angular momentum the semi-major axis a of a binary changes according to $a \propto (M_A M_B)^{-2}$ (Equation (6) in Paczynski, 1971); thus a $q = 1$ binary would widen its orbit when its mass ratio becomes less than unity after mass exchange (by a factor ~ 2, if it ends up with $M_A : M_B = 1 : 3$).

One might speculate that the same effect operates when a binary system first forms. That would mean that $q = 1$ binaries should have systematically smaller separations than binaries with a small mass ratio but the same total mass.

---

* It is not obvious, however, how a recent new discovery fits in here; namely, the discovery that the frequency distribution of separations of early (B)-type Main-Sequence stars with visual low mass secondaries does *not* increase all the way towards smaller separations but reaches a maximum and decreases again (Lindroos, 1982).

Another puzzling fact is the discovery that close early (O)-type SB Main-Sequence stars lack mass ratios smaller than $q = 0.3$ (Garmany and Conti, 1980; Abt, 1983).

2.3. MULTIPLE SYSTEMS

We will not discuss triple systems but concentrate on quadruple systems only. (Dynamically, triple systems have much in common with quadruple systems, anyway)*.

Two distinct types of quadruple systems exist (Batten, 1973):

(a) hierarchical systems (Evans, 1968);

(b) trapezium-like systems (Ambartsumian, 1955).

Hierarchical systems (like Capella) consist of two close pairs being in a wide orbit around each other. Trapezium-like systems (as in Orion) share the property that all the member stars have roughly equal distances from each other. Recent observational work on hierarchical systems has been done by Fekel (1981) with special emphasis on the period ratio between the long and the short period, the question of coplanarity, and the mass ratios of the close pairs. The results were as follows:

For 25 systems with a long period of about 100 yr or less, the mean ratio of long to short period is roughly 3000, a factor ~ 12 higher than given in Batten (1973). A third out of 21 orbital pairs are not coplanar, for the rest coplanarity is a permitted possibility. Of the 25 short-period pairs whose mass ratio are known, 18 have mass ratios greater than 0.6. It is also noteworthy that two substantially different mass ratios can occur in the same system as is the case for the quadruple μ Ori.

Observations of Trapezium-like systems have recently been reported by Salukvadze (1980a, b) concentrating on the young T-associations in Orion and Tau/Aur. It is very interesting that Trapezium-like systems are not only common for massive OB-stars but also for low-mass T-Tauri stars. Of course, Trapezium-like systems have to be extremely young, because these systems are known to be very unstable ($\sim 10^6$ yr, see Allen and Poveda, 1974). Beichman *et al.* (1979) have found multiple compact infrared sources in molecular cluds which seem to be precursors to Trapezium-like systems of OB stars. The mean separation of these sources is 0.17 ± 0.04 pc based on a total of 14 systems which is very similar to the corresponding number for the mean separation of the stars in 31 Trapezium-type OB-clusters (0.12 ± 0.01 pc). Considerably more than half of the compact infrared sources come in double or multiple systems (Wynn-Williams, 1982). It is perhaps not surprising that the galactic distribution of Trapezium-type systems as derived from a total of 915 visual systems is strongly concentrated to the galactic plane (Allen *et al.*, 1977), since these systems are known to be young.

---

* In order to be stable, triple systems like quadruples have to be hierarchical, i.e. the separation of the third star C to the center of mass of the binary AB has to be large enough (see the discussion and the reference in Szebehely, 1977; Fekel, 1981).

A classical example for a stable triple system is Algol = β Persei at a distance of 30 pc. The long-period system Algol AB-C ($P = 1.862$ yr) is nearly coplanar with the eclipsing binary Algol AB ($P = 2.867$ days). Masses newly determined by speckle-interferometric observations (Bonneau, 1979) are: $M_A = 0.73 \pm 0.12 M_\odot$, $M_B = 3.4 \pm 0.6 M_\odot$, and $M_C = 1.7 \pm 0.2 M_\odot$.

2.4. YOUNG DOUBLE STARS (Pop. I)

2.4.1. *Protostars and pre-Main-Sequence Objects*

With the advent of modern observing techniques – such as infrared speckle interferometry – it has become possible to catch a couple of double stars still associated with their mother molecular cloud which gave birth to them. We have listed them in Table IV including tentative parameters as far as these are known. Note that the famous BN-object in the Orion molecular cloud appears to be a single star (Foy *et al.*, 1979), while the most powerful embedded Orion source IRc2 is probably double with a separation of 350 AU (Chelli *et al.*, 1983).

TABLE IV

Young objects discovered to be double

Name	Distance	Mass	Separation	Ref.	Comments
T-Tau	150 pc	$3 M_\odot$	145 AU	1, 2	$\Delta m_{3.8\mu} = 1.47$
MonR2 IRS3	950 pc	$8 M_\odot$	830 AU	3, 5, 6	triple?
W3 IRS5	2300 pc	$25 M_\odot$	3000 AU	3, 4, 6	member of a small cluster

Key to the references:

1 = Dyck *et al.* (1982a).
2 = Hanson *et al.* (1983).
3 = Dyck and Howell (1982).

4 = Howell *et al.* (19881).
5 = McCarthy (1982).
6 = Wynn-Williams (1982).

In addition to the young objects listed in Table IV, the first pre-Main-Sequence spectroscopic binary (named X-ray 1 or E0429 + 1755) has recently been discovered in the Taurus-Auriga star formation complex (Mundt *et al.*, 1983). It is a non-eclipsing SB2 with mass ratio unity and period 4 days.

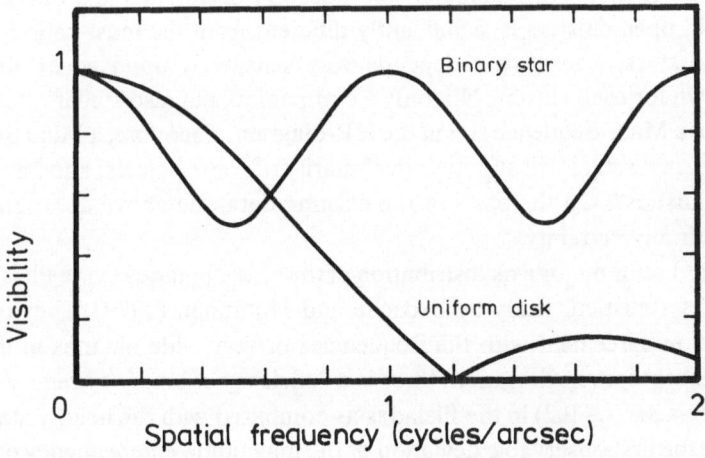

Fig. 3.   Typical result of an infrared speckle observation: square root of object power spectrum (visibility I) as a function of spatial frequency, for a binary system and a uniform disk. The visibility function is defined in the usual way that $(I_{max} - I_{min})/(I_{max} + I_{min})$ equals the brightness ratio of the binary components (1 : 3 here). The separation of the binary components is given by the inverse of the period of the spatial frequency (1″ here). (This figure has been kindly provided by M. Dyck.)

In the southern hemisphere S CrA has long been known to be a visual pair of pre-Main-Sequence stars (separation 1″, period ~ 1000 yr).

Finally, we draw attention to the list of pre-Main-Sequence stars compiled by Cohen and Kuhi (1979). A fair fraction of their optical pairs may well be physical pairs, i.e. visual binaries.

The infrared speckle technique allows are to distinguish a young binary system from a protostellar disk (see Figure 3). It also allows the determination of the binary's angular separation as well as the brightness ratio. Therefore, this new observational method holds great promise for the extension of binary statistics to star formation regions.

2.4.2. *Open Clusters*

It is clearly very important to ask whether binary statistics in open clusters are different from the binary statistics of field stars addressed so far in the previous sections. The general impression that prevails in the literature is that there is no striking difference between the total binary frequency in open clusters as compared to the field. For example, Jaschek (1976) gives 35% as the minimum average percentage of binaries for 4 well-studied clusters (Pleiades, Praesepe, Coma Berenices, and α Persei). For the Hyades cluster, the corresponding number is 40% (Carney, 1982). This fraction is slightly higher, presumably because the photometric identification was extended from the optical to the near infrared yielding a few binary candidates among the cluster dwarfs. Possible cluster-to-cluster variations in binary content have been discussed by Abt and Sanders (1973) who also claimed the existence of an anticorrelation between the axial rotational velocity of cluster stars and the percentage of cluster binaries (see also Abt, 1979; Levato and Morrell, 1983). A particular controversial case is the cluster IC 4665 (Abt and Snowden, 1964; Crampton *et al.*, 1976).

Trimble and Ostriker (1978) examined the question of whether the mass ratio in binary systems in open clusters is significantly different from the mass-ratio in binary systems for field stars. The answer depends very sensitively upon where the Main Sequence is drawn for each cluster. Not only a companion, but also stellar rotation can lift the position of a Main-Sequence star in the HR-diagram. Therefore, a Main Sequence fitted to rotating single stars will suppress the binarity of some objects, and hence serve to distort the statistics*. On the basis of the existing data, the above question cannot be answered with any certainty.

The number and semi-major axis distribution of the visual binaries in the Pleiades was the objective of a statistical study by Brosche and Hoffmann (1979). They obtained results that were in agreement with the frequencies of very wide binaries in the solar neighborhood, except for one deviation: there is a deficiency of binary systems with faint components ($6.5 < M_{vis} \leq 9.2$) in the Pleiades as compared with the nearby stars. This might constitute the first observable deviation of the magnitude independency of double star statistics.

---

* There is also the danger of confusion between single pre-Main-Sequence objects and Main-Sequence binaries (see Stauffer, 1982, for the Pleiades cluster).

Another major issue connected with binary stars in open clusters is the following: Do the orbital planes of cluster binaries exhibit a preferential orientation or do they not? Kraft (1965) investigating the Hyades and the Coma cluster came up with the conclusion that the orbital planes are randomly distributed, although he also stated that one cannot exclude a weak correlation. Note that Huang and Wade (1966) could not find a preferred orientation of the orbital planes of field star eclipsing binaries with the galactic plane. Finally, knowing that some clusters tend to have their fastest rotating stars towards the projected cluster center (Abt, 1970), it would be of great interest to get information about the spatial distribution of binary systems inside the clusters. Extensive speckle observations of clusters members may be a suitable way to approach this problem.

2.5. OLD DOUBLE STARS (Pop. II)

Was the dynamics of star formation different in the early epoch when the galactic disk was not yet formed? One approach to deal with this issue is to investigate the binary frequency among the metal-poor high-velocity stars which belong to the halo population (Pop. II). If the binary frequency among those stars turns out to be significantly different from the binary frequency among the low-velocity disk stars, then we have some evidence how important the large-scale dynamics of the collapse of the proto-Galaxy and perhaps metallicity were to the small-scale dynamics of star formation. Searches for halo binaries have been undertaken in both domains in which halo stars are found: in the field and in globular clusters. The 'primordial' binary fraction in globular clusters is difficult to determine, because of continuing binary disruption and formation (Hut, 1983), but the binary fraction plays a crucial role in the dynamical evolution of a globular cluster (Dokuchaev and Ozernoy, 1978; Spitzer and Mathieu, 1980). Trimble (1980) reviewed the interplay between theory and observations of binaries in globular clusters, and we will not discuss it further, except to say that the common belief that there are no close binaries in the globular clusters is almost certainly going to be challenged in the course of future observations (see Alexander and Budding, 1979, for a discussion of the selection effects).

Among the field halo stars, results of systematic searches for radial velocity variability have been published by Abt and Levy (1969) and by Crampton and Hartwick (1972). Abt and Levy concluded that the frequency of short-period spectroscopic binaries among Pop. II stars was lower than among stars of Pop. I. This conclusion was confirmed by Crampton and Hartwick in an enlarged sample of extremely metal-deficient subdwarfs. The Abt and Levy sample included 68 F- and G-type high-velocity dwarfs, and it was compared with a sample of 42 low-velocity dwarfs of about the same spectral types.

As far as the long-period or visual binaries are concerned, the frequency among Population II dwarfs may be similar to the frequency among Population I dwarfs. Observational data on this problem comes from the stars within 20 pc of the Sun (Gliese's catalogue). These stars can be separated into two categories according to their space velocities, and the binary fraction in the high-velocity Pop. II and the low-velocity Pop. I can be compared. In making such a comparison, Partridge (1967) was the first

to address the question of binarity among the halo field population. His result was: 18%
out of 127 stars with high (> 70 km s^{-1}) total space velocity are binaries, while out of
275   stars with low (< 40 km s^{-1}) total space velocity 23% are binaries, ~ 7% of
which are spectroscopic binaries. The validity of these percentages depends, of course,
on the absence of strong selection effects. On one hand, there may be an observational
bias against the detection of Pop. II binaries, since both members of such a pair will
necessarily be faint. Moreover, many Pop. II binary systems are expected to contain a
white dwarf, making the detection even more difficult. On the other hand, there is the
possibility that two nearby stars moving with the same high space velocity would be
more readily identified as members of a binary system than two stars with lower space
velocites.

Worley (1969) has discussed the duplicity characteristics of high velocity subdwarfs
(mostly spectral type F or G) finding at least 15 of 127 systems which were positively
identified as subdwarfs to be double, some of them close visual or spectroscopic pairs.
Remembering the observational selection effects, Worley concludes that it has not been
proved that binaries are any less frequent among Pop. II stars than amng younger
objects (see also Gehren *et al.*, 1981).

A fresh attempt to search for binaries in the halo dwarf stars is reported by Carney
(1983). The technique is based on *uvbyUBVRIJHK* photometry and involves the
color excess method ($B - V$ vs $V - K$ colors). The sample includes 71 stars and is free
from post-Main-Sequence objects. A binary can be identified via its flux distribution.
As a result, Carney estimates that the halo dwarf binary frequency may be as high as
20–25%.

New radial velocity work is also in progress. There is an ongoing radial velocity survey
of Lowell proper motion stars, about 500 with $7 \leq V \leq 13$ so far, with an extension
planned to $V \sim 15$. Repeated radial velocity measurements (accuracy < 1 km s^{-1}) have
been made of all stars with radial velocities exceeding 100 km s^{-1}. It appears that the
halo binary frequency is some 10% at least, but the analysis is still preliminary (Latham,
Stefanik, and Carney, private communication).

If the apparent lack of close binaries in the halo (in the field as well as in globular
clusters) is largely due to observational selection, the star formation processes in the
early stages of the Galaxy need not be different from those acting at present in the
galactic disk. However, Abt (1979) conjectured that metal poor stars are deficient in
close binaries, because for low metallicity a contracting protostar will have a greater
difficulty in radiating away excess energy and therefore will contract more slowly – slow
enough probably that it may be able to shed the excess angular momentum without
bifurcating (in this context see also Barry (1977) who studied binarity as a function of
metallicity for disk stars in the solar neighborhood).

Since wide binaries are as frequent for halo field stars as for disk field stars, there
is no reason to claim that halo stars form a low-angular momentum population (as stated
in the literature now and then). If anything, a possible deficiency of close binaries,
together with the normal incidence of wide binaries may be taken to indicate a high-
angular momentum population (presumably due to the highly turbulent velocity field
generated during an irregular protogalactic collapse).

2.6. BINARIES IN OTHER GALAXIES

Almost nothing is known about binaries in other galaxies. One thing we do know (Kopal, private communication) is the lack of bright eclipsing binaries in the spiral arms of M31: the absolute magnitude of the brightest eclipsing binaries in M31 is about 2 magnitudes fainter than in our own Galaxy. Another fact is the discovery of an eclipsing binary in Ursa Minor, a metal-poor dwarf elliptical galaxy in the Local Group (Webbink, 1980).

3. On the Origin of Binary and Multiple Systems

In this section we attempt to interpret the present statistical data in an effort to provide a framework for future observations related to binary statistics. Firstly, we introduce the angular momentum problem in star formation; secondly we will contrast four existing theories of binary formation with each other. Thirdly, we will consider the implications of present and future data for attempts to discriminate between the various binary formation theories.

3.1. THE ANGULAR MOMENTUM PROBLEM IN STAR FORMATION

3.1.1. *An Example*

Stars are formed mainly, though perhaps not entirely, in molecular clouds. Molecular clouds tend to be clumpy on all scales that have been resolved up to now (see, for instance, Walmsley's 1982 good overview of molecular clouds and star formation). Clumps having masses of the order of a solar mass corresponding to typical sizes ~ 0.1 pc in which the ambient gas density is $\sim 10^{-19}$ g cm^{-3} seem to be quite common. Although those appear to be rather quiescent (in the sense that their molecular linewidths are nearly thermal, i.e. 0.1 km s^{-1} at typical temperatures ~ 10 K), we estimate that the specific angular momentum of such a clump – a fair fraction of the product of the linewidth times the size – amounts to as much as $\sim 10^{21}$ cm^2 s^{-1}, a factor 10^4 higher than the specific angular momentum of the solar system. If this clump is to form a binary system with two $0.5 M_\odot$ components orbiting each other, and without losing angular momentum, their separation would have to be 5×10^3 AU, extremely wide indeed. The calculated separation is also in agreement with the expected shrinkage of a uniformly rotating sphere whose ratio $r = $ rot. energy/grav. energy in our case is $\approx 6\%$ (shrinkage = $3r$). The example that we have chosen here is realistic enough to demonstrate the angular momentum problem in star formation. Asking how it might be possible to obtain a binary system with a smaller separation from the above initial conditions is asking how to solve or to circumvent the angular momentum problem. The former implies transport of angular momentum (either local or global transport) while the clump contracts, the latter implies segregation of angular momentum, i.e. the specific angular momentum of each fluid element is conserved (at least up to the stage of fragmentation) yet only the low angular momentum material near the rotation axis finally ends up in a pair of protostars. Thus depending on transport or conservation of angular momentum the outcome of the collapse may be a binary with component masses $\sim 0.5\,M_\odot$ or $\sim 0.1\,M_\odot$ each. In either case, conversion of spin to orbital angular momentum is required to form a binary. Moreover, even after conversion, the spin of the fragments orbiting each other may still be too large to allow for further dynamical collapse of the fragments towards the high density which is characteristic for the stellar interior ($\sim 1\ \mathrm{g\ cm^{-3}}$). Hence, the fragments would remain fragments and could not become stars unless again transport of angular momentum is operative. Thus, strictly speaking, transport of angular momentum is absolutely essential for star formation, even for binary stars for which much of the angular momentum is stored in orbital motion. Hierarchical fragmentation without any transport of angular momentum would lead to substellar masses (Mestel, 1965; von Hoerner, 1968).

3.1.2. *The Collapse of a Rotating Cloud – General Results*

Let us follow the collapse of a rotating cloud in more detail in order to get an idea of how a binary system may actually be forming. Although fragmentation is fundamentally a three-dimensional problem, many of the results that we have come from two dimensional, i.e. axisymmetric numerical collapse calculations (for a recent review of numerical collapse calculations with emphasis on the effects of rotation see Bodenheimer (1981)). We shall distinguish between calculations based on conservation and transport of angular momentum, respectively.

(i) *Conservation of Angular Momentum*

Provided that the distribution of specific angular momentum is normal, people now agree that the axisymmetric collapse of a rotating cloud results in a toroidal density maximum ('ring structure') around the cloud center rather than in a central condensation (e.g. Tscharnuter, 1980). The reality of this feature was a matter of strong debate in the past. It was believed that artificial angular momentum transport towards the cloud center (caused by the numerical method) could be the reason for the appearance of the ring. Meanwhile comparative studies of different numerical codes and a higher order numerical scheme have confirmed the physical nature of the ring; in addition, convincing analytical arguments have been given (Tohline, 1980; Norman, 1980; Boss, 1980) showing that ring formation results from the dynamical competition between centrifugal and gravitational forces: while the collapse proceeds the fluid elements near the cloud center overshoot the centrifugal barrier and rebound in the radial direction colliding with the still infalling outer layers. This excites a toroidal density wave (cf. Bodenheimer, 1981, p. 24). The initial distribution of specific angular momentum within the cloud as well as the initial cloud density profile affect the position and the size of the ring (Tohline, 1980). Once the ring is formed, not only does it accrete mass but it also accretes angular momentum. Therefore the ring-like density maximum moves outward away from the rotation axis (the ring diameter grows). When the ring mass grows, the gravitational potential minimum moves into the ring, and the ring approaches a stage of hydrostatic gravo-centrifugal equilibrium (cf. Ostriker, 1964). At that point the ring mass (M), the ring diameter (D), and the specific angular momentum of the ring (J/M) are related approximately as $D \propto M$ and $(J/M) \propto M$.

As more mass is added to the ring, the ring may start collapsing on itself. With non-axisymmetric azimuthal density perturbations imposed on the ring $(\cos(m\phi)$-perturbations, $m = 2, 3, 4, \ldots)$, the ring will fragment so that two or more orbiting fragments will emerge (see Norman and Wilson, 1978, for isothermal rings; Cook and Harlow, 1978, for adiabatic rings; see also Lucy, 1981, whose method avoids the adoption of an initial perturbation mode). Since $(J/M) \propto M$ and M, the ring mass, is typically an order of magnitude less than the cloud mass, only 10% of the original specific spin angular momentum of the cloud show up in the orbital plus spin angular momentum of the fragments (the spin of each fragment is in turn of the order of 10% of the orbital angular momentum). If a binary system is formed, the mass ratio could be $q = 1$ but there is no a priori reason against a smaller mass ratio (Lucy, 1981). If a triple system is formed (see Boss, 1982), the final configuration might be a binary with the lightest member being ejected. In this case we expect the mass ratio to be closer to unity.

(ii) *Transport of Angular Momentum*

– *Local transport*. Turbulent friction has long been known to cause redistribution of angular momentum in rotating disks (von Weizsäcker, 1948; Lüst, 1952; see also Lynden-Bell and Pringle, 1974). Angular momentum flows to the outer parts of the disk and this enables mass to flow to the inner parts and to form a central condensation. In recent times, axisymmetric 2D numerical collapse calculations including turbulent friction (Regev and Shaviv, 1981; Tscharnuter, 1981) have shown that even small amounts of turbulent friction prevent toroidal structures from forming, but lead to a central stellar object surrounded by a disk in approximately centrifugal equilibrium. The bulge-to-disk mass ratio after the dynamical phase depends on the efficiency of the redistribution of angular momentum. In 2D, most of the disk gas will be accreted onto the single, central object but, in 3D, the possibility of disk fragmentation remains, especially for cold disks (Quirk, 1973; Genkin and Safronov, 1975; Schmitz, 1983). For dynamical and geometrical reasons it appears likely that the mass ratio in binaries resulting from disk fragmentation would tend to be rather small, although this is only a guess which calls for confirmation by detailed calculations. In 3D, gravitational torques from a central bar-like or triaxial structure (Wood, 1981) or from spiral density waves (Larson, 1983) may allow much of the mass to fall to the center, and the formation of binary or multiple systems may be due to independent condensation of the components, similar to what Larson (1978) finds in his numerical calculations of cloud fragmentation.

– *Global transport*. The magnetic field permeating a rotating fragment can carry a substantial amount of the fragment's angular momentum to the external cloud medium, provided that the magnetic field in the fragment is connected with the external medium (see Mestel, 1965). The reason for such efficient 'magnetic braking' lies in the fact that the moment of inertia of the external medium is large. Matter in the external medium near the fragment is pulled by the field lines in the direction of motion of the fragment;

a torque is imparted to it at the expense of the rotational motion of the fragment which therefore most rotate slower than before (Mouschovias, 1981; review paper). In the case of magnetic field lines perpendicular to the axis of rotation the braking is more efficient (typically by an order of magnitude) than for the aligned rotator where magnetic field and rotation vectors are parallel to each other. The perpendicular case can even result in retrograde rotation of the fragment (Mouschovias and Paleologou, 1980; Dorfi, 1982). Ambipolar diffusion, in other words the gradual separation between the ionised matter component of the fragment and its neutral matter component at high gas density (Black and Scott, 1982) ultimately limits the efficiency of this angular momentum transport mechanism. For a discussion of the time scales of ambipolar diffusion and of magnetic braking (more or less of the order of the free-fall time) we refer again to Mouschovias (1981). In summary, as stated by Mouschovias and Paleologou (1980), the magnetic field seems to remain frozen in the matter long enough to resolve much of the angular momentum problem, and decouples from the matter rapidly enough for solar-type stars to form within about 3×10^7 yr, as evidently required by observations of the spatial separation of young stars and the apparent location of a spiral shock wave.

3.2. THEORIES OF THE FORMATION OF BINARY STARS

There are four rival theories of the formation of binary stars in the literature which, in turn, may be subdivided into two groups roughly related to the observational classification into spectroscopic binaries (SB) and visual binaries (VB). These basic binary formation mechanisms are displayed in Table V.

TABLE V

The 4 basic binary formation mechanisms

SB (close pairs)	VB (wide pairs)
Fragmentation	Capture (indep. condens.)
Fission	Disintegration (small clusters)

The physical difference between fragmentation and fission is explained in the following way (cf. Lucy, 1981): Fragmentation results from the break-up of a rotating protostellar cloud into two or more pieces during or immediately following a phase of dynamical collapse. Fission results from the bifurcation of a rotating protostar during its quasi-static pre-Main-Sequence Kelvin–Helmholtz contraction, if the ratio of the rotational to the gravitational energy density exceeds 0.25 (dynamical instability) or 0.14 (secular instability) according to Ostriker and Bodenheimer (1973).

– *Fragmentation.* The current idea of fragmentation differs from the former picture in which fragments spontaneously appear when their masses are greater than the local Jeans mass (Hoyle, 1953; Hunter, 1962). Density perturbations can, in fact, damp initially due to pressure effects, unless the perturbation amplitude is high enough. After about one initial free-fall time, when the cloud collapse is slowed down by pressure effects parallel to the rotation axis and primarily rotational effects perpendicular to the axis, the fragmentation begins. Fragmentation can occur either directly as a consequence of the initial perturbation imposed on the cloud, or through an intermediate ring stage depending on the thermal energy of the cloud measured in units of the gravitational energy. The dominant mode of fragmentation seems to be the binary mode but in many calculations this mode is imposed on the cloud from the beginning. A low thermal energy favors the formation of multiple systems. A comprehensive summary of all these results about fragmentation of an isothermal cloud is given in a paper by Bodenheimer *et al.* (1980). The properties of the fragments in the isothermal case are such that they are unstable to further collapse. The fragments form in the innermost part of the cloud which has a lower angular momentum per unit mass than the average for the cloud. This effect, combined with the conversion of spin to orbital motion, results in a reduction of spin angular momentum per unit mass by a factor 10 to 20 from that of the initial cloud. Thus, after a series of several collapses and fragmentations the specific angular momentum as well as the fragment masses can be reduced by considerable factors. Bodenheimer (1978) extrapolated that such a hierarchical process could result in direct evolution from a massive interstellar cloud to Main-Sequence binary and multiple systems within the observed range of masses* and orbital angular momenta.

Larson (1978) approached the fragmentation problem with a different numerical method, simulating hydrodynamics by a coarse fluid particle code. This method implies efficient transport of angular momentum, and fragmentation turns out to be quite direct rather than hierarchical with several steps. Multiple systems or small star clusters seem to be the outcome for clouds comprising several Jeans masses (i.e. clouds with low thermal energy at the onset of their collapse). It is also noteworthy that the fraction of cloud mass ending up in protostellar objects is higher by far in Larson's scheme compared with Bodenheimer's which leads to a fraction less than 1%.

– *Fission.* The outcome of numerical fission calculations for optically thick, rotating protostars is currently uncertain. Following the pioneering 3D numerical approach to the fission problem (Lucy, 1977, for apolytropic index 0.5 and a uniformly rotating initial model), further similar 3D numerical experiments of differentially rotating polytropes (Gingold and Monaghan, 1978, for polytropic indices 0.5 and 1.5) have been performed. The results of these calculations were diverse: fission accompanied by mass shedding (40%), fission without mass loss, and no fission but mass shedding (25%), respectively, were found. According to Lucy (1977) and Lucy and Ricco (1979) fission results from the instability of the third harmonic of an elongated triaxial rotating ellipsoid yielding a mass ratio of the components of about one third. In contrast with this result, Hachisu and Eriguchi (1982) obtained fission with mass ratio unity for the case of incompressible polytropes (i.e. zero polytropic index). They investigated the dumbbell- or bone-shaped equilibrium sequence of rigidly rotating polytropes. Durison and Tohline (1981) also

---

* In fact, such a *multiplicative* star formation process is able to explain the shape and the dispersion of the Miller/Scalo-IMF (Zinnecker, 1981).

studied the fission hypothesis for rotating polytropes of polytropic index 1.5 whose angular momentum distribution was that of a rigidly rotating uniform sphere. They imposed density perturbations which evolved into a dumbbell configuration for very rapid rotation. This configuration develops inside corotation while outside corotation material is ejected in the form of two spiral arms which wrap due to differential rotation, emerge into a detached disk, and eventually narrow into a radially expanding ring. The ring contains 16% of the mass but more than half the angular momentum. A similar process had been suggested earlier by Drobyshevski (1974). His idea was as follows: After collapse, convection is set up quickly in the protostar. Convection will make rotation uniform, even if it was initially differential, so that the balance between centrifugal and gravitational forces in the outer layers will be destroyed. These layers will then be thrown off forming a ring around the star. The mass loss results in an incrase of the convective zone in extent so that transfer of matter to the ring will proceed until all the convective envelope is lost. The ring is unstable and will probably form a second component. The mass ratio of the new system is therefore determined by the mass ratio of the convective zone and the stable core of the protostar. However, irrespective of the initial mechanism of separation of the components, equalisation of their masses may occur in the course of subsequent disc accretion of matter onto the system (cf. McCrea, 1956).

Finally, concerning fission, we would like to refer to an earlier review paper of Ostriker (1970).

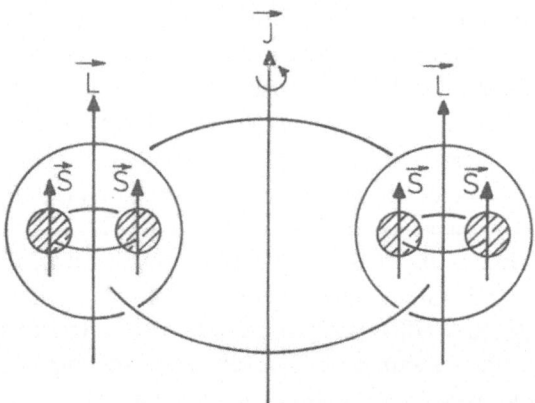

Fig. 4a.   Bodenheimer's scheme of hierarchical ring fragmentation.

Fig. 4b.   Dumbell-equilibrium sequence and fission of a protostar (from Hachisu and Eriguchi, 1982).

The fragmentation and the fission process are illustrated in Figures 4a and 4b, respectively. After such a long discussion of fragmentation and fission theory (warranted because it is the close binaries whose origin is the big issue) the two theories for the origin of wide binaries (capture theory and cluster disintegration theory) will be dealt with more briefly.

– *Capture.* Capture theory assumes stars to originate from single independent nearby condensations which subsequently get bound into a wide binary when fragments (proto-stellar cores with extended accretion envelopes) collide with non-zero impact parameters (greater than the sum of the core sizes), and viscous interaction dissipates the kinetic energy of the relative motion of the fragments (Silk, 1978; Silk and Takahashi, 1979). Capture theory implies random pairing in a protocluster cloud. The chances of capture through three-body encounters of galactic field stars are very small, although this kind of capture does occur in dense star clusters like globular clusters. Random pairing during star formation according to a stellar mass spectrum increasing towards low masses implies that the distribution of the mass ratio increases towards low mass ratios. A Salpeter-type mass spectrum would, however, be too steep at the low mass end to be consistent with the distribution of the mass ratio given in Figure 2 (Warner, 1962a, b). A Miller/Scalo-type mass spectrum which increases only moderately towards lower masses is likely to be consistent (cf. Abt, 1978). The distribution of orbital periods expected from capture theory may explain the observed period distribution for periods exceeding 100 yr (see Silk, 1978). An intriguing test of the capture theory for wide binaries would be to observe the spin vectors of the binary component (via the method proposed in Strittmatter, 1981). If the spin vectors are poorly or even randomly orientated to each other, this might be taken as evidence in favor of capture theory. Similarly the absence of any correlation between spin and orbital angular momentum would support capture theory.

– *Cluster disintegration.* Cluster disintegration theory assumes stars to form in small bound clusters (see e.g. Figure 4 in Herbig, 1977) which subsequently lose most of their members by close two-body encounters leaving behind a wide binary which absorbs and carries the initial binding energy of the whole cluster (van Albada, 1968a, b). The left over binary most likely consists of the two heaviest members of the cluster, so the statistical distribution of the mass ratio should increase towards unity if this case prevails. If the cluster initial had n stellar members with mean separation $\bar{a}$ (0.01 pc), then, for equal masses, from the principle of energy conservation we have $\bar{a} = a_{**}n(n-1)/2$, a_{**} being the separation of the remaining binary (e.g. $a_{**} = 0.1\bar{a}$ for $n = 5$ or $a_{**} = 0.02\bar{a}$ for $n = 11$). Since during the condensation of the cluster stars neighboring protostars had a larger separation, tidal disruption during the formation stage was less of a problem than it would be if the binary components formed with the closer separation right away (Kumar, 1972). The disadvantage with the cluster disintegration theory is that it is not a very prolific process of binary formation. However, cluster disintegration can obviously account for the formation of multiple systems (Aarseth, 1977) including Trapezium-type systems (Allen and Poveda, 1974).

3.3. IMPLICATIONS FROM BINARY STATISTICS

3.3.1. *Mass Ratios*

The all important question is whether the true frequency distribution of the mass ratio in close binaries increases or decreases towards small mass ratios. A steady increase towards small mass ratios would neither be consistent with fragmentation theory nor with fission theory. These theories predict component masses of comparable size. While ring fragmentation might prefer $q = 1$ for $m = 2$ perturbations, ring fragmentation cannot be said to lead to $q = 1$ for the most general perturbations (Lucy, 1981). On the other hand, if ring fragmentation leads to $q \neq 1$, subsequent disk accretion onto the binary may equalise the component masses. The same is true for fission if it results in $q \neq 1$ in the first place. Mass exchange during the contact phase might change the fission mass ratio to $q = 1$ (Lucy, 1977). However, the fission process itself has not been studied extensively enough to claim that it does not produce $q = 1$ in many cases. The conclusion from these considerations is that the present theory for the formation of close binaries is in trouble if future observations do indeed reveal that the frequency of the mass ratio in close binaries is not peaked near $q = 1$ but increases for $q \to 0$. In that case the dominant formation process could be the fragmentation of a gaseous disk surrounding the central condensation. As far as the mass ratio in wide binaries is concerned, capture theory better accounts for the observed frequency of the mass ratio than does cluster disintegration theory. This frequency increases towards small mass ratios, while cluster disintegration theory would predict the opposite trend. This, however, does not render cluster disintegration theory obsolete. The existence of very wide binaries, triple systems with a distant faint third body, and Trapezium systems may well require the disintegration process.

3.3.2. *Orbital Periods*

The crucial question here is whether a single process of star formation can explain the broad unimodal (roughly log-normal) frequency distribution. The wide spread in binary periods or equivalently in orbital angular momentum per unit mass could either be attributed to various mechanisms that change the separation after the formation of binaries by a single process (Kuiper, 1955) or to several formation processes (like those discussed previously in Section 3.2). (An incisive discussion of this problem is given by Huang, 1977.) Moreover, the wide spread of binary orbital angular momentum could have been imposed on the protostellar clouds before cloud collapse and fragmentation into a binary got underway, for example by the process of magnetic braking (Mouschovias, 1977), by the processes of hydrodynamical turbulence (Woolfson, 1978; Larson, 1981), or due to cloud-cloud collisions (Horedt, 1982).

Hierarchical fragmentation à la Bodenheimer (1978) can, in principle, serve as a mechanism to explain the distribution of orbital angular momentum if the number of steps in the hierarchy follow a statistical distribution centered on a most likely number of steps equal to 3 within a range 3 ± 2. We recall that each step corresponds to an order of magnitude in orbital angular momentum per unit mass (see the fragmentation table in Bodenheimer, 1978). The problem is how a very wide binary having gone through only 1 or 2 steps can achieve its very low ratio of spin to orbital angular momentum, while for very close binaries the same ratio is relatively high. This shows that for wide binaries additional processes must be operating to brake the spin of the two subcondensations. Magnetic braking is a good candidate process, since it is efficient primarily during the earlier stages of a contracting protostellar cloud. However, at every step in the hierarchy there is a chance to redistribute angular momentum, and that is why we have invoked a statistical distribution in the number of steps (if there is no redistribution at any step, the hierarchical fragmentation requires the full number of five steps as in Bodenheimer (1978)).

The observed period ratios between the long and the short period in hierarchical quadruple systems (of the order of 1000) is naturally explained in terms of two steps of the hierarchical ring fragmentation scheme (see again the fragmentation table in Bodenheimer, 1978), although hierarchical quadruple systems could also result from the dynamical decay of Trapezium-type quadruple systems. Trapezium-type quadruple systems themselves may either be the result of cluster disintegration (see above) or fragmentation through a $m = 4$ perturbation for very cold rings (Norman and Wilson, 1978; Rozyczka et al., 1980). The coplanar orbits in some of the observed systems allow us to conclude that rotational hierarchical fragmentation does in fact occur, rarely perhaps, but at least in some favorable cases.

4. Conclusions and Suggestions

4.1. CONCLUSIONS

(1) Various binary formation mechanisms exist. These include:
   (a) fragmentation of a collapsing protostellar cloud;
   (b) fission during pre-Main Sequence contraction;
   (c) capture after independent condensation;
   (d) disintegration of small star clusters.
(2) The initial frequency distribution of the mass ratio, once it is established from observations free from selection effects, contains the outstanding information about the dominant binary formation mechanism(s). If largely unequal component masses where the rule rather than the exception for close binaries (an unpopular view at present), a new additional formation mechanism would be required ('the fifth mechanism'; possibly the fragmentation of a gaseous disk surrounding a centrally condensed protostar).
(3) The four basic mechanisms (a) through (d) are able to provide complementary coverage of a wide range of binary separations. The multimodal origin of binaries coupled with a mixture of initial conditions and other processes such as transport of angular momentum is likely to account for the broad, unimodal frequency distribution of the orbital periods. The period ratio $P(\text{long})/P(\text{short})$ in hierarchical

quadruple systems is well explained by hierarchical ring fragmentation during the rotating collapse.

(4) Transport of angular momentum is vital not only for single star formation but also for the formation of binary stars. Magnetic braking is an efficient transport mechanism up to a moderately high gas density.

(5) Binary statistics in open clusters is probably not different from that of Pop. I field stars, implying that star formation processes are not altogether different in both categories. The similar statistics may also imply that single stars are not predominantly the ejecta of unstable triple systems.

(6) More and more Pop. II field stars (high velocity subdwarfs) are now found to be close binaries, challenging previously reported results. Selection effects may operate against the detection of eclipsing binaries in globular clusters. Thus it appears premature to claim that Pop. II forms a low angular momentum population.

4.2. SUGGESTIONS

(a) *Concerning theory:*

(1) Develop a fluid particle code with $N > 10^4$ particles for a Cray-1 computer to simulate the 3D hydrodynamics of fission and fragmentation starting with random initial conditions.

(2) Try to model $q = 1$ ring fragmentation analytically. Calculate how the specific ring angular momentum is split into orbital and spin angular momentum. Check whether the fragments so formed rotate slowly enough to undergo substantial contraction.

(b) *Concerning observation:*

(1) Free the SB-catalogues from eclipsing variables and evaluate the data anew.

(2) Check whether newer data reduce the dispersion around the $q = 1$ peak of the frequency distribution of the mass ratio in Lucy and Ricco's work on SB2's.

(3) Check the SB2-data with respect to a possible correlation between the mass ratio and the period for a given total mass of the system (see Section 2.2.3).

(4) Make maximum use of the exciting possibilities of speckle interferometry in observational binary star research. Discover nearby ($\lesssim 100$ pc) double stars down to separations of about 0.025 arc sec (the diffraction limit of a 4 m optical telescope); this would correspond to linear separations less than 2.5 AU, i.e. to close binaries (visual rather than spectroscopic). By measuring the intensity ratio of the components at several wavelengths it may be possible to infer the mass ratio. In that way we could approach the real frequency distribution of the mass ratio for close binaries.

The speckle method should also be used to study the separations in the difficult range from 1 to 40 AU for a well-defined sample such as the members of nearby open star clusters (e.g. the Hyades and the Pleiades).

(c) *The future in binary star research:*

It is our belief that the future in binary star research belongs to speckle observations (optical and infrared), although competitive developments are also in progress (better radial velocity spectrometers, fuller exploitation of lunar occultations). Since the speckle technique can in principle distinguish between binary stars and single stars surrounded by circumstellar disks, the improvement of this technique may ultimately result in an understanding of the bifurcation between the formation of a binary system and the formation of a planetary system associated with a single star. Meanwhile the answer to that problem is up to theory.

Acknowledgments

I am deeply indebted to F. Gieseking and C. Rucinski for illuminating discussions on selection effects in binary star research. Helpful comments from L. Lucy on binary formation are also appreciated. Furthermore, thanks go to B. Carney who has supplied me with new information on Pop. II binaries, and to M. Dyck who provided me with Figure 3. S. Shore was a big psychological help when this work was written up during the Les Houches Summer School 1983, and thanks to S. Heathcote, B. Guthrie, and again to F. Gieseking for their suggestions after reading the manuscript.

It is a pleasure to thank B. Hidayat and J. Rahe for their kind invitation to come to Indonesia. Finally, I thank the Deutsche Forschungsgemeinschaft for a travel grant which enabled me to attend the Bandung IAU-Colloquium No. 80 (and, after that, to watch the June 11 total solar eclipse). The patience of the editor, Z. Kopal, with the manuscript is gratefully acknowledged.

Appendix A: The Detection of Binary Systems

(a) *Visual binaries*

Visual binaries typically have mean separations of 3–5 arc sec depending on luminosity. These separations translate into semi-major axes ranging from 5 AU (for KM dwarf pairs) up to 100 AU (for OB star pairs). The difference arises, because the intrinsically fainter binaries are statistically less distant from the Sun than the intrinsically bright binaries. The recent development and application of the speckle interferometric technique represent a major break-through in visual binary star research (see McAllister, 1977; Morgan *et al.*, 1978; Bonneau *et al.*, 1980; Bonneau and Foy, 1980). An angular resolution of the order of $0''.1$ (or even less) can be achieved and the data are just about to incrase by a great deal (McAllister *et al.*, 1983). Interestingly enough, the speckle masking method (Weigelt and Wirnitzer, 1983) is capable of measuring the intensity ratio of SB components to an accuracy of 25% (at 5000 Å).

(b) *Spectroscopic binaries*

The threshold in the projected radial velocities of the orbital motion about the center of mass of a binary system for detection of SB1's is typically 3 km s^{-1} (or somewhat less), whereas that for detection of SB2's is typically 10 km s^{-1} (or somewhat more). These numbers hold for intermediate spectral types (A0–K0). For very early spectral types (OB) and very late spectral types (KM) the velocity threshold is higher. For

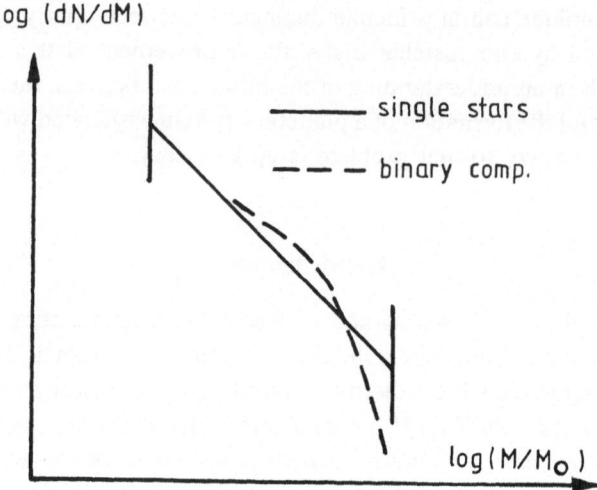

Fig. 5.   Model stellar mass spectrum ($M_{min} \leq M \leq M_{max}$) illustrating the effect of the presence of binary systems.

OB-binaries this is due to rotational broadening of the lines caused by the rapid rotation of the individual components about their spin axes; for KM-binaries it is due to the appearance of molecular bands.

By virtue of Kepler's 3rd law the above thresholds for intermediate spectral types correspond to semi-major axes typically less than 1 AU for SB2's and typically less than 10 AU for SB1's. As a rule of thumb, SB1's are binary systems in which the brightness difference between the two components is more than 1 mag while it is less than 1 mag for SB2's. It is important to keep this in mind.

Appendix B: The Problem of Correcting the IMF for Unresolved Binary Systems

In the definition of the Initial Mass Function (IMF) it is implicitly assumed that the fraction of stars which are double is independent of mass ('random pairing'). Otherwise one would have to distinguish between the IMF of single stars (IMF$_0$) and the IMF of (1) the primaries of binary stars (IMF$_1$) and (2) the secondaries of binary stars (IMF$_2$). This problem which is usually ignored (Hartmann, 1970) may be investigated qualitatively by a simple thought experiment: Suppose there would be only single stars with a mass distribution which we may denote IMF$_0'$ (e.g. a power law). Now, let a fraction of single stars split into binary stars with the mass of the primary component being M_1 and the mass of the secondary component being M_2 ($M_0 = M_1 + M_2$). If there is preferential splitting of the higher mass stars (a likely situation, because these are faster rotators, as is known), the effect will be a depletion of the high mass end of the total IMF (IMF$_{tot}$ = IMF$_0$ + IMF$_1$ + IMF$_2$) compared to IMF$_0'$, i.e. a steepening.

By mass conservation the intermediate masses will be more frequently populated in the distribution function IMF$_{tot}$ (see Figure 5). We suggest to call IMF$_{tot}$ the true IMF.

Since the observed IMF should correspond to a superposition of IMF_0 and IMF_1 ($IMF_{obs} = IMF_0 + IMF_1$) it is obvious that the observed IMF is not really the true IMF, since it is not corrected for the fainter unresolved companions of the primaries of binary systems. We conclude that the true IMF is somewhat steeper at the high mass end than the observed IMF, if the binary fraction is larger for the higher mass stars (cf. Vanbeveren, 1982).

References

Aarseth, S.: 1977, *Revista Mexicana Astron. Astrofiz.* **3**, Secial Issue (*IAU Colloq.* **33**), 199.

Abt, H. A.: 1970, in A. Slettebak (ed.), *Stellar Rotation*, D. Reidel Publ. Co., Dordrecht, Holland, p. 193.

Abt, H. A.: 1977, *Revista Mexicana Astron. Astrofiz.* **3**, Special Issue (*IAU Colloq.* **33**), 47.

Abt, H. A.: 1978, in T. Gehrels (ed.), *Protostars and Planets*, Univ. of Arizona Press, p. 323.

Abt, H. A.: 1979, *Astron. J.* **89**, 1519.

Abt, H. A.: 1983, *Ann. Rev. Astron. Astrophys.* **21**, 343.

Abt, H. A. and Levy, S. G.: 1969, *Astron. J.* **74**, 908.

Abt, H. A. and Levy, S. G.: 1976, *Astrophys. J. Suppl.* **30**, 273.

Abt, H. A. and Sanders, W. L.: 1973, *Astrophys. J.* **186**, 177.

Abt, H. A. and Snowden, M. S.: 1964, *Astrophys. J.* **139**, 1139.

Alexander, M. E. and Budding, E.: 1979, *Astron. Astrophys.* **73**, 227.

Allen, C. and Poveda, A.: 1974, in Y. Kozai (ed.), 'The Stability of the Solar System and the Small Stellar Systems', *IAU Symp.* **62**, 239.

Allen, C., Tapia, M., and Parrao, L.: 1977, *Revista Mexicana Astron. Astrofiz.* **3**, 119.

Ambartsumian, V. A.: 1955, *Observatory* **75**, 22.

Barry, D. C.: 1977, *Nature* **268**, 509.

Batten, A. H.: 1973, *Binary and Multiple Systems of Stars*, Pergamon Press, New York.

Batten, A. H., Fletcher, J. M., and Mann, P. J.: 1978, *Publ. Dom. Astrophys. Obs.* **15**, 121.

Beichmann, C. A., Becklin, E. E., and Wynn Williams, C. G.: 1979, *Astrophys. J.* **232**, L47.

Blaauw, A.: 1981, Proc. ESO Conference *Scientific Importance of High Angular Resolution at Infrared and Optical Wavelengths*, p. 391.

Black, D. C. and Scott, E. H.: 1982, *Astrophys. J.* **263**, 696.

Bodenheimer, P.: 1978, *Astrophys. J.* **224**, 488.

Bodenheimer, P.: 1981, in D. Sugimoto, D. Q. Lamb, and D. N. Schramm (eds.), 'Fundamental Problems in the Theory of Stellar Evolution', *IAU Symp.* **93**, 5.

Bodenheimer, P., Tohline, J. E., and Black, D. C.: 1980, *Astrophys. J.* **242**, 209.

Bonneau, D.: 1979, *Astron. Astrophys.* **80**, L11.

Bonneau, D. and Foy, R.: 1980, *Astron. Astrophys.* **86**, 295.

Bonneau, D., Blazit, A., Foy, R., and Labeyrie, A.: 1980, *Astron. Astrophys. Suppl.* **42**, 185.

Brosche, P.: 1964, *Astron. Nachrichten* **288**, 33.

Brosche, P. and Hoffmann, M.: 1979, *Astrophys. Space Sci.* **63**, 467.

Boss, P. A.: 1980, *Astrophys. J.* **237**, 866.

Boss, P. A.: 1982, *Icarus* **51**, 623.

Carney, B. W.: 1982, *Astron. J.* **87**, 1527.

Carney, B. W.: 1983, *Astron. J.* **88**, 623.

Chelli, A., Perrier, C., and Lena, P.: 1983, ESO-preprint (submitted to *Astrophys. J.*).

Cohen, M. and Kuhi, L. V.: 1979, *Astrophys. J. Suppl.* **41**, 743.

Cook, T. L. and Harlow, F. H.: 1978, *Astrophys. J.* **225**, 1005.

Crampton, D. and Hartwick, F. D. A.: 1972, *Astron. J.* **77**, 590.

Crampton, D., Hill, G., and Fisher, W. A.: 1976, *Astrophys. J.* **204**, 502.

Dokuchaev, V. I. and Ozernoy, L. M.: 1978, *Soviet Astron.* **22**, 15.

Dorfi, E.: 1982, *Astron. Astrophys.* **114**, 151.

Drobyshevski, E. M.: 1974, *Astron. Astrophys.* **36**, 409.

Ducati, J. R. and Jaschek, C.: 1982, preprint.

Durison, R. H. and Tohline, J. E.: 1981, in D. Sugimoto, D. Q. Lamb, and D. N. Schramm (eds.), 'Fundamental Problems in the Theory of Stellar Evolution', *IAU Symp.* **93**, 109.

Dyck, H. M. and Howell, R. R.: 1982, *Astron. J.* **87**, 400.

Dyck, H. M., Simon, T., and Zuckerman, B.: 1982, *Astrophys. J.* **255**, L103.

Evans, D. S.: 1968, *Quart. J. Roy. Astron. Soc.* **9**, 388.

Fekel, F.: 1981, *Astrophys. J.* **246**, 879.

Fleck, R. C.: 1982, *Astrophys. J.* **261**, 631.

Foy, R., Chelli, A., Lena, P., and Sibille, F.: 1979, *Astron. Astrophys.* **79**, L5.

Franco, J.: 1983, *Astrophys. J.* **264**, 508.

Garmany, C. and Conti, P.: 1980, in M. J. Plavec, D. M. Popper, and R. K. Ulrich (eds.), 'Close Binary Stars: Observations and Interpretation', *IAU Symp.* **88**, 163.

Gehren, T., Hippelein, H., and Münch, G.: 1981, *Mitt. Astron. Ges.* **52**, 68.

Genkin, I. L. and Safronov, V. S.: 1975, *Soviet Astron.* **19**, 189.

Gieseking, F.: 1983, *Comptes Rendus sur les Journées de Strasbourg*, 5ème Reunion, p. 4.

Gingold, R. A. and Monaghan, J. J.: 1978, *Monthly Notices Roy. Astron. Soc.* **184**, 481.

Giuricin, G., Mardirossian, F., and Mezetti, M.: 1983, *Astrophys. J. Suppl.* **52**, 35.

Gray, F.: 1982, *Astrophys. J.* **261**, 259.

Guthrie, B. N. G.: 1983, in A. Maeder and A. Renzini (eds.), 'Observational Tests of the Stellar Evolution Theory', *IAU Symp.* **105**, in press.

Hachisu, I. and Eriguchi, Y.: 1982, *Prog. Theor. Phys.* **68**, 206.

Hanson, R. B., Jones, B. F., and Lin, D. N. C.: 1983, *Astrophys. J.* **270**, L27.

Harris, D. L., III, Strand, K. A., and Worley, C. E.: 1963, in K. A. Strand (ed.), *Stars and Stellar Systems III*, University of Chicago Press, Chapter 15, p. 273.

Hartmann, W. K.: 1970, *Mem. Soc. Roy. Sci. Liège* **14**, 49.

Heintz, W. D.: 1969, *J. Roy. Astron. Soc. Canada* **63**(6), 275.

Herbig, G. H.: 1977, *Astrophys. J.* **214**, 747.

Horedt, G. P.: 1982, *Astron. Astrophys.* **106**, 29.

Howell, R. R., McCarthy, D. W., and Low, F. J.: 1981, *Astrophys. J.* **251**, L21.

Hoyle, F.: 1953, *Astrophys. J.* **118**, 513.

Huang, S. S.: 1977, *Revista Mexicana Astron. Astrofiz.* **3**, Special Issue (*IAU Colloq.* **33**), 175.

Huang, S. S. and Wade, C., Jr.: 1966, *Astrophys. J.* **143**, 146.

Hunter, C.: 1962, *Astrophys. J.* **136**, 594.

Hut, P.: 1983, *Astrophys. J.* **272**, L29.

Jaschek, C.: 1976, *Astron. Astrophys.* **50**, 185.

Jaschek, C. and Ferrer, O.: 1972, *Publ. Astron. Soc. Pacific* **84**, 292.

Jeffers, H. M., van den Bos, W. H., and Greeby, F. M.: 1963, *Index Catalogue of Visual Double Stars*, Mt. Hamilton, Lick Observatory.

Kraft, R. P.: 1965, *Astrophys. J.* **142**, 681.

Kraicheva, Z. T., Popova, E. I., Tutukov, A. V., and Yungelson, L. R.: 1979, *Soviet Astron.* **23**, 290 and **22**, 670.

Kruszewski, A.: 1967, *Adv. Astron. Astrophys.* **4**, 233.

Kuiper, G. P.: 1935, *Publ. Astron. Soc. Pacific* **47**, 15 and 121.

Kuiper, G. P.: 1955, *Publ. Astron. Soc. Pacific* **67**, 387.

Kumar, S.: 1972, *Astrophys. Space Sci.* **17**, 453.

Larson, R. B.: 1972, *Monthly Notices Roy. Astron. Soc.* **156**, 437.

Larson, R. B.: 1978, *Monthly Notices Roy. Astron. Soc.* **184**, 69.

Larson, R. B.: 1981, *Monthly Notices Roy. Astron. Soc.* **194**, 809.

Larson, R. B.: 1984, *Monthly Notices Roy. Astron. Soc.* **206**, 197.

Levato, H. and Morrell, N.: 1983, *Astrophys. Letters* **23**, 183.

Lindroos, P.: 1982, *The Messenger (ESO)* **27**, 4.

Lucy, L. B.: 1977, *Astron. J.* **82**, 1013.

Lucy, L. B.: 1981, in D. Sugimoto, D. Q. Lamb, and D. N. Schramm (eds.), 'Fundamental Problems in the Theory of Stellar Evolution', *IAU Symp.* **93**, 75.

Lucy, L. B. and Ricco, E.: 1979, *Astron J.* **84**, 401.

Lüst, R.: 1952, *Z. Naturf.* **7a**, 87.

Lynden-Bell, D. and Pringle, J.: 1974, *Monthly Notices Roy. Astron. Soc.* **168**, 603.

McAlister, H. A.: 1977, *Astrophys. J.* **215**, 159.

McAlister *et al.*: 1983, *Astrophys. J. Suppl.* **51**, 309.

McCarthy, D. W.: 1982, *Astrophys. J.* **257**, L93.

McCarthy, D. W.: 1983, in A. R. Upgren (ed.), 'The Nearby Stars and the Stellar Luminosity Function', *IAU Colloq.* **76**, in press.

McCrea, W. H.: 1956, *Astrophys. J.* **124**, 461.

Mestel, L.: 1965, *J. Roy. Astron. Soc.* **6**, 161.

Morgan, B. L., Beddoes, D. R., Dainty, J. V., and Scaddan, R. J.: 1978, *Monthly Notices Roy. Astron. Soc.* **183**, 701.

Mouschovias, T.: 1977, *Astrophys. J.* **211**, 147.

Mouschovias, T.: 1981, in D. Sugimoto, D. Q. Lamb, and D. N. Schramm (eds.), 'Fundamental Problems in the Theory of Stellar Evolution', *IAU Symp.* **93**, 27.

Mouschovias, T. and Paleologou, V. E.: 1980, *Moon and Planets* **22**, 31.

Mundt, R., Walter, F. M., Feigelson, E. D., Finkenzeller, U., Herbig, G. H., and Odell, A. P.: 1983, *Astrophys. J.* **269**, 229.

Norman, M. L.: 1980, Ph.D. Diss., Univ. California at Davis.

Norman, M. L. and Wilson, J. R.: 1978, *Astrophys. J.* **224**, 497.

Ostriker, J. P.: 1964, *Astrophys. J.* **140**, 1067.

Ostriker, J. P.: 1970, in A. Sletteback (ed.), *Stellar Rotation*, D. Reidel Publ. Co., Dordrecht, Holland, p. 147.

Ostriker, J. P. and Bodenheimer, P.: 1973, *Astrophys. J.* **180**, 171.

Paczynski, B.: 1971, *Ann. Rev. Astron. Astrophys.* **9**, 183.

Partridge, R. B.: 1967, *Astron. J.* **72**, 713.

Popper, D. M.: 1967, *Ann. Rev. Astron. Astrophys.* **5**, 85.

Popper, D. M.: 1980, *Ann. Rev. Astron. Astrophys.* **18**, 115.

Popov, M. V.: 1970, *Peremennye Zvezdy* **17**, 412.

Poveda, A., Allen, C., and Parrao, L.: 1982, *Astrophys. J.* **258**, 589.

Quirk, W. J.: 1973, *Bull. Am. Astron. Soc.* **5**, 9.

Retterer, J. H. and King, I. V.: 1982, *Astrophys. J.* **254**, 214.

Regev, O. and Shaviv, G.: 1981, *Astrophys. J.* **245**, 934.

Rozyczka, M., Tscharnuter, W. M., Winkler, K.-H., and Yorke, H. W.: 1980, *Astron. Astrophys.* **83**, 118.

Salukvadze, G. N.: 1980, *Astrofisika* **16**, 505 and 687.

Schmitz, F.: 1983, *Astron. Astrophys.* **125**, 333.

Silk, J.: 1978, in T. Gehrels (ed.), *Protostars and Planets*, University of Arizona Press, p. 172.

Silk, J. and Takahashi, T.: 1979, *Astrophys. J.* **229**, 242.

Shu, F. H. and Lubow, S. H.: 1981, *Ann. Rev. Astron. Astrophys.* **19**, 277.

Spitzer, L., Jr. and Mathieu, R. D.: 1980, *Astrophys. J.* **241**, 618.

Staniucha, M.: 1979, *Acta Astron.* **29**, 587.

Stauffer, J. R.: 1982, *Astron. J.* **87**, 1507.

Strittmatter, P. A.: 1981, in Ulrich and Kjar (eds.), *The Scientific Importance of High Angular Resolution and Infrared and Optical Wavelengths*, ESO Workshop, p. 359.

Szebehely, V.: 1977, *Revista Mexicana Astron. Astrofiz.* **3**, Special Issue (*IAU Colloq.* **33**), 145.

Thomas, H. C.: 1977, *Ann. Rev. Astron. Astrophys.* **15**, 127.

Tohline, J. E.: 1980, *Astrophys. J.* **236**, 160.

Trimble, V.: 1974, *Astron. J.* **79**, 967.

Trimble, V.: 1978, *Observatory* **98**, 163.

Trimble, V.: 1980, in J. E. Hesser (ed.), 'Star Clusters', *IAU Symp.* **85**, 259.

Trimble, V.: 1983, *Nature* **303**, 137.

Trimble, V. and Ostriker, J.: 1978, *Astron. Astrophys.* **63**, 433.

Tscharnuter, W. M.: 1980, *Space Sci. Rev.* **27**, 235.

Tscharnuter, W. M.: 1981, in D. Sugimoto, D. Q. Lamb, and D. N. Schramm (eds.), 'Fundamental Problems in the Theory of Stellar Evolution', *IAU Symp.* **93**, 105.

van Albada, T. S. and Blaauw, A.: 1967, in *On the Evolution of Double Stars*, Observatoire Royal de Belgique, Communications, Ser. B, No. 47, p. 44.

van Albada, T. S.: 1968, *Astron. Inst. Neth.* **20**, 57 and **19**, 479.

van Beveren, D.: 1982, *Astron. Astrophys.* **115**, 65.

van de Kamp, P.: 1971, *Ann. Rev. Astron.* **9**, 103.

Vogel, S. N. and Kuhi, L. V.: 1981, *Astrophys. J.* **245**, 960.

von Hoerner, S.: 1968, in Y. Terzian (ed.), *Interstellar Ionized Hydrogen*, Benjamin Inc., New York, p. 101.

von Weizsäcker, C. F.: 1948, *Z. Naturf.* **3a**, 524.

Walmsley, C. M.: 1982, *Irish. Astron. J.* **15**, 161.

Warner, B.: 1962a, *Ann. Astrophys.* **25**, 94.

Warner, B.: 1962b, *Publ. Astron. Soc. Pacific* **73**, 439.

Webbink, R. F.: 1980, in M. J. Plavec, D. M. Popper, and R. K. Ulrich (eds.), 'Close Binary Stars: Observations and Interpretation', *IAU Symp.* **88**, 561.

Weigelt, G. and Wirnitzer, B.: 1983, *Optics Letters* **8**, 389.

Wesson, P. S.: 1979, *Astron. Astrophys.* **80**, 296.

Wolff, S. C., Edwards, S., and Preston, G. W.: 1982, *Astrophys. J.* **252**, 322.

Wood, D.: 1981, *Monthly Notices Roy. Astron. Soc.* **199**, 331.

Woolfson, M. M.: 1978, in S. F. Dermott (ed.), *The Origin of the Solar System*, Wiley Publ., New York.

Worley, Ch. E.: 1969, in S. S. Kumar (ed.), *Low Luminosity Stars*, Gordon and Breach, New York, p. 117.

Wynn-Williams, G.: 1982, *Ann. Rev. Astron. Astrophys.* **20**, 587.

Zinnecker, H.: 1981, Ph.D. Diss. Techn. Univ. München, Report MPI-PAE Extraterr. Physik 167.

FISSION SEQUENCE AND EQUILIBRIUM MODELS OF RIGIDITY ROTATING POLYTROPES*

IZUMI HACHISU and YOSHIHARU ERIGUCHI**

Department of Aeronautical Engineering, Kyoto University, Kyoto, Japan

(Received 4 June, 1983)

Abstract. Equilibrium sequences of self-gravitating and polytropic stars including binary stars are computed for various polytropic indexes. We find a fission sequence from an ellipsoidal configuration to a binary by way of dumb-bell equilibrium.

1. 3D Hydrostatic Equilibriums

It is required to obtain accurate equilibrium models of rotating stars including binary stars from the following points of view: (1) reexamination of the fission theory; (2) checking the results of three-dimensional (3D) hydrodynamical calculation; (3) proposing proper model of mass-exchange or mass-losing close binary systems.

Recently, we have developed methods for obtaining 3D hydrostatic equilibrium of rotating polytropes, and succeeded in calculating various sequences. In this paper we will summarize our results and discuss the physical nature of the sequences.

2. Equilibrium Sequences

2.1. INCOMPRESSIBLE CASE (POLYTROPIC INDEX $n = 0$)

As many bifurcation points have been found on the Maclaurin and the Jacobi sequences, the calculations are started from these points and the equilibriums are obtained step by step. These sequences are shown in Figure 1.

(A) Pear-Shaped: this bifurcates from the Jacobi sequence but it terminates shortly by mass-shedding (this denotes that the gravity balances with or is less than the centrifugal force at the surface). The shapes just before mass-shedding are shown in Figure 2.

(B) Dumb-Bell: this also bifurcates from the Jacobi sequence and continues to binary sequence with mass ratio $q = 1$. This is a *fission* sequence.

(C) Binary sequences: the less massive component necessarily fills up its Roche lobe first and the sequences terminate.

---

* Paper presented at the Lembang-Bamberg IAU Colloquium No. 80 on 'Double Stars: Physical Properties and Generic Relations', held at Bandung, Indonesia, 3–7 June, 1983.
** Department of Earth Science and Astronomy, College of Arts and Sciences, University of Tokyo, Japan.

Astrophysics and Space Science **99** (1984) 71–74. 0004–640X/84/0991–0071$00.60.

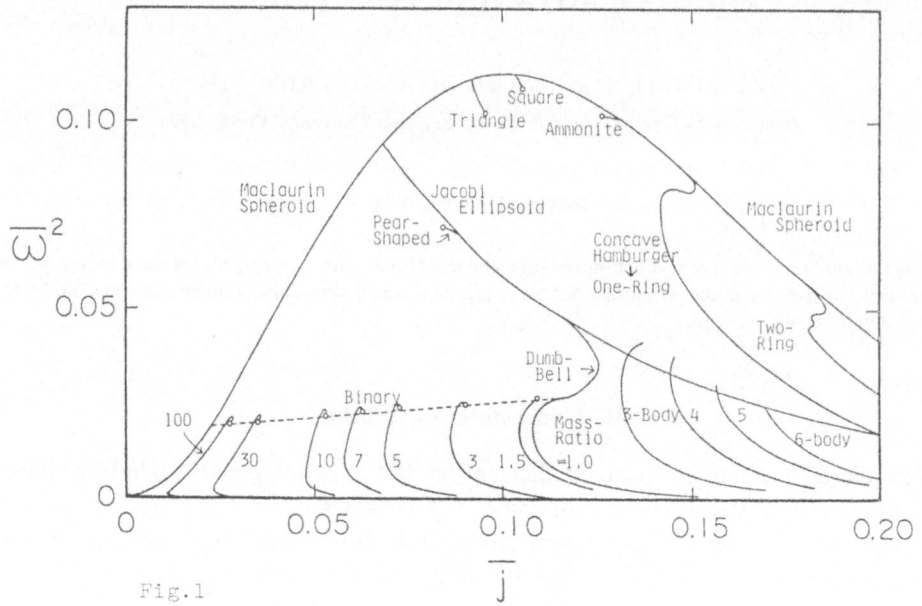

Fig. 1.   Angular velocity $\overline{\omega}^2$-angular momentum $\overline{j}$ diagram.

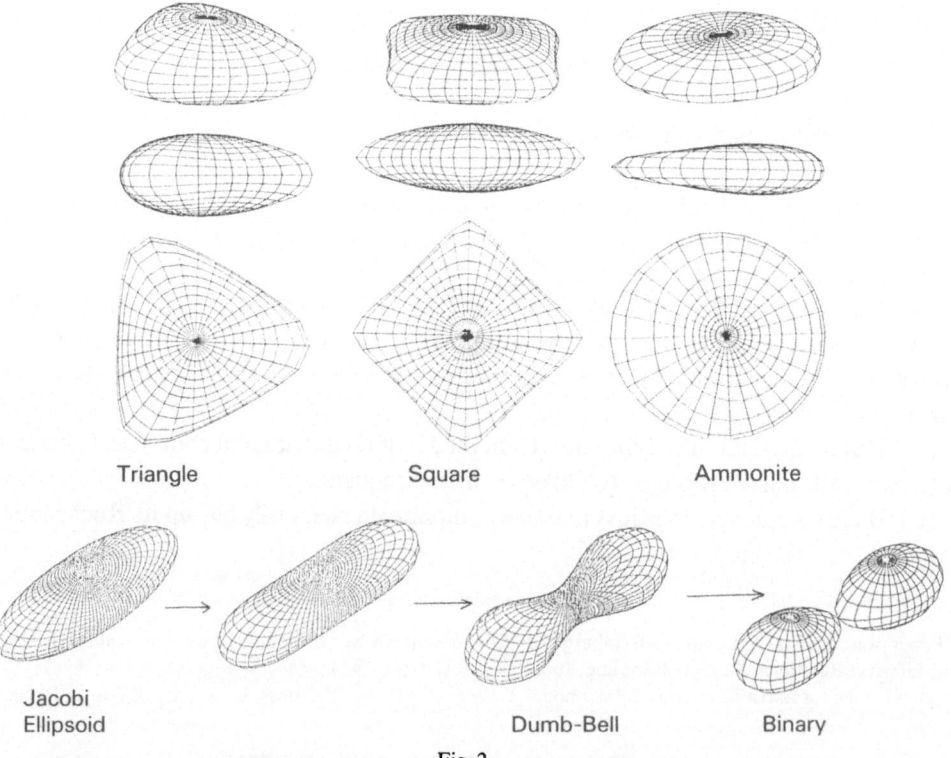

Fig. 2.

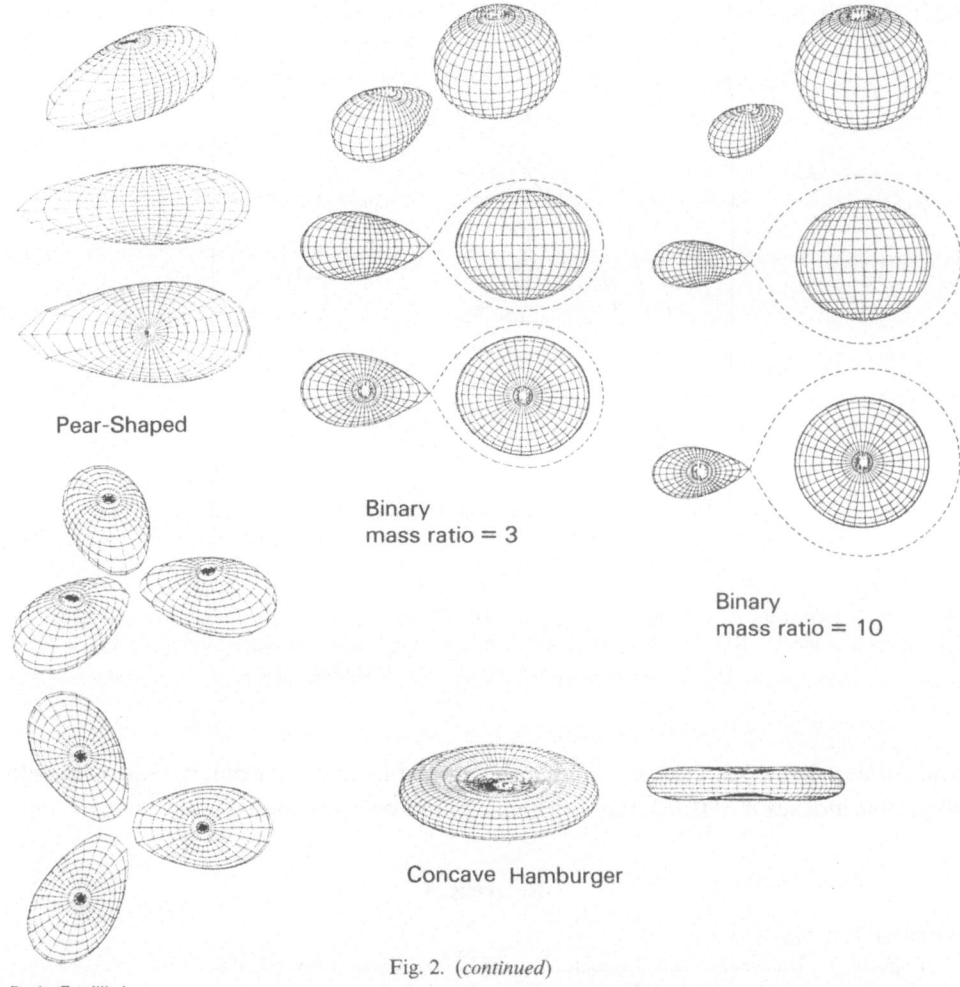

Pear-Shaped

Binary
mass ratio = 3

Binary
mass ratio = 10

Concave Hamburger

3-Body Equilibrium

Fig. 2. (*continued*)

(D) Triangle: this bifurcates from the Maclaurin sequence. This sequence also shortly terminates because of mass-shedding.

(E) Square.

(F) Ammonite.

(G) Concave Hamburger: this also bifurcates from the Maclaurin sequence and continues to the Dyson-Wong toroid. We call this sequence One-Ring sequence.

(H) Two-Ring: this bifurcates from the Maclaurin sequence.

(J) Multi-Body: we can calculate multi-body sequences.

2.2. COMPRESSIBLE CASE (POLYTROPIC INDEX $n \neq 0$)

Various sequences with various polytropic indexes have been computed. However, there are no thorough sequences thoroughly from spheroid-like configuration to binary

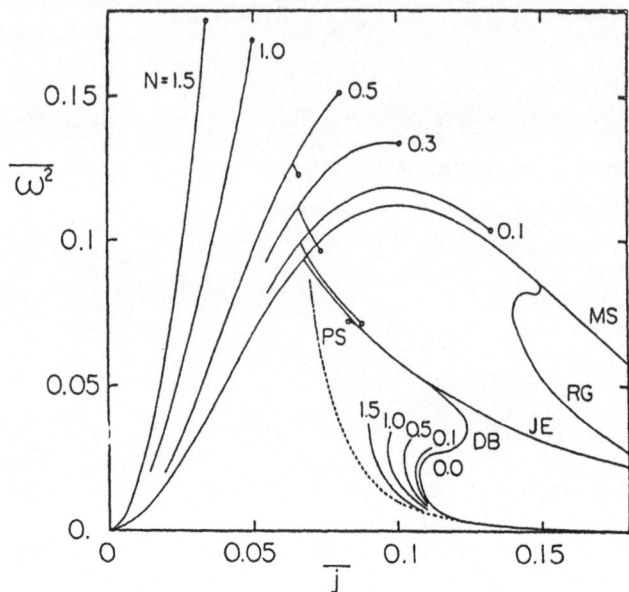

Fig. 3. $\overline{\omega}^2 - \overline{j}$ diagram. MS: Maclaurin sequence. JE: Jacobi sequence. DB: Dumb-Bell sequence. PS: Pear-Shaped sequence. RG: One-Ring sequence. And two-point mass sequence (dotted) is added. Open circle denotes the point where mass-shedding occurs.

configuration because of mass-shedding as seen in Figure 3. Five binary sequences with polytropic indexes $N = 0, 0.1, 0.5, 1.0,$ and 1.5 have been computed to the contact stage.

References

(A) Dumb-Bell: Pear-Shaped;
  Eriguchi, Y., Hachisu, I., and Sugimoto, D.: 1982, *Prog. Theor. Phys.* **67**, 1068.
  Hachisu, I. and Eriguchi, Y.: 1982, *Prog. Theor. Phys.* **68**, 206.
(B) Binary;
  Hachisu, I. and Eriguchi, Y.: 1983a, submitted to *Publ. Astron. Soc. Japan* (Compressible case).
  Hachisu, I. and Eriguchi, Y.: 1983b, submitted to *Publ. Astron. Soc. Japan* (Incompressible case).
  Hachisu, I. and Eriguchi, Y.: 1983c, submitted to *Publ. Astron. Soc. Japan* (Incompressible case, planet and satellite system).
(C) Triangle: Square: Ammonite: Two-Ring;
  Eriguchi, Y. and Hachisu, I.: 1982, *Prog. Theor. Phys.* **67**, 844.
  Eriguchi, Y. and Hachisu, I.: 1983, *Prog. Theor. Phys.* **69**, 1131.
(D) Concave Hamburger: One-Ring;
  Eriguchi, Y. and Sugimoto, D.: 1981, *Prog. Theor. Phys.* **65**, 1870 (Incompressible case).
  Hachisu, I. and Eriguchi, Y.: 1982, *Prog. Theor. Phys.* **68**, 191 (Compressible case).
(E) Multi-Body;
  Eriguchi, Y. and Hachisu, I.: 1983, submitted to *Prog. Theor. Phys.*

FISSION CANDIDATES AMONG DETACHED CLOSE BINARIES*

H.-A. OTT

Astronomisches Institut der Universität Münster, F.R.G.

(Received 26 July, 1983)

Abstract. A comparison between results of model calculations and observed properties of close, but detached low mass binaries with Main Sequence primaries shows statistical trends, which seem to support fission origin for some of these objects: the mass-momentum relation, the relation between mass ratio and separation and the relation between mass ratio and synchronisation speak in favour of close initial separations and small mass ratios of the components.

1. Introduction

Fission, i.e. splitting of a dynamically collapsed object (pre-Main Sequence star) by rotational instability, is often proposed as a mechanism of forming close binary systems (CBS), see e.g. Lucy (1981) and other articles in *IAU Symp.* **93** (1981). Though the theory of rotating polytropes provides us with a rough criterion for instability, that can be applied to models of low mass protostars (Roxburgh, 1966), the physics of the fission process itself and the characteristic properties remain largely unknown. The rationale of this work is to estimate these properties.

Two basic assumptions are: (1) fission is a process which occurs in single rotating objects not able to store large amounts of momentum; (2) fission forms two gravitationally bound stars which star their dynamical evolution from a very small initial separation.

2. A Survey of the Literature

Some theoretical results and statistical analyses of CBS observations constitute the background for these studies and are summarized in Table I. Throughout this work we use the following units and definitions:

Mass, distance, time in solar units: solar mass, astronomical unit, year.

Momentum units can easily be derived.

Separation: in units of sum of radii of the two components.

Mass ratio: = mass (secondary)/mass (primary).

Synchronisation: = orbital frequency/rotational frequency (primary).

Assuming that Roxburgh's theory gives the correct momentum $H_0(M)$ for stellar masses smaller than 5, we have to define parametrically the status of a close binary

---

* Paper presented at the Lembang-Bamberg IAU Colloquium No. 80 on 'Double Stars: Physical Properties and Generic Relations', held at Bandung, Indonesia, 3–7 June, 1983.

Astrophysics and Space Science **99** (1984) 75–83. 0004–640X/84/0991-0075$01.35.

TABLE I

Fission – parameters in theory and observation
SB 1 – single line spectroscopic binary
SB 2 – double line spectroscopic binary

Total mass (M)	Mass ratio (Q)	Momentum (H), separation (A) period (P)
Theory:		
Roxburgh (1966): pre-Main Sequence stars (convective-radiative phase): Fission requires $M \gtreqqless 0.8$ $M > 4$ result in detached systems	Lucy (1981): N-particle simulations favour small Q (0.2–0.3) Generally accepted: $Q \ll 1$ (no serious theoretical foundation)	Roxburgh (1966): $H = H_0(M)$ computed for mass range $0.8 \leqq M \leqq 5$ Van 't Veer (1979): strong H loss in contact systems results in evolution to smaller Q. Form all contact systems with Q near 1?
Observation:		
Staniucha (1979): statistical analysis of the catalogue by Batten et al. (1978): * selection effects strong for M < 1 * peak of distribution near M = 1.2 for SB 2 De Grève and Vanbeveren (1980): primary masses of near B(V) stars show bimodal distribution (peak near M = 7 and rise for M < 4)	Staniucha (1979): Q distribution bimodal (peak near 0.2 for SB 1 and 1 for SB 2) Trimble (1978): non-evolved systems show the peak near 0.2 (i.e. no evolution effect)	Staniucha (1979): distributions of P (peak near 1.5 days) and Asini (peak near 17 $R_\odot$) strongly biased, no significance for star formation Mochnacki (1981): contact systems (WUMa) often show too much H/M (theory incorrect?) Farinella et al. (1979): from late B to G stars: common average geometry, separation about 3 (sum of radii of stars)

after the (hypothetical) fission. Hints from the literature and simple momentum considerations lead to the following plausible scenario: the phase shortly after fission is characterized by a low separation ($A < 3$) and a small mass ratio ($Q < 0.3$) together with a rotational frequency of the primary, which is not much higher than the orbital frequency of the companion ($S > 0.2$). Possible mass and momentum loss during the fission process is not included in our preliminary computations.

3. Orbital Evolution in Detached Close Binaries

An answer to the question which observed CBS are likely to be fission candidates, is only possible, if the dynamical evolution of the systems can be traced over a long time interval, e.g. the main sequence lifetime of the primary. In order to understand some statistical trends in our sample of observed CBS (see next section), we here present some results of model calculations concerning the evolution of separation and synchronisation with time.

The dynamical equations for tidal dissipation in stars with convective envelopes (Zahn, 1977) have been integrated numerically under the assumptions of: (i) insignificant mass loss or transfer; (ii) circular orbits; (iii) possible steady momentum loss of the primary (magnetic braking of a stellar wind). Fixed parameters are: total momentum $H = H_0(M)$ from Roxburgh and initial separation $A = 1$. Depending on the chosen total mass (M) and mass ratio (Q) the calculations trace the orbital evolution with or without momentum loss. The upper time limit is the main sequence lifetime of the primary $(6 \times 10^9 M^{-2})$.

Figure 1 shows the final separation as a function of model mass and mass ratio, all without momentum loss. Figure 2 shows the final synchronisation for the same models as in Figure 1. The curves represent upper limits for empirical CBS, because the ages of observed systems with main sequence primaries fall well within the given evolutionary limit.

The results of calculations including momentum loss are more complex and difficult to interpret. In general they show smaller final separations, as expected, but the distance between the components can reach a maximum, before it slowly approaches a final value. Synchronisation sometimes reaches unity very early, but goes through prolonged oscillatory phases afterwards – orbital evolution continues!

We will not discuss these results here, because presently available empirical data

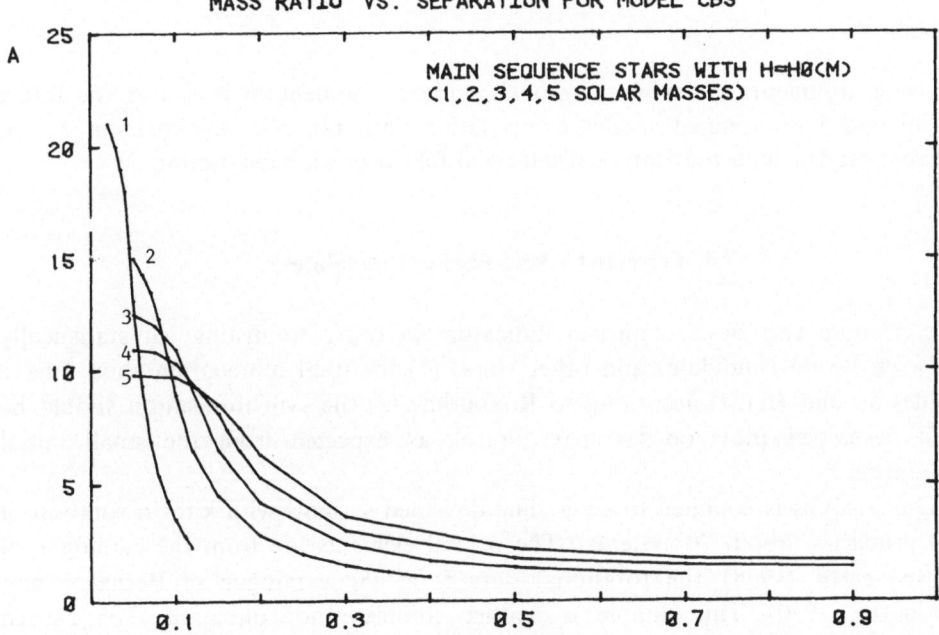

Fig. 1. System configuration with respect to separation (A) after one Main Sequence lifetime of the primary.

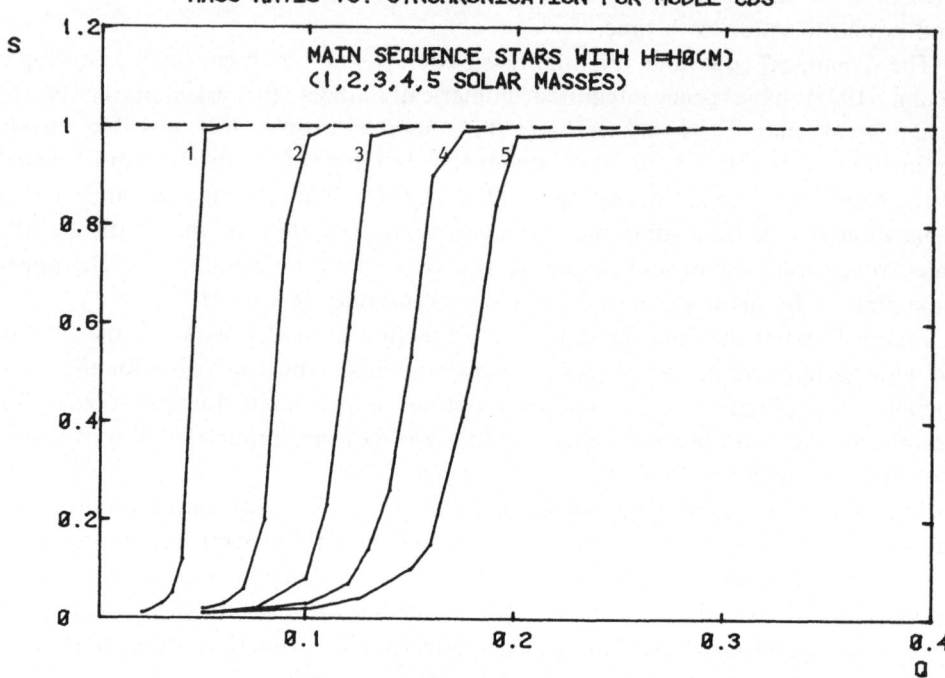

Fig. 2.   System configuration with respect to synchronisation (S) after one Main Sequence lifetime of
the primary.

provide arguments in favour of models without momentum loss. For the future
more and more refined model computations are planned. Comparison of the
present models with the empirical data will follow in the next section.

4. Observed CBS: Fission Candidates?

We propose two basic empirical indicators in order to distinguish statistically
between fission candidates and other stars: (1) the total momentum should have
values around $H_0(M)$ according to Roxburgh; (2) the synchronisation should be
well developed (not too far from unity), as expected from the small initial
separation.

Our analysis is confined to single line detached systems with known rotation of
the primaries (about 70 systems). The orbital elements are from the catalogue of
Batten *et al.* (1978), the rotational data from the catalogue of Bernacca and
Perinotto (1970). This sample is neither complete nor unbiased. The chosen
indicators however should be applicable to this inhomogenous sample, an
assumption which will be tested by Monte-Carlo simulations.

4.1. Mass-momentum relation (M/H)

According to Brosche (1963) a universal relation exists between momentum and mass ($H \sim M^2$), for double stars with a larger internal scatter than for single stars on the main sequence. If this is correct, fission objects – formed from single stars – should show a better defined M/H relation than systems with components of independent origin.

Taking synchronisation as fission indicator, this expectation is confirmed by Figures 3 and 4. Most of the data points lie within reasonable limits around Roxburgh's results, for small values of Q (0 to 0.3) the M/H relation is better defined than for larger Q values, Systems with known inclination angles i display the relation even better.

4.2. Mass ratio and separation (Q/A)

Our model calculations show a clear correlation between mass ratio and final separation for Main Sequence stars with Roxburgh's momentum (Figure 1). The rapid stellar evolution of the more massive primaries does not allow the separation to reach the synchronous stage during the Main Sequence lifetime. This trend has a good empirical counterpart in the diagram of Figure 6 (synchronous systems with momentum near $H_0(M)$), but non-synchronous CBS or such with excess

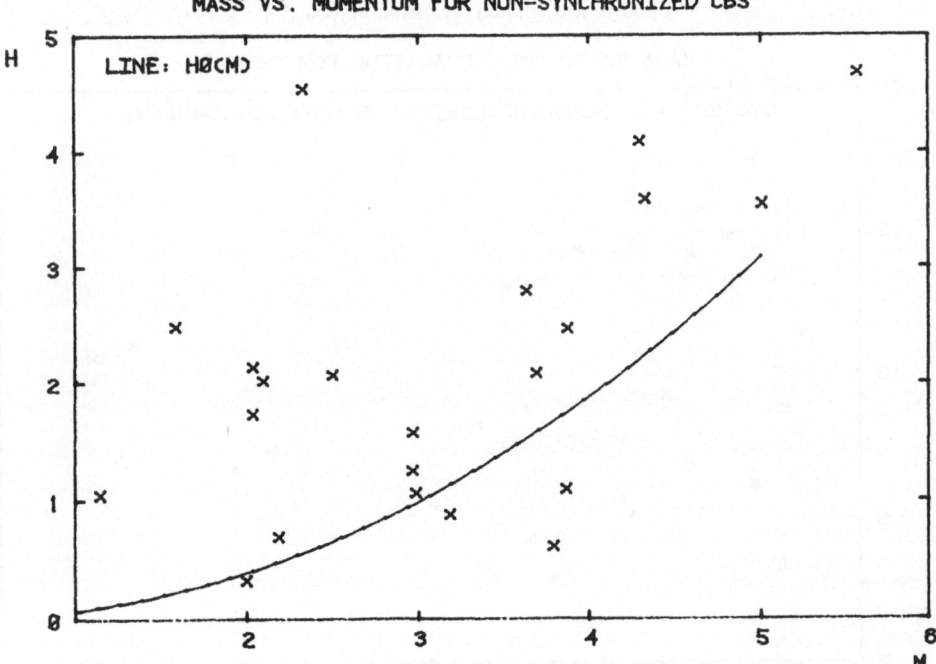

Fig. 3. Observed mass-momentum relation for non-synchronized CBS. Line represents Roxburgh's theoretical relation for fission objects.

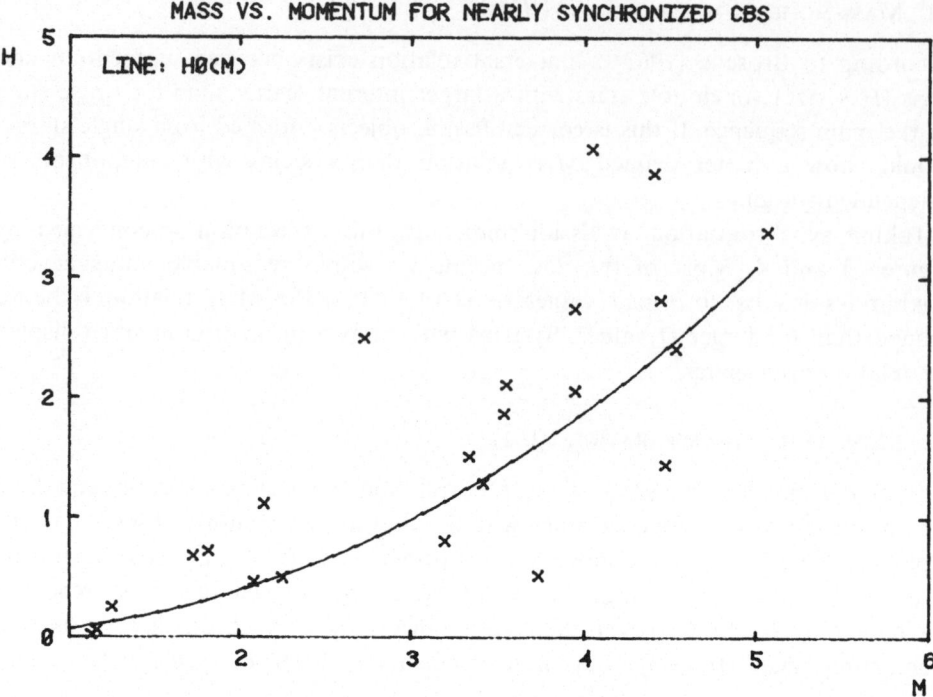

Fig. 4. Observed mass-momentum relation for nearly synchronized CBS. Line represents Roxburgh's theoretical relation for fission objects.

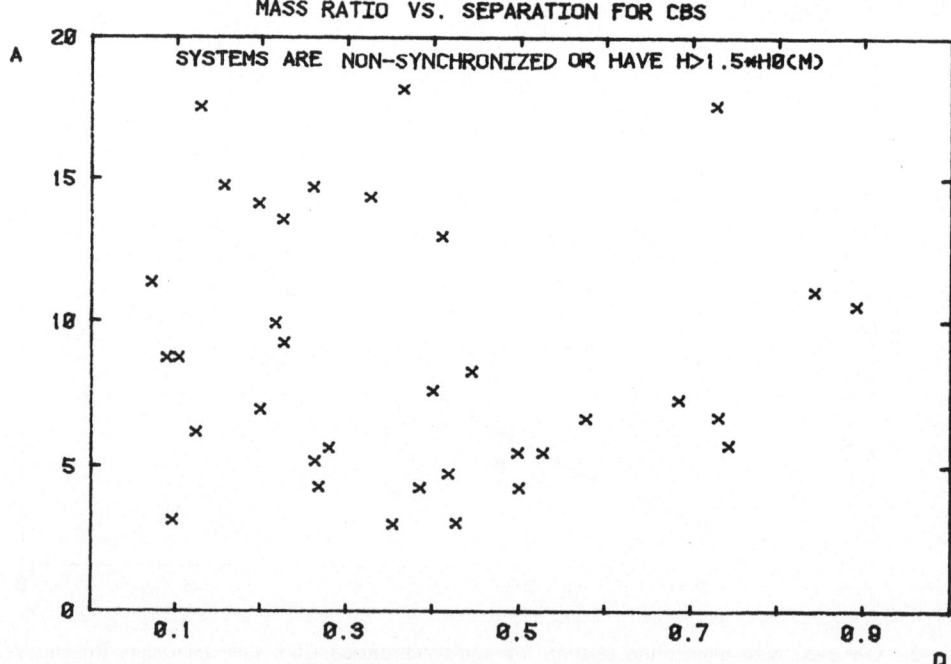

Fig. 5. Observed combinations of system parameters for non-synchronized or high momentum CBS. Compare with model results of Figure 1.

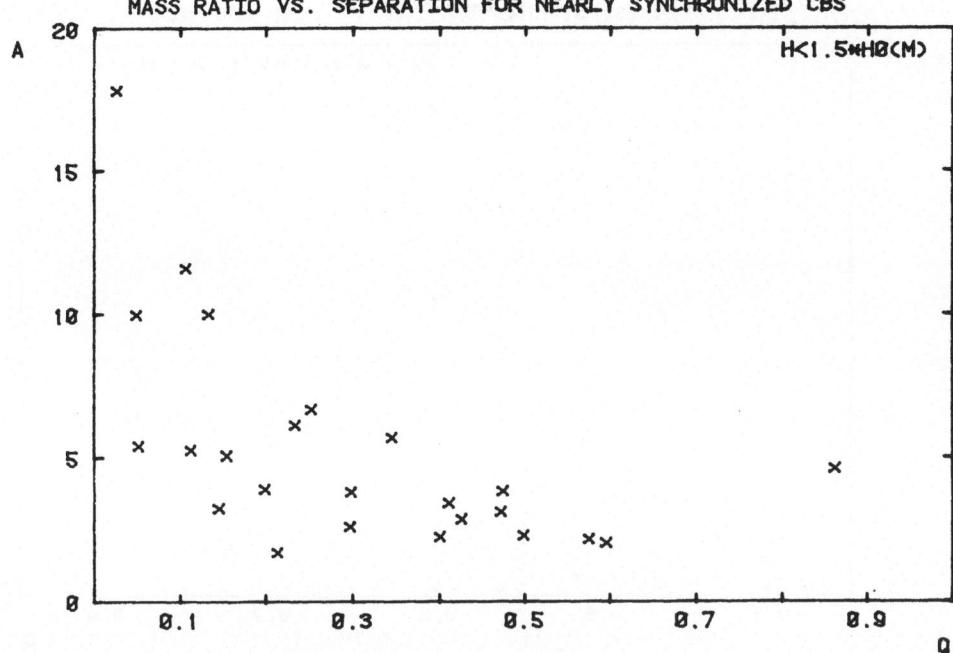

Fig. 6. Observed combinations of system parameters for nearly synchronized and low momentum CBS. Compare with model results of Figure 1.

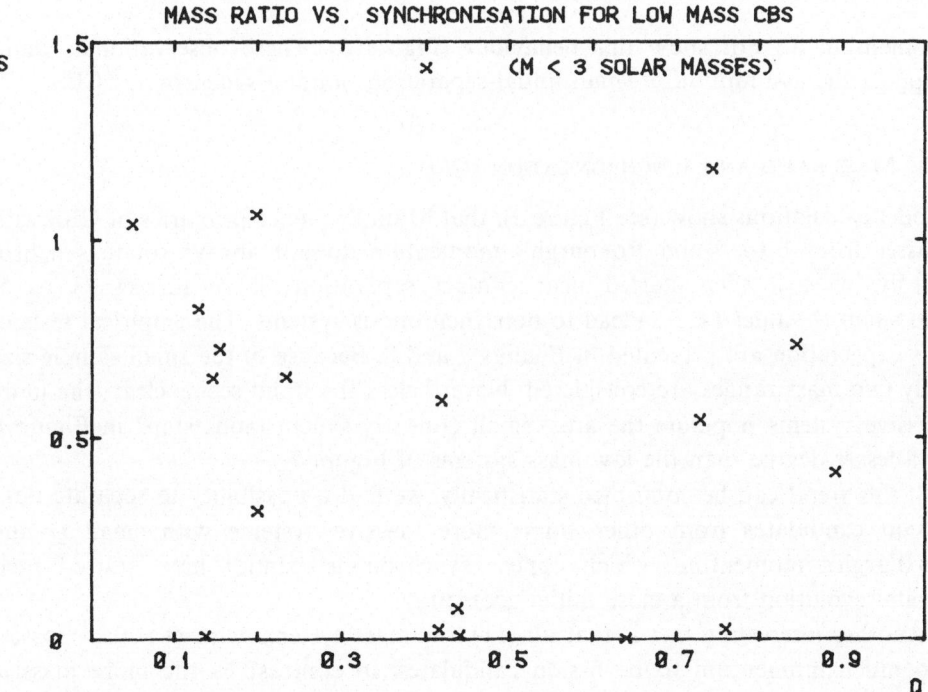

Fig. 7. Observed combinations of mass ratio and synchronisation for low mass CBS. All evolved fission candidates should have reached synchronisation, viz. Figure 2.

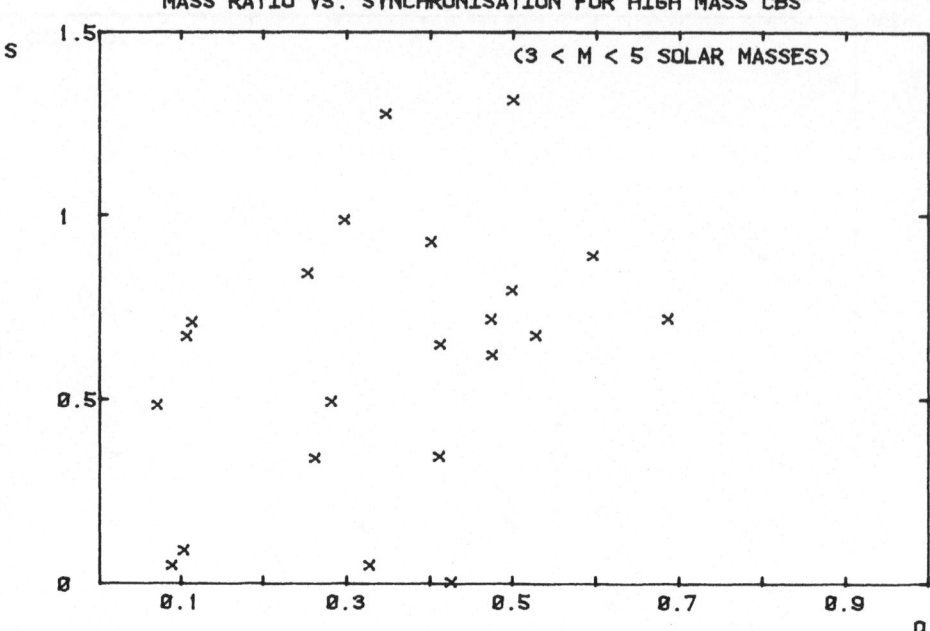

Fig. 8. Observed combinations of mass ratio and synchronisation for high mass CBS. Only evolved
fission candidates with large Q values should have reached synchronisation, viz. Figure 2.

momentum do not show this behaviour (Figure 5). These observational results
support the assumption of small initial separations among synchronous CBS.

4.3. Mass ratio and synchronisation (Q/S)

Model calculations show (see Figure 2), that Main Sequence primaries of CBS with
masses from 1 to 5 and Roxburgh's momentum do not always rotate synchronously, even if they started near contact separation; large masses (3 to 5)
and small Q values (< 0.2) lead to nonsynchronous systems. The empirical tests of
this expectation are presented in Figures 7 and 8. Because of the small sample size
only two mass ranges are considered. Nevertheless the trend seems clear: the more
massive systems populate the area 'small Q/nearly synchronous state' in Figure 8
to a lesser degree than the low mass systems of Figure 7.

If this trend can be confirmed statistically, we find a possibility to separate non-
fission candidates from other stars: more massive systems with small Q and
Roxburgh's momentum, which appear synchronous, cannot have started their
orbital evolution from a close initial geometry.

Another interesting fact is that all (11) low mass systems with $Q > 0.25$ possess
too much momentum to be fission candidates, in contrast to the more massive
systems, among which only 7 out of 17 (41 %) have larger momentum values. This

and the above remark on the M/H relation, concerning Q, speak in favour of small Q values to be characteristic of fission candidates.

5. Summary

Our comparison between results of model calculations without momentum loss and observed properties of detached low mass CBS with main sequence primaries shows some statistical trends, which seem to support fission origin for some of these objects. If Roxburgh predicts the correct momentum values for fission of protostars, the *mass-momentum* relation, the relation between *mass ratio and separation* and the relation between *mass ratio and synchronisation* speak in favour of: (1) a very close initial separation of fission candidates; and (2) small mass ratios of the binary components.

Of course, more detailed models and further statistical confirmation are needed to strengthen the argument for fission origin of some close binary stars.

References

Batten, A. H., Fletcher, J. M., and Mann, P. J.: 1978, *Publ. Dominion Astrophys. Obs.* 15, No. 5
Bernacca, P. L. and Perinotto, M.: 1970, *Contr. Oss. Astrofiz. Padova*, No. 239.
Brosche, P.: 1963, *Z. Astrophys.* 57, 143.
Cassinelli, J. P.: 1979, *Ann. Rev. Astron. Astrophys.* 17, 275.
De Grève, J. P. and Vanbeveren, D.: 1980, *Astrophys. Space. Sci.* 68, 433.
Farinella, P., Luzny, F., Mantegazza, L., and Paolicchi, P.: 1979, *Astrophys. J.* 234, 973.
Lucy, L. B.: 1981, in D. Sugimoto, D. Q. Lamb, and D. N. Schramm (eds.), 'Fundamental Problems in the Theory of Stellar Evolution', *IAU Symp.* 93, 75.
Mochnacki, S. W.: 1981, *Astrophys. J.* 245, 650.
Roxburgh, I. W.: 1966, *Astrophys. J.* 143, 111.
Staniucha, M.: 1979, *Acta Astron.* 29, No. 4, 587.
Trimble, V.: 1978, *Observatory*, 98, 163.
Van 't Veer, F.: 1979, *Astron. Astrophys.* 80, 287.
Zahn, J.-P.: 1977, *Astron. Astrophys.* 57, 383.

UV OBSERVATIONS OF THREE CLASSICAL NOVAE DURING EARLY STAGES OF DECLINE*

H. DRECHSEL, W. WARGAU and J. RAHE

Dr Remeis Sternwarte Bamberg, Astronomisches Institut Universität Erlangen Nürnberg, F.R.G.

(Received 23 August, 1983)

Abstract. We present IUE UV spectra of the three recent classical novae N Sgr 1982, N Ser 1983, and N Mus 1983, obtained during early decline, transitional stage, and nebular phase, respectively. The line spectra are discussed. Preliminary conclusions concerning CNO over-abundances, shell geometry, and speed classes are drawn.

1. Introduction

IUE observations of the three recent novae N Sagittarii 1982, N Serpentis 1983, and N Muscae 1983 have been collected between October, 1982 and March, 1983.

N Sgr 1982 was discovered by Honda (1982) on 4 October, 1982 at a visual brightness of $9^m\!\!.0$. The absolute visual maximum of $8^m\!\!.0$ was reported for 14 October, 1982. Subsequently, the brightness decreased steeply to $10^m\!\!.6$ on 25 October, 1982, and afterwards increased again to $8^m\!\!.5$ on 9 November, 1982.

The outburst of N Ser 1983 has been reported by Wakuda (1983) on 21 February, 1983, at a visual brightness of $11^m\!\!.0$. Soon after discovery, a peak brightness of $7^m\!\!.7$ was reached on 22 February, 1983, followed by a very steep decline of about 4 mag within only 6 days.

The outburst of N Mus 1983 was discovered by Liller (1983), at a visual brightness of $7^m\!\!.2$ on 18 January, 1983. After the brightness maximum of about 7^m on 21 January, the object declined by about 2 mag within a few days. Subsequent fluctuations around 9^m were reported until 22 February. Later, the nova declined only gradually, and still was at a visual brightness of $10^m\!\!.9$ at the beginning of 1983 August (Cassatella, 1983).

Except for N Mus 1983, our measurements represent the first UV observations of these objects. The satellite observations were carried out between 3 and 44 days after the visual brightness maxima. Since the three classical novae belong to different speed classes, the observing epochs coincide with different stages of the nova events.

2. IUE Data

The journal of UV observations of the three novae N Sgr 1982, N Ser 1983, and N Mus 1983 is presented in Table I, containing the dates of observations, image

* Paper presented at the Lembang-Bamberg IAU Colloquium No. 80 on 'Double Stars: Physical Properties and Generic Relations', held at Bandung, Indonesia, 3–7 June, 1983.

Astrophysics and Space Science **99** (1984) 85–91. 0004–640X/84/0991–0085$01.05.

TABLE I

Journal of UV Observations

Object	Date	(UT)	Camera	Image no.	Disp.	Exp. time (min.)
N Sgr 1982	1982, Oct.	18.637	SWP	18323	low	5
		18.644	LWR	14433	low	5
		18.665	SWP	18324	low	60
		18.710	LWR	14434	low	20
		18.729	SWP	18325	low	30
	1982, Oct.	19.632	SWP	18329	low	40
		19.665	LWR	14441	low	25
		19.684	SWP	18330	low	30
		19.707	LWR	14442	low	6
	1982, Oct.,	31.631	SWP	18439	low	30
		31.655	LWR	14530	low	30
		31.678	SWP	18440	low	60
		31.703	LWR	14531	low	5
N Mus 1983	1983, March	4.176	SWP	19383	low	20
		4.192	LWR	15421	low	5
		4.211	SWP	19384	low	5
		4.232	SWP	19385	low	2
		4.250	SWP	19386	high	120
		4.356	LWR	15422	low	0.5
		4.358	SWP	19387	high	20
		4.397	SWP	19388	low	0.5
		4.406	LWR	15423	high	45
		4.441	SWP	19389	high	12
N Ser 1983	1983, March	5.252	LWR	15427	low	40
		5.283	SWP	19394	low	150
		5.412	SWP	19395	low	54

numbers, dispersion modes, and exposure times. All spectra were taken through the large aperture. A total of 13, 10, and 3 spectra were sampled of N Sgr 1982, N Mus 1983, and N Ser 1983, respectively, at the ESA-Vilspa ground station. For N Mus 1983, 4 high dispersion spectra were obtained, while the other exposures are all in low dispersion.

3. Results

3.1. Nova sagittarii 1982

IUE low dispersion spectra were taken 3, 4, and 16 days after the visual brightness maximum. The early spectra of 18 and 19 October, 1982, were obtained only very shortly after the maximum, while the observations of 31 October were made in an ascending part of the light curve following an intermediate brightness minimum. The spectra taken on two consecutive days are very similar, and reveal only minor

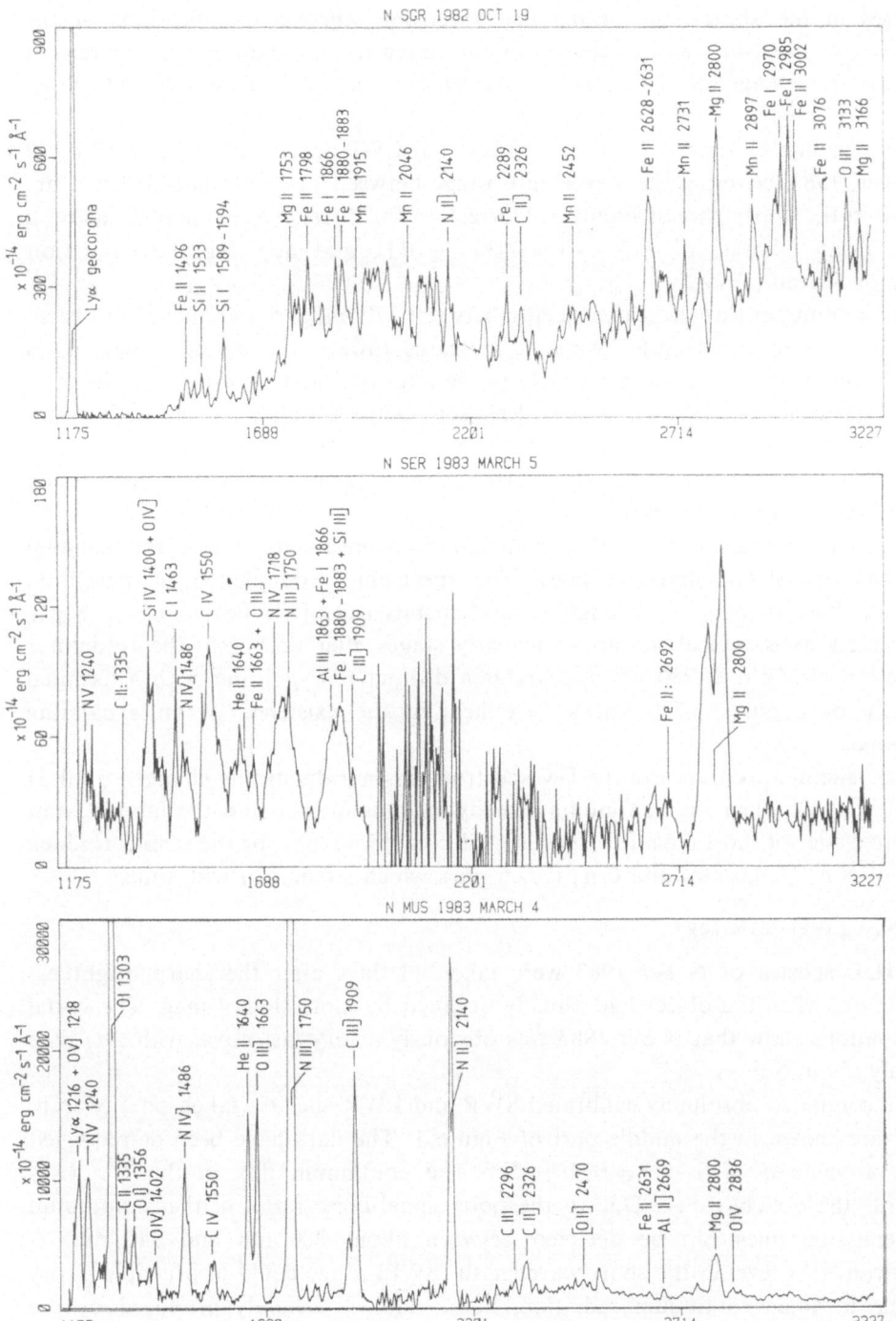

Fig. 1. IUE UV spectra of the three classical novae N Sgr 1982, N Ser 1983, and N Mus 1983 (from top to bottom); the spectra are absolutely calibrated, and have been de-reddened using $E(B-V) = 0.^m30$, $0.^m40$, and $0.^m50$ ($\pm 0.^m1$), respectively.

changes in the appearance of the line spectrum, whereas modifications of the
relative line intensities and of the continuum intensity distribution are apparent at
the later nova stage on 31 October (cf. Drechsel *et al.*, 1982; Blades and Drechsel,
1982).

The absolutely calibrated combined SWP and LWR spectra of N Sgr 1982 of 18
October, 1982, covering the wavelength range between 117.5 nm and 320 nm, are
shown in the upper part of Figure 1. Correction for interstellar extinction has been
made using a value of $E(B-V) = 0\overset{m}{.}30$ ($\pm 0\overset{m}{.}1$), and applying the extinction
curve of Seaton (1979).

The continuum flux decreases steeply between 170 nm and 160 nm toward zero-
level at shorter wavelengths, while it increases toward the optical range, as is
usually observed for novae in an early phase after outburst maximum; earlier UV
observations of the classical novae FH Ser 1970 (Andersen *et al.*, 1971), and
N Cyg 1978 (Stickland *et al.*, 1979) showed that the maximum of intensity
distribution gradually shifts toward shorter wavelengths in the UV with progressing
evolutionary stage of the nova.

The hump between about 170 nm and 220 nm is presumably due to the blending
of numerous strong emission lines. The spectrum is dominated by permitted
emission lines of neutral or singly-ionized metals (e.g., Fe I, Fe II, Si I, Si II, Mg II,
Mn II, etc.), as is typical for novae in early stages after outburst. The resonance
transition of Mg II at 280 nm appears as a distinct P Cygni line, with a terminal
velocity of about -1700 km s^{-1}, indicating the existence of an expanding
envelope.

The general appearance of the UV spectrum has not drastically changed until 31
October 1982, when the continuum already extends further into the far UV, and
the intensities of most emission lines have decreased, except for the semi-forbidden
lines of N II] (214.0 nm) and C II] (232.6 nm), which strengthen with time.

3.2. NOVA SERPENTIS 1983

The IUE spectra of N Ser 1983 were taken 11 days after the sharp brightness
maximum, when the object had already declined by more than 4 mag. The visual
observations show that N Ser 1983 was obviously a very fast nova, with a t_3-time
of only about 5 days.

The combined absolutely calibrated SWP and LWR spectra, taken on 5 March,
1983, are shown in the middle part of Figure 1. The data have been de-reddened
using a value of $E(B-V) = 0\overset{m}{.}4(\pm 0\overset{m}{.}1)$. The continuum flux in the UV rises
towards the optical range. Due to the poor signal/noise ratio, neither continuum
nor emission lines can be detected between about 200 nm and 240 nm. A
significant flux level in the short wavelength (SWP) range could in principle partly
be due to some continuum radiation, but is almost certainly produced by the
blending of numerous emission lines. A wealth of strong and broad emission lines
is present: resonance lines of moderate to high ionization levels (e.g., of Al III,
C IV, Si IV, and N V); semi-forbidden intercombination lines (e.g., of C III], N III], or

O III]); as well as permitted lines of neutral and singly-ionized metals such as Fe I, Fe II, or C I; the recombination line of He II at 164 nm is weak. The most prominent emission feature in the long wavelength (LWR) range is due to the resonance doublet of Mg II at 280 nm. It is composed of two distinct (blue- and red-shifted) components with different strengths and peak radial velocities of -1520 and $+2250$ km s^{-1}, respectively, which indicate the expansion of a non-spherically symmetric shell expelled by the nova. The amazingly intense Mg II-line shows a distinct P Cygni-type profile, with a terminal velocity – corresponding to the shortward absorption edge – as high as about -11000 km s^{-1}. The C IV (155 nm) line also exhibits a P Cygni profile with a terminal velocity of about -9000 km s^{-1}. The general appearance of the short wavelength IUE spectrum resembles that of Nova Aquilae 1982 (Snijders *et al.*, 1982).

The line spectrum is composed of lines with a wide range in ionization potential, and consists of permitted lines, which are typical for the early decline stage, as well as of semi-forbidden lines usually observed during the nebular phase. Our UV observations were therefore probably made in the transitional stage of the nova.

3.3. Nova Muscae 1983

The high and low dispersion spectra of N Mus 1983 were obtained 44 days after the visual brightness maximum, when the nova had declined by about 2.5 mag. The visual observations show that N Mus 1983 is probably a slow nova. After a relatively steep decline by about 2 mag within a few days, the object remained about constant in brightness at about 9^m, for more than 30 days. At the time of our observations, its brightness was $9^m.6$.

The combined low dispersion SWP and LWR spectra, taken on 4 March, 1983, are shown in the lower part of Figure 1. The data have been de-reddened according to $E(B-V) = 0^m.5(\pm0^m.1)$. Almost no continuum flux is detectable. The spectrum is dominated by semi-forbidden emission lines of C II], C III], N II], N III], N IV], and O III]. Lines with relatively high ionization level such as He II, C IV, N V, O IV], and O V] are also present. The large number and relative strength of the inter-combination lines suggest that the nova had already entered its nebular stage. Comparison can be made with Nova Cygni 1978: the spectral evolution of this nova was followed with IUE from very early phases into the nebular stage, where the appearance of its line spectrum was strikingly similar to that of N Mus 1983 (see, e.g., Figure 1a of Stickland *et al.*, 1981).

The relative strength, especially of the nitrogen lines, but also of those of carbon and oxygen, suggests very high abundances of these elements, as can be achieved through the operation of the CNO cycle, and in accordance with theoretical models of the nova outburst. A detailed analysis of the high dispersion spectra will yield quantitative abundance results. The overabundance of C, N, and O with respect to solar values is estimated to be at least of 1 or 2 orders of magnitude.

4. Discussion

The UV spectra of the three novae N Sgr 1982, N Ser 1983, and N Mus 1983 are compared in Figure 1. The spectral appearance together with the photometric evolution allows a classification of the objects according to their speed class.

The spectrum of N Sgr 1982 shown in the upper part of the figure is typical for early stages after the (visual) outburst maximum: a wealth of neutral and singly-ionized metal lines are seen; the continuum decreases toward shorter wavelengths, and drops to zero already in the mid-UV; while the UV spectra of 1982, 18 and 19 October (about 14 days after maximum) are very similar, the one of 31 October shows already a definite shift of the maximum of intensity distribution toward far UV wavelengths. Comparison with the UV spectroscopic evolution of the fast Nova Cygni 1978 (Stickland *et al.*, 1979) suggests that N Sgr 1982 belongs to the class of classical fast novae, and was observed during its early decline prior to the nebular phase.

The spectrum of N Ser 1983 in the middle part of Figure 1 was probably taken during the transitional stage. Lines of low and high-ionization levels as well as resonance, permitted and semi-forbidden lines do coexist. Since this spectrum was taken only 11 days after the sharp visual brightness maximum, N Ser 1983 is probably a very fast nova; this can also be anticipated from the exceptionally short t_3-time of less than 5 days.

The P Cygni profile of Mg II (280 nm) indicates a very strong and fast wind (~ 10000 km s^{-1}) with asymmetric structure. Two blobs or clouds of material contained in the expanding shell can possibly cause the two emission components with a peak separation of several thousand km s^{-1}. The existence of Mg II together with high ionization species such as C IV, N IV, or Si IV implies that these ions are formed in spatially separated parts of the wind, which is optically thin and has a large temperature gradient. The high-ionization lines have no double-peaked emission comparable to that of Mg II. They are conceivably formed in deeper layers of the expelled envelope, which do not reflect the asymmetric structure of the outer shell.

The UV spectrum of N Mus 1983 in the lower part of Figure 1 was obtained 44 days after visual maximum light. The spectral appearance is typical for the nebular phase, as one can readily derive from the line spectrum and from a comparison with the spectral evolution of N Cyg 1978 in the UV range (Stickland *et al.*, 1979, 1981). Though the relatively early onset of the nebular phase would lend support to the classification of N Mus 1983 as a moderately fast nova, this is in contradiction with the phtometric development: in early August 1983, some 200 days after maximum, the visual brightness was still at 10^{m}9, which is less than 4 mag below the maximum brightness, and corresponds to a decline rate typical for a slow nova.

Another striking feature of N Mus 1983 is the exceptional strength of the nitrogen lines, and – to somewhat less extent – also of the carbon and oxygen lines. The enhancement of nitrogen relative to the Sun, amounts to more than two orders of magnitude, and indicates that the CNO process was certainly important in the thermonuclear runaway event.

A more detailed analysis of all three novae, for which in part simultaneous optical spectroscopy and high resolution IUE spectra have been obtained, is in progress and will be published elsewhere.

Acknowledgements

Special thanks are due to Drs G. Shaviv and A. Cassatella for helpful comments. We appreciate the competent support of the staff of the ESA-Vilspa IUE ground station in Villafranca. This research was partly supported by the Deutsche Forschungsgemeinschaft grants Ra 136/10–2 and Ra 136/11–1.

References

Andersen, P. H., Borra, E. F., and Dubas, O. V.: 1971, *Publ. Astron. Soc. Pacific* **83**, 5.
Blades, J. C. and Drechsel, H.: 1982, *IAU Circ.*, No. 3741.
Cassatella, A.: 1983, private communication.
Drechsel, H., Rahe, J., Wargau, W., Blades, J. C., Cacciari, C., and Wamsteker, W.: 1982, *IAU Circ.*, No. 3736.
Honda, M.: 1982, *IAU Circ.*, No. 3733.
Liller, W.: 1983, *IAU Circ.*, No. 3764.
Seaton, M. J.: 1979, *Monthly Notices Roy. Astron. Soc.* **187**, 73.
Snijders, M. A. J., Seaton, M. J., and Blades, J. C.: 1982, in *Proc. Third European IUE Conference*, Madrid, ESA SP-176, p. 177
Stickland, D. J., Penn, C. J., Seaton, M. J., Snijders, M. A. J., Storey, P. J., and Kitchin, C. R.: 1979, in *Proc. of The First Year of IUE*, Londen, p. 63.
Stickland, D. J., Penn, C. J., Seaton, M. J., Snijders, M. A. J., and Storey, P. J.: 1981, *Monthly Notices Roy. Astron. Soc.* **197**, 107.

A CATALOGUE AND FINDING LIST OF GALACTIC NOVAE*

H. W. DUERBECK

Observatorium Hoher List der Universitäts-Sternwarte Bonn, Daun, F.R.G.

(Received 1 August, 1983)

Abstract. A catalogue of galactic novae and an atlas of finding charts are under preparation and will be published in 1984. The status of the project is described.

The most complete lists of galactic novae been compiled by Payne-Gaposchkin (1957, 1977). These catalogues serve a statistical purpose, but lack useful information for the active observer, such as precise positions and finding charts. Major finding lists and charts were published by Khatisov (1971) and Wyckoff and Wehinger (1978) and contain 42 and 25 objects, respectively. In the former list, not all identifications are correct.

Since many novae at minimum light are within the reach of modern equipment attached to medium-size telescopes, an extended collection of nova data would now be appropriate for observers. Also, accurate positions are vital for observations in other wavelength regions, especially those with satellites.

The new catalogue and atlas of finding charts now under preparation will comprise all novae for which sufficient data are available or can be derived from a study of plate collections. It will contain:

(1) positions (equinox 1950.0 and 2000.0), accurate to at least $1''$;
(2) apparent magnitudes at maximum and minimum light;
(3) light curve types (according to the classification of Duerbeck (1981));
(4) brief notes of spectral appearances;
(5) other minimum data (e.g., orbital periods);
(6) finding charts, generally prepared from sky atlas plates.

The present status of the project is as follows: the literature search for data of 245 novae (some of them suspected ones) is essentially completed. It yielded accurate positions for 35% of the objects. For the remainder it is planned to evaluate the plates of several observatory archives with the aim of increasing the positional accuracies by factors 10 or more, as compared with previously available data. This will faciliate an identification of these novae at minimum light.

Finding charts for about 55% of the novae could be located. Most of them are insufficient due to the fact that they were intended for use during the maximum state. New standardized charts will be prepared from sky atlas plates.

The publishing data of the catalogue and atlas in 1984.

* Paper presented at the Lembang-Bamberg IAU Colloquium No. 80 on 'Double Stars: Physical Properties and Generic Relations', held at Bandung, Indonesia, 3–7 June, 1983.

For the very faint novae – e.g., in the Sgr-Sco region – which are generally poorly documented and whose minimum brightness lies in most cases below the plate limit of the sky atlases, the use of large telescopes is needed both for identifications and for further study.

Any information about unpublished identifications is welcome, also suggestions concerning the contents of the catalogue. Should, e.g., all stars once classified as novae be included even if they are now listed as other types of variable stars?

Acknowledgement

I thank the Deutsche Forschungsgemeinschaft, Bonn-Bad Godesberg, for a travel grant (477/832/83(2)).

References

Duerbeck, H. W.: 1981, *Publ. Astron. Soc. Pacific* **93**, 165.
Khatisov, A. Sh.: 1971, *Bull. Abastumani Astrofiz. Obs.* **40**.
Payne-Gaposchkin, C. H.: 1957, *The Galactic Novae*, North-Holland Publ. Co., Amsterdam, Holland.
Payne-Gaposchkin, C. H.: 1977, in M. Friedjung (ed.), *Novae and Related Stars*, D. Reidel Publ. Co., Dordrecht, Holland, p. 1.
Wyckoff, S. and Wehinger, P. A.: 1978, *Publ. Astron. Soc. Pacific* **90**, 557.

THE OLD NOVA BT MONOCEROTIS*

W. C. SEITTER**

Astronomisches Institut der Universität Münster, Münster, F.R.G.

(Received 30 August, 1983)

Abstract. Spectroscopic observations through most of the eclipse cycle of BT Mon reveal the presence of both low and high velocity gas streams. Acceleration through a Laval-nozzle-effect at the inner Lagrangian point of the system and powering of the emission lines through kinetic energy losses of Coriolis deflected and subsequently colliding gas streams are considered as possible mechanisms at work in the system.

1. Introduction

The old nova BT Mon (Nova Monocerotis 1939) deserves and has recently received much attention. Photometric investigations have yielded one of the deepest eclipses found in any postnova so far (Robinson *et al.*, 1982). Our own spectroscopic investigations have left us puzzled.

2. The Data

BT Mon was observed on 7 nights in December, 1982, with the IDS spectrograph at the 1.52 m telescope at ESO, La Silla. The 54 spectrograms cover 82% of the eclipse cycle. The wavelength range of most observations was 3800–7300 Å with a dispersion of 172 Å mm^{-1}. Several observations are of higher dispersion and correspondingly smaller wavelength regions. The data were reduced at the ESO reduction facility in Garching and at the newly installed system in Münster.

3. The Light Curve

The calibrated spectra allow us to measure fluxes in the lines as well as in the continuum and to follow them through the eclipse cycle. Figure 1 shows the continuum variation at 5000 Å.

From several observed eclipses a mean minimum time was derived,

$$\text{JD hel. (min)} = 2\,445\,323.6875\,,$$

which fits very well into Robinson *et al.*'s ephemeris. No indication of a period change could be detected.

---

* Paper presented at the Lembang-Bamberg IAU Colloquium No. 80 on 'Double Stars: Physical Properties and Generic Relations', held at Bandung, Indonesia, 3–7 June, 1983.
** Based on observations obtained at the European Southern Observatory, La Silla, Chile.

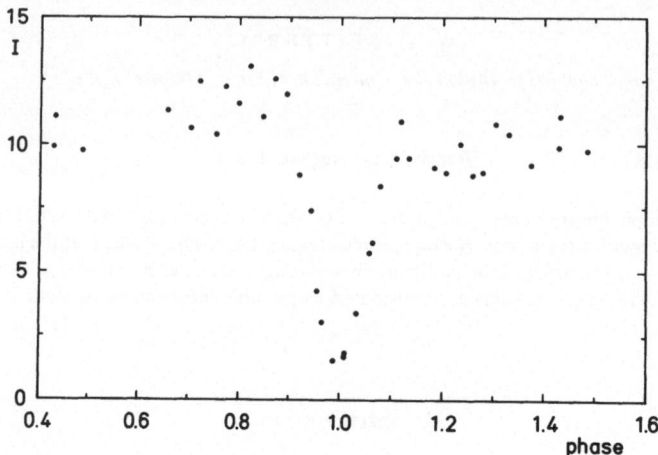

Fig. 1.   Light curve of BT Mon at 5000 Å; arbitrary intensity units.

4. The Radial Velocity Curves

Radial velocity curves were obtained through various procedures. Mean RV-curves were derived from cross correlations of the spectra. Detailed RV-curves were obtained from Gaussian fits to the emission lines He II, Hα, and Hβ. Figure 2 shows a mean RV-curve. It is at once apparent that the curve has a hitherto unknown peculiarity, a high velocity spike at the time of eclipse. While it is fairly obvious that the spike must be associated with the contribution of high velocity gas dominant during eclipse, the details of the model are not simple. A more subtle but no less disturbing phenomenon becomes obvious when instead of using a mean RV-curve, individual curves for the

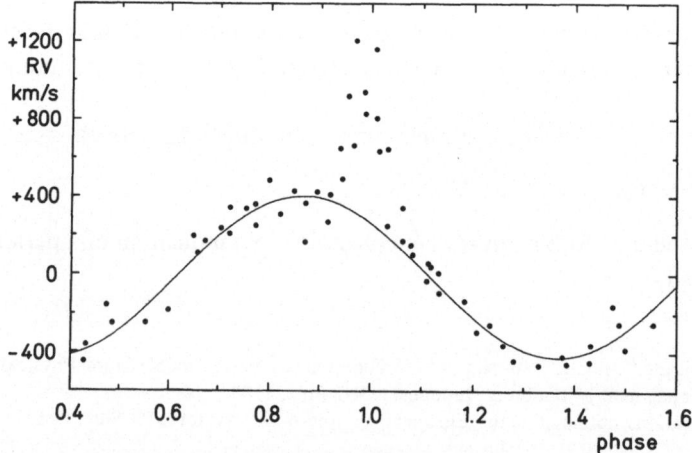

Fig. 2.   Radial velocity curve of BT Mon, derived from cross-correlation of spectra.

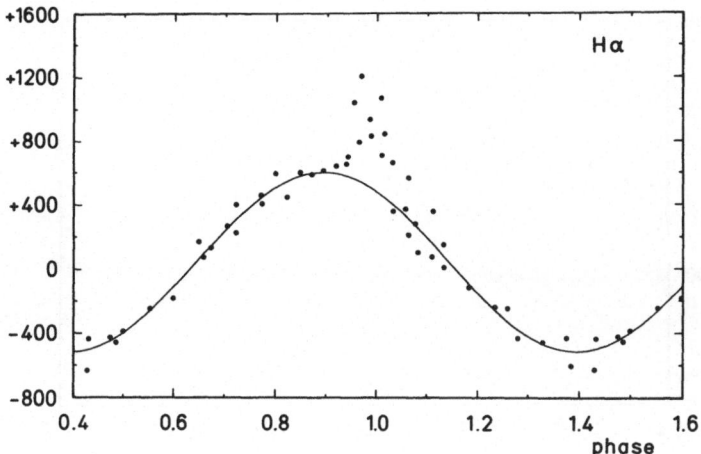

Fig. 3. Radial velocity curve of BT Mon, based on the Doppler displacements of the Hα line.

different lines are plotted (Figure 3: Hα RV-curve). The differences in amplitudes are significant (Hα: 557 km s^{-1}, Hβ: 402 km s^{-1}, HeII: 303 km s^{-1}) while the peak velocities of the spikes seem to be approximately the same, if one allows for a short eclipse of the HeII contribution which will be justified later on. The most important feature, however, is the noticeable phase shift from phase zero (eclipse) as a function of amplitude. The simple interpretation that the eclipse is that of the hot spot rather than the eclipse of a luminous disc can be ruled out. While the presence of a region of higher luminosity is indicated in Figure 1 at phase 0.8, it is far too weak to account for the decline through 2.7 mag. during minimum, and furthermore, it is not at the right location.

Apparently, we observe not the RV-curve due to orbital motion alone but the superposition of different velocities whose resultants are shown in Figure 3. Assuming that the observed curves result from adding different sine curves at different phase shifts for the three lines Hα, Hβ, and HeII to a fixed curve representing the unknown orbital motion, it is possible to derive the true velocities. The K_1 value is 248 km s^{-1}, the added values are 105 km s^{-1} for HeII, 266 km s^{-1} for Hβ and 489 km s^{-1} for Hα. It is now possible to determine the zero points of the added velocities and thus the directions of the additional motions, if the zero points for the orbital motion are assumed to occur at eclipse and at 180° from eclipse.

5. The Line Fluxes

Figures 4 and 5 show the variations of emission line fluxes of Hα and HeII during the eclipse cycle. A strenghtening of the lines is observed before eclipse with a peak probably around phase 0.75. The lines appear systematically weaker after eclipse. The variations for most of the lines differ not greatly from a sinusoidal variation, which makes them likely to be due to the view of a hot region from different aspects during the cycle.

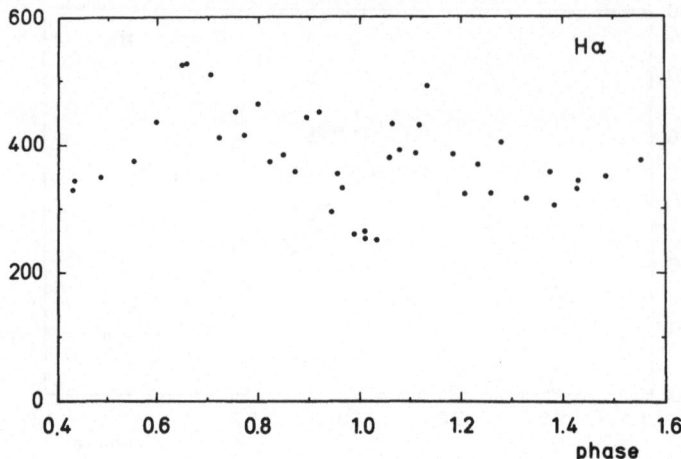

Fig. 4.   Line fluxes of the Hα emission line; arbitrary intensity units. The photometric minimum is only weakly indicated.

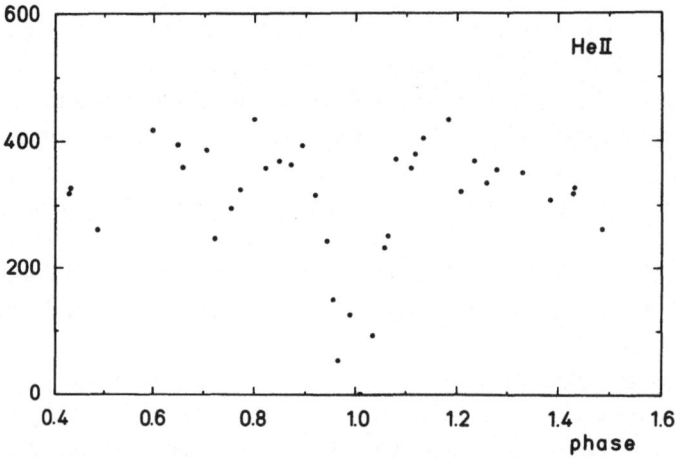

Fig. 5.   Line fluxes of the He II 4648 line; arbitrary intensity units. The minimum is very pronounced.

The most surprising result, however, is the fact that the strength of Hα shows a decline by only $\frac{1}{3}$ of the maximum value during eclipse, while C III disappears completely, He II and Hβ for short intervals during the eclipse.

6. The Model

As often in astronomy, data are not sufficiently scarce to permit a simple model, but not sufficiently complete to reveal the full detail of a more complicated picture. There are, however, a few preliminary conclusions which can be drawn: Assuming that the correct RV of the orbit has been derived, taking the secondary to be of type K5–7 as seems to be indicated from 21 identified weak absorption lines and blends visible mostly during minimum, furthermore, using Robinson *et al.*'s relation between q and $\cos i$, we find no acceptable solution for the masses.

However, an orbital velocity of 190 km s^{-1} together with $i = 79°$, yields a solution $\log q = 0.17$, $M_1 = 0.48 M_\odot$, $M_2 = 0.72 M_\odot$. The dimensions of the system derived from these data allow a purely gravitational acceleration to about 500 km s^{-1}. If, however, the energy radiated in the emission lines derives from collisions of gas streams of different directions, the original velocities must be higher by at least a factor two. The presence of high velocity gas in the system is corroborated by Boroson-Greenstein diagrams, especially for Hα. Here, velocities up to 3000 km s^{-1} are observed, velocities which clearly do not result from gravitational acceleration at the location of origin of the line.

We thus propose an acceleration mechanism already at the location where the gas stream leaves the secondary. Such a mechanism could exist in the form of a Laval nozzle which increases the sound velocity of some 10 km s^{-1} at which the gas enters the region around the inner Lagrangian point to supersonic velocities. The wide spread of velocities in the system then requires either a self-regulating system supplying a range of different velocities or noticeable contributions from the velocities around the Mach disks forming at some distance from the nozzle.

If a simplified two-gas stream model is invoked, we can reproduce the observed features with a low-velocity stream showing a larger Coriolis deflection, which shows sufficient bending towards the primary to eventually collide with a high-velocity gas stream of small Coriolis deflection and a velocity which at all locations within the Roche lobe of the primary exceeds the velocity of escape, such that this stream leaves the system almost undeflected. The interaction of the two streams marks the region of highest radiative energy whereby the outer parts contribute mostly Hα radiation while regions closer to the primary produce more Hβ in accordance with a flatter Balmer decrement for hotter regions, in the innermost region He II is generated. Ionization and excitation are the results of collisions of the gas stream particles which loose considerable velocity as suggested by the increasingly lower gas velocities observed for lines of higher energy.

So far we can deduce only this rather qualitative model which has been checked for consistency only. Besides this, two rather firm conclusions can be drawn: ex-novae may well be more complicated than has hitherto been assumed, and, more observations are needed.

Reference

Robinson, E. L., Nather, R. E., and Kepler, S. O.: 1982, *Astrophys. J.* **254**, 646.

SYMBIOTIC STARS*

DAVID A. ALLEN

Anglo-Australian Observatory, Epping, N.S.W., Australia

(Received 9 July, 1983)

Abstract. I review our current knowledge of symbiotic stars. A great many papers have graced the literature in the fifty years of their study, and many data are available on the spectral variations at optical wavelengths these stars undergo. I do not give extensive references to those data, for previous reviews have done so quite adequately. Rather, I concentrate on the extensive widening of the wavebands within which symbiotic stars have been studied over the past few years, and attempt to synthesise the data into a coherent picture.

Symbiotic stars are most readily explained as interacting binaries, though single star models may still be tenable for some systems. They are made much more complex than most other interacting binaries by the variety of accreting stars, and because gas flows may be highly structured. Moreover, their study is more difficult than that of dwarf novae because the orbital periods are long compared to the activity cycles of the accretion phenomena.

Our data base has expanded enormously with our present spectral catholicism. But there remains much valuable work to be done with even simple equipment on small telescopes. I suggest in a final section areas for future work.

1. What are symbiotic stars?

This review was presented at a conference devoted to binary stars. I shall therefore open with the blunt dogma that symbiotic stars are interacting binaries, in which a late-type giant sheds material onto a more compact companion. In adopting this definition I am patently doing injustice to the many papers in which highly credible models have been constructed describing symbiotic stars as single objects. It would be foolish to dismiss these models so casually, and below I shall refer to some of them more specifically. It would also be as well for the reader to recall that the very term symbiotic star, whilst implying the presence of two interdependent components, has also become synonymous with an object that cannot be dropped into any other descriptive bin. (Nussbaumer, 1982). Some of the objects we so designate may indeed be single stars. Certain too is that many are interacting binaries: over the last few years the proportion for which this diagnosis is irrefutable has increased imposingly, whilst our understanding of how the interaction can generate what we call the symbiotic phenomenon has blossomed.

What do we mean when we speak of a symbiotic star? The definition has broadened of late, and now allows inclusion of objects that might not have been considered fifty years ago when the first specimens were found by Mount Wilson

---

* Invited paper presented at the Lembang-Bamberg IAU Colloquium No. 80 on 'Double Stars: Physical Properties and Generic Relations', held at Bandung, Indonesia 3–7 June, 1983.

Astrophysics and Space Science **99** (1984) 101–125. 0004–640X/84/0991–0101$03.75.

observers. In Allen (1979) I proposed a set of defining criteria that allowed a catalogue of slightly over 100 examples to be drawn up. Some revision of both the definition and the catalogue has been enforced by data from the International Ultraviolet Explorer (IUE). The present number in my private catalogue is 128.

A current definition must relate to the simultaneous presence of two grossly different temperature regimes – the symbiosis. The cooler of these is a giant of spectral type G or later; the hotter is expressed by emission lines of high excitation, such as He II and [Fe II]. In addition, variability of a nova-like nature has often been considered a defining characteristic, but this I reject for reasons that will become apparent below. Rather, to those stars which undergo frequent outbursts of a few magnitudes on time-scale of a few months, I give the epithet classical symbiotic stars. The principal members of this group are (in right ascension order) AX Per, RW Hya, AG Dra, BF Cyg, CI Cyg, AG Peg, and Z And.

This definition is unsatisfactory because of the looseness of the term 'high excitation'. My 1979 requirement of He II emission (λ 4686) was proved inadequate when IUE began to reveal lines of C IV, He II, etc. in the ultraviolet spectra of stars (such as R Aqr) which did not from optical data alone fulfill my criteria. Unfortunately, there are many late-type stars which exhibit Balmer emission lines, but which certainly do not merit being called symbiotic. Exactly where in this continuum of objects one draws his dividing line is unclear. I prefer to leave the definition vague at the moment, in the belief that within a few years we will be able to use instead the dogma with which I opened this section. Inasmuch as symbiotic stars are interacting binaries, they are relatively low-mass specimens, and are distinct from systems which produce Wolf–Rayet stars. This fact should be borne in mind, for Wolf–Rayet spectral features will be mentioned below.

In what follows I shall not use the terms primary or secondary to describe the components of the symbiotic stars. In systems where mass transfer occurs, this is inevitably confusing. Moreover, as Plavec (1980) comments, different authors mean different things by the terms. In symbiotic stars the distinction is made easier by the two temperature regimes. I refer to the hot star and the cool star. Plavec would call these the gainer and the loser respectively.

2. Previous Reviews

The field of symbiotic stars has become particularly active over the last five years, and reviews dating from the 1970's are already outdated in large part. Of course, they nicely enshrine the history of thinking on the subject. Space does not permit the indulgence of a detailed historical summary, and I merely draw attention to the reviews of Sahade (1965, 1976), and Boyarchuk (1970a, 1975). The first partial review that took significant note of data outside the visible waveband was that of Allen (1979). For the most comprehensive review of these stars the reader is referred to the proceedings of *IAU Colloq.* **70** (Friedjung and Viotti, 1982); there he will find brief reviews by specialists on various aspects of their study. The present paper hopefully presents a more coherent picture than can be achieved in a major meeting such as *IAU Colloq.* **70**, but suffers from the inevitable bias introduced by a single author.

Many papers have been written on individual stars; I have eschewed a complete bibliography, giving references to specific cases of interest to the theme of this review. I apologise to those whom I have neglected: the list of references is already lengthy.

3. Methods of Study

More than any other, the lesson we have learnt from the last five years is that symbiotic stars are a phenomenon of almost the entire electromagnetic spectrum. To restrict one's attention to the optical waveband is to be an ostrich. I now summarise the range of wavelengths at which useful data can and have been obtained.

3.1. X-RAY

Only a few symbiotic stars are X-ray sources. The most prominent example is V2116 Oph, a star discovered initially by its X-radiation and formerly known as GX1 + 4, or 3U 1728-24 (Davidsen *et al.*, 1977). It is the only hard X-ray source known amongst the symbiotic stars, and is further distinguished by its pulsations of period near 2 min. (4 min.?) and large period derivative. Analysis of these pulsations shows the system to contain an accreting neutron star (Mason, 1977), whilst the optical data reveal an M giant.

Soft X-ray sources might be expected in many symbiotic stars, where there is emission from species with ionization potentials exceeding 100 eV. That this is not the case may be attributed to two factors. First, there is a large column depth of neutral hydrogen in front of many; second, the X-ray source has a very steep spectral gradient on the short-wavelength side, a fact suggesting that the X-rays are thermal and emanate from a black-body of temperature a few hundred thousand Kelvin. To date the following have been detected: AG Dra (Anderson *et al.*, 1981), T CrB (Cordova *et al.*, 1981), HM Sge, V1016 Cyg, RR Tel, and possibly AS 295B (Allen, 1981).

From the X-ray data we can determine an equivalent angular size to the emitting object. If the distance is taken from the infrared (see below), then the hot component is typically the size of a white dwarf star.

3.2. ULTRAVIOLET

Despite the pioneering work by Gallagher *et al.* (1979) using the OAO2 satellite on AG Peg, it is IUE that has really taught us about the hot gas and continuum in symbiotic stars. Several groups have been active in this field. A summary of the results is attempted below; Table I gives the major references to ultraviolet data.

Early data from IUE seemed to show the simple continuum from a black body

TABLE I

The major references to ultraviolet data on symbiotic stars

CH Cyg	Hack (1979)
HM Sge, V1016 Cyg	Flower et al. (1979)
R W Hya	Kafatos et al. (1980)
AG Peg	Keyes and Plavec (1980)
R Aqr	Michalitsianos et al. (1980)
EG And	Stencel and Sahade (1980)
Z And	Altamore et al. (1981)
T CrB	Krautter et al. (1981)
V1016 Cyg	Nussbaumer and Schild (1981)
AG Dra	Altamore et al. (1982)
HM Sge, V1016 Cyg, V1329 Cyg	Feibelman (1982)
RX Pup	Kafatos et al. (1982)
V1329 Cyg	Kindl and Nussbaumer (1982)
BX Mon, SY Mus (in eclipse?). CL Sco, YY Her	Michalitsianos et al. (1982a)
SY Mus (in outburst?)	Michalitsianos et al. (1982b)
12 systems	Slovak and Lambert (1982)
CI Cyg	Stencel et al. (1982)
V1329 Cyg	Nussbaumer and Schmutz (1983)
RR Tel	Penston et al. (1983)
AG Dra (in outburst)	Viotti et al. (1983)

of a few 10^5 K underlying emission lines formed in a high-density gas. As more stars were studied, the picture muddied impressively. In many cases the continuum is not well represented by any credible combination of hot star and gaseous continuum. The variability of some systems may provide a way of discriminating various components which, on present data, appear to behave independently (e.g., AG Dra in outburst, references in Table I).

Further, whilst the emission lines have proved a common feature of symbiotic stars, they are in some cases broad and structured and in others unresolved. AG Peg is a prime example of a system with broad lines, the velocities in this case being several hundred kilometres per second. These greatly exceed the orbital velocities, and indicate that gas streams flow between the stars. In most such cases, some evidence of broad emission lines (mostly He II $\lambda4686$) can be seen in the optical. These lines have often been taken to imply the existence of a Wolf–Rayet star in the system. However, the IUE line profiles do not resemble those of Wolf–Rayet stars. Moreover, as noted above, symbiotic stars have lower masses and luminosities that the classic Population I Wolf–Rayet stars.

Unquestionably the ultraviolet has contributed significantly to the data bank on symbiotic stars, but it would be an exaggeration to state that it has provided as many answers. The emission-line results have been valuable; a convenient summary of the use to which they can be put is given by Nussbaumer (1982). The difficulty of understanding the continuum data can be appreciated from the following list of potential contributing components: the hot star; an accretion disk, generally of unknown inclination; the boundary between disk and star; free-free, free-bound, and two-photon continua from a stratified nebula. of hydrogen and helium; blended, weak emission lines; scattering of light by dust grains; reddening, part of which may be circumstellar and therefore different from the normal reddening law; radiation from the heated side of the cool star.

The most comprehensive attempt at unravelling the ultraviolet continua of symbiotic stars is that of Kenyon and Webbink (1983).

3.3. OPTICAL

Most data on symbiotic stars have, of course, accumulated in the optical domain. The picture they suggest, nicely summarised by the work of Boyarchuk (1966, 1967a, b, 1969), combines a cool giant with a hot star and the emission nebula the latter excites. The nova-like outbursts are seen to be events centred on the hot component. There is convincing evidence that the bolometric luminosity of the hot star remains fairly constant during such an outburst, whilst the temperature fluctuates.

Several of the symbiotic stars are eclipsing systems, and therefore indisputably binary. The first such to be studied was AR Pav (Thackeray and Hutchings, 1974), and this analysis neatly confirmed Boyarchuk's model of other systems, whilst further demonstrating that a stream of gas flows from the cool star towards the hot companion. More recently, CI Cygni has been shown to undergo eclipses (Pucinskas, 1970; Belyakina, 1979), and it seems likely that SY Mus is another example (Uitterdijk, 1934; Michalitsianos et al., 1982a, b).

Radial velocity data do not assist in detecting orbital motions of these systems. The spectral features of late-type giants are insufficiently sharp to permit easy velocity measurement, whilst the emission lines indicate the complex streaming of the gas rather than any motion of the hot star. The difficulty of deriving even simple evidence of binary motion is poignantly highlighted by the stars AG Peg and V1329 Cyg. In the latter, a binary orbit has indeed been deduced (Grygar et al., 1979), but the mass function, 25 $M_\odot$, is inadmissibly high. Ijima and Mammano (1981) have suggested that the 950 day period is one of outburst rather than orbital motion. In AG Peg the velocity curves do not close around the orbit, and in the analysis of Hutchings et al. (1975) gas appeared to be flowing up a potential gradient from the hot star to the cool. In both these cases it is the complex nature of the gas streams that undoubtedly confuses the analysis, a fact noted in the case of AG Peg by Sahade (1976).

The stratified nebula also complicates analysis of optical and ultraviolet emission-line ratios: the H II regions are sufficiently structured that unique values of electron temperature and density are inappropriate. In many studies of physical conditions and elemental abundances the authors have been forced to assume homogeneous, spherical nebulae, and the results of these analyses are questionable.

If estimates of electron temperatures (usually around 10000–15000 K) are meaningful, they indicate photoionization to be more important than shock or other ionization mechanisms.

The majority of objects now classified as symbiotic stars are not known to partake in the nova-like variability considered by Boyarchuk to be a pre-requisite for inclusion in the class. And a few, referred to as the slow novae, do so in a more protracted way, and with much greater increase in brightness. AG Peg, mentioned above, erupted in about 1850, and has not yet returned to its pre-outburst magnitude. RR Tel is progressing at a comparable rate.

Yet, if we were to take single snapshots, these systems would resemble most of the classical symbiotic stars in their optical and ultraviolet spectra, and in many other ways. It seems unrealistic to restrict the definition of symbiotic stars to those whose variability occurs on a convenient timescale. Figure 1 illustrates this point: in it are uncalibrated spectra of two objects secured on the same night with the same equipment. One of these, RT Ser, is the prototype slow nova; the other, lying only a few degrees away in the sky, was until recently classified as a planetary nebula, and is not a known variable. The spectra scarcely differ. The main features of symbiotic spectra are well represented: TiO absorption in the cool continuum of an M giant, a blue continuum due to free-free and free-bound emission from the nebula, and emission lines covering a range of excitations. It would not be difficult to present an essentially identical spectrum of a classical symbiotic star at a suitable phase.

Broad emission lines are a characteristic feature of more than half of the known symbiotic stars. In a few cases the lines are Wolf–Rayet-like. The luminosities are much lower than those of Population I Wolf–Rayet stars; moreover, the rapid evolution of RR Tel's broad-line spectrum (Thackeray and Webster, 1974) argues against the presence of a true Wolf–Rayet star in the system. It is interesting that this progression, towards higher ionization subclass, exactly parallels the evolution that is now believed to occur in population I Wolf–Rayet stars due to the shedding of mass (Moffat, 1981). We should not consider these features as evidence of true Wolf–Rayet stars; rather a condition comparable to that in the atmospheres of the more massive Wolf–Rayet stars is simulated near the hot component.

In RR Tel, the Wolf–Rayet features faded at the same time that the $\lambda6830$ band appeared. This, and its weaker satellite at $\lambda7088$, are unidentified features. They are found only in symbiotic stars (and in particular not in Population I Wolf–Rayet stars), and clearly originate at high excitation. The presence of such emission bands suggests the existence of a compact star with escape velocity of order 1000 km s^{-1} or more (Allen, 1980a).

A recent review of optical data is given by Ciatti (1982) in *IAU Colloq.* **70**, and of the photographic infrared in the same publication by Andrillat (1982).

3.4. INFRARED

Early infrared observations (Swings and Allen, 1972; Glass and Webster, 1973; Szkody, 1977) served to show that symbiotic stars are merely late-type giants in

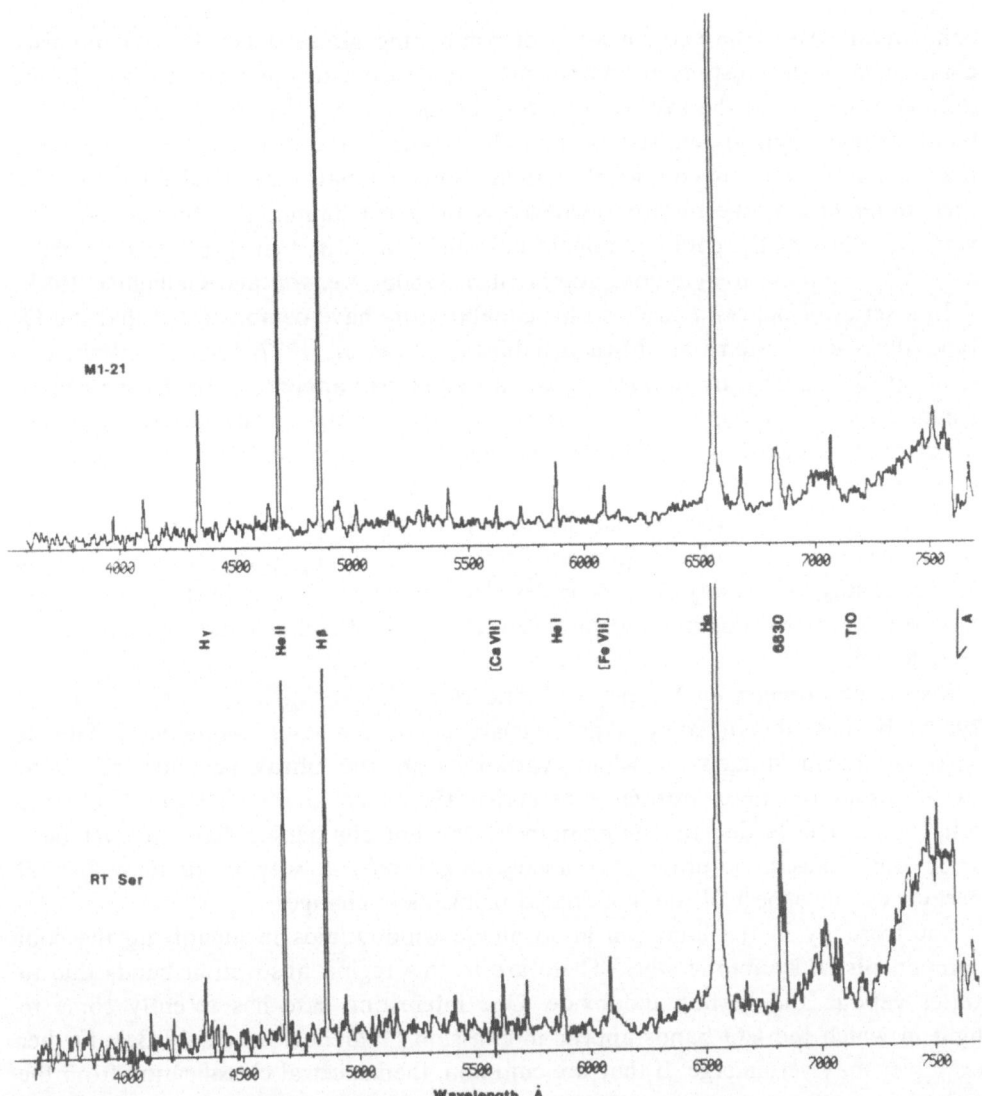

Fig. 1. Low-dispersion optical spectra of two symbiotic stars, M1–21 (originally classified as a planetary nebula) and RT Serpentis (the prototype slow nova). The ordinate on these plots is uncalibrated photon rate. These spectra clearly reveal the high-excitation forbidden lines, the gaseous continuum and the presence of an M giant.

this waveband. Subsequent data showed a subset of them to have circumstellar dust emission in the 1–4 μm range (Webster and Allen, 1975). The distinction into two classes, S and D according to whether the cool star or dust dominate, remains relevant now almost a decade after its discovery.

The dust itself resembles that found in single late-type giants except that the black-body component is hotter. Roche *et al.* (1983) suggest that the hot companion causes the extra heating of conducting grains which form a minority component of the dust. That component may be iron. In this respect alone do the infrared data reveal the symbiotic nature of the systems. In all other respects the late-type giant appears entirely normal. A number of recent papers have suggested that symbiotic stars are cool giants evolving towards planetary nebulae by shedding their outer layers; the infrared evidence is the most damning of that model. The very normality of the cool component not only proves its reality, but also provides a distance estimate to symbiotic stars (Allen, 1980b; Kenyon and Gallagher, 1983).

In a series of papers, Feast and his collaborators have demonstrated that the D-type (dust-rich) systems are Mira variables (Feast *et al.*, 1977, 1983a, b; Whitelock *et al.*, 1983a, b). The division into two infrared classes appears to be determined by whether the cool component is or is not a Mira variable, whilst the much higher mass loss seen in Miras accounts for the existence of good correlations between the infrared class and other parameters (Allen, 1979).

Of particular relevance is the demonstration by Feast *et al.* (1983a) that the Mira variable in RR Tel was quite unaffected by the 7 magnitude flare seen in 1944 and persisting to this day. This provides clear evidence that the outburst in RR Tel is associated with a companion star, even though no direct evidence of binarity has been found.

Recent photometry of S-type symbiotic stars (Taranova and Yudin, 1981a, b, 1982a, b) has shown some slight variability of the cool component. This is encouraging to models in which variations in the ultraviolet flux relate to variations in the mass transfer rate across the inner Lagrangian point. On the other hand, effects due to illumination by the hot companion have not yet been ruled out. Indeed, accurate photometry might offer a way of deriving orbital periods as our aspect of the illuminated hemisphere changes.

Spectroscopy in the 2-micron atmospheric window aids in identifying the cool components of symbiotic stars. There are in this region absorption bands due to water vapour and carbon monoxide. One interesting case has recently come to light in which the CO bands appear in emission (Whitelock *et al.*, 1983c): further examples may yet emerge. If they are common, then spectral classification from the CO bands will cease to be practicable, for fear of infilling by emission.

The late-type star is revealed by the infrared data even if it cannot be seen optically. Some examples have recently been found of systems in which the cool star is highly extinguished. This point will be discussed further in the section on H1-36, below.

Infrared data are summarised by Allen (1982) in *IAU Colloq.* **70**.

3.5. RADIO

A continuum survey by Wright and Allen (1978) showed 10 % of symbiotic stars to be detectable with a single dish telescope. Arrays (especially the VLA) overcome the confusion in the galactic plane which so troubles single-dish work, so the proportion detected can be expected to rise.

Almost all detected cases show a spectral index of about $+1$, intermediate between optically thin (-0.1) and thick $(+2)$ cases. The radio data are interpreted in terms of prolonged mass loss (Wright and Barlow, 1975; Panagia and Felli, 1975; Olnon, 1975). It is significant that all but one of Wright and Allen's detections were classed as D-type in the infrared: such systems can be expected to have undergone considerable mass loss from the cool star.

Some of the radio spectra of symbiotic stars turn over at high frequency to become optically thin. This was attributed by Marsh (1975) to the presence of a neutral cavity generated as a result of spasmodic mass loss. The lack of temporal evolution as the cavity expanded has caused some embarrassment to this model, and a new interpretation was proposed by Allen (1983a, b). This will be enlarged upon below.

Temporal variation has been seen to occur in some systems. Seaquist (1977) found RX Pup to vary rapidly, and no entirely satisfactory explanation of the changes have been forthcoming. More gradual secular changes in the radio flux from some symbiotic stars have also been recorded, as reported in Kwok (1982). The most notable of these has been the fading of V1329 Cyg by an order of magnitude or more between the first detection by Altenhoff and Wendker (1973) and the recent upper limit of Kwok et al. (1981).

Of interest also is the recently discovered optical/radio jet in R Aqr (Sopka et al., 1982). As pointed out by Kafatos et al. (1983a), the jet appears to have been ejected from the system, and its provocative alignment with another radio source suggests previous ejection episodes. Again, a satisfactory explanation is lacking.

The radio offers in addition to continuum coverage the possibility of detecting orbital motion in systems which emit in one of the maser lines. Several searches have been conducted for OH masers (Lépine and Nguyen-Quang-Rieu, 1974;) Brocka, 1979; Cohen and Ghigo, 1980; Michalitsianos, private communication; Allen and Caswell, unpublished), but without success. An SiO maser is known only in the star R Aqr (Lépine et al., 1978; Zuckerman, 1979). Unfortunately, in this system the orbital period is likely to be very long. The absence of OH emission is surprising in some of the very late-type systems, and probably reflects the influence of the hot companion.

Radio data are also reviewed by Kwok (1982) in *IAU Colloq.* **70**.

4. The Nature of the Hot Star and its Outbursts

As far as I know, the first attempt at explaining the idiosyncracies of the hot star in symbiotic systems was that of Tutukov and Yungel'son (1976), who argued that accretion of gas from the cool star's wind provided a variable ultraviolet source. They in turn were doubtless influenced by observational evidence of accretion, especially in the system AR Pav (Thackeray and Hutchings, 1974). A more detailed analysis of the mechanism for a Main-Sequence star was given by Bath (1977). Later, Paczyński and Rudak (1980) divided the symbiotic stars into two groups according to their optical variability, and related these groups to the accretion rate onto a white dwarf.

In discussing the accretion process it is necessary to have some idea of the type of accreting star. The absence of hard X-rays in all but V2116 Oph precludes a neutron star or black hole as the accretor, and the debate therefore ranges around main sequence stars, subdwarfs or white dwarfs. The latter raises the possibility of a thermonuclear flash, for accreted hydrogen-rich material will spontaneously ignite once its temperature becomes sufficiently high. It should be remembered that conventional novae are currently explained as flashes on white dwarfs accreting gas from a Main-Sequence star. Moreover, detailed modelling of flashes on cold and hot white dwarfs (Fujimoto, 1982a, b; Iben, 1982 and references therein) indicates that the interflash period and characteristic flash time-scale are both functions of the accretion rate. Thus, as suggested by Paczyński and Żytkow (1978), and by Gallagher *et al.* (1979) for the specific case of AG Peg, slow novae may simply be conventional novae governed by slightly different conditions. Further, as I have pointed out (Allen, 1979, 1983b), there seems no reason to deny the possibility of flashes persisting for decades or even centuries. I therefore see no difficulty in the hypothesis that many of the symbiotic stars which have shown no dramatic optical variability are in the throes of a shell flash. This may be a more comfortable explanation of their steady light than the release of gravitational energy from accretion, since accretion tends to be a highly unstable phenomenon.

If a white dwarf accretes hydrogen-rich material at a sufficiently high rate, continuous nuclear burning will occur, and the resulting object will be considerably larger. Plavec (1982) has referred to such stars as subdwarfs, though they are not the conventional Population II subdwarfs. Plavec argued that we may indeed require subdwarfs in order not to release too much accretion energy. Only minor thermonuclear flashes occur on white dwarfs which are burning their accreta steadily.

It is certain that an accretion disk with significant luminosity surrounds the hot component of some symbiotic stars. The flickering seen in optical photometry of CH Cyg (Slovak and Africano, 1978) indicates a disk to be present, and a recent analysis by Duschl (1983) invokes a disk of marginal stability in this system. In CI Cygni (see section 7) the existence of an accretion disk is certainly implied by the eclipse light curve.

We must not be beguiled into believing that accretion is essential to a symbiotic star. There are a few planetary nebulae in which the ionization is as high as exhibited by the symbiotic stars. NGC 6302 and 7027 come to mind, and I know of no claims that either is a binary with accretion flows (though the bipolar morphology of NGC 6302 may suggest this configuration). It is possible that the same evolutionary trend can occur in a binary system, and therefore that a planetary nebula may be forming alongside a red giant. Simple statistical arguments suggest that only a very small proportion of symbiotic stars are such systems, however.

Current thinking amongst the majority of students of symbiotic stars (as expressed by contributors of Friedjung and Viotti (1982), and to the proceedings of a workshop held some months earlier on the other side of the Atlantic: Stencel (1981)) is clearly directed towards accretion phenomena in binary systems. One star, the loser, overfills the tidal lobe generated by the presence of its companion, and so transfers mass to the latter. Although the dynamic tidal lobe will not in general equate with the static Roche lobe, it is appropriate to give credit to Roche's work in this area by referring to the tidal lobe as the Roche lobe. In so doing, I follow popular convention. The details that remain to be settled are:

(i) the nature of the accreting star;

(ii) the accretion dynamics: Roche lobe overflow or wind accretion;

(iii) the role (if any) of thermonuclear flashes;

(iv) the structure and stability of any accretion disk.

In a recent study of the ultraviolet continua of 18 symbiotic stars, Kenyon and Webbink (1983) found 5 to involve Main-Sequence accretors. Many of the remainder appeared to be small, hot stars which may have been white dwarfs accreting from the stellar wind.

5. V1016 Cygni: Single Star Concepts

But it would be unreasonable to ignore those models of symbiotic stars that require only single objects. Although many have been proposed, it is particularly convenient here to consider those related to the star V1016 Cyg. In so doing, I set the pattern for the next few sections, in which I will examine a particular star or group of stars to illustrate how our thinking may be directed by specific cases.

In 1965, V1016 Cyg flared into prominence when it brightened by several magnitudes from a faint emission-line star (AS 373: Merrill and Burwell, 1950) to a system likened by Fitzgerald et al. (1966) to the symbiotic stars. The outburst resembled those of novae except that the time-scale was very much longer, and there was no observed optically thick phase, during which time only a cool shell absorption spectrum would have been expected. The subsequent development of the spectrum and light curve also shows similarities to novae, but on a much more protracted timescale. V1016 Cyg is, in fact, a slow nova.

Although the symbiotic binary interpretation was championed by several authors, notably Boyarchuk (1968), Mammano and Ciatti (1975), and Taranova and Yudin (1983), proponents of the single-star hypothesis have provided very convincing arguments. Their principal thesis is that we are here viewing a planetary nebula in its formative stages. The most important of these single star models were those of Baratta et al. (1974), Ahern et al. (1977), Kwok (1977), Flower et al. (1979), Nussbaumer and Schild (1981), and Kindl et al. (1982). Kwok developed a model in which a fast wind from the newly exposed hot nucleus ripped through the more gradually expanding shell of the former M star. Kindl et al.

(1982), analysing IUE data, postulated a spicular wind from the hot star as a means of explaining the observed line profiles.

There is only one feature of these models which I find unsatisfactory: the fact that there is convincing evidence for the continuing presence of a Mira variable in V1016 Cyg. The variability was first noted at infrared wavelengths by Harvey (1974), and it persists, according to Taranova and Yudin (1983). The 2-micron spectrum also shows the CO absorption of a Mira variable (Puetter *et al.*, 1978). Since this is an overtone band, it cannot easily arise in a tenuous nebula, but requires photospheric conditions. If this star has evolved to the point of revealing its core, then it cannot possibly still resemble a Mira.

This argument does not preclude the alternative, though less popular, interpretation in which a cool giant has a particularly active corona. Aller (1954), Gauzit (1955), and Wood (1974) have discussed such hypotheses. Boyarchuk (e.g., 1970a) has argued forcibly against the coronal model, demonstrating that so energetic a corona cannot be maintained. He ignores the electrical discharge theory of Bruce (1975) which might provide an alternative mechanism whereby a corona could be supported. However, I am aware of no way in which any of these coronal models can account for the intense ultraviolet continuum found in all studied symbiotic stars, a continuum which provides adequate radiative ionization to power the observed emission.

There remains one difficulty with the binary interpretation of V1016 Cyg. The distance implied by the infrared luminosity is much greater than the distance inferred from the interstellar reddening, as pointed out by Nussbaumer and Schild (1981). This probably reflects the difficulty of determining the true contribution of the *M* component in a dust-rich system where both stars contribute to the dust heating.

6. RX Puppis: Symbiotic Stars Are Not Always What They Seem

I use the star RX Pup to illustrate how easy it is to be misled by the optical spectrum. At the time of the Henry Draper catalogue, and later when studied by Swings and Struve (1941), RX Pup showed a high-excitation emission-line spectrum very reminiscent of the symbiotic stars, but lacking the cool giant's continuum in the red. When Sanduleak and Stephenson (1973) undertook their objective-prism survey of the southern Milky Way, about 1967, the star showed a strong, blue continuum with relatively few, low-excitation emission lines. This condition persisted into the 1970's, and led Klutz *et al.* (1978) to interpret the system as a single B supergiant. That this was not the case was again demonstrated in the infrared, where a Mira variable is seen (Barton *et al.*, 1979; Whitelock *et al.*, 1983), but more dramatically by the reversion to its former spectrum as the 1980's dawned (Klutz and Swings, 1981; Andrillat, 1982).

Another example, though less well documented, is BI Cru, which showed a high-excitation spectrum in 1962 (Henize and Carlson, 1980), but more recently has been a low-excitation object (Allen, 1974). The fact that the M star remains visible in BI Cru demonstrates that the varying component is associated with the hot companion. Other stars have undergone the same transformations during rapid, large outbursts (Z And: Swings and Struve, 1940; AX Per: Gauzit, 1955). And an optically thick shell dominated RR Tel early in its outburst (Thackeray, 1950; Pottasch and Varsavsky, 1960), and AG Peg around the turn of the century (Merrill, 1929).

The lesson to be learned is clear. Conditions around the hot star can mimic a cooler object. The false photosphere in RX Pup could have been that of an accretion disk, or of a dense wind from the hot component. Neither can yet be ruled out.

The lesson must be applied with caution to other systems. Zipoy (1975) did indeed do so for the star M1–2 (which he considered to be a single star). But we should also question the reality of the apparently cool star in such systems as HD 330 036 (Webster, 1966; Lutz, 1977, 1983), AS 210 (Wilde, 1965) and He2–467 (Lutz *et al.*, 1976; Lutz, 1977). And is this the explanation of Herbig's (1960) classification of the cool star in CM Aql as M4, whereas on spectra I secured with the Anglo-Australian Telescope in 1977 the continuum is more like that of an F star?

We should also be alert to the possibility that other systems which currently show scant similarity to symbiotic stars may be different manifestations of the same phenomenon.

7. CI Cygni: Accretion Onto a Main-Sequence Star

One of the classical symbiotic stars is CI Cygni, a system first noted by Merrill and Humason (1932); its optical spectrum was described in detail by Merrill (1933, 1950), Swings and Struve (1940), Tcheng and Bloch (1954), Fehrenbach and Huang (1981) and Iijima (1981). The variability of this star could fuel an unending succession of papers that would not greatly improve our understanding of it. Progress became possible when CI Cyg was discovered to be an eclipsing binary. This fact was first hinted at by Whitney (see Aller, 1954), Hoffleit (1968), and Pucinskas (1972), but became clear in the data of Belyakina (1976, 1979).

With this vital fact at their fingertips, several authors were able to model CI Cygni in considerable detail. Kenyon *et al.* (1982) and Iijima (1982) showed that a bright M4 giant sheds material onto a Main-Sequence star, and that temperatures up to 160 000 K are formed, probably at the boundary layer where the accretion disk abuts against the stellar surface. The disk itself has been modelled by Bath and Pringle (1982), who were able to reproduce the optical light curve over several outbursts by suitable mass transfer events. The disk is physically thick, so that simple descriptions as befit dwarf novae are not entirely appropriate.

The rapid rise and slow fall of the outburst luminosity is readily explained by the sporadic formation and subsequent evolution of the disk. The same characteristic variations are seen in many classical symbiotic stars.

Although CI Cygni must be regarded as the most satisfactorily explained symbiotic star at the time of writing, we should not ignore the work on T CrB a full quarter of a century ago. Despite the absence of eclipses in this system, it was shown spectroscopically to be a binary involving a 1.9 $M_\odot$ Main-Sequence star orbiting a red giant by Kraft (1958; modified by Paczyński, 1965). Later, Webbink (1976) demonstrated that its occasional outbursts can be attributed to accretion events on the Main-Sequence star. T CrB is called a recurrent nova, a designation which belies its similarity to the symbiotic stars during the outbursts. Between outbursts, however, T CrB resides in a much more quiescent, low-excitation state.

The structure of the H II region has been studied in the optical by Mikołajewska and Mikołajewska (1982) and Oliversen and Anderson (1983), and in the ultraviolet by Stencel et al. (1982). During eclipse of the hot star and its accretion disk, the H II region is not fully eclipsed. Neither the majority of the forbidden lines nor the high-excitation resonance lines are weakened by eclipse, so these must arise in the outer parts. A complete explanation is not yet available, and it may be that the latter are formed in a shock-excited region where winds from the two stars collide. In this respect, the formulation of Kwok (1977; see also Kwok et al., 1978; Kwok and Purton, 1979) is particularly relevant, even though it presumes a single star interpretation of the slow novae, and hence, a single centre of mass loss. Alternatively, there may be polar plumes where radiatively-driven mass-loss from the hot star is able to escape the dominance of the accretion disk. The narrowness of N v emission in many symbiotic stars also suggests an origin some distance from the hot star.

In Z And, Altamore et al. (1981) argue that some of the highest-excitation lines are formed around the cool star. Their arguments rely on a demonstration that the N v emission region has relatively small linear thickness. But this might also pertain to bipolar plumes or to a shock front where winds collide. A mechanism whereby the cool star's coronal activity is enhanced by forced rotation at the orbital period is explored by Friedjung et al. (1983). Electron temperatures in the high-excitation regions are presently though to be sufficiently low that only photoionization can sustain the nebula, and this favours an origin as near as possible to the hot star (but see caveat in Section 3.3).

There would be considerable interest in seeking bipolar structure in CI Cyg and other symbiotic systems. Recently, a VLA radio map of V1016 Cyg by Newell (see Hjellming and Bignell, 1982) suggested a bipolar morphology. Direct optical confirmation of this structure was made by Solf (1983). Of course the extensive nebulosity around R Aqr also has a bipolar morphology. The bipolar nebula surrounding He2–104 may be another manifestation of the effect (Allen, 1979).

8. H1-36: A Thermonuclear Flash on a White Dwarf?

The case of H1–36 has been analysed by Allen (1983a, b). In this star the only evidence for the presence of a cool companion is the near-infrared spectrum. The M star is of quite late type, and is probably a Mira variable of long period. It is reddened by about 20 mag. at V, whereas the emission nebula has a scant 2.2 mag. extinction.

Similar, but less extreme instances have been found by Bregman (1982: HM Sge) and Taranova and Yudin (1983: HM Sge and V1016 Cyg). It naturally follows that the Mira variable in these systems is enveloped in its own dust cloud. In H1–36 we can estimate the size of this cloud by the following approach, which is equally applicable to HM Sge and V1016 Cyg, although in their cases not all the relevant data are to hand.

The radio spectral index of H1–36 indicates that prolonged mass loss has occurred. Additionally, the flattening of the radio spectrum above 10 GHz (Purton et al., 1977) shows a neutral cavity to reside within this pattern of outflow. Since there is no temporal evolution of the radio spectrum, the cavity is reasonably static. An interpretation of the optical emission spectrum requires extensive neutral portions of the nebula in order to enhance the low-excitation lines. A central ionizing source in a uniform outflow nebula cannot produce the type of optical spectrum seen in H1–36. All these features are explained by placing the source of ionization outside the centre of mass loss. An inner region around the cool star remains neutral and can shield the dust. From the radio spectrum and an adopted distance of 4.5 kpc (based on the apparent luminosity of the Mira component) can be derived the distance from the cool star to the ionization inner boundary, and hence the separation of the stars. The value I deduce for the latter is 3×10^{16} cm, or 2000 AU. The figure exceeds the limiting separation derived by Tutukov and Yungel'son (1982), who assumed that dust absorption would obscure the entire symbiotic phenomenon rather than just the Mira variable. It also represents about the largest separation known for binary stars.

The Mira by no means fills its Roche-lobe, and accretion must be from the wind. At so large a separation, the accretion rate cannot exceed $10^{-10} M_\odot \text{ yr}^{-1}$. Hence not even a white dwarf will liberate sufficient accretion energy to power the observed H II region.

The most attractive alternative is a thermonuclear flash on a white dwarf that has been accreting for about 10^4 yr. Because the Mira will shed about 0.1 $M_\odot$ between flashes (unless the orbit is very elliptical), this is probably the first flash to have occurred on this particular dwarf, and it is likely to be quite long-lived. The fact that no photometric variations are known in H1–36 need not therefore distract from this argument.

The parallels with the slow novae are very evident here. HM Sge, RR Tel, and V1016 Cyg, all slow novae, show strong similarities to H1–36 with the exception that their outbursts have been observed to occur. In particular, the radio turnover is observed in V1016 Cyg (whereas in the other two stars it must exist, but at a frequency not yet studied), and the known parameters are sufficiently comparable that the argument can be carried almost verbatim to that system, provided that it is indeed a binary. Recently, Nussbaumer and Schmutz (1983) lent their support to a similar model for the dust-free slow Nova V1329 Cyg.

9. HD 330036: Cool Dust

The dust in H1–36 and its ilk lie quite close to the Mira, and presumably protected by the neutral cavity. Indeed, it is difficult to conceive of a site for dust formation better than the circumstellar cavity. But there are a few symbiotic stars in which the dust is cool, so that its presence is apparent only in data beyond 3 μm wavelength. I have distinguished these as class D' (Allen, 1982). The dust must be no hotter than 500 K. In these systems the cool component is not a Mira variable, rather a hotter star of spectral type F–K (these systems are sometimes called the yellow symbiotic stars: Glass and Webster (1973); but see the caveat in Section 6).

Where does the dust lie in these systems? Again our inclination is to locate it in neutral regions shielded from the hard ultraviolet radiation. But it must occupy a niche considerably more distant from the relatively warm giant than the hotter dust we find surrounding cooler Miras. We do not know the orbital periods of these stars, or indeed whether they really are binaries. However, it would seem unlikely that an F or G giant could eject sufficient matter for wind accretion to be effective, so they are probably examples of Roche lobe overflow. A suitable location for the dust remains elusive, and it may be that shielding within the orbital plane offers the only explanation.

The D' systems appear not to show silicate emission, in contrast to the majority of the Mira systems (Roche *et al.*, 1983). HD 330036 itself is unique amongst the known symbiotic stars in exhibiting infrared emission bands at 3.3 and 11.3 μm, otherwise known only in (probably carbon-rich) planetary nebulae and H II regions (Allen *et al.*, 1982; Roche *et al.*, 1983). This and other considerations have prompted Lutz (1983) to classify it as a dense planetary nebula forming in a binary system with an F5 giant. In view of the low excitation of HD 330036, such a classification is quite acceptable.

10. Luminosities: Symbiotic Stars in the Magellanic Clouds

For very few of the symbiotic stars do we have a reliable distance. We may estimate this vital parameter by using the late-type component, but we must then assume that the star is a normal giant. The assumption may be unreliable, and considerable discussion has devolved upon this point, particularly for AG Peg (Gallagher *et al.*, 1979; Keyes and Plavec, 1980). In the case of this star, the M giant can fill its Roche lobe only if it is of luminosity class II; the hot star is then likely near its Eddington luminosity, but the height of the system above the galactic plane becomes uncomfortably large. It is the lengthy orbital period (820 days) that forces the dilemma. We may have to face this problem, for in the eclipsing system AR Pav we can be certain that the height above the galactic plane is nearly 2 kpc. For CI Cyg the period is 855 days, but the smaller galactic latitude allows the greater distance, and a bright (i.e., class II) giant is acceptable. Indeed, Plavec (1982) has cogently argued that interacting binaries will most often involve bright (asymptotic branch) giants rather than stars on their first ascent of the giant branch.

We should like to know the luminosities of more of these systems. One obvious approach is to study specimens in the Magellanic Clouds. True, we will tend to locate first the most luminous examples therein, but even they will offer considerable help.

Feast and Webster (1974) first drew attention to some possible symbiotic stars in the Large Magellanic Cloud. Further study of these, together with some candidates kindly provided by Sanduleak, allowed Allen (1980c) to confirm three. Subsequent study suggests that one of these, S18 in the Small Cloud, may not be symbiotic (Shore, private communication). More recently, Walker (1983) has catalogued three more in the Small Cloud.

As so often happens, complications arise in the study of these stars. In half of them the cool components are carbon-rich rather than M-type, so the derived luminosity is not relevant to the Galactic specimens. In another (Sanduleak's object in the LMC) the infrared data indicate that a cool dust shell envelopes a star hotter than 4000 K. Infrared observations of Walker's stars have yet to be made.

Ultraviolet observations of the two LMC specimens were made by Kafatos et al. (1983b), but the objects are really too faint for a good continuum shape to be derived. Coupled with uncertainties in the reddening, it is not yet possible to infer a reliable luminosity for the hot components. For S63 Kafatos et al. (1983b) find the luminosity of the hot component to approach the Eddington luminosity of a $1\,M_\odot$ star.

The LMC star HD 269227 may be closely related to the symbiotics. It combines the emission lines of a WN star with a cool companion (Allen and Glass, 1976; Andrillat et al., 1982), though there is considerable doubt about the validity of a Wolf–Rayet classification. The luminosity of the cool star places it in the supergiant class, and thus HD 269227 is not here considered to be a symbiotic system.

Further work on the Magellanic Cloud symbiotic stars is clearly required.

The very fact that half of the examples in the Magellanic Clouds involve carbon stars is of interest. In our Galaxy only two stars out of more than one hundred are carbon-rich (UV Aur: Sanford, 1944, 1949; UKS-Cel: Longmore and Allen, 1977). Although the numbers are small, there is a clear indication that carbon symbiotic stars are commoner in the Magellanic Clouds. This fact exactly parallels the greater proportion of carbon field stars in the Magellanic Clouds (Blanco et al.,

1978). Hence, we infer that symbiosis is a catholic phenomenon, one that can afflict a giant star irrespective of its chemisty. The M components of symbiotic stars are weighted much more towards the later subdivisions of the class than are field stars (Allen, 1980b). We therefore realise that one pre-requisite for a symbiotic star is a cool component capable of shedding mass at a high rate.

11. The λ6830 Band

The unidentified band at λ6830 in about 50 % of symbiotic stars is a sure indication that velocities approaching 1000 km s^{-1} exist (Allen, 1980a). Its structure is complex, often double- or triple-peaked, but not clearly that of a rotating accretion disk (as modelled by Smak, 1981). The implied velocity is marginally consistent with Friedjung's (1981) prediction for an optically thick wind driven by an object exceeding the Eddington luminosity.

If the velocities are circular, as in an accretion disk, they must be lower than the escape velocity at the surface of the hot star. If the velocities represent outflow in a radiation-driven wind, they will tend slightly to exceed the escape velocity (Cassinelli and Castor, 1973; Castor et al., 1975). In either case we may tentatively conclude that the escape velocities of the hot components are of this order. This rules out Main Sequence stars, and leaves the choice of subdwarfs (using Plavec's definition: see section 4) or white dwarfs. The choice between a white dwarf and a subdwarf is dictated by the accretion conditions.

A prediction is that neither the broad λ6830 band nor any Wolf–Rayet-like emission line will be seen in systems involving Main-Sequence accretors. The recent claim by Blair et al. (1983) that the λ6830 band is seen in CI Cygni would appear to violate the prediction. Examination of their data suggests that the feature may not be real, confusion having been caused by the steep continuum of the M giant and the neighbouring atmospheric (B-band) absorption.

12. Evolutionary Considerations

The hot components of symbiotic systems may be either Main-Sequence stars or more compact objects. We are therefore dealing not with a single phase in the evolution of some (maybe all) binary stars, but with at least two stages. In one, the more massive of a pair of stars has recently evolved onto the giant branch, and has expanded sufficiently that it either fills its Roche lobe or is otherwise capable of depositing gas onto its companion at a substantial rate. This phase is typified by CI Cygni.

In examples of the other phase, both stars have left the Main Sequence. One has evolved to the white dwarf stage whilst the other is now a giant. In reaching this stage of evolution the system has passed through an earlier mass-transfer episode which may have influenced the evolution (e.g., Rudak, 1982). For a system as widely separated as H1–36, unless the eccentricity is high or there has been a significant increase in the orbital separation, the first mass transfer may have had slight effect. In closer systems the evolutionary history will have been complex.

Models of the mass transfer relevant to CI Cygni and its ilk have been computed by Lauterborn (1970) and by Plavec *et al.* (1973). In these models a class II giant is the mass donor, and eventually a massive white dwarf results. The subsequent evolution is computationally tedious to follow. Webbink (1979a, b) has sketched the likely course of events, with particular relevance (1979b) to the formation of white dwarfs and (1979a) to the recurrent nova T CrB, which (together with RS Oph) can be considered closely related to the symbiotic stars. It is possible that those symbiotic stars involving white dwarf accretors are precursors to dwarf novae. In his excellent review, Plavec (1982) compared the symbiotic stars to several other interacting binaries including Algol systems. The evolutionary sequence that he sketches forms an elaboration of Webbink's ideas. More recently, Trimble (1983) has summarised our current thinking on the evolution of interacting binaries.

The sole example, V2116 Oph, in which accretion is onto a neutron star represents a possible third evolutionary state in which a supernova explosion occurred non-disruptively in a binary system. It should be recalled that a white dwarf can, on accreting sufficient mass, become a supernova (Sugimoto and Nomoto, 1980).

Until we have a more complete understanding of the exact nature of the accreting components, it is unlikely that a definitive evolutionary scheme can be proposed. It may eventually be demonstrated that the symbiotic phase is an important (though short-lived) one in the evolution of many binary systems.

13. Future Prospects

In astronomy, changing fashions cause classes of objects to be studied in depth for periods of typically a decade, after which they are largely ignored until some breakthrough in understanding or instrumentation forces a reappraisal of the scene. Symbiotic stars, largely ignored in the 1960's and early 1970's are, once more, fashionable. In part this is due to the impetus provided by infrared techniques, which allow study of the cool component; in part the opening up of the ultraviolet domain (principally by IUE) is responsible. Important, too, has been our improved observational and theoretical understanding of accretion phenomena, primarily through study of dwarf novae. As the Space Telescope becomes operational, it will permit a new assault on the ultraviolet spectroscopy of symbiotic stars.

But we should not rest on our laurels awaiting the Space Telescope. In this section I give some guidelines to areas I consider ripe for study.

If indeed symbiotic stars are binaries, and if the cool star is often Roche-lobe filling, then as many as one in three should exhibit eclipses. The eclipses may be of the hot star, or merely of part of the emission nebula. It is vital to discover and study all the eclipsing systems. We now have a good understanding of CI Cygni, but only because of the recent discovery of its eclipses. We can do the same with other systems. AX Per and BF Cyg may be other examples (Kenyon, 1982; Oliversen and Anderson, 1983; Pucinskas, 1970). The orbital periods are long, typically a few hundred days, so that occasional observations are all that we require to discover more eclipsing examples. These may be photometric or spectroscopic, and need only a modest telescope. *UBV* photometry on a 0.5 m telescope is a potent tool. It will require many orbits, and hence many years, to disentangle eclipses from the stochastic variations and nova-like outbursts these systems also exhibit. In some cases the relevant data may already exist, in the form of observations by amateur networks such as the AAVSO and the Variable Star Section of the Royal Astronomical Society of New Zealand, or in archival plate material. There are more than one hundred systems out there just waiting for someone to begin monitoring them.

The eclipses do not provide answers to all the questions. High-resolution spectroscopy is essential to define the gas streams. CI Cygni is a case in point: the fact that some of the high-excitation gas is not eclipsed tells us that the nebula is complex. Studies of the velocity structure of a variety of emission lines around the orbit will help to disentangle the complexity. This is also true of systems not known to eclipse. Recall that the binary nature of T CrB and AG Peg were sleuthed spectroscopically (Kraft 1958; Cowley and Stencel, 1973). The programme of echelle spectroscopy by Anderson *et al.* (1980) and Oliversen and Anderson (1982) is a beginning to this work.

Again, in searching for binary motion, we should not ignore such tools as polarimetry, which can reveal orbital motion and determine the orbital inclination (as for the Wolf–Rayet binary HD 50896: McLean, 1980a, b). Variable linear polarization has been found in HM Sge (Efimov, 1979), R Aqr (Nikitin and Khudyakova, 1979) and CH Cyg (Piirola, 1982), though in the latter two it may arise in the M star (Svatoš and Šolc, 1981) rather than in the gas or dust as required if McLean's approach is to reach fruition.

If we can detect SiO maser emission in more of these systems, then the motion of the M star can be monitored with high precision by radio techniques, something which should be regarded as a luxury by optical observers. It appears also that the lines of Fe II and [Fe II], often so numerous in symbiotic stars, arise in the vicinity of the cool star (Boyarchuk, 1970a; and references therein), and so provide another handle on its motion. The advent of CCD detecors may permit better radial velocities of the M components directly from cross-correlation analyses of their TiO bands.

The binary periods are in themselves of interest. Those currently known are so long that in most systems a cool giant could not fill its Roche lobe. Even an asymptotic giant branch star may not do so. Determination of the luminosity of the cool star is of great interest to evolutionary studied. Attempts in the infrared (Kenyon and Gallagher, 1983) are not obviously going to succeed. The 8125 Å CN band is a luminosity discriminant, but is also sensitive to abundances: perhaps a study in which modelling of the H II region yields the abundance will permit luminosity classification from the CN band.

The whole question of accretion from an M giant's wind is raised. It is tempting to suggest that systems which show few or no forbidden lines in their optical spectra are examples of Roche-lobe overflow, so that most of the gas lies in a stream of high density (as in dwarf novae). But this may be wrong. And is the standard accretion model of Bondi and Hoyle (1944) applicable when the stream is neither cold nor collimated, and the M star's mass loss wind crosses the inner Lagrangian point? Or are we in fact viewing systems in which wind accretion is enhanced by partial streaming of the gas, and in which enough angular momentum is transferred with the gas to create a disk-like accretion cloud even in the absence of direct Roche-lobe overflow?

Since the hot star may be driving mass loss radiatively, we seem to require a disk-like accretion flow and two bipolar outflowing lobes, except in cases where thermonuclear shell flashes causes occasional disruption of the steady-state accretion mechanism. Do many of these systems have bipolar symmetry? This question could be answered by rather difficult observations of the type undertaken by Solf (1983), by speckle interferometry, or by aperture synthesis at frequencies of order 10 GHz of the radio-emitting region using the VLA or the University of Manchester's MERLIN facility.

Magnetic fields are important in accretion onto white dwarfs: dwarf novae and similar cataclysmic variables are now subdivided according to whether the field is strong (AM Her systems), weak (intermediate polars) or irrelevant. It would be of interest to seek evidence for magnetic influence in symbiotic stars. This could manifest itself as a cyclotron accretion column, or as modification to the gas flows. If, in fact, most white dwarfs either have stable shells or are undergoing a shell flash, then the effects of magnetic fields may be very hard to detect.

Variability is probably the key to understanding all the contributing components of the ultraviolet continuum, as I indicated above. Further monitoring is certainly needed. Concurrent soft X-ray data would help to define temperatures, but the sensitivity required is greater than offered by EXOSAT. It is, needless to say, unfortunate that we cannot secure observations in the wavelength region 100–900 Å. Variability studies in the infrared also seem of value. Now that we have good data on the Mira variables in D-type systems, we should be turning our attention to the S-types. Can we detect changes with orbital phase due to ellipsoidal deformation and/or heating of one hemisphere of the cool star by its hyperactive companion; or will its own random fluctuations defeat us?

Finally, the field of interacting winds from the two stars needs more thought. It is perhaps early yet to expect wind models to be applied to such complex systems. If so, then we should concentrate on observational attempts to locate the gas emitting both N v and the mysterious $\lambda 6830$ band.

Let these thoughts stimulate my readers to activity!

References

Ahern, F. J., FitzGerald, M. P., Marsh, K. A., and Purton, C. R.: 1977, *Astron. Astrophys.* **58**, 35.
Allen, D. A.: 1974, *Inf. Bull. Var. Stars*, No. 911.
Allen, D. A.: 1979, *IAU Colloq.* **46**, 125.
Allen, D. A.: 1980a, *Monthly Notices Roy. Astron. Soc.* **190**, 75.
Allen, D. A. 1980b, *Monthly Notices Roy. Astron. Soc.* **192**, 521.
Allen, D. A.: 1980c, *Astrophys. Letters* **20**, 131.
Allen, D. A.: 1981, *Monthly Notices Roy. Astron. Soc.* **197**, 739.
Allen, D. A.: 1982, *IAU Colloq.* **70**, 27.
Allen, D. A.: 1983a, *Monthly Notices Roy. Astron. Soc.* **204**, 113.
Allen, D. A.: 1983b, *Proc. Astron. Soc. Australia*, in press.
Allen, D. A. and Glass, I. S.: 1976, *Astrophys. J.* **210**, 666.
Allen, D. A., Baines, D. W. T., Blades, J. C. and Whittet, D. C. B.: 1982, *Monthly Notices Roy. Astron. Soc.* **199**, 1017.
Aller, L. H.: 1954, *Publ. Dominion Astrophys. Obs.* **9**, 321.
Altamore, A., Baratta, G. B., Cassatella, A., Friedjung, M., Giangrande, A., Ricciardi, O., and Viotti, R.: 1981, *Astrophys. J.* **245**, 630.
Altamore, A., Baratta, G. B., Cassatella, A., Giangrande, A., Ponz, D., Ricciardi, O., and Viotti, R.: 1982, *IAU Colloq.* **70**, 183.
Altenhoff, W. J. and Wendker, H. J.: 1973, *Nature* **241**, 37.
Anderson, C. M., Cassinelli, J. P., and Sanders, W. T.: 1981, *Astrophys. J.* **247**, L127.
Anderson, C. M., Oliversen, N. A., and Nordsieck, K. H.: 1980, *Astrophys. J.* **242**, 188.
Andrillat, Y.: 1982, *IAU Colloq.* **70**, 47.
Andrillat, Y., Dennefeld, M., and Vreux, J. M.: 1982, in C. W. H. de Loore and A. J. Willis (eds.), 'Wolf-Rayet Stars: Observations, Physics, Evolution', *IAU Symp.* **99**, 527.
Baratta, G. B., Cassatella, A., and Viotti, R.: 1974, *Astrophys. J.* **187**, 651.
Barton, J. R., Phillips, B. A., and Allen, D. A.: 1979, *Monthly Notices Roy. Astron. Soc.* **187**, 813.
Bath, G. T.: 1977, *Monthly Notices Roy. Astron. Soc.* **178**, 203.
Bath, G. T. and Pringle, J. E.: 1982, *Monthly Notices Roy. Astron. Soc.* **201**, 345.
Belyakina, T. S.: 1976, *Inf. Bull. Var. Stars*, No. 1169.
Belyakina, T. S.: 1979, *Izv. Krymsk. Astrofiz. Obs.* **59**, 133 (in Russian).
Blair, W. P., Stencel, R. E., Feibelman, W. A., and Michalitsianos, A. G.: 1983, *Astrophys. J. Suppl.* (in press).
Blanco, B. M., Blanco, V. M., and McCarthy, M. F.: 1978, *Nature* **271**, 638.
Bondi, H. and Hoyle, F.: 1944, *Monthly Notices Roy. Astron. Soc.* **104**, 273.
Boyarchuk, A. A.: 1966, *Astrophys.* **2**, 50.
Boyarchuk, A. A.: 1967a, *Izv. Krymsk. Astrofiz. Obs.* **38**, 155 (in Russian).
Boyarchuk, A. A.: 1967b, *Soviet Astron.* **10**, 783.
Boyarchuk, A. A.: 1968, *Astrophysics* **4**, 109.
Boyarchuk, A. A.: 1969, *Izv. Krymsk. Astrofiz. Obs.* **39**, 124 (in Russian).
Boyarchuk, A. A.: 1970a, in A. A. Boyarchuk and R. E. Gershberga (eds.), *Eruptive Stars*, Academy of Sciences, Moscow ch. 3 (in Russian).
Boyarchuk, A. A.: 1970b, *Izv. Krymsk. Astrofiz. Obs.* **41–42**, 264 (in Russian).
Boyarchuk, A. A.: 1975, in V. E. Sherwood and L. Plaut (eds.), 'Variable Stars and Stellar Evolution', *IAU Symp.* **67**, 377.
Bregman, J. D.: 1982, *Bull. Am. Astron. Soc.* **14**, 982 (abstract only).
Brocka, B.: 1979, *Publ. Astron. Soc. Pacific* **91**, 519.
Bruce, C. E. R.: 1975, *Observatory* **95**, 204.
Cassinelli, J. P. and Castor, J. I.: 1973, *Astrophys. J.* **179**, 189.
Castor, J. I., Abott, D. C., and Klein, R. I.: 1975, *Astrophys. J.* **195**, 157.
Ciatti, F.: 1982, *IAU Colloq.* **70**, 61.
Cohen, N. L. and Ghigo, F. D.: 1980, *Astron. J.* **85**, 451.
Cordova, F. A., Mason, K. O., and Nelson, J. E.: 1981, *Astrophys. J.* **245**, 609.
Cowley, A. P. and Stencel, R. E.: 1973, *Astrophys. J.* **184**, 687.
Davidsen, A., Malina, R., and Bowyer, S. : 1977, *Astrophys. J.* **211**, 866.

Duschl, W. J.: 1983, *Astron. Astrophys.* **119**, 248.

Efimov, Yu. S.: 1979, *Soviet Astrophys. Letters* **5**, 352.

Feast, M. W., Robertson, B. S. C., and Catchpole, R. M.: 1977, *Monthly Notices Roy. Astron. Soc.* **179**, 499.

Feast, M. W. and Webster, B. L.: 1974, *Monthly Notices Roy. Astron. Soc.* **168**, 31P.

Feast, M. W., Whitelock, P. A., Catchpole, R. M., Roberts, G., and Carter, B. S.: 1983a, *Monthly Notices Roy. Astron. Soc.* **202**, 951.

Feast, M. W., Catchpole, R. M., Whitelock, P. A., Carter, B. S., and Roberts, G.: 1983b, *Monthly Notices Roy. Astron. Soc.* **203**, 373.

Fehrenbach, C. and Huang, C. C.: 1981, *Astron. Astrophys. Suppl.* **46**, 257 (in French).

Feibelman, W. A.: 1982, *Astrophys. J.* **258**, 548.

Fitzgerald, M. P., Houk, N., McCuskey, S. W., and Hoffleit, D.: 1966, *Astrophys. J.* **144**, 1135.

Flower, D. R., Nussbaumer, H., and Schild, H.: 1979, *Astron. Astrophys.* **72**, L1.

Friedjung, M.: 1981, *Acta Astron.* **31**, 373; and errata in (1982) *ibid.* **32**, 446.

Friedjung, M. and Viotti, R. (eds.): 1982, *IAU Colloq.* **70**.

Friedjung, M., Stencel, R. E., and Viotti, R.: 1983, in preparation.

Fujimoto, M. Y.: 1982a, *Astrophys. J.* **257**, 752.

Fujimoto, M. Y.: 1982b, *Astrophys. J.* **257**, 767.

Gallagher, J. S., Holm, A. V., Anderson, C. M., and Webbink, R. F.: 1979, *Astrophys. J.* **229**, 994.

Gauzit, J.: 1955, *Ann. Astrophys.* **18**, 354 (in French).

Glass. I. S. and Webster, B. L.: 1973, *Monthly Notices Roy. Astron. Soc.* **165**, 77.

Grygar, J., Hřič, L., Chochol, D., and Mammano, A.: 1970, *Bull. Astron. Inst. Czech.* **30**, 308.

Hack, M.: 1979, *Nature* **279**, 305.

Harvey, P. M.: 1974, *Astrophys. J.* **188**, 95.

Henize, K. G. and Carlson, E. D.: 1980, *Publ. Astron. Soc. Pacific* **92**, 479.

Herbig, G. H.: 1960, *Astrophys. J.* **131**, 632.

Hjellming, R. M. and Bignell, R. C.: 1982, *Science* **216**, 1279.

Hoffleit, D.: 1968, *Irish Astron. J.* **8**, 149.

Hutchings, J. B., Cowley, A. P., and Redman, R. O.: 1975, *Astrophys. J.* **201**, 404.

Iben, I.: 1982, *Astrophys. J.* **259**, 244.

Iijima, T.: 1981, *Astron. Astrophys.* **94**, 290.

Iijima, T.: 1982, *Astron. Astrophys.* **116**, 210.

Iijima, T. and Mammano, A.: 1981, *Astrophys. Space Sci.* **79**, 55.

Kafatos, M., Hollis, J. M., and Michalitsianos, A. G.: 1983a, *Astrophys. J.* **267**, L103.

Kafatos, M., Michalitsianos, A. G., Allen, D. A., and Stencel, R. E.: 1983b, *Astrophys. J.*, in press.

Kafatos, M., Michalitsianos, A. G., and Feibelman, W. A.: 1982, *Astrophys. J.* **257**, 204.

Kafatos, M., Michalitsianos, A. G., and Hobbs, R. W.: 1980, *Astrophys. J.* **240**, 114.

Kenyon, S. J.: 1982, *Publ. Astron. Soc. Pacific* **94**, 165.

Kenyon, S. J. and Gallagher, J. S.: 1983, *Astron. J.* **88**, 666.

Kenyon, S. J. and Webbink, R. F.: 1983, *Astrophys. J.*, in press.

Kenyon, S. J., Webbink, R. F., Gallagher, J. S., and Truran, J. W.: 1982, *Astron. Astrophys.* **106**, 109.

Keyes, C. D. and Plavec, M. J.: 1980, in M. J. Plavec, D. M. Popper, and R. K. Ulrich (eds.), 'Close Binary Stars: Observations and Interpretation', *IAU Symp.* **88**, 535.

Kindl, C. and Nussbaumer, H.: 1982, *IAU Colloq.* **70**, 175.

Kindl, C., Marxer, N., and Nussbaumer, H.: 1982, *Astron. Astrophys.* **116**, 265.

Klutz, M. and Swings, J. P.: 1981, *Astron. Astrophys.* **96**, 406.

Klutz, M., Simonetto, O., and Swings, J. P.: 1978, *Astron. Astrophys.* **66**, 283.

Krautter, J., Klare, G., Wolf, B., Duerbeck, H. W. Rahe, J., Vogt, N., and Wargau, W.: 1981, *Astron. Astrophys.* **102**, 337.

Kraft, R. F.: 1958, *Astrophys. J.* **127**, 625.

Kwok, S.: 1977, *Astrophys. J.* **214**, 437.

Kwok, S.: 1982, *IAU Colloq.* **70**, 17.

Kwok, S. and Purton, C. R.: 1979, *Astrophys. J.* **229**, 187.

Kwok, S., Purton, C. R., and Fitzgerald, M. P.: 1978, *Astrophys. J.* **219**, L125.

Kwok, S., Purton, C. R., and Keenan, D. W.: 1981, *Astrophys. J.* **250**, 232.

Lauterborn, D.: 1970, *Astron. Astrophys.* **7**, 150.

Lépine, J. R. D., Le Squeren, A. M., and Scalise, E.: 1978, *Astrophys. J.* **225**, 869.

Lépine, J. R. D. and Nguyen-Quang-Rieu: 1974, *Astron. Astrophys.* **36**, 469.

Longmore, A. J. and Allen, D. A.: 1977, *Astrophys. Letters* **18**, 159.

Lutz, J. H.: 1977, *Astron. Astrophys.* **60**, 93.

Lutz, J. H.: 1983, *Astrophys. J.*, in press.

Lutz, J. H., Lutz, T. E., Kaler, J. B., Osterbrock, D. E., and Gregory, S. A.: 1976, *Astrophys. J.* **203**, 481.

Mammano, A. and Ciatti, F.: 1975, *Astron. Astrophys.* **39**, 405.

Marsh, K. A.: 1975, *Astrophys. J.* **201**, 190.

Mason, K. O.: 1977, *Monthly Notices Roy. Astron. Soc.* **178**, 81P.

McLean, I. S.: 1980a, *Astrophys. J.* **236**, L149.

McLean, I. S.: 1980b, in M. J. Plavec, D. M. Popper, and R. K. Ulrich (eds.), 'Close Binary Stars: Observations and Interpretation', *IAU Symp.* **88**, 65.

Merrill, P. W.: 1929, *Astrophys. J.* **69**, 330.

Merrill, P. W.: 1933, *Astrophys. J.* **77**, 44.

Merrill, P. W.: 1950, *Astrophys. J.* **111**, 484.

Merrill, P. W. and Burwell, C. G.: 1950, *Astrophys. J.* **112**, 72.

Merrill, P. W. and Humason, M. L.: 1932, *Publ. Astron. Soc. Pacific* **44**, 56.

Michalitsianos, A. G., Kafatos, M., and Hobbs, R. W.: 1980, *Astrophys. J.* **237**, 506.

Michalitsianos, A. G., Kafatos, M., Feibelman, W. A., and Hobbs, R. W.: 1982a, *Astrophys. J.* **253**, 735.

Michalitsianos, A. G., Kafatos, M., Feibelman, W. A., and Wallerstein, G.: 1982b, *Astron. Astrophys.* **109**, 136.

Mikołajewska, J. and Mikołajewska, M.: 1982, *IAU Colloq.* **70**, 147.

Moffat, A. F. J.: 1981, *IAU Colloq.* **59**, 301.

Nikitin, S. N. and Khudyakova, T. N.: 1979, *Pis'ma Astron. Zh.* **5**, 611 (in Russian).

Nussbaumer, H.: 1982, *IAU Colloq.* **70**, 85.

Nussbaumer, H. and Schild, H.: 1981, *Astron. Astrophys.* **101**, 118.

Nussbaumer, H. and Schmutz, W.: 1983, *Astron. Astrophys.* (in press).

Oliversen, N. A. and Anderson, C. M.: 1982, *IAU Colloq.* **70**, 71.

Oliversen, N. A. and Anderson, C. M.: 1983, *Astrophys. J.* **268**, 250.

Olnon, F. M.: 1975, *Astron. Astrophys.* **39**, 217.

Paczyński, B.: 1965, *Acta Astron.* **15**, 197.

Paczyński, B. and Rudak, B.: 1980, *Astron. Astrophys.* **82**, 349.

Paczyński, B. and Żytkow, A. N.: 1978, *Astrophys. J.* **222**, 604.

Panagia, N. and Felli, M.: 1975, *Astron. Astrophys.* **39**, 1.

Penston, M. V., Benvenuti, P., Cassatella, A., Heck, A., Selvelli, P., Macchetto, F., Ponz, D., Jordan, C., Cramer, N., Rufener, F., and Manfroid, J.: 1983, *Monthly Notices Roy. Astron. Soc.* **202**, 833.

Piirola, V.: 1982, *IAU Colloq.* **70**, 139.

Plavec, M. J.: 1980, in M. J. Plavec, D. M. Popper, and R. K. Ulrich (eds.), 'Close Binary Stars: Observations and Interpretation', *IAU Symp.* **88**, 3.

Plavec, M. J.: 1982, *IAU Colloq.* **70**, 231.

Plavec, M. J., Ulrich, R. K., and Polidan, R. S.: 1973, *Publ. Astron. Soc. Pacific* **85**, 769.

Pottasch, S. R. and Varsavsky, C. M.: 1960, *Ann. Astrophys.* **23**, 516.

Pucinskas, A.: 1970, *Bull. Vilnius Astron. Obs.* **27**, 24 [in Russian].

Pucinskas, A.: 1972, *Bull. Vilnius Astron. Obs.* **33**, 50 [in Russian].

Puetter, R. C., Russel, R. W., Soifer, B. T., and Willner, S. P.: 1978, *Astrophys. J.* **223**, L93.

Purton, C. R., Allen, D. A. Feldman, P. A., and Wright, A. E.: 1977, *Monthly Notices Roy. Astron. Soc.* **180**, 97P.

Roche, P. F., Allen, D. A., and Aitken, D. K.: 1983, *Monthly Notices Roy. Astron. Soc.*, in press.

Rudak, B.: 1982, *IAU Colloq.* **70**, 275.

Sahade, J.: 1965, *IAU 3rd, Colloq. Var. Stars, Bamberg*, p. 140.

Sahade, J.: 1976, *Mem. Soc. Roy. Sci. Liège* **9**, 303.

Sanduleak, N. and Stephenson, C. B.: 1973, *Astrophys. J.* **185**, 899.

Sanford, R. F.: 1944, *Publ. Astron. Soc. Pacific* **56**, 122.

Sanford, R. F.: 1949, *Publ. Astron. Soc. Pacific* **61**, 261.

Seaquist, E. R.: 1977, *Astrophys. J.* **211**, 547.

Slovak, M. H. and Africano, J.: 1978, *Monthly Notices Roy. Astron. Soc.* **185**, 591.

Slovak, M. H. and Lambert, D. L.: 1982, *IAU Colloq.* **70**, 103..

Smak, J.: 1981, *Acta Astron.* **31**, 395.

Solf, J.: 1983, *Astrophys. J.* **266**, L113.

Sopka, R. J., Herbig, G. H., Kafatos, M., and Michalitsianos, A. G.: 1982, *Astrophys. J.* **285**, L35.

Stencel, R. E. (ed.): 1981, *Proc. N. Am. Workshop Symbiotic Stars*, University of Colorado, Boulder.

Stencel, R. E. and Sahade, J.: 1980, *Astrophys. J.* **238**, 929.

Stencel, R. E., Michalitsianos, A. G., Kafatos, M., and Boyarchuk, A. A.: 1982, *Astrophys. J.* **253**, L77.

Sugimoto, D. and Nomoto, K.: 1980, *Space Sci. Rev.* **25**, 155.

Svatoš, J. and Šolc, M.: 1981, *Astrophys. Space Sci.* **78**, 503.

Swings, J. P. and Allen, D. A.: 1972, *Publ. Astron. Soc. Pacific* **84**, 523.

Swings, P. and Struve, O.: 1940, *Astrophys. J.* **91**, 546.

Swings, P. and Struve, O.: 1941, *Astrophys. J.* **94**, 291.

Szkody, P.: 1977, *Astrophys. J.* **217**, 140.

Taranova, O. G. and Yudin, B. F.: 1981a, *Soviet Astron.* **25**, 598.

Taranova, O. G. and Yudin, B. F.: 1981b, *Soviet Astron.* **25**, 710.

Taranova, O. G. and Yudin, B. F.: 1982a, *Soviet Astron.* **26**, 57.

Taranova, O. G. and Yudin, B. F.: 1982b, *Soviet Astron. Letters* **8**, 90.

Taranova, O. G. and Yudin, B. F.: 1983, *Astron. Astrophys.* **117**, 209.

Tcheng M.-L. and Bloch, M.: 1954, *Ann. Astrophys.* **17**, 6 (in French).

Thackeray, A. D.: 1950, *Monthly Notices Roy. Astron. Soc.* **110**, 45.

Thackeray, A. D. and Hutchings, J. B.: 1974, *Monthly Notices Roy. Astron. Soc.* **167**, 319.

Thackeray, A. D. and Webster, B. L.: 1974, *Monthly Notices Roy. Astron. Soc.* **168**, 101.

Trimble, V.: 1983, *Nature* **303**, 137.

Tutukov, A. V. and Yungel'son, L. R.: 1976, *Astrophys.* **12**, 342.

Tutukov, A. V. and Yungel'son, L. R.: 1982, *IAU Colloq.* **70**, 283.

Uitterdijk, J.: 1934, *Bull. Astron. Inst. Neth.* **7**, 177.

Viotti, R., Ricciardi, O. Ponz, D., Giangrande, A., Friedjung, M., Cassatella, A., Baratta, G. B., and Altamore, A.: 1983, *Astron. Astrophys.*, in press.

Walker, A. R.: 1983, *Monthly Notices Roy. Astron. Soc.* **203**, 25.

Webbink, R. F.: 1976, *Nature* **262**, 271.

Webbink, R. F.: 1979a, *IAU Colloq.* **46**, 102.

Webbink, R. F.: 1979b, *IAU Colloq.* **53**, 426.

Webster, B. L.: 1966, *Publ. Astron. Soc. Pacific* **78**, 136.

Webster, B. L. and Allen, D. A.: 1975, *Monthly Notices Roy. Astron. Soc.* **171**, 171.

Whitelock, P. A., Feast, M. W. Catchpole, R. M., Carter, B. S., and Roberts, G.: 1983a, *Monthly Notices Roy. Astron. Soc.* **203**, 351.

Whitelock, P.A., Catchpole, R. M., Feast, M. W., Roberts, G., and Carter, B. S.: 1983b, *Monthly Notices Roy. Astron. Soc.* **203**, 363.

Whitelock, P. A., Feast, M. W., Roberts, G., Carter, B. S. and Catchpole, R. M.: 1983c, *Monthly Notices Roy. Astron. Soc.*, in press.

Wilde, K.: 1965, *Publ. Astron. Soc. Pacific* **77**, 208.

Wood, P. R.: 1974, *Astrophys. J.* **190**, 609.

Wright, A. E. and Allen, D. A.: 1978, *Monthly Notices Roy. Astron. Soc.* **184**, 893.

Wright, A. E. and Barlow, M. J.: 1975, *Monthly Notices Roy. Astron. Soc.* **170**, 41.

Zipoy, D. M.: 1975, *Astrophys. J.* **201**, 397.

Zuckerman, B.: 1979, *Astrophys. J.* **230**, 442.

CATACLYSMIC VARIABLE STARS*

G. T. BATH

Department of Astrophysics, Oxford University, Oxford, England

(Received 29 July, 1983)

Abstract. The properties of cataclysmic variable outbursts are reviewed, and interpreted in terms of dynamical instabilities by the red component, and subsequent evolution of the white dwarf accretion disc subject to a mass transfer burst in the transfer stream.

1. Introduction

During the past two decades many classes of eruptive star have been shown to be interacting binaries in which mass exchange onto an accreting, relatively compact companion is taking place. In these variable stars the energy liberated by accretion is commonly the dominant emission process in the system. The spectral region where the bulk of the flux is emitted depends on the mass and size of the accreting component, the rate of accretion, and the role of viscosity and angular momentum in generating an accretion disc. The cause of eruptions in the emitted flux has been ascribed to modulated accretion due either to dynamical instabilities in the cool, mass transfering star, or to intrinsic instabilities within the accretion disk.

A recent re-examination (Bath and van Paradijs, 1983) of the outburst behaviour of SS Cygni confirms that it is a relaxation oscillator. The behaviour is incompatible with simple disc-instability models, but in agreement with simple expectations of overflow instabilities by the companion. G, K, and M spectral-class companions are predicted to be susceptible to dynamical instabilities during mass exchange (Bath, 1975; Papaloizou and Bath, 1975). Time-dependent studies of the evolution of viscous discs indicate that the resulting overflow instabilities accurately model the eruptions of dwarf novae and certain symbiotics, and possibly also transient X-ray sources containing late spectral-type companions. The spectral evolution, eruption decay-rate and stream-impact behaviour are in accord with theories of bursting mass transfer by the companion (Mantle and Bath, 1983; Bath et al., 1983; Bath, 1983). Intrinsic disc instabilities, due to the existence of multiple solutions to the local structure at ionization temperatures, do not normally produce outbursts, but rather a rapid oscillating wave which may be related to the observed flickering in dwarf novae light-curves, but does not appear related to the outbursts.

* Paper presented at the Lembang-Bamberg IAU Colloquium No. 80 on 'Double Stars: Physical Properties and Generic Relations', held at Bandung, Indonesia, 3–7 June, 1983.

Astrophysics and Space Science **99** (1984) 127–137. 0004–640X/84/0991–0127$01.65.

2. Accretion as an Energy Source

Accretion, and the release of gravitational potential energy liberated by infall, plays a
major role in theoretical studies of a whole range of astrophysical problems. Accretion
is commonly invoked in many models of quasars and active galactic nuclei (Lynden-
Bell, 1969; Lynden-Bell and Pringle, 1974; Abramowicz *et al.*, 1980) as the powerhouse
which fuels these most energetic radiation sources in the Universe. It is also well
established as the power source of X-ray sources (Shakura and Sunyaev, 1973; Novikov
and Thorne, 1973; Rees, 1976) and, through accretion onto white dwarfs rather than
neutron stars as the dominant energy generation mechanism in cataclysmic variables
(Bath *et al.*, 1974; Bath and Pringle, 1981), though now only at luminosities of the order
of the Sun.

The cataclysmic variable stars are well established interacting binary stars containing
accreting white dwarfs (or in some cases Main-Sequence stars) encircled by angular
momentum supported discs. In this review I shall describe the statistical properties of
the outbursts of one of the best observed systems, discuss past work on the stability of
semi-detached binary systems and recent work on disc structure in cataclysmic variables,
and try to illustrate the rich rewards theoretical models may contribute to our understanding of stellar structure and of accretion disc physics.

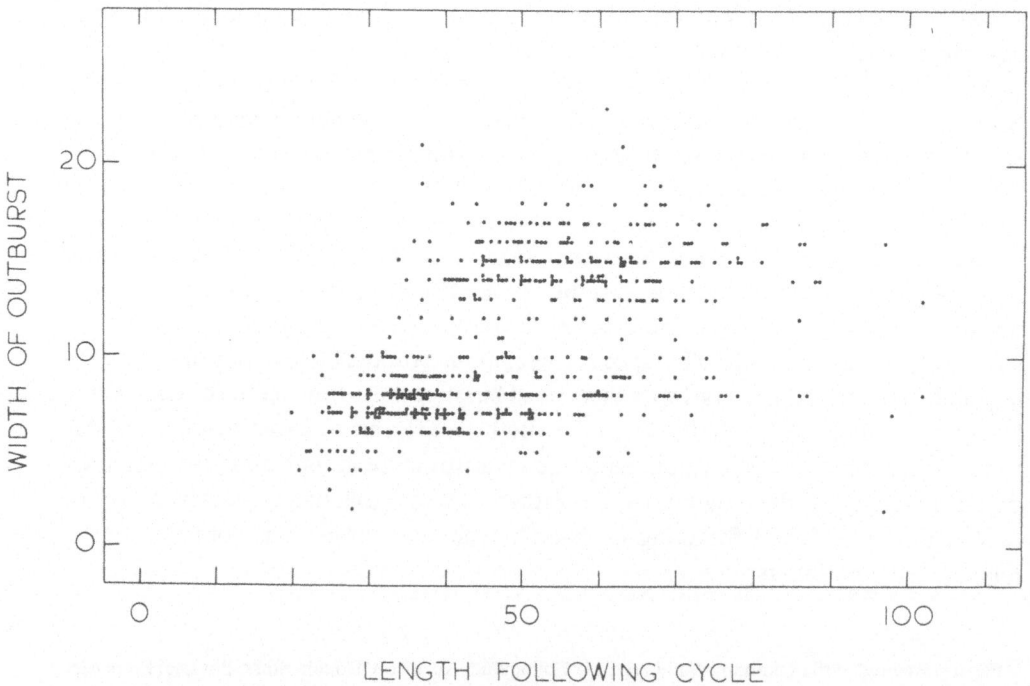

Fig. 1.   Width-of-outburst vs outburst cycle-time of the following quiescent period. Note evidence for a
relaxation cycle, and a bimodal distribution of burst widths.

3. Outbursts of SS Cygni

Dwarf novae are commonly considered to conform to a relation between outburst amplitude and period. Payne-Gaposchkin (1977) pointed out that no striking relation for dwarf novae as a class exists but that there is a larger proportion of large ranges for longer cycles.

Within any individual system, the range of outburst behaviour permits one to search for a similar relation, but now obeyed by individual outbursts. The main question is then whether any correlation exists between the previous cycle-time and the outburst energy (storage process), or between the following cycle-time and outburst energy (relaxation oscillator). Examination of the light curve of SS Cygni indicates a clear correlation of the latter-type (Kruytbosch, 1928; Sterne and Campbell, 1934; Brecher *et al.*, 1977; Bath and van Paradijs, 1983). There is no correlation of the former type. The behaviour is illustrated in Figures 1 and 2 where outburst energy is measured by the outburst width in days at a magnitude level of 10.0 mag., and the outburst cycle-length is measured as the time between successive outburst rises at the same magnitude level. Covering the period 1897 to 1979, with a gap in the data between 1933 and 1940, the correlation coefficients are 0.516 ± 0.031 between width and following cycle-time but only 0.089 ± 0.047 between width and preceding cycle-time. The outburst behaviour is, on average, a relaxation process in which the energy liberated within the accretion disc at outburst is approximately related to the time which must elapse before a second outburst can occur.

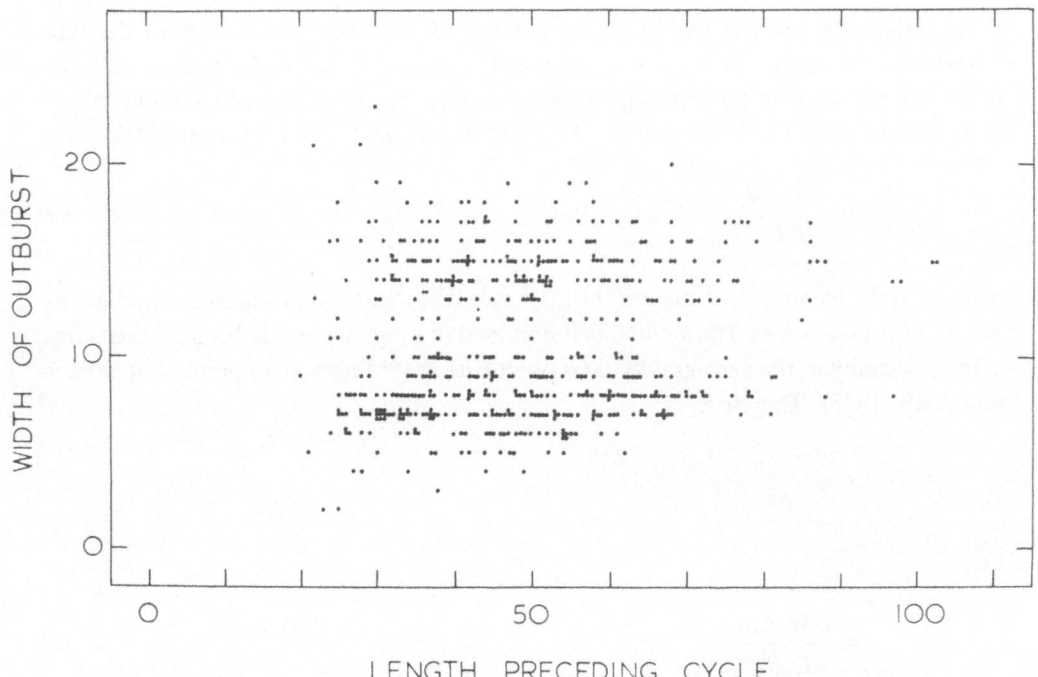

Fig. 2. Width-of-outburst vs outburst cycle-time of the previous quiescent period.

This is in direct conflict with the assumptions of simple disc instability models (e.g., Osaki, 1974). Osaki takes as his starting point the assumed existence of a correlation between outburst amplitude and cycle-length *of the preceding cycle*. Matter stored in the disc in the quiescent inter-outburst state is assumed to be dumped down onto the white dwarf in the succeeding outburst. The main assumption of this class of model is that the mass exchange rate is, at least roughly, constant. Then the preceding cycle time τ is related to the transfer rate $\dot{m}$ and the mass ΔM stored in time τ by

$$\tau = \frac{\Delta M}{\dot{m}} . \tag{1}$$

The outburst energy is

$$E = \frac{GM_1 \Delta M}{R_1} , \tag{2}$$

where the symbols have their usual meaning and the subscripts 1 refers to the white dwarf and 2 to the companion. Osaki points out that such a model obeys a period/amplitude relation, but the fact that it is not a relaxation process is not recognised. Osaki's relation is

$$\tau(\text{previous}) = \left(\frac{R_1}{GM_1 \dot{m}} \right) E . \tag{3}$$

An analogous relation to (3) can be derived for overflow instabilities of the type described by Bath (1977). Following an overflow instability driven by ionization zones in the red component, the envelope must relax and recover thermal equilibrium before a second instability can take place. The thermal relaxation time between outbursts is

$$\tau = \frac{\mathcal{R}\bar{T}\Delta M}{\Delta L} , \tag{4}$$

where $\bar{T}$ is the mean temperature of the thermally disturbed region suffering a luminosity deficit ΔL, of mass ΔM. The instability is confined to a region of scale-height dimensions z, in the vicinity of the zero-gravity tidal point (the inner Lagrangian point; Papaloizou and Bath, 1975). Therefore,

$$\Delta L \approx \frac{L_2 z^2}{R_2^2} \approx \left(\frac{\mathcal{R}\bar{T} L_2 R_2^2}{GM_2} \right),$$

i.e.,

$$\tau \approx \frac{GM_2 \Delta M}{L_2 R_2} . \tag{5}$$

On the average, ΔM will equal the mass transfered in the previous burst and models indicate that $\Delta L \simeq L_2$. Thus we obtain a cycle-time/outburst energy relation

$$\tau(\text{following}) \approx \left(\frac{M_2}{M_1}\frac{R_1}{R_2}\right)\frac{E}{L_2}. \tag{6}$$

Expression (6) is in quantitative agreement with the behaviour of SS Cygni. The known distance and orbital elements of SS Cygni allow E, M_1, and M_2 to be measured. If we assume white-dwarf and Main-Sequence structure for stars 1 and 2, respectively, then R_1, R_2, and L_2 can be estimated. The deduced range of outburst cycle times is 20–100 days, as observed.

4. Stellar Stability in Semi-Detached Systems

Two unique features of cataclysmic variables affect the stability properties of the mass-losing component. The first is the gravitational potential in which the star is distributed, and the second is its spectral class, or temperature.

The red component is well established to be filling its tidal, or Roche-lobe, and transfering material, both in the quiescent state and at outburst, toward the white dwarf companion. The stability of this star must be discussed in the appropriate potential, with the effect of the gravitational saddle-point at the inner Lagrangian point included. The turnover in potential leads to mass loss and the generation of a free-fall stream following expansion into contact with the Roche-surface. This condition of unbinding of matter at the inner $\mathscr{L}_1$ point is equivalent to the condition that the energy of escape of matter within the envelope (along the line of centes of the binary) is the energy required to lift it to the inner Lagrangian point and *not*, as in an isolated single star, the energy required to lift it to infinity. As Papaloizou and Bath (1975) have explicitly shown, the stability of the companion is severely affected by this major difference in the form of the potential surfaces as compared to an isolated star.

A star approaching filling it's Roche, or tidal, potential, is structurally affected by the presence of the companion. Perturbation of the configuration within the potential results in different behaviour of the eigenfunctions, as a consequence of weakening and eventual turnover of the gravitational field at the inner Lagrangian point. If the star is cool enough to possess, ionization zones, i.e., G, K, or M spectral class, then these can act to dynamically destabilize the envelope through the release of recombination energy. Destabilization is concentrated along the line of centres, affecting a region within at least a scale height of $\mathscr{L}_1$ (Bath, 1972, 1975).

A simply analogy with this problem is given by the following simple fluid model, shown in Figure 3. A cylindrical jar of water, area A, is suspended by elastic. A filled syphon, with an outlet bend as illustrated, is inserted until the water levels between the cylinder and the outlet are aligned. The water is analogous to the stellar envelope, with a 'leak' at a fixed point in the potential (analogous to the Roche potential and the inner Lagrangian point). The elastic, and its associated stored energy, is analogous to the influence of the ionization zones in the stellar envelope.

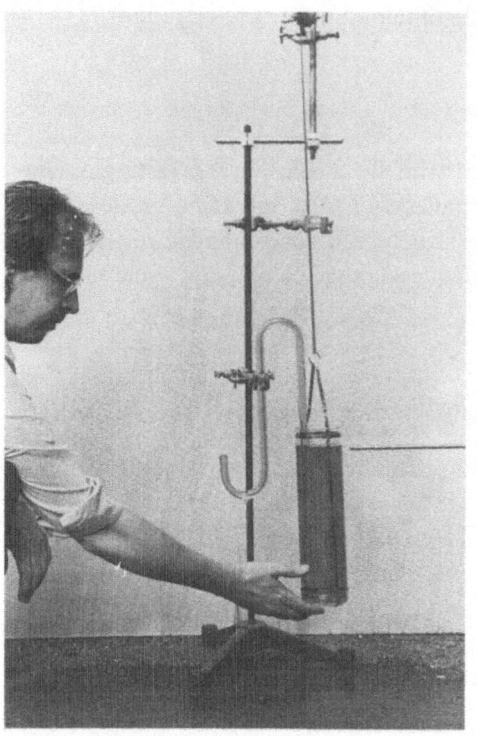

Fig. 3.   Simple experimental demonstration of dynamical instability in a system with close analogy to a cool ($T_e < 10\,000$ K) star filling its Roche-lobe. The glass cylinder is suspended by elastic from the clamp at the top edge of the picture. Perturbation of the initial equilibrium configuration (left) leads to an exponentially growing instability (right) with water loss driven through the syphon by release of stored elastic energy. If the elastic is instead held rigidly by the central clamp, the elastic energy is insufficient to drive dynamical instability. This is analogous to the stability of hotter stars. In that case the cylinder oscillates until following viscous damping of the oscillations a new equilibrium with water level below the syphon outlet is achieved.

If the cylinder is perturbed upwards water will leak out of the syphon. The subsequent behaviour then depends on the stored elastic energy. If this is insufficient to lift the water over the gravitational potential the cylinder oscillates, and eventually settles down to a new equilibrium level with the water level below the end of the outlet.

However, if the stored elastic energy is sufficient, then the cylinder starts to rise, rising exponentially against gravity. Water is expelled with increasing force through the syphon outlet. The condition for instability is that if the associated elastic stretch length, x, is initially (in the equilibrium state) longer than the height of the water column, h, then the system is dynamically unstable to water loss. In practice a finite perturbation is required, but only in order to overcome surface tension effects.

The cause of instability is easily seen. If, following water loss from the syphon, and hence, the cylinder, the new equilibrium level of the water is higher in the Earth's gravitational potential than it was before water removal, then it will be above the syphon outlet, and further water loss will take place.

The important factor is whether the water level rises or falls as water is poured into the cylinder. If the level falls as it is filled, the cylinder will be dynamically unstable to water loss. The process is closely analogous, both physically and mathematically, to the stability problem of semi-detached binaries. However, in stellar instabilities of the type experienced by cataclysmic variables the instability is eventually damped by the changing thermal conditions in the envelope. These are only restored to their initial state following energy transfer from the interior and a subsequent eruption may then take place. The energy transfer must be sufficient to refill depleted energy resources in the envelope, as described in the previous section.

In the stellar context the elastic property of the ionisation zones is an intrinsic property of the fluid. Their effect is also apparent in the context of Cepheid pulsations. Stellar pulsations are the equivalent of oscillations of a free cylinder with no syphon, or with the syphon continuing sufficiently above the outlet bend that there is no water loss following perturbation. Cataclysmic variables are a consequence of a leak through a 'hole' in the gravitational potential, equivalent to instability of the cylinder when there is a cut-off in the syphon.

5. Disc Structure and Evolution

The response of the accretion disc to sudden fluctuations in the mass transfer rate requires the development of time-dependent disc evolution models. The general evolution equation of a time-dependent, thin, viscous, axially-symmetric, Keplerian disc, is given by

$$\frac{\partial \Sigma}{\partial t} = \frac{3}{R} \frac{\partial}{\partial R} \left(R^{1/2} \frac{\partial}{\partial R} (\nu \Sigma R^{1/2}) \right) + \frac{1}{2\pi R} \frac{\partial \dot{m}}{\partial R} +$$

$$+ \frac{1}{\pi R} \frac{\partial}{\partial R} \left(R \left(1 - \frac{R_k^{1/2}}{R^{1/2}} \right) \frac{\partial \dot{m}}{\partial R} \right). \tag{7}$$

The associated vertical structure may be described by one zone, vertically averaged, structure equations (Lightman, 1974; Bath and Pringle, 1981). In this expression ν is the vertically averaged kinematic viscosity; Σ, the surface density; $\partial \dot{m}/\partial r$, the rate of matter supply as a function of radius due to stream/disc collision; and R_k, the circular Keplerian radius at which the stream would orbit if there were no viscous angular momentum transfer within the disc.

The first term is the diffusion term describing matter and angular momentum redistribution through viscous stress. This term causes a disc configuration to be formed on a viscous evolution time-scale, in which matter spirals inward, losing angular momentum to more slowly rotating outer regions, and radiating the energy dissipated as a result of the viscous stress through the top and bottom of the disc. In α-viscosity discs ν is parametrized as $\nu = \alpha c H$. $\alpha \simeq 1$ is the largest value of the viscosity parameter compatible with subsonic turbulence, c is the sound speed, and H the disc thickness.

The second term describes the way in which matter is inserted into disc material already in Keplerian motion. $\partial \dot{m}/\partial R$ measures the rate of input of matter into the disc and the degree of stream penetration. This may be parametrized in terms of the momentum available in disc material for acceleration of stream material into disc orbits.

$$\frac{\partial \dot{m}}{\partial R} = \beta \Sigma \left(\frac{GM_1}{R} \right)^{1/2}. \tag{8}$$

If $\beta = 1$ all the momentum of disc material is absorbed by the stream and used in deflecting the maximum flux of matter out of the stream into the disc. $\beta = 1$ is an absolute upper limit on the stream injection rate. A more realistic upper limit, corresponding to equipartition of energy, is $\beta = 0.5$.

The third term describes the tendency of new material with angular momentum appropriate to a Keplerian orbit at radius R_k, to squeeze the disc into an annulus orbiting at that radius. R_k is less than the outer disc radius, R_{out}. In typical cataclysmic models $1 \times 10^{10} \lesssim R_k \lesssim 2 \times 10^{10}$ cm and $2 \times 10^{10} \lesssim R_{out} \lesssim 8 \times 10^{10}$ cm.

The inclusion of Cox–Stewart opacities leads to multiple solutions for disc structure at ionization temperature of $\simeq 10\,000$ K. In this region the kinematic viscosity is no longer a single valued function of the surface density, but shows a characteristic S-band with triple solutions. This leads to the possibility of limit cycle behaviour, with the structure oscillating between the upper and lower stable branches. The middle branch is unstable (Bath and Pringle, 1982).

The model described below allows such behaviour. In addition the ionization state of hydrogen and helium is computed assuming the validity of Saha's equation. The optically thin regions of the disc are computed using a time-dependent structural procedure similar to the approach of Tylenda (1981). No treatment of vertical convection is included, since the convective turnover times are typically longer than the viscous radial transport time, and convection is therefore inefficient.

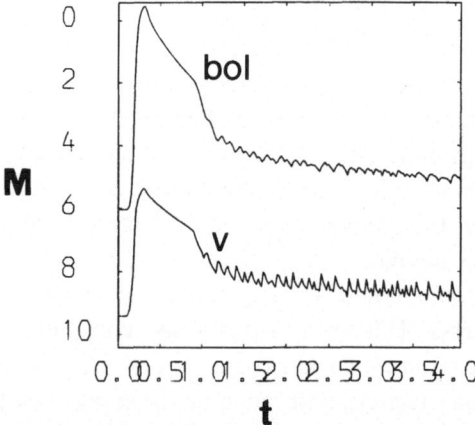

Fig. 4.   Outburst bolometric and visual light curves of disc models with multiple solution regions giving disc instabilities (small oscillations) together with the normal outburst due to a mass transfer burst. t is the time in units of 10^6 s.

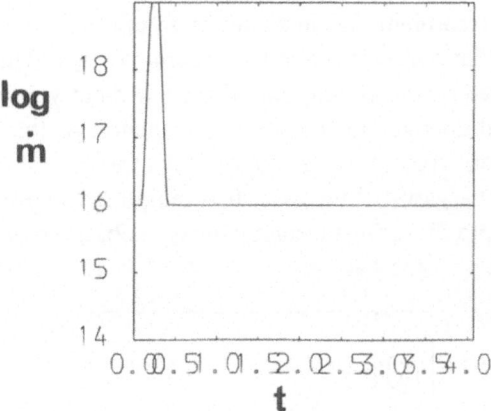

Fig. 5. Mass transfer burst assumed as initial input.

In Figure 4, the light curve of a typical model following a mass-transfer burst is shown, with burst form as in Figure 5. At outburst the whole disc is optically thick, hotter than 10 000 K, and the limit cycle instabilities disappear. They return after outburst, as the accretion flow dies down to the quiescent level, and the disc cools. A new unstable region then migrates back into the disc from the outside edge.

This change in structure is illustrated in Figure 6. Oscillations only appear with the reappearance of the optically thin exterior. Also shown in this figure are the positions (at increasing radius) of points with ionization fractions = 0.5 of He II, He I, and H. The depth of penetration of the stream, and associated shrinkage of the disc during the mass transfer burst is also shown.

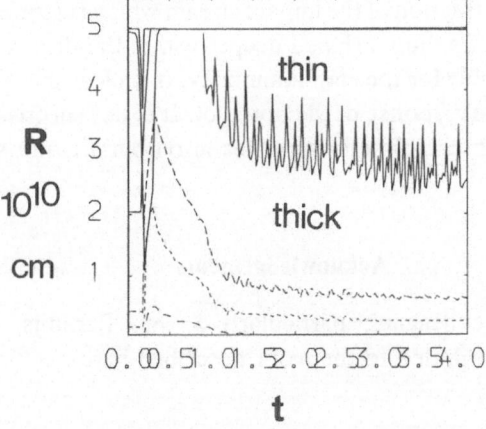

Fig. 6. Change in disc structure associated with the burst input. In the burst the disk shrinks to 3.8×10^{10} cm and the stream penetrates to 1.1×10^{10} cm. Following the burst the disc becomes optically thick out to the outermost radius at 5×10^{10} cm, and the unstable disc region is expelled. Later, following viscous evolution the optically thin outer region is re-established, and associated disc instabilities return. The positions of partial ionization of hydrogen, He I, and He II are shown by broken lines at successively smaller radii. Note that the disc instabilities lie on top of the global outburst behaviour introduced by the mass transfer burst.

We find that with our treatment disc instabilities do not produce eruptive behaviour but rather irregular oscillations with period of hours to days. The properties of the oscillations are determined by the assumption of axial symmetry. In realistic conditions in which large azimuthal changes in the shear, and hence, in the viscous stress, are produced due to deviations from circular motion, then the instabilities are likely to be confined to discrete local regions of the disc. It is tempting to speculate that the final effect may be produce rapid flickering of exactly the type observed in the quiescent state in cataclysmic variables.

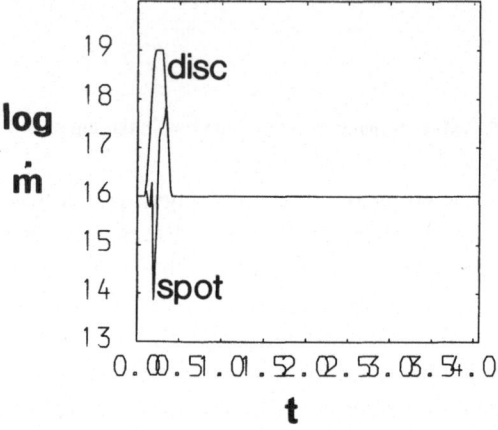

Fig. 7.   The fraction of stream material, $\dot{m}$, which is stripped in the outer spot region, and that stripped deeper within the disc. Note the delay in the spot flux, similar to the observed luminosity delay.

Finally, in Figure 7, the fraction of the impact stream which is stripped at the hot-spot, on the disc edge, and the fraction stripped deeper within the disc is shown. The mass flux in the spot is responsible for the spot luminosity. It is clear that stream penetration effects cause a delay in the response of the hot-spot. It is not surprising, therefore, that the change in the hump in cataclysmic variables at outburst is not simultaneous with the outburst itself.

Acknowledgements

I should like to thank colleagues, particularly J. van Paradijs, V. J. Mantle, and A. C. Edwards, who contributed to the work described here.

References

Abramowicz, M. A., Calvani, M., and Nobili, L.: 1980, *Astrophys. J.* **242**, 772.
Bath, G. T.: 1972, *Astrophys. J.* **173**, 121.
Bath, G. T.: 1975, *Monthly Notices Roy. Astron. Soc.* **171**, 311.
Bath, G. T.: 1977, in M. Friedjung (ed.), *Novae and Related Stars*, D. Reidel Publ. Co., Dordrecht, Holland.
Bath, G. T.: 1983, *Physica Scripta* (in press).
Bath, G. T. and Pringle, J. E.: 1981, *Monthly Notices Roy. Astron. Soc.* **194**, 967.

Bath, G. T. and Pringle, J. E.: 1982, *Monthly Notices Roy. Astron. Soc.* **199**, 267.

Bath, G. T. and van Paradijs, J.: 1983, *Nature* **305**, 33.

Bath, G. T., Edwards, A. C., and Mantle, V. J.: 1983, *Monthly Notices Roy. Astron. Soc.* (in press).

Bath, G. T., Evans, W. D., Papaloizou, J. C. B., and Pringle, J. E.: 1974, *Monthly Notices Roy. Astron. Soc.* **169**, 447.

Brecher, K., Morrison, P., and Sadun, A.: 1977, *Astrophys. J.* **217**, L139.

Kruytbosch, W. E.: 1928, *Bull. Astron. Inst. Neth.* **144**, 145.

Lightman, A. P.: 1974, *Astrophys. J.* **194**, 419.

Lynden-Bell, D.: 1969, *Nature* **223**, 690.

Lynden-Bell, D. and Pringle, J. E.: 1974, *Monthly Notices Roy. Astron. Soc.* **168**, 603.

Mantle, V. J. and Bath, G. T.: 1983, *Monthly Notices Roy. Astron. Soc.* **202**, 151.

Novikov, I. D. and Thorne, K. S.: 1973, in DeWitt and DeWitt (eds.), *Black Holes*, Gordon and Breach, New York.

Osaki, Y.: 1974, *Publ. Astron. Soc. Japan* **26**, 429.

Papaloizou, J. C. B. and Bath, G. T.: 1975, *Monthly Notices Roy. Astron. Soc.* **172**, 339.

Payne-Gaposchkin, C.: 1977, in M. Friedjung (ed.), *Novae and Related Stars*, D. Reidel Publ. Co., Dordrecht, Holland.

Rees, M. J.: 1976, in P. Eggleton, S. Mitton, and J. Whelan (eds.), 'Structure and Evolution of Close Binary Systems', *IAU Symp.* **73**, 225.

Shakura, N. I. and Sunyaev, R. A.: 1973, *Astron. Astrophys.* **24**, 337.

Sterne, T. E. and Campbell, L.: 1934, *Ann. Harv. Coll. Obs.* **90**, 189.

Tylenda, R.: 1981, *Acta Astron.* **31**, 147.

RADIAL VELOCITY AND PROFILE VARIATIONS OF THE ULTRAVIOLET CIRCUMSTELLAR LINES IN ζ TAURI*

D. N. DAWANAS

Bosscha Observatory, Department of Astronomy, Bandung Institute of Technology, Indonesia

and

Department of Astronomy, University of Kyoto, Japan

and

R. HIRATA**

Department of Astronomy, University of Kyoto, Japan

(Received 28 July, 1983)

Abstract. The radial velocity and profile variations of UV lines of the shell star ζ Tau have been examined in the IUE spectra obtained in 1978–1982. The neutral atoms, and once or twice-ionized ions (except C II, Al III, Si III resonance lines) follow the same velocity variations as in the visual spectra, while the Si IV and C IV resonance lines show a constant negative velocity (~ -50 km s^{-1} at the core). The Al III, C II resonance lines and probably Fe III (mult. No. 34) are formed in both regions, i.e., in lowly-ionized and highly-ionized regions and the Si III resonance line is formed in a highly-ionized region.

1. Introduction

The shell star ζ Tau (HD 37202, B2IIIp) is known as a single-line spectroscopic binary with a period of 132.91 days (Hynek and Struve, 1942; Underhill, 1952). In the visual spectra, this star shows a long-term cyclic velocity and V/R variation with periods of 7–4–7 years (Delplace and Chambon, 1976).

Hubert-Delplace *et al.* (1982, hereafter referred to as Paper I) reported that the radial velocities of low-ionization lines in the UV region follow the same velocity variations as in the visual region, while those of Si IV and C IV resonance lines are always negative. They ruled out a simple one-dimensional arrangement of two regions from the fact of the co-existence of positive and negative velocities, and they proposed a two-dimensional model for the circumstellar envelope of this star, which consists of highly-ionized spherical expanding region and a cool equatorial disk. As an extension of Paper I, we report here our further examination of IUE spectra, considering earlier IUE spectra and our new IUE observations of 1982.

* Paper presented at the Lembang-Bamberg IAU Colloquium No. 80 on 'Double Stars: Physical Properties and Generic Relations', held at Bandung, Indonesia, 3–7 June, 1983.
** Guest observer with the International Ultraviolet Explorer Satellite.

Astrophysics and Space Science **99** (1984) 139–144. 0004–640X/84/0991–0139$00.90.

2. Observational Material

In Table I are presented the relevant data of high-dispersion IUE spectra analysed in this study. Spectra obtained in 1978, 1979 were collected through the World Data Center A for Rockets and Satellites. The spectra in September 1982 were obtained by one of us (R.H.) at the Goddard Space Flight Center. The image numbers with asterisk were analysed in Paper I, but are also analysed in this study for the consistency of measurement. In the visual region, the radial velocities of the circumstellar lines, referred to the star, were positive in 1978, and about zero at the end of 1979 (Paper I). The epoch of last IUE observation (September, 1982) corresponds to the second zero-velocity phase after the negative velocity phase, which was disclosed by our present analysis.

TABLE I

The IUE spectra used

SWP 2356*	LWR 2136*	1978.233.21.09	1978.233.22.09	210.58
SWP 2387		1978.236.12.58		210.60
SWP 2682		1978.261.14.42		210.79
SWP 3138	LWR 2707	1978.298.11.48	1978.298.11.51	211.07
SWP 3506*	LWR 3084*	1078.335.16.10	1978.335.16.15	211.35
	LWR 3085		1978.335.17.35	211.35
SWP 6930*	LWR 5888*	1979.293.15.30	1979.293.15.54	213.78
SWP 18114	LWR 14266	1982.269.09.52	1982.269.09.56	221.84
SWP 18115	LWR 14267	1982.269.10.49	1982.269.10.53	221.84

Column 1: Image number of SWP spectra ($\lambda\lambda$ 1150–2100 Å, resolution 0.15 Å). Column 2: Image number of LWR spectra ($\lambda\lambda$ 1850–3200 Å, resolution 0.20 Å). Column 3 (for SWP spectra) and column 4 (for LWR spectra): Epoch of observation in UT, year, day, hour, and min. Column 5: Phase of the 132.91 days cycle.

3. Radial Velocity Measurements

The line identification was made by using the table of Kelly and Palumbo (1973), Kelly (1979), and a table of expected shell line intensity which was generated with the curve-of-growth technique (see Hirata *et al.*, 1982). We selected all possible unblended lines and measured the radial velocities of the centroid of the absorption lines in the core part of the profiles. In the case of broad lines we took the mean value of central positions at several depth points in the line.

After correction for satellite and earth motion, we also corrected the instrumental systimatic shift in wavelength by using the interstellar lines, which were easily detected as a secondary absorption in the spectra obtained in 1978. The velocity of the instellar lines was assumed to be $+20$ km s^{-1} as in Paper I. We corrected also the stellar velocity (the center-of-mass velocity and binary motion) which was taken from Underhill (1952). Thus we reduced all measured values to those referred to the star. The accuracy of our measurement is about 10 km s^{-1} for narrow lines and about 30 km s^{-1} for broad lines.

4. Results and Discussion

Figure 1 shows the radial velocities of different ions in UV and visible regions, where the latter was taken from Paper I. The radial velocities of neutral atoms, once and twice ionized species (except C II, Al III, Si III resonance lines and Fe III mult. No. 34) vary in the same manner as in the visual region, while those of the resonance lines of highly-ionized species, such as Si IV and C IV are always negative and seem to be constant. Thus we can discriminate two regions, i.e., lowly-ionized and highly-ionized regions which are dynamically uncoupled with each other.

The radial velocities of Al III resonance lines are intermediate between those of lowly-ionized and highly-ionized species and follow the same variation pattern as of lowly-ionized species. From Figures 2a and 2b we see that the blue wing part of these lines did not change so greatly, while the red wing part displaced towards the longer wavelength in 1978 (positive velocity phase) compared to those in 1979 and 1982 (zero velocity phase). This variation cannot be attributed to the contamination by other shell lines and should be interpreted as a true variation of

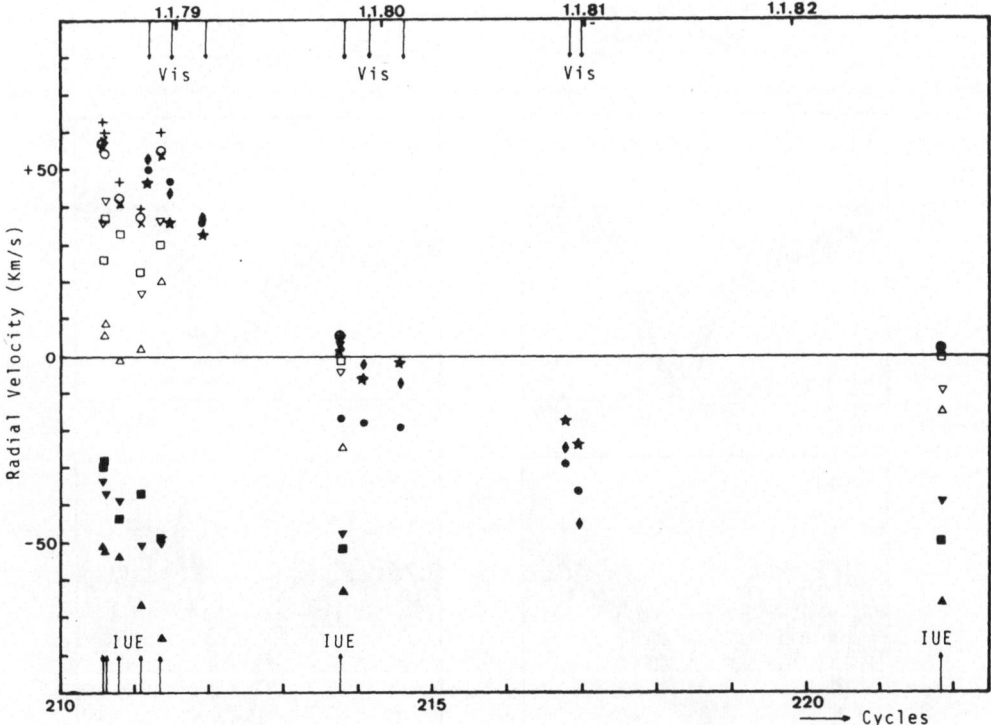

Fig. 1.   Radial velocity (referred to the star) of UV and visible lines of ζ Tau in 1978–1982. Symbols have the following meanings, for UV lines, ▲; Si III, ▲; Si IV, ■; C IV, △; Al III resonance lines, □; C II resonance and non-resonance lines, ▽; Fe III (34), ×; neutral atoms, +; once and ○; twice-ionized species (except Al III resonance lines. C II resonance and non-resonance lines and Fe III mult. No. 34).
For visible lines (Paper I),   ★; HγHδHε, ◆; H$_{19}$H$_{20}$H$_{21}$, and ●; Fe II.

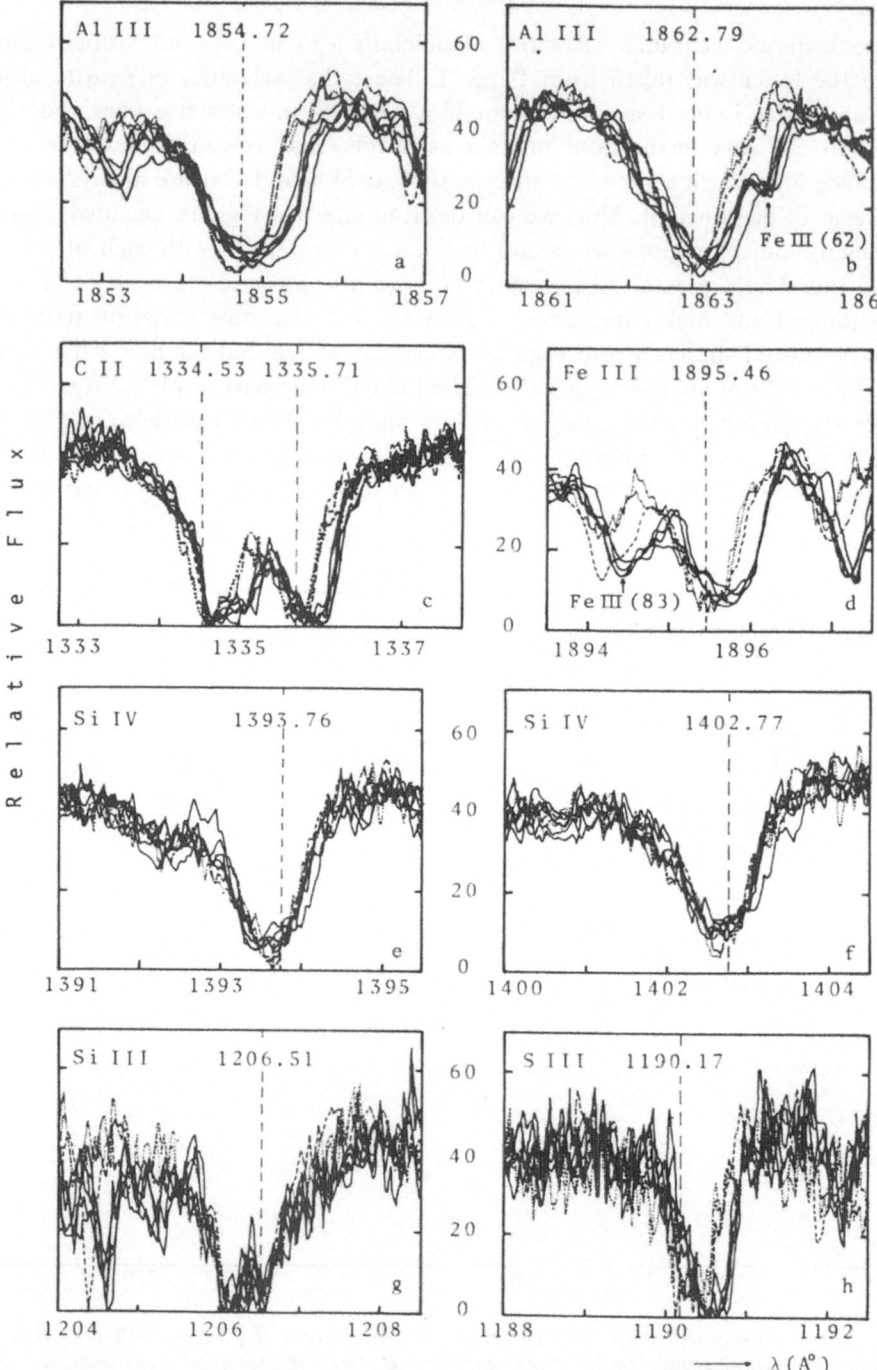

Fig. 2.   Variation of line profiles of ζ Tau in 1978, 1979, and 1982. The solid lines represent the profiles
observed in 1978, the dashed line in 1979 and the dotted lines in 1982.

the Al III resonance lines because, first, the same variation is seen in both resonance lines and, second, similar wing profiles are observed at two zero-velocity phases with strong shell lines (1979) and week shell lines (1982). From the characteristics of the profile variation of these lines mentioned above and from the fact that centroid velocities have just the mean value of those of lowly-ionized and highly-ionized species, we conclude that the Al III resonance lines are formed in both regions and that the contributions of both regions are approximately same.

The same profile variations are also found in the C II resonance and non-resonance lines ($\lambda\lambda$ 1334.53 and 1335.71 Å, see Figure 2c) and in Fe III mult. No. 34 ($\lambda\lambda$ 1895.46, 1914.06, and 1926.30 Å, Figure 2d is an example for $\lambda\lambda$ 1895.46 Å). Thus C II and probably Fe III (34) lines are formed also in both regions, but the contribution from the lowly-ionized region is greater than from highly-ionized region, because we see in Figure 1 that their radial velocities are intermediate between those of lowly-ionized species and Al III resonance lines, and blue wing part is displaced in the same sense as the red wing part, but with less extent.

The Si IV resonance lines are always strong, blue-shifted and blue-winged, indicating an expansion (Figures 2e and 2f). Both resonance lines have quite similar profiles. The core became narrower and deeper in 1982 than in 1978 and 1979.

The radial velocity of Si III resonance line has a constant negative value which is smaller than those of Si IV and C IV resonance lines. But the velocity difference between Si III and Si IV lies in the range of uncertainty of measurement. The profile of Si III resonance line did not change so greatly in 1978–1982 (Figure 2g). Therefore, we interpret that Si III resonance line is formed in the highly-ionized region.

The S III resonance line has the same veloctiy as those of low-ionization. From Figure 2f we can see that the blue and red wing parts of the profile are displaced towards longer wavelengths in 1978, in comparison with those in 1979 and 1982, with the same degree. From this fact we conclude that the S III resonance line is formed in the region of low ionization.

5. Conclusions

From the results given above, we can distinguish three groups of the circumstellar lines in the envelope of ζ Tau, first, the lines which are formed in the region of low ionization (cool equatorial disk), such as N I, O I, Ni II, Si II, Fe II, Al II, Ni III, Cr III, Si III, Fe III, etc., second, formed in the highly-ionized region such as Si III, Si IV, and C IV resonance lines, and third, formed in both regions such as Al III, C II resonance lines and probably Fe III mult. No. 34. The situation will become much clearer if we can observe this star at the negative velocity phase. It is highly probable that the other resonance lines of doubly-ionized species such as Fe III, C III, and N III are also formed in both regions, though the IUE spectrum does not cover those lines.

Acknowledgments

The authors wish to thank the IUE Observatory staff for their help in acquisation and reduction of the data and also thank the World Data Center A for Rockets and Satellite for providing us with the released IUE data. One of us (D.N.D.) would like to thank the Japan Society for Promotion of Science for the Research Fellowship. He also thanks Prof. T. Kogure and staff members of the Department of Astronomy, University of Kyoto, for the hospitality, R. H. express his heartly gratitude for grands from Kajima Foundation which made his travel to Greenbelt possible.

References

Delplace, A. M. and Chambon, M. Th.: 1976, in A. Slettebak (ed.), 'Be and Shell Stars', *IAU Symp.* **70**, 79.

Hirata, R., Katahira, J., and Jugaku, J.: 1982, in M. Jaschek and H. G. Groth (eds.), 'Be Stars', *IAU Symp.* **98**, 161.

Hubert-Delplace, A. M., Mon, M., Ungerer, V., Hirata, R., Paterson-Beeckmans, F., Hubert, H., and Baade, D.: 1982, *Astron. Astrophys.* **121**, 174.

Hynek, J. A. and Struve, O.: 1942, *Astrophys. J.* **96**, 425.

Kelly, R. L.: 1979, NASA TM 80268, Goddard Space Flight Center, Greenbelt, Maryland.

Kelly, R. L. and Palumbo, L. J.: 1973, NRL Report 7599, Washington, Naval Research Laboratory.

Underhill, A.B.: 1952, *Publ. Dominion Astrophys. Obs.* **9**, 138.

SPECTROSCOPY AND OPTICAL/IR PHOTOMETRY OF THE CATACLYSMIC VARIABLE CPD −48°1577 * **

W. WARGAU, A. BRUCH[a], H. DRECHSEL, J. RAHE, and R. SCHOEMBS[b]

Dr. Remeis Sternwarte Bamberg, Astronomisches Institut der Universität Erlangen, Nürnberg, F.R.G.

(Received 23 August, 1983)

Abstract. The photometric variability of CPD −48°1577 in the optical and IR ranges is discussed. The structure and variation of prominent emission line profiles are investigated. An estimate of the distance is given.

1. Introduction

CPD −48°1577 was recently discovered as a cataclysmic variable by Garrison *et al.* (1982). The optical spectra show extremely broad and shallow hydrogen absorption lines with emission cores. In this respect it resembles other novalike systems such as TT Ari (Cowley *et al.*, 1975), VY Scl (Burrel and Mould, 1973) or the old nova DI Lac (Kraft, 1964). The color indices and flickering behavior are also typical for cataclysmic systems. The UV spectra obtained by Böhnhardt *et al.* (1982) confirm the classification of CPD −48°1577 as a novalike system. The visual magnitude of 9^m.8 (Garrison *et al.*, 1982) makes this star the brightest member of its class.

First spectrophotometric observations of CPD −48°1577 were discussed by Wargau *et al.* (1983; hereafter referred to as Paper I). From radial velocity measurements of the emission cores of the hydrogen lines, a tentative orbital period of 0.187 days was derived.

In the following we study the long-term photometric variability of CPD −48°1577, present results of first photometric measurements in the infrared, and discuss results of the spectroscopic observations published in Paper I in more detail. Finally, an estimate of the distance of CPD −48°1577 is given.

2. Photometric Observations

2.1. THE LONG-TERM VARIABILITY

The long-term photographic variability of CPD −48°1577 was determined from measurements of 92 plates of the Bamberg sky-survey series sampled between December 1963 and January 1973. An iris-diaphragm-photometer was used to measure the density of the stellar image. Seven SAO stars were used as calibration

---

* Paper presented at the Lembang-Bamberg IAU Colloquium No. 80 on 'Double Stars: Physical Properties and Generic Relations', held at Bandung, Indonesia, 3–7 June, 1983.
** Based on data collected at the European Southern Observatory in Chile.
ᵃ Astronomisches Institut der Westfälischen Wilhelms-Universität Münster, F.R.G.
ᵇ Institut für Astronomie und Astrophysik der Universität München, F.R.G.

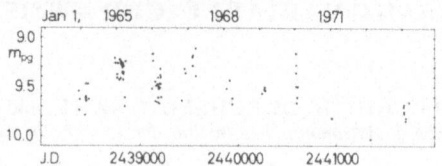

Fig. 1.   Long-term variability of CPD−48°1577, as determined from photographic plates of the Bamberg sky-survey series, obtained between December 1963 and January 1973.

standards. The mean error for a single measurement can be estimated to be a few hundredths of a magnitude. The light curve is shown in Figure 1. During the nine years covered by the plates, CPD−48°1577 varies in the range between $m_{pg} = 9\overset{m}{.}1$ and $10\overset{m}{.}0$ around the mean value of $m_{pg} = 9\overset{m}{.}5$. A systematic trend is not detectable. The observed irregular long-term light variations are a common feature for many novalike systems. In particular, no indication for an outburst or for a temporary light depression like those of TT Ari (Krautter *et al.*, 1980), MV Lyr (Robinson *et al.*, 1981), or of some AM Her-type stars could be found for CPD−48°1577.

2.2. INFRARED PHOTOMETRY

Infrared observations of CPD−48°1577 were carried out between 25 January and 1 February 1983, with the 1 m-telescope of the European Southern Observatory at La Silla, Chile. The telescope was equipped with an InSb-photometer. The measurements were taken through the J (1.25 μm, H (1.65 μm), K (2.2 μm), and L (3.4 μm) filters. The integration time of a single measurement was 20 s; in each night about 4 *JHKL* measurements were obtained.

Table I gives the mean J-magnitude and the (J–H), (H–K), and (K–L) colors for each night. During the observing period, CPD−48°1577 was essentially constant in the infrared light. In order to check a possible dependence of the IR brightness and colors on the orbital phase, the light and color curves are plotted versus orbital

TABLE I

Mean J brightness and IR colors of CPD−48°1577

Date 1983	J	$(J–H)$	$(H–K)$	$(K–L)$
26 Jan.	$9\overset{m}{.}199 \pm 034$	$0\overset{m}{.}208 \pm 019$	$0\overset{m}{.}084 \pm 017$	$0\overset{m}{.}322 \pm 217$
27 Jan.	$9\overset{m}{.}402 \pm 057$	$0\overset{m}{.}285 \pm 067$	$0\overset{m}{.}145 \pm 052$	$0\overset{m}{.}332 \pm 278$
29 Jan.	$9\overset{m}{.}313$	$0\overset{m}{.}267$	$0\overset{m}{.}120$	$0\overset{m}{.}421$
30 Jan.	$9\overset{m}{.}180 \pm 042$	$0\overset{m}{.}191 \pm 035$	$0\overset{m}{.}130 \pm 012$	$0\overset{m}{.}155 \pm 184$
31 Jan.	$9\overset{m}{.}379 \pm 134$	$0\overset{m}{.}346 \pm 108$	$0\overset{m}{.}127 \pm 013$	$0\overset{m}{.}323 \pm 097$
1 Feb.	$9\overset{m}{.}293 \pm 050$	$0\overset{m}{.}297 \pm 070$	$0\overset{m}{.}085 \pm 071$	$0\overset{m}{.}080 \pm 278$
Mean	$9\overset{m}{.}29 \pm 11$	$0\overset{m}{.}26 \pm 08$	$0\overset{m}{.}12 \pm 04$	$0\overset{m}{.}26 \pm 21$

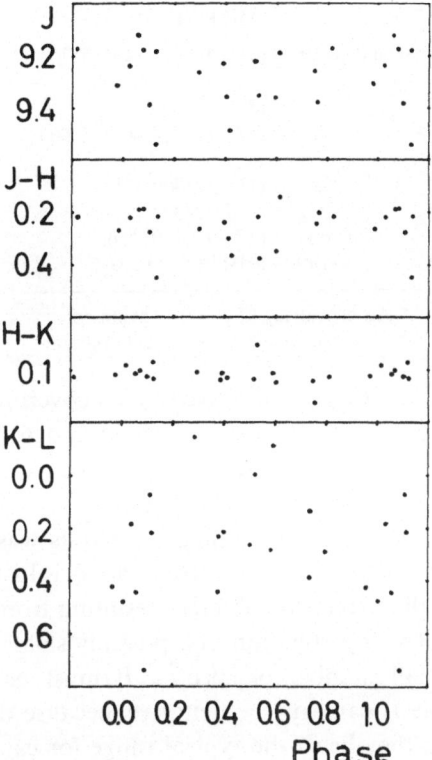

Fig. 2.   Infrared light and color curves. The J (1.25 μm) filter curve together with the color curves $(J-H)$, $(H-K)$, and $(K-L)$ are drawn as a function of orbital phase.

phase in Figure 2. The appearance of these curves suggests that the system shows no occultation or eclipse effects, though this statement is somewhat uncertain due to the large amount of scattering. An upper limit of roughly 65–70° can be put on the orbital inclination of the system. This is consistent with the estimate of $i = 63°$ given in Paper I on the basis of the cataclysmic variable model.

The infrared colors of CPD$-48°1577$ are similar to those of other novalike variables such as UX UMa $((J-K) = 0.^{m}50$: Szkody (1977)), and RW Tri $((J-K) = 0.^{m}5$: Longmore *et al.* (1981)); whereas the colors of most dwarf novae at minimum are considerably redder (SS Cyg: $J-K = 0.^{m}76$; RX And: $J-K = 0.^{m}77$; AH Her: $J-K = 0.^{m}93$: Szkody (1977); OY Car: $J-K = 0.^{m}75$: Sherrington *et al.* (1982)), which might reflect different contributions to the IR light by accretion discs in nova-like systems and dwarf novae.

3. Optical Spectroscopy

Spectroscopic observations of CPD$-48°1577$ have been obtained on 1982, 30 and 31 December with the 1.5 m-telescope of the European Southern Observatory at La Silla,

TABLE II

Orbital parameters of CPD $-48°1577$ [a]

$$P = 0\overset{d}{.}187 \pm 0\overset{d}{.}002$$
$$T_0\,^b = JD\,2\,445\,334\overset{d}{.}609 \pm 0\overset{d}{.}003$$

$$K_1 = (135 \pm 8)\,\text{km s}^{-1}$$
$$\gamma_0 = (-6 \pm 7)\,\text{km s}^{-1}$$
$$a_1 \sin i = (3.5 \pm 0.2)\,10^{10}\,\text{cm}$$
$$f(m) = (0.047 \pm 0.006)\,M_\odot$$

[a] An eccentricy of $e = 0$ is assumed.
[b] T_0 is the time of inferior conjunction.

Chile, using the Image Dissector Scanner. Five spectra covering the wavelength range from 4080 to 5260 Å, and 2 spectra between 4450 and 6730 Å were taken at a dispersion of 59 Å mm^{-1} and 114 Å mm^{-1}, respectively. For details of the observations and reduction procedure, see Paper I.

The most important result of the preliminary analysis was the determination of the orbital period which could be derived from radial velocity measurements (see Paper I). Table II gives the orbital parameters resulting from a least squares fit to the radial velocity data. The investigation of a possibly small eccentricity is difficult due to the relatively small number of spectra. It must be emphasized that the parameters given in Table II can only be tentative, because they were derived from only 7 spectra. However, they lie in the typical range for cataclysmic variables.

The appearance of the spectral lines is described in Paper I. The most prominent features are the emission cores in the center of the hydrogen absorption troughs. The structure of these lines is strongly variable. Figure 3 shows tracings of Hβ and Hγ of the 5 spectra taken on 30 December 1982, ordered with orbital phase. The line profiles contain different peaks with irregularly variable relative intensities. In particular, no systematic variations with phase can be inferred from these data.

The intensities of the Hα, Hβ, Hγ, and He II ($\lambda\,4686$Å) emission lines and the equivalent widths of the hydrogen absorption troughs have been measured and are given in Table III. The errors quoted result from the deviations of the individual measurements from the mean value. The actual errors (in particular of the equivalent widths) may be larger, because the determination of the continuum in the spectra of CPD $-48°1577$ is rather difficult. The line intensities of the hydrogen emission components confirm the shallow Balmer decrement derived by Garrison et al. (1982).

4. Distance

Bailey (1981) derived a method to determine the distance of cataclysmic variables. His approach makes use of an empirical relation between the K-surface brightness and $(V - K)$ colors for nearby late-type stars, and anticipates that the red component of cataclysmic systems contributes most of the light in the K (2.2 μm) filter.

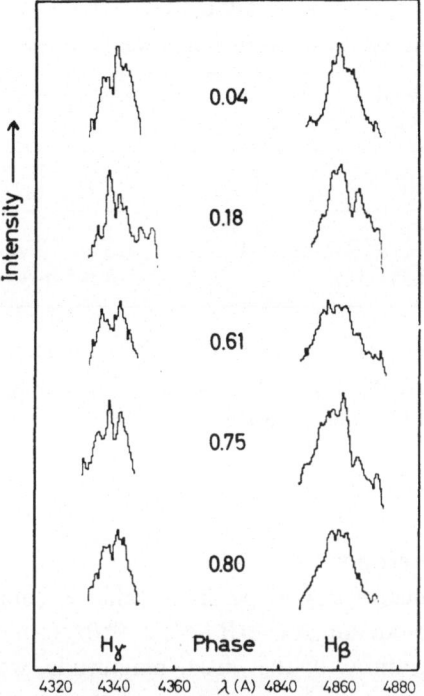

Fig. 3. Emission line profiles of Hβ and Hγ. Tracings of 5 spectra taken on 30 December, 1982, are shown; corresponding orbital phases are given.

Bailey (1981) expressed the K-surface brightness $S(K)$ by

$$S(K) = K + 5 \log\left(\frac{R_2}{R_\odot}\right) - 5 \log d + 5,$$

where K is the observed magnitude in the K filter, R_2 is the radius of the late-type component, and d is the distance of the system in parsecs.

The calibration of the K-surface brightness as a function of $(V-K)$-colors, which represents an effective temperature sequence, gives two line segments of different slopes, separated by $(V-K) = 3\overset{m}{.}5$. The spectral type of the late component of CPD $-48°1577$, which would immediately yield a $(V-K)$ value, is not known. On the other hand, we know the almost constant K-brightness of CPD $-48°1577$ during our observations, which is $8\overset{m}{.}91 \pm 0\overset{m}{.}04$. For the V-magnitude we use the value of $V = 9\overset{m}{.}5 \pm 0\overset{m}{.}2$ by Böhnhardt et al. (1982), which is in agreement with the mean brightness of the historical light curve. This gives a $(V-K)$ value of $0\overset{m}{.}59$. From Bailey's (1981) relation, we obtain

$$S(K) = 2.56 + 0.508(V-K) \quad \text{(valid for } (V-K) < 3\overset{m}{.}5\text{)}.$$

The radius R_2 of the secondary can be approximated from Kepler's third law and from the geometry of the Roche model, assuming that the secondary fills its lobe

TABLE III

Emission line intensities and equivalent widths of absorption lines

Date 1982	Emission line intensities ($\times 10^{-12}$ erg cm^{-2} s^{-1}Å^{-1})				Equivalent widths of absorption lines (Å)	
	Hα	Hβ	Hγ	He II	Hβ	Hγ
30 Dec.	–	1.6 $\pm$0.3	1.1$\pm$0.2	1.1 $\pm$0.1	1.4$\pm$0.4	2.2$\pm$0.5
31 Dec.	1.4$\pm$2	1.40$\pm$0.02	–	0.57$\pm$0.03	3.4$\pm$0.3	

(Paczynski, 1971), and from the mass-radius relationship for the lower Main-Sequence stars (Lacy, 1977). Then we find

$$\frac{R_2}{R_\odot} = 1.970 \times 10^{-5}\, P^{1.0474},$$

where P is expressed in seconds.

This leads to a distance of $d = 82$ pc. This value is compatible with the small amount of interstellar reddening of $E(B-V) = 0^m\!\!.02$ found by Böhnhardt et al. (1982). Allowing for variations of the visual magnitude, we derived the distances corresponding to the upper and lower bounds of the photographic light curve of Figure 1. The results are $d = 90$ pc and 73 pc, respectively.

The main uncertainty of the distance determination arises from the $(V-K)$ value. In particular, the V magnitude does not represent the secondary star, since it is dominated by the light of the accretion disc and hot spot. Wade (1979) showed that the red component in the U Gem system, which has a period of $P = 0^d\!\!.177$, dominates the radiation of the system at wavelengths longer than about 0.7 μm. Fortunately, Bailey's method is not very sensitive to errors in the $(V-K)$ value, i.e., an error of 1 mag in $(V-K)$ only leads to an 25% error in distance.

5. Conclusions

During a time interval of 9 y, the photographic brightness of CPD-48°1577 showed irregular variations with a maximum amplitude of about 1 mag with undetermined time-scale. IR photometric measurements revealed essentially constant brightness in the J-, H-, K-, L-bands during the observing time. The IR color indices lie in a typical range of novalike systems. The Balmer line profiles are composed of broad absorption throughs with central emission cores which show irregular structure and intensity variations. Making use of the K-magnitude of the secondary, a distance of 82 pc is derived. The absolute visual magnitude of $M_V = 4^m\!\!.9$ is typical for a nova-like variable. The mean visual brightness of $9^m\!\!.5$ makes CPD $-48°1577$ the brightest known cataclysmic system. Due to the relatively small distance, an attempt to measure the parallax appears promising.

Acknowledgements

This research was supported in part by the Deutsche Forschungsgemeinschaft grants Ra 136/10–2 and Dr 131/3–1.

References

Bailey, J.: 1981, *Monthly Notices Roy. Astron. Soc.* **197,** 31.

Böhnhardt, H., Drechsel, H., Rahe, J., Wargau, W., Klare, G., Stahl, O., Wolf, B., and Krautter, J.: 1982, *IAU Circ.*, No. 3749.

Burrel, J. F. and Mould, J. R.: 1973, *Publ. Astron. Soc. Pacific* **85,** 627.

Cowley, A. P., Crampton, D., Hutchings, J. B., and Marlborough, J. M.: 1975, *Astrophys. J.* **195,** 413.

Garrison, R. F., Hiltner, W. A., and Schild, R. E.: 1982, *IAU Circ.*, No. 3730.

Kraft, R. P.: 1964, *Astrophys. J.* **139,** 457.

Krautter, J., Klare, G., Wolf, B., Vogt, N., Rahe, J., and Wargau, W.: 1980, *IAU Circ.*, No. 3541.

Lacy, C. H.: 1977, *Astrophys. J. Suppl.* **34,** 479.

Longmore, A. J., Lee, T. J., Allen, A. D., and Adams, D. J.: 1981, *Monthly Notices Roy. Astron. Soc.* **195,** 825.

Paczynski, B.: 1971, *Ann. Rev. Astron. Astrophys.* **9,** 183.

Robinson, E. L., Barker, E. S., Cochran, A. L., Cochran, W. D., and Nather, R. E.: 1981, *Astrophys. J.* **251,** 611.

Sherrington, M. R., Jameson, R. F., Bailey, J., and Giles, A. B.: 1982, *Monthly Notices Roy. Astron. Soc.* **200,** 861.

Szkody, P.: 1977, *Astrophys. J.* **217,** 140.

Wade, R. A.: 1979, *Astron. J.* **84,** 562.

Wargau, W., Bruch, A., Drechsel, H., and Rahe, J.: 1983, *Monthly Notices Roy. Astron. Soc.* **204,** 35 P.

THE Hα LINE PROFILE IN EARLY TYPE-BINARY SYSTEMS HD 47129 AND γ VELORUM*

R. RAJAMOHAN

Indian Institute of Astrophysics, Bangalore, India

(Received 1 August, 1983)

Abstract. A brief description of the variation of the Hα Line profile in γ Velorum and HD 47129 is given.

The radial velocity curves of binary systems which contain massive early-type components are generally distorted. Whereas the primary velocity curve is fairly well defined, the velocity of the secondary shows large and erratic fluctuations. Batten (1973) remarks "In some systems, such as Plaskett's star (HD 47129) and HD 190967, the secondary spectrum seems to arise at least partly from a nonstellar source. No matter how well it may be observed, it will be impossible to obtain accurate masses for these systems".

In spite of the complexities of these systems, where one has to propose phase-dependent mass loss and stellar wind (Hutchings, 1976) gas streams and expanding clouds of gases (Sahade, 1962) for an interpretation of the observed spectroscopic behaviour, we plan to observe such systems to see if we can unravel at least a part of the mystery that shrouds their spectroscopic behaviour. These observations are being obtained to separate the phase-dependent regular changes from those which are sporadic in nature. We hope that in the not so long a future, such periodic changes if any can be interpreted.

I present here our preliminary results on the behaviour of the Hα line in HD 47129. I have also included here the profile of the Hα line in γ₂ Velorum from a few selected plates. This Wolf–Rayet binary has been extensively observed by Dr Bappu and myself from $\lambda\lambda 3700$–6700 Å.

γ Velorum: These profiles indicate that there is not much of a change associated with phase. We find a sharp fairly stationary emission component superposed over a broad flat-topped profile. There seems to be a periodic change in the intensity of the violet and red wings. No strong changes are apparent as a function of phase where as in the blue region, the violet shifted He I 3888 Å line is very strong and sharp close to zero phase (Wolf–Rayet star behind) and strong but split near phase 0.25.

HD 47129: The profile of the Hα line determined from various spectra that were obtained (listed in Table I) with the one-meter telescope at Kavalur are shown in Figure 1. The phase was calculated from the time of periastron passage T_0 given by Abhyankar (1959). The general behaviour of the Hα line is similar to that described by Struve *et al.* (1958). However, two interesting points are to be noted.

* Paper presented at the Lembang-Bamberg IAU Colloquium No. 80 on 'Double Stars: Physical Properties and Generic Relations', held at Bandung, Indonesia, 3–7 June, 1983.

TABLE I

List of observations

Plate number	Date	Mid Exp UT	Plate	Dispersion Å mm^{-1}	Phase (period)
HD 47129					
δ 1365	Feb. 1, 1981	15^{h}11^m	O9802	45	0.98
1437	19, 1981	14 10	IIa–D + IT	22	0.23
1448	20, 1981	13 34	IIa–D + IT	22	0.30
1508	Mar. 3, 1981	18 40	IIa–D + IT	22	0.14
1696	Nov. 19, 1981	18 55	O9802	45	0.21
1704	20, 1981	18 31	O9802	45	0.28
1713	21, 1981	18 22	O9802	45	0.35
1721	22, 1981	18 39	O9802	45	0.42
1806	Feb. 2, 1982	15 10	IIa–D + IT	22	0.41
γ Vel					
δ 1301	Jan. 17, 1981	22 35	IIIa–F	22	0.99
1369	Feb. 1, 1981	19 32	IIIa–F	22	0.18
1389	5, 1981	15 57	IIIa–F	22	0.23
1399	7, 1981	19 38	IIIa–F	22	0.25
1469	22, 1981	15 20	IIIa–F	22	0.44
1497	26, 1981	19 16	IIIa–F	22	0.50
1530	Mar. 15, 1981	17 12	IIIa–F	22	0.71
1543	17, 1981	17 10	IIIa–F	22	0.73

IT = Image Tube (Varo).

The strength of the emission component seems to be associated with the radial velocity of the primary component. In plate δ 1721 (phase 0.42) when the primary has maximum velocity of approach, the violet component is stronger and in plate δ 1365 (phase 0.98) when the primary has large velocity of recession, the red component is stronger. At phase 0.23 (δ 1437) close to conjunction, the primary is behind the secondary and the violet emission component is almost absent. Unfortunately, we have not taken so far a spectra at the other conjunction. However, in the illustration of the Hα spectra of this star by Struve *et al.* (1957) we can notice that at the other conjunction, when the secondary is behind the primary, the red emission component is extremely weak. At phase 0.14, just before conjunction both emission feature show self reversal, with the violet absorption being much stronger.

It seems probable that the mass loss and stellar wind from the components is not spherically symmetric and that such phenomenon are stronger on the hemisphere away from the common centre of mass. Such a cometary tail of expanding material at each component due to radiation pressure may be responsible for the changes in the line spectrum due to changing geometry as the components revolve around their common centre of mass.

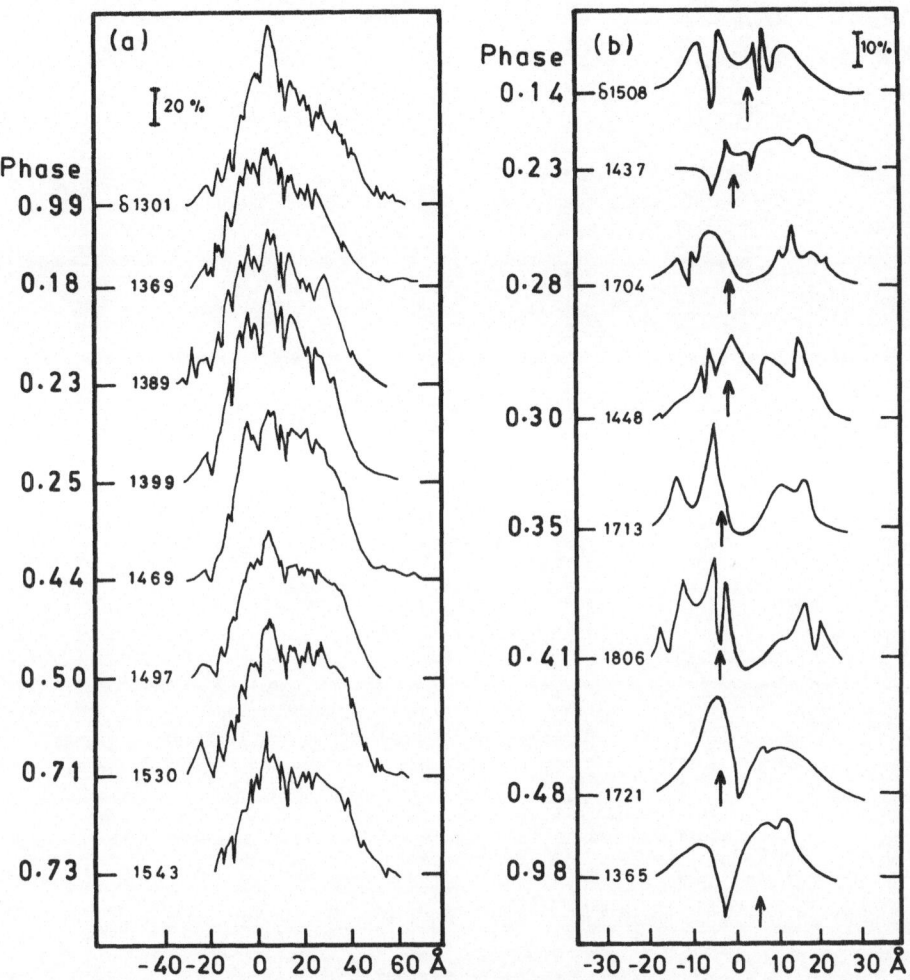

Fig. 1. The profile of Hα line in γ Velorum (a) and HD 47129 (b). Arrow indicated the expected position of Hα due to the primary component alone.

Acknowledgement

I am thankful to Mr K. Kuppuswamy for help in the reduction of the spectra.

References

Abhyankar, K. D.: 1959, *Astrophys. J. Suppl.* **4**, 147.
Batten, A. H.: 1973, *Binary and Multiple Systems of Stars*, Pergamon Press, p. 112.
Hutchings, J. B.: 1976, *Astrophys. J.* **206**, 490.
Sahade, J.: 1962, in J. Sahade (ed.), *Symposium on Stellar Evolution*, La Plata Observatory, p. 185.
Struve, O., Sahade, J., and Huang, S. S.: 1958, *Astrophys. J.* **127**, 148.

THE CONTACT BINARY AE PHOENICIS – AN ANALYSIS
FROM DECONVOLUTED SPECTRA*

H. J. BECKER

Universitäts-Sternwarte Bonn, F.R.G.

(Received 30 August, 1983)

Abstract. A method is described for an analysis of deconvoluted stellar spectra, and applied to new spectroscopic observations of the close eclipsing system AE Phuenicis, obtained with the 1.52 m telescope of the Southern European Observatory at La Silla, Chile. Results are presented for the rotational velocities of this variable, together with a proposed model of the system.

60 spectrograms of the W UMa-type eclipsing binary AE Phe were obtained with the coudé spectrograph at the 1.52 m telescope of the European Southern Observatory, La Silla, Chile. The emulsion is IIa–0, the dispersion 20 Å mm^{-1}. The spectrograms were digitized with the PDS 1010A microdensitometer of the Astronomical Institutes, University of Bonn.

We observe the intensity distribution

$$I(\lambda) = A(\lambda) * B(\lambda) * C(\lambda) ,$$

where $A(\lambda)$ is the original spectrum without rotational broadening; $B(\lambda)$ is the instrument profile; $C(\lambda)$ is the broadening function due to rotation. The symbol $*$ denotes convolution. This assumes that the usual 'uniform profile approximation' is justified.

The broadening function $C(\lambda)$ can be derived from the observed $I(\lambda)$ by deconvolution. For its derivation, we make use of the convolution theorem

$$\tilde{I}(v) = \tilde{A}(v)\tilde{B}(v)\tilde{C}(v) .$$

Furthermore, we can use a standard spectrum of the same spectral type (κ Cet, $v \sin i = 0$ km s^{-1}) for $A(\lambda)$. For κ Cet, we have

$$D(\lambda) = A(\lambda) * B(\lambda) ,$$

where $B(\lambda)$ is the same instrumental profile as above. Then,

$$\tilde{I}(v) = \tilde{D}(v)\tilde{C}(v) \quad \text{or} \quad \tilde{C}(v) = \tilde{I}(v)/\tilde{D}(v) ,$$

i.e., division of the Fourier transforms. Transforming the result back into the spatial domain yields $C(\lambda)$. Before transforming back, a filter must be applied to suppress high Fourier frequencies (noise).

---

* Paper presented at the Lembang-Bamberg IAU Colloquium No. 80 on 'Double Stars: Physical Properties and Generic Relations', held at Bandung, Indonesia, 3–7 June, 1983.

Astrophysics and Space Science **99** (1984) 157–161. 0004–640X/84/0991–0157$00.75.

158   H. J. BECKER

The practical procedure is as follows:

– digitization of the spectra of the standard star (κ Cet);

– digitization of the spectra of the program star (AE Phe);

– conversion from x to λ (dispersion curve), then to $\log \lambda$ (the radial velocity introduces a constant shift for the complete data interval);

– determination of the λ-dependent characteristic curve and conversion density → intensity;

– FFT after usual treatment of data arrays;

– division, filter function;

– FFT.

We have investigated the two spectral intervals $\lambda\lambda\,4000\ldots4080$ (mainly Fe I lines) and $\lambda\lambda\,4080\ldots4120$ (Hδ). Results are shown in Figures 1 to 6.

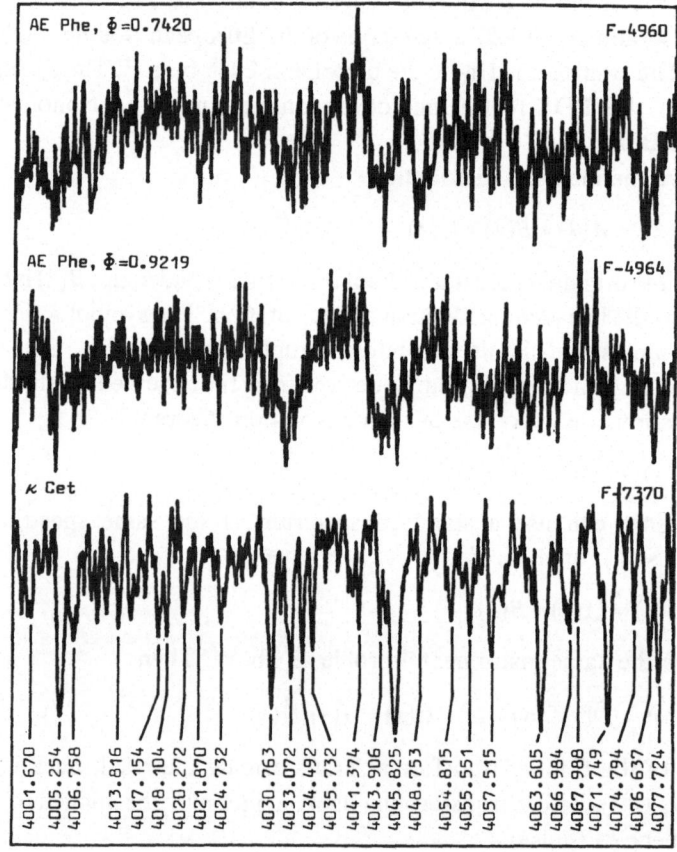

Fig. 1. $I(\lambda)$ in the interval $\lambda\lambda\,4000\ldots4080$, AE Phe at different phases, and κ Cet.

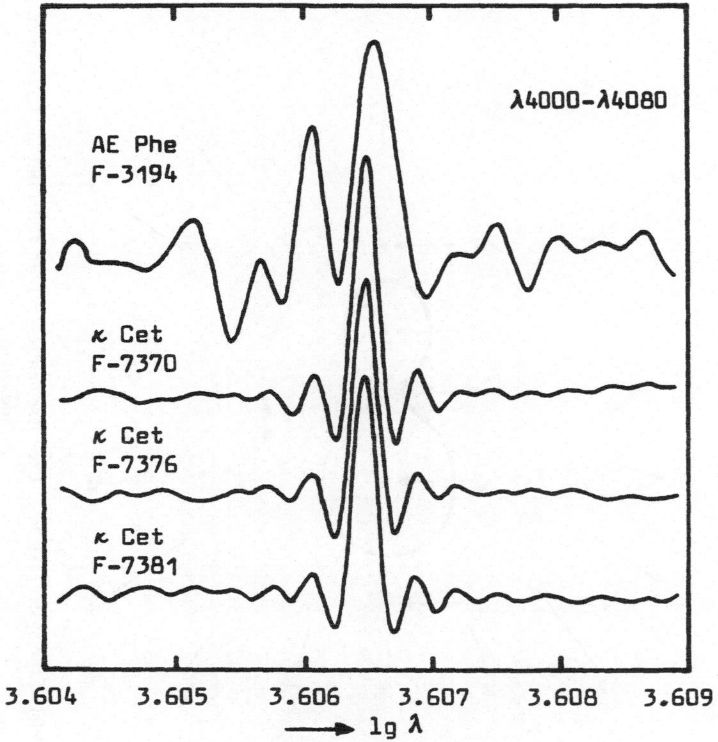

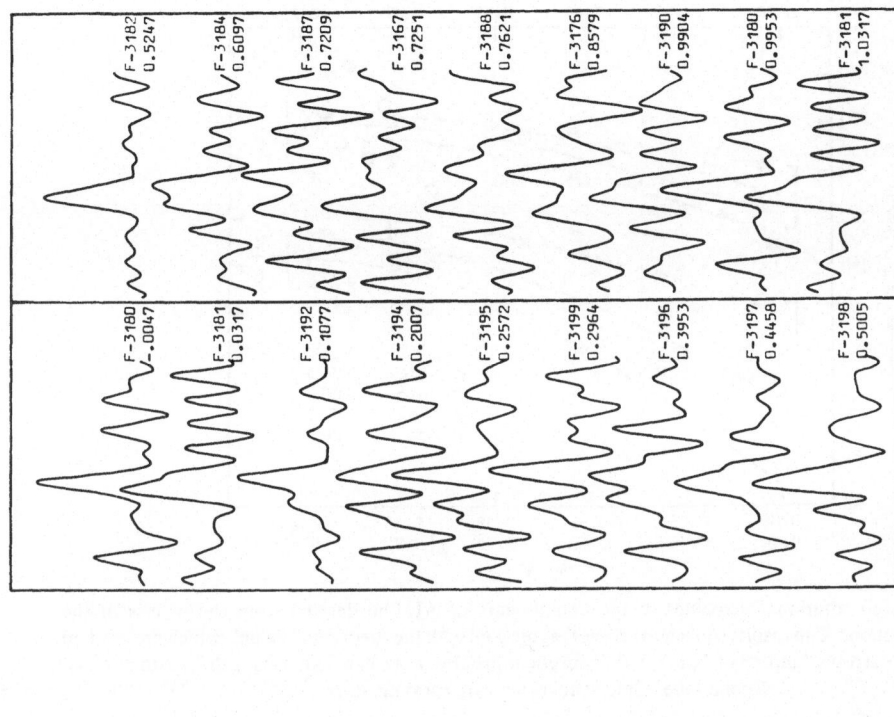

Fig. 3. Broadening functions $C(\lambda)$ of AE Phe at different orbital phases. The interval is $\lambda\lambda 4000 \ldots 4080$. The abscissa has a length $\Delta \log \lambda = 0.005$.

Fig. 2. Broadening function of κ Cet, deconvoluted by the mean spectrum of κ Cet, and broadening function of AE Phe, obtained in the same way. Wavelength interval $\lambda\lambda 4000 \ldots 4080$. The 'emission' and 'absorption' lobes to the left and the right of the broadening function of κ Cet are an artefact of the filtering process.

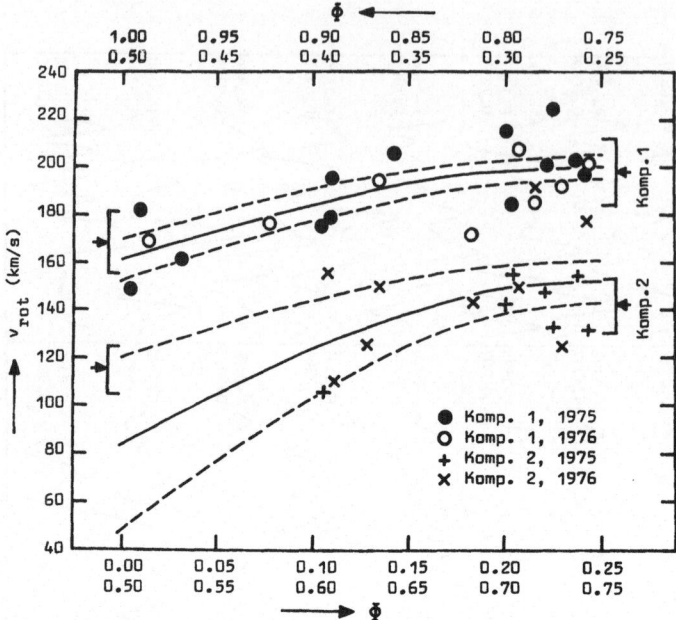

Fig. 4. Measured rotational velocities of the components of AE Phe derived from the widths of the broadening functions. The results (dots and crosses) agree well with the theoretical values, which are marked at the edges by arrows (and error bars). The theoretical profiles were calculated with the assumption of Roche lobe filling synchronously rotating stars.

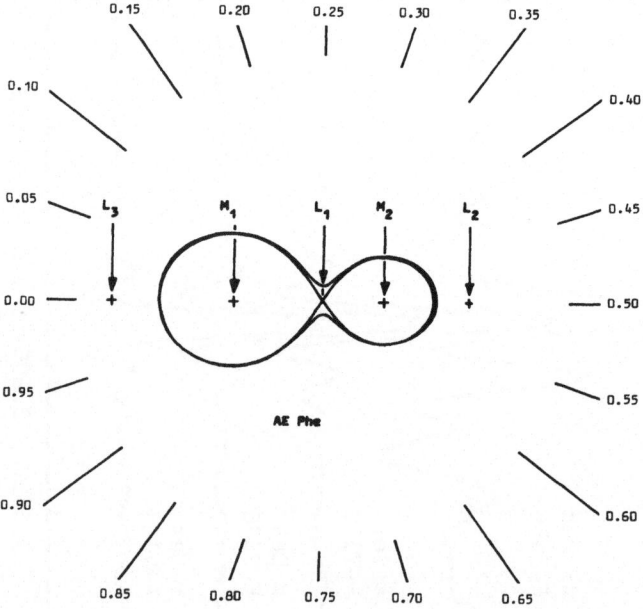

Fig. 5. Model of the system AE Phe. The mass centres of both stars and the Lagrangian points are indicated.

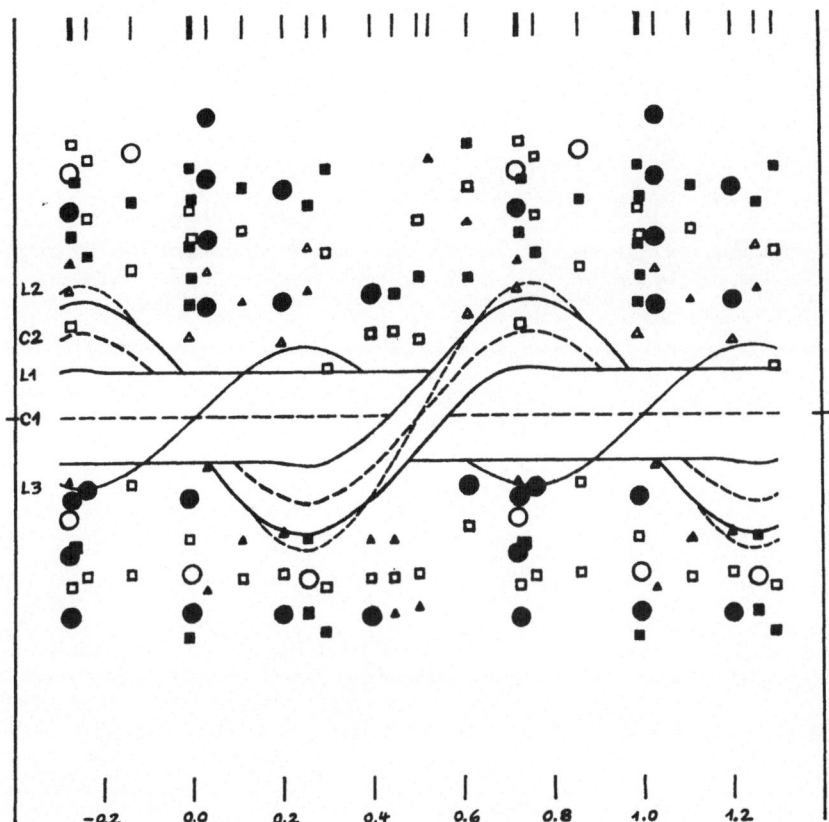

Fig. 6. The kinematics of special points of the system, as given in Figure 5, and referred to M_1, the mass centre of the primary component. Besides the two principal maxima of the broadening function, which indicate the motions of the two components, more maxima (absorption lines) and minima (emission lines) of the broadening function are indicated in Figure 4, with can hitherto not be explained (chromospheric lines?). These velocity components are marked in the figure as dots (absorption lines) and circles (emission lines).

VV ORIONIS: A WELL-BEHAVED EARLY-TYPE ECLIPSING
BINARY SYSTEM*

CARLSON R. CHAMBLISS

Kutztown University, Kutztown, Pennsylvania, U.S.A.

(Received 27 July, 1983)

Abstract. VV Orionis is a detached eclipsing binary system consisting of stars of spectral types B1 V and B4 V with a period of $1.^{d}4854$. There is also a third component whose orbital period about the eclipsing pair is about 120 days. In recent years there have been several new photoelectric and spectroscopic investigations of this system, and the results of these are compared. Both the geometric and photometric orbital elements appear to be well determined. VV Ori A appears to be of normal size and mass for its spectral class, but VV Ori B seems to be somewhat smaller than would be expected for a normal B4 V star. Linear limb-darkening coefficients are derived for VV Ori A, and these are in good agreement with theory. This system is of particular importance, because it is only one of very few early-type systems for which reliable limb-darkening coefficients can be expected to be obtained. The contribution of the light of the third component to the system has also been determined, and it appears most likely that VV Ori C is a star of spectral type A3 V.

1. Introduction

VV Orionis (HD 1868) is a detached eclipsing binary system having a period of $1^{d}.4854$. The eclipses are complete, the primary is a transit, and both components are Main-Sequence stars of spectral type B. The details of the earlier history of this system are summarized by Wood (1946). More recent photoelectric investigations include those made by Huffer and Kopal (1951), Bolokadze (1953), and by Atkins (1971). Orbital elements were calculated by each of these investigators, and Kopal and Shapley (1956) obtained an additional set of orbital elements based upon observations made by Huffer in blue and in yellow light.

The earliest radial velocity curve of VV Ori was obtained by Daniel (1916), who was also able to show that periodic variations in the residuals of the radial velocities could be interpreted as due to the presence of a third body in the system. He estimated the period of the third component to be about 120 days. Although the mass of this component has been approximately determined from spectroscopic analysis, its contribution to the light of the system has either been ignored or has been only crudely approximated in most photometric analyses of this system.

It has long been recognized that VV Ori should be capable of yielding reliable

* Paper presented at the Lembang-Bamberg IAU Colloquium No. 80 on 'Double Stars: Physical Properties and Generic Relations', held at Bandung, Indonesia, 3–7 June, 1983.

Astrophysics and Space Science **99** (1984) 163–170. 0004–640X/84/0991–0163$01.20.
© 1984 *by D. Reidel Publishing Company.*

limb-darkening coefficients (Koch *et al.*, 1963). In fact, apart from SZ Cam, VV Ori is practically the only early-type eclipsing binary system for which highly reliable limb-darkening coefficients can presently be expected to be determined. Most photometric analyses of VV Ori, however, have concentrated on the geometrical elements of the system, and only assumed or approximate values of the limb darkening have generally been used or reported.

2. Recent Investigations of VV Orionis

Recent photometric investigations of VV Ori include those by Duerbeck (1975), Eaton (1975), and by Chambliss (cf. Chambliss and Leung, 1982). The observations of Duerbeck were obtained at the ESO in one season during 1973–74 in *UBV* and in Hβ. The observations of Eaton are from the OAO-2 and are in eight bandpasses ranging from 1330 Å to 4250 Å. The observations of Chambliss were obtained at KPNO in four seasons during 1975–79 in *UBV* and in *uvby*.

Several analyses have been published from these investigations. Duerbeck analyzed his *UBV* data by means of the method of Russell and Merrill (1952) and the iterative procedure of Irwin (1947). Eaton analyzed five of his light curves with the use of the method of Wilson and Devinney (1971). A unique geometry was assumed for all five light curves. A similar approach was used by Leung in analyzing light curves formed from the normal points of Chambliss. All seven light curves were analyzed simultaneously, and a unique geometry was assumed.

Other investigations have also been made using these sets of data. The data of Duerbeck were analyzed by Budding and Najim (1980) using a frequency-domain analysis procedure. The independent treatment of L_1 and L_2 permitted the small contribution of the light of the third component to be taken into consideration. Cester *et al.* (1978) reanalyzed the *UBV* light curves of Duerbeck and the 3320 Å and 4250 Å light curves using the model of Wood (1971, 1972). The most recent photometric analyses of VV Ori have been those by Giuricin *et al.* (1983) and by Chambliss (1983) using the Wood model on the observations of Chambliss.

As might be expected, the different methods of analysis used by different investigators on different bodies of data do not all produce the same results. As was noted by Giuricin *et al.* (1983) and by Chambliss (1983), the attempt by Leung to fit all seven light curves of Chambliss to a unique geometry resulted in a compromised solution. The derived inclination ($i = 90°\!.0$) was plainly too high, as all other recent investigations of VV Ori have yielded $i = 85°$ or $86°$. Other criticisms of this solution are also given in the two later papers.

Only the Trieste astronomers have analyzed all three of these sets of photometric data using the same method: i.e., the Wood model. The agreement between the different sets of elements obtained by them is very good. In fact, the agreement between the orbital elements derived from the three independent sets of data but analyzed with the same method is much better than between the orbital elements reported by Chambliss and Leung and those reported by Giuricin *et al.*, both of which used the normal points of Chambliss in their analysis. Thus it appears that the computed orbital elements of a well-observed eclipsing binary seem to depend more on the method of analysis chosen than they do on whose data base is utilized.

The formal errors quoted by Chambliss and by Giuricin *et al.* average about ± 0.002 for the parameters r_1 and k. It appears that this is about the highest precision that one can expect to obtain for these parameters by using the Wood model. Although the errors for r_1 and r_2 quoted in Leung's solution, which made use of the Wilson–Devinney method, are lower than this, it does not seem likely that the derived parameters are reliable to the precision implied by these formal errors (± 0.0004). It is interesting to note that the errors for r_1 and k of VV Ori quoted in the solution tabulated in the paper by Kopal and Shapley (1956) are about ± 0.003. In this paper the iterative method of Kopal was used to solve the light curves, and it would appear that this method is capable of yielding nearly as satisfactory solutions for systems such as VV Ori as is the Wood model.

Of the various investigations only those by Kopal and Shapley, Duerbeck, Eaton, and Chambliss attempted to determine limb darkening coefficients empirically. Duerbeck reported values of $u = 0.07$, 0.14, and 0.17 for VV Ori A in V, B, and U, respectively. These values are much lower than the predicted theoretical values, but Duerbeck notes they may be the consequence of the asymmetries which he observed in the annular portion of the primary eclipse. The value of 0.5 ± 0.1 quoted by Kopal and Shapley for VV Ori A in blue light is higher than one would predict from theory. Eaton's values are more in line with the theoretical values for the limb darkening coefficients, but most of his bandpasses lie well outside the optical range. The results obtained by Chambliss are summarized in the next section.

Only the papers by Budding and Najim and by Chambliss attempted to empirically determine the contribution of the third component to the light of the system. It is the belief of this author that the values of L_3 which were obtained by Budding and Najim are too low.

There have been two recent spectroscopic investigations of VV Ori. These are by Beltrami and Galeotti (1970) and by Duerbeck (1975). The former investigation was made at a lower dispersion, but the masses reported for the components ($\mathfrak{M}_1 = 10.2\ \mathfrak{M}_\odot$, $\mathfrak{M}_2 = 4.5\ \mathfrak{M}_\odot$) are realistic for stars of spectral types B1 V and B4 V, respectively, while those reported by Duerbeck ($\mathfrak{M}_1 = 7.60\ \mathfrak{M}_\odot$, $\mathfrak{M}_2 = 3.42\ \mathfrak{M}_\odot$) appear to be much too low.

3. The Physical Nature of VV Orionis

The photometric properties of VV Ori are summarized in the papers of Duerbeck and of Chambliss and Leung. In the latter paper the following magnitudes are given:

	Maximum	Primary	Secondary
V	+5.33	+5.66	+5.50
$B-V$	−0.19	−0.18	−0.19
$U-B$	−0.92	−0.90	−0.93

Chambliss estimates $E(B-V)$ as +0.07 implying that $(B-V)_0 = -0.26$ and $(U-B)_0 = -0.97$ for VV Ori. This star is a member of the I Orion association, and its distance is estimated to be about 400 parsecs.

All published times of minimum light observed for this system are listed in these two papers, and the period is given in both as $1^d.485378$. All available data indicate that the period of VV Ori has remained constant for the past 70 y. The third component will introduce a small light-time effect – a total amplitude of approximately two minutes – on the times of minimum light, but it has not proven possible to extract the orbital period of VV Ori C from variations in the observed times of minimum light of the eclipsing pair.

The light curves of Chambliss do not show some of the peculiarities reported by Duerbeck for his light curves, and this author doubts that most of these effects are regularly present in the light curves of VV Ori. Most notable of these is the dip at about 0.65 phase present in all three of Duerbeck's light curves. In the light curves of Chambliss only the intermediate-band u curve shows this effect and only to a much smaller extent than is the case with the light curves observed by Duerbeck. Duerbeck's model of two gas streams as a possible explanation for this effect raises questions as to how such streams could arise from both components of a dynamically detached system.

Unlike those of Duerbeck, the light curves of Chambliss are symmetrical in the annular phase about primary minimum. Were this not the case, reliable limb darkening coefficients for the primary component could not be obtained, and the low values for u_1 obtained by Duerbeck are probably attributable to the peculiarities which he observed at this phase of the light cycle.

The geometrical elements reported in the various analyses of VV Ori are in substantial agreement with each other. The following are taken from the 1983 paper of Chambliss:

i	$85°.60 \pm 0°.17$	k	0.4864 ± 0.0025
r_1	0.3627 ± 0.0017	r_2	0.1764 ± 0.0014
a_1	0.3830	a_2	0.1796
b_1	0.3713	b_2	0.1770
c_1	0.3588	c_2	0.1755

The system is detached. The oblateness of the primary component is about 6 %, while that of the secondary is only about 2 %.

One of the main reasons for studying VV Ori is to obtain reliable empirical limb darkening coefficients. As has already been noted, VV Ori is one of the very few early-type systems for which this is possible. It is really only possible to obtain limb darkening coefficients for the primary component, since the secondary component contributes less than 10 % of the light of the system. Furthermore the portion of the light curves which would be most sensitive to the value of u_2 would be the secondary eclipse. This eclipse, however, is a complete occultation, and hence, observations made in the total portion of secondary minimum would contain no information at all about u_2. Listed below are theoretical values for the limb darkening (u_1^*, u_2^*) taken from the data given by Al-Naimiy (1978) and the empirical values for u_1 adopted by Chambliss in his 1983 paper:

λ	u_1^*	u_1	u_2^*
V	0.28	0.26 ± 0.05	0.32
B	0.35	0.32 ± 0.04	0.41
U	0.33	0.35 ± 0.05	0.35
y	0.28	0.29 ± 0.03	0.32
b	0.34	0.25 ± 0.04	0.39
v	0.37	0.29 ± 0.05	0.42
u	0.34	0.27 ± 0.04	0.36

The agreement between theory and observation in this case is remarkably good.

Attempts were also made by Chambliss to derive non-linear limb darkening coefficients from his VV Ori observations. The formula treated in the Wood model is that of Klinglesmith and Sobieski (1970), namely:

$$I(u) = I_0 \left(1 - A(1 - \mu) - B\mu \log \mu \right)$$

where

$$\mu = \cos \gamma.$$

All attempts to obtain reliable values for A and B proved unsuccessful, however, and this supports the contention of Grygar et al. (1972) who stated that non-linear limb-darkening laws are beyong the reach of the presently available observations of close eclipsing binary stars – even if these are well-behaved in most other respects.

Of the recent papers on VV Ori only two, those of Budding and of Najim (1980) and of Chambliss (1983), have attempted to empirically determine the light of the third component. The values of L_3 derived in the former are too low to be compatible with the existing spectroscopic data, but the latter indicates that the third component is most likely a star of spectral type A3 V. The fractional luminosities adopted by this investigator are as follows:

	L_1	L_2	L_3
V	0.891	0.092	0.017
B	0.895	0.091	0.014
U	0.925	0.070	0.005
y	0.894	0.089	0.017
b	0.899	0.086	0.015
v	0.907	0.081	0.012
u	0.929	0.066	0.005

On the basis of the photometric data a temperature of 25 000 K has been assigned to the primary component. In the Wood model the temperature of the secondary component is usually treated as an independent variable while that of the primary is held constant. A mean temperature of $15\,650 \pm 90$ K for T_2 is derived. These temperatures are compatible with spectral types of B1 V and B4 V, respectively. Both Eaton and Leung used values of 21 000 K for T_1 in their solutions. This value is not compatible, however, with the presently accepted temperature scales for stars having the spectroscopic and photometric properties of VV Ori A.

As has already been noted, there is still much controversy over the spectroscopic orbital elements of VV Ori. Duerbeck describes VV Ori as a double-lined spectroscopic binary, but this contention has been challenged by Andersen (1976), whose comments have in turn been questioned by Duerbeck (1976). VV Ori B contributes only $\sim 10\,\%$ of the light of the system (somewhat less than this according to Chambliss, Leung, Giuricin, and Kopal and Shapley; somewhat more than this according to Duerbeck). Thus it is difficult to see how this component can display spectral features over almost all orbital phases. Furthermore the masses of the components as reported by Duerbeck are decidedly low. The values given by Beltrami and Galeotti (1970) imply that the two components are of fairly normal size and mass, but Duerbeck's spectroscopic orbital elements imply components which are severely undersized and undermassive for their luminosities. These two papers agree on the orbital period of the third components ($115^d.9$ and $119^d.1$, respectively), but the former assigns a mass of 2.3 $\mathfrak{M}_\odot$ to this component, while the latter estimates that VV Ori C has a mass of 1.6 $\mathfrak{M}_\odot$. According to this investigator, the spectral type of this component is A3 V, and the normal mass for a star of this type is 2.1 $\mathfrak{M}_\odot$.

The following are believed to be the most realistic estimates for the absolute dimensions of the components of VV Ori:

R_1	4.94 $\odot$	R_2	2.36 $\odot$
$\mathfrak{M}_1$	10.2 $\odot$	$\mathfrak{M}_2$	4.5 $\odot$
L_1	8425 $\odot$	L_2	260 $\odot$
M_{bol1}	-5.01	M_{bol2}	-1.24

The larger component appears to be normal for a star of spectral type B1 V and is found to be fairly close to the ZAMS. The secondary component, however, appears to be unusually small for its mass.

4. Conclusions

Although many of the problems concerning VV Ori appear to have been solved, a new radial velocity curve is definitely needed. Even if VV Ori does prove to be only a single-lined system, new values for K_1 and $f(\mathfrak{M})$ are still urgently needed.

The peculiarities of the light curve of VV Ori discussed by Duerbeck have not been confirmed by Chambliss: and it seems that the light variations of this system can be described as being essentially without complications.

The limb darkening results for VV Ori A are particularly important, since they are the most precisely determined linear limb darkening coefficients which have been found thus far for an early-type star. They are essentially in good agreement with the values determined in recent theoretical studies, and this fact should inspire confidence in those researchers who use the results of such studies when deriving the other orbital elements of eclipsing binary systems. The failure of attempts to obtain reliable empirical non-linear limb darkening coefficients supports the view that non-linear limb darkening laws are beyond the reach of the presently available observations of eclipsing binary stars.

VV Ori is a detached system, and the components are sufficiently far apart for the close proximity effects to be relatively weak. Although VV Ori appears to be small for its mass, VV Ori A is an archetypal early B-type Main-Sequence star.

References

Al-Naimiy, H. M.: 1978, *Astrophys. Space Sci.* **53**, 181.
Andersen, J.: 1976, *Astron. Astrophys.* **47**, 467.
Atkins, H. L.: 1971, M.S. Thesis, University of Georgia.
Beltrami, G. and Galeotti, P.: 1970, *Mem. Soc. Astron. Ital.* **41**, 167.
Bolokadze, R. D.: 1953, *Perem. Zvezdy* **9**, 379.
Budding, E. and Najim, N. N.: 1980, *Astrophys. Space Sci.* **72**, 369.
Cester, B., Fedel, B., Giuricin, G., Mardirossian, F., and Mazzetti, M.: 1978, *Astron. Astrophys. Suppl.* **33**, 91.
Chambliss, C. R.: 1983, *Astrophys. Space Sci.* **89**, 15.
Chambliss, C. R. and Leung, K. C.: 1982, *Astrophys. J. Suppl.* **49**, 531.
Daniel, Z.: 1916, *Publ. Allegheny Obs.* **3**, 179.
Duerbeck, H. W.: 1975, *Astron. Astrophys. Suppl.* **22**, 19.
Duerbeck, H. W.: 1976, *Astron. Astrophys.* **47**, 471.
Eaton, J. A.: 1975, *Astrophys. J.* **197**, 379.
Giuricin, G. Mardirossian, F., Mazzetti, M., and Chambliss, C. R.: 1983, *Astron. Astrophys. Suppl.* **51**, 111.
Grygar, J., Cooper, M. L., and Jurkevich, I.: 1972, *Bull. Astron. Inst. Czech.* **23**, 147.
Huffer, C. M. and Kopal, Z.: 1951, *Astrophys. J.* **114**, 297.
Irwin, J. B.: 1947, *Astrophys. J.* **106**, 380.
Klinglesmith, D. A. and Sobieski, S.: 1970, *Astron. J.* **75**, 175.
Koch, R. H., Sobieski, S., and Wood, F. B.: 1963, *Publ. Univ. Penna. Astron. Ser.* **9**, 32.

Kopal, Z. and Shapley, M. B.: 1956, *Jodrell Bank Ann.* **1**, 141.
Russell, H. N. and Merrill, J. E.: 1952, *Contr. Princeton Univ. Obs.*, No. 26.
Wilson, R. E. and Devinney, E. J.: 1971, *Astrophys. J.* **166**, 304.
Wood, F. B.: 1946, *Contr. Princeton Univ. Obs.* **21**, 25.
Wood, D. B.: 1971, *Astron. J.* **76**, 701.
Wood, D. B.: 1972, 'A Computer Program for Modeling Non-Spherical Eclipsing Binary Star Systems', NASA Publ. X 110 72 473.

THE LIGHT VARIABILITY OF BD + 25°2511*

A. V. RAVEENDRAN

Indian Institute of Astrophysics Bangalore, India

(Received 12 July, 1983)

Abstract. Photometry of the spectroscopic binary BD+25°2511 obtained during the 1981, 1982, and 1983 observing seasons shows that it is a variable with a period of about 3.5 days.

BD + 25°2511 is a spectroscopic binary with double Ca II H and K emissions, and it is most likely a member of the Coma Berenices cluster (Wilson, 1963; Trumpler, 1938). Both Wilson (1963) and Trumpler (1938) assign BD + 25°2511 a spectral type close to dG8. Photometry of this object was begun in 1981 as part of a programme on late-type emission binaries to study their photometric behaviour and chromospheric activity.

Observations were made with the 34-cm Cassegrain reflector of the Kavalur Observatory through standard B and V filters. All the measurements were made with respect to the near by comparison star HD 108 806. As a check on the photometric constancy of this comparison, HD 108576 was also observed on several nights. Table I gives the Julian day intervals covered by the observations (ΔT), the number of nights observed during each interval (n) and the corresponding total range in the observed visual magnitudes (Δv). It is clear from the Table I that BD + 25°2511 is a variable and thus confirms its suspected light variability reported by Eggen (1978).

In order to determine the period of light variation, the period finding technique as outlined in Raveendran *et al.* (1982) was employed. All the three data sets were independently subjected to a period analysis. The 1981 observations yielded a 3.540 days period, while the period given by the 1982 data is 3.475 days. Since the two sets of observations cover only a few photometric cycles, the period determinations

TABLE I
Total range in the observed visual magnitudes and the derived periods

Year	ΔT	n	Δv	Period (day)
1981	2 444 617–4678	22	0.08	3.540
1982	2 444 984–5026	24	0.11	3.475
1983	2 445 366–5412	14	0.02	–

* Paper presented at the Lembang-Bamberg IAU Colloquium No. 80 on 'Double Stars: Physical Properties and Generic Relations', held at Bandung, Indonesia, 3–7 June, 1983.

are not sufficiently accurate to decide definitely whether the difference in the
derived periods is real. The 1983 observations did not yield any satisfactory period.
We attribute this to the small range in the observed visual magnitudes, which is
comparable to the uncertainty (~ 0.01 mag) in the measurements.

The Julian days of observation were converted to photometric phases by use of
the following equations:

(1) Phase = JD $2\,444\,617.494 + 3^{d}.540E$, for the 1981 data, and

(2) Phase = JD $2444\,984.474 + 3^{d}.475E$, for the 1982 data.

The initial epoch in each case corresponds to the time of the first observation. In
Figure 1, the differential magnitudes (in the sense, BD$+25°2511$–HD 108 806)
obtained during the 1981 and 1982 observing seasons are plotted separately. Each
point is a mean of 3–4 independant measurements. It is clearly seen that the light
curves are nearly sinusoidal and the amplitude is variable. During the 1981 season
the amplitude was ~ 0.07 mag while during 1982 observing run it was ~ 0.11 mag.
The recent photometry shows that the amplitude of light variation has decreased
to about 0.015 mag.

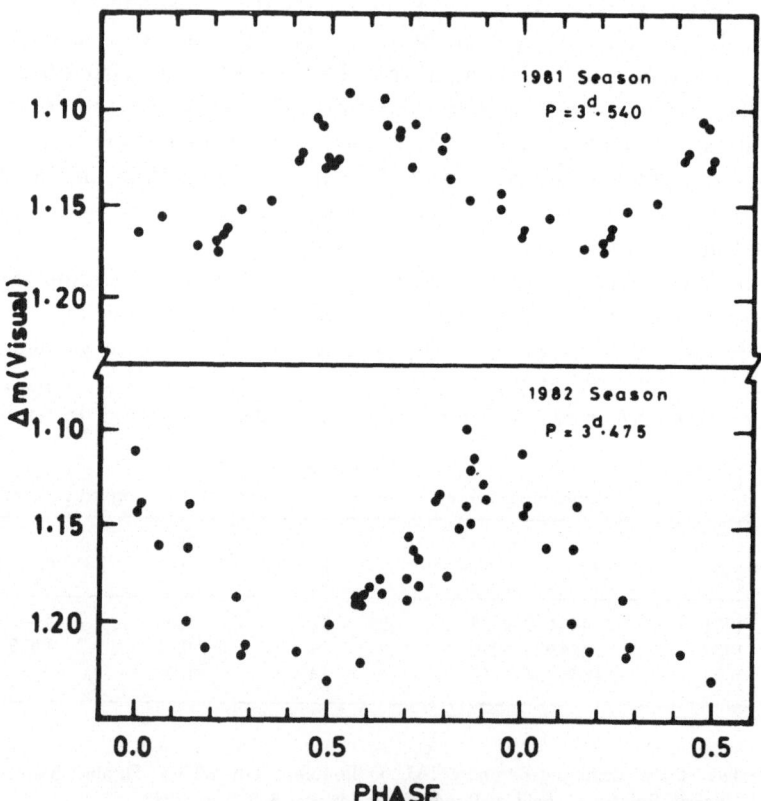

Fig. 1.   Light curves of BD$+25°2511$ obtained during the 1981 and 1982 seasons.

The nature of the light curves shows that the cause for the observed light variation is not geometrical eclipses. Most likely, in BD+25°2511 we are seeing the type of 'activity' exhibited by the well-known RS Canum Venaticorum and BY Draconis variables, where the photometric variation is attributed to the presence of 'starspots' which rotationally modulate the observed flux. A detailed analysis of the available data including the spectrophotometric scans obtained with the automated spectrum scanner attached to the 102-cm Cassegrain reflector of the Kavalur Observatory is in progress and will be published elsewhere.

Acknowledgement

I am thankful to Mr. K. Jayakumar and Mr. M. J. Rosario for their help in collecting the photometric data.

References

Eggen, O. J.: 1978, *Inf. Bull. Var. Stars*, No. 1426.
Raveendran, A. V., Mekkaden, M. V., and Mohin, S.: 1982, *Monthly Notices Roy. Astron. Soc.* **199**, 707.
Trumpler, R. J.: 1938, *Lick Obs. Bull.*, No. 494.
Wilson, O. C.: 1963, *Astrophys. J.* **138**, 832.

WOLF–RAYET BINARIES:
EVOLUTIONARY CAUSES FOR THEIR DISTRIBUTION
IN THE GALAXY***

BAMBANG HIDAYAT and A. GUNAWAN ADMIRANTO

Observatorium Bosscha of the Institut Teknologi Bandung, Lembang, Indonesia

and

KAREL A. VAN DER HUCHT

Space Research Laboratory of The Astronomical Institute, Utrecht, The Netherlands

(Received 22 August, 1983)

Abstract. On the basis of the most recent data, the fraction of known Wolf–Rayet binaries is 0.22. In the solar neighbourhood ($d < 2.5$ kpc) this fraction is 0.34

In order to assess the relative importance of massive binary evolution as one of the ways to produce WR stars, the galactic distribution of WR binaries is compared with that of single WR stars using improved intrinsic parameters and new data for the fainter WR stars.

In the galactic plane the increase of the binary frequency with galactocentric distance is confirmed.

In a direction perpendicular to the galactic plane it is demonstrated at all distances from the Sun that the single-line spectroscopic WR binaries with small mass functions have definitely larger $|z|$-distances than the 'single' WR stars and the WR binaries with massive companions. This is consistent with the evolutionary scenario for massive binaries summarized by van den Heuvel (1976). Among the 'single' WR stars the fraction of those with large $|z|$-distances is increasing with galactocentric distance, like the fraction of the known binaries. This implies that among the high-$|z|$ 'single' WR stars as well as among the WR stars with lower $|z|$-values many binaries are still to be discovered.

The total WR binary frequency in the Galaxy could be well above 50 %.

1. Introduction

Population I WR stars are evolved massive stars, primarily in their core He-burning phase, their progenitors being O-type stars (e.g., Conti, 1982). Due to removal of their outer hydrogen layers by mass loss and mass transfer, these stars reveal in their emission spectra overabundances of nucleosynthesis products like He + N (WN stars), He + C (WC stars) or He + C + O (WO stars), as elaborated on by Smith and Willis (1982) and Barlow and Hummer (1982).

Before the importance of mass loss was realized, it was thought that only mass transfer in a binary system could remove sufficient material from a massive star to reveal core H-burning and core He-burning products at its surface. And thus it was thought that all WR stars were produced in binary systems. In recent years

* Contribution from the Bosscha Observatory No. 79.
** Invited paper presented at the Lembang-Bamberg IAU Colloquium No. 80 on 'Double Stars: Physical Properties and Generic Relations', held at Bandung, Indonesia, 3–7 June, 1983.

Astrophysics and Space Science **99** (1984) 175–190. 0004–640X/84/0991–0175$02.40.
© 1984 *by D. Reidel Publishing Company.*

it has been shown (e.g., Maeder, 1982) that mass loss and internal mixing can cause a single O-type star with $\mathscr{M} \geqslant 23 \, \mathscr{M}_\odot$ (Maeder and Lequeux, 1982) to evolve into a WR star, and that generally an O → (RSG →)WN → WC scheme can be expected. During the O-phase the mass loss rate is of the order of $\dot{\mathscr{M}} = 5 \times 10^{-6} \, \mathscr{M}_\odot$; during the WR phase $\dot{\mathscr{M}} = 2.8 \times 10^{-5} \, \mathscr{M}_\odot$ (Abbott, 1982). In the latter phase this mass is lost in a dense, optically thick stellar wind with typically $T_{\mathrm{eff}} = 2{-}6 \times 10^4 \, \mathrm{K}$ and $n = 10^{12}{-}10^{14} \, \mathrm{cm}^{-3}$ (e.g., van der Hucht, 1982). Due to this mass loss a massive star is stripped of its outer layers. This process can be accompanied by internal mixing of various kinds (Maeder, 1982), which makes it even easier to bring N, C, and O to the surface and thus to cause the WR phenomenon.

WR star masses in binary systems have been determined between 5 and 50 $\mathscr{M}_\odot$ (Massey, 1981b).

The WR lifetime is of the order of $3{-}6 \times 10^5$ yr (Maeder and Lequeux, 1982).

In the past few years, there has been an increasing interest in the study of the galactic distribution of Wolf–Rayet stars, in response to the recognition that WR stars are descendants of massive early-type stars, and that their distribution may give clues to their origin and evolution.

Recent studies by Moffat and Isserstedt (1980), Hidayat et al. (1982), Garmany et al. (1982), and Conti et al. (1983) each used improved parameters over previous work.

Thanks to the work of Lundström and Stenholm (1983), Hidayat et al. (1984) could use improved intrinsic parameters and new colours and spectral types for many of the fainter stars, in order to reinvestigate the galactic distribution of Population I Wolf–Rayet stars. In the present paper emphasis is put on the binaries versus the single stars.

2. Known WR Binaries

In our Galaxy, 159 Population I Wolf–Rayet stars are known (van der Hucht et al., 1981; Hidayat et al., 1984). Table I gives the distribution between single WR stars, single WR stars with intrinsic absorption lines, double-line spectroscopic binaries, and single-line spectroscopic binaries. Stars are called single if no orbit solution is known. Duplicity is published of 35 WR stars, i.e., 22%.

Double-line spectroscopic WR binaries (SB2) have been discussed extensively by Massey (1982), single-line spectroscopic WR binaries (SB1) have been discussed extensively by Moffat (1982). Since then a few more cases have been discovered. Table II lists the WR SB2's arranged by spectral type, Table III lists the WR SB1 systems also arranged by spectral type, and subdivided between SB1's with small mass functions $(f(\mathscr{M}) < 0.3 \, \mathscr{M}_\odot)$, henceforth labeled as lmSB1, and SB1's with large mass functions $(f(\mathscr{M}) > 0.3 \, \mathscr{M}_\odot)$, henceforth labeled as mSB1.

In these tables the absolute visual magnitude M_v, the heliocentric distance d, the galactocentric distance r, and the separation from the galactic plane z are from Hidayat et al. (1984) and based on the intrinsic parameters given in Table IV

TABLE I

Distribution of galactic Wolf–Rayet stars in subclasses

WR	Single		Double		Total
Subclass	Single	Single + abs.	SB2	SB1	
WN2	1				1
WN3	3	1	1		5
WN4	8		3	1	12
WN4.5	5		1		6
WN5	4		2	1	7
WN6	13		2	6	21
WN7	9	3		4	16
WN8	7			4	11
WN9	1				1
WN10	1				1
unclassified WN	1				1
Subtotal WN	53	4	9	16	82
WC4	3				3
WC5	12		1		13
WC6	11	3	2		16
WC7	4	2	3		9
WC8	5	1	2		8
WC8.5	5				5
WC9	13			1	14
WC10	1				1
Subtotal WC	54	6	8	1	69
WO1	1				1
WO2	1				1
WN + WC	2			1	3
unclassified WR	2	1			3
Grand total	113	11	17	18	159

(see next section). For the SB2 systems the M_v values of the individual binary components have been determined from data in the literature (see notes to Table II).

It should be noted that some of the more recent duplicity determinations would be served by confirmative studies. At least one case, i.e., WR 140 in Table II, is subject to controversy: Conti (1983, private communication) did not find an orbit solution from his data.

Among the 18 SB1 systems in Table III are 13 lmSB1's, in which the unseen component may be a compact star. The listed masses are calculated by assuming that $i = 60°$ and $\mathcal{M}$ (unseen companion) $= 1.6 \, \mathcal{M}_\odot$, following Moffat (1982).

Both the SB2 and SB1 systems can be identified with links in the evolutionary

TABLE II

Double-line spectroscopic Wolf–Rayet binaries (SB2) (17)

WR	HD/Name	Spectral type	Period (days)	$\mathcal{M}_{WR}\sin^3 i$ ($\mathcal{M}_\odot$)	$\mathcal{M}_{WR}/\mathcal{M}_O$	i	$\mathcal{M}_{WR}$ ($\mathcal{M}_\odot$)	Refs.	$(b-v)_0$ (system)	M_v (WR)	M_v (O)	M_v (system)	d (kpc)	r (kpc)	z (pc)
97	E320102	WN3+O5–7	8.83	1.8	0.33	30°	11	1	−0.33	−3.6	−5.1	−5.3	2.95	7.07	−58
21	90657	WN4+O4−6	8.2	8–11	0.52	50°	18	2	−0.33	−4.0	−5.2	−5.5	3.51	9.70	−55
31	94546	WN4+O7	4.9	8	0.34	–	≥8	3	−0.33	−4.0	−4.9	−5.3	4.96	9.65	+2
127	186943·	WN4+O9.5V	9.5548	9–11	0.52	70°	13	4	−0.21	−3.6	−4.0	−4.6	4.41	8.99	+133
133	190918	WN4.5+O9.5 Ia	112.8	0.7	0.26	25°	9	5	−0.33	−4.3	−6.0	−6.2	2.09	9.59	+75
139	193576	WN5+O6	4.21238 (e)	9.3	0.40	55°	17	6,7	−0.30	−4.8	−5.1	−5.7	1.74	9.74	+43
151	CX Cep	WN5+O8V	2.1267 (e)	5	0.43	≥50°	5–11	8	−0.33	−4.6	−4.6	−5.4	5.08	12.17	+123
47	E311884	WN6+O5V	6.34	40	0.84	70°	50	9	−0.30	−5.3	−5.1	−6.0	3.80	8.61	−15
153	211853	WN6+O	6.6884 (e)	–	≥0.22	≥50°	10–25	4	−0.33	−5.3	−4.8	◄6.2	3.56	11.33	−40
9	63099	WC5+O7	27.63	17	0.16	–	–	1	−0.33	−4.0	−4.5	−5.0	2.08	10.91	−174
30	94305	WC6+O6–8	18.82	15	0.48	–	–	10	−0.33	−4.4	−5.0	−5.5	8.99	11.00	−410
48	113904	WC6+O9.5I	18.34	–	–	–	–	11	−0.33	−4.4	−6.0	−6.2	1.59	9.19	−69
42	97152	WC7+O7V	7.886	3.6	0.59	35°	20	12	−0.33	−4.7	−4.9	−5.6	3.47	9.34	−30
79	152270	WC7+O5–8	8.893	1.8	0.36	25°	20	13	−0.35	−4.7	−5.6	−6.0	2.00	8.10	+40
140	193793	WC7+O4–5	1085	11	0.22	74°	13	14	−0.33	−4.7	−5.5	−5.9	1.43	9.88	+104
11	68273	WC8+O9I	78.5002	17	0.54	70°	20	15	−0.32	−4.9	−6.4	−6.6	0.48	10.07	−64
113	168206	WC8+O8–9III–V	29.707	11	0.48	70°	13	16	−0.39	−4.8	−4.8	−5.5	2.00	8.13	+61

Note: (e) = eclipsing system.

References:
(1) Niemela (1982).
(2) Niemela and Moffat (1982).
(3) Niemela (1980).
(4) Massey (1981a).
(5) Fraquelli and Horn (1983).
(6) Ganesh et al. (1967).
(7) Münch (1950).
(8) Massey and Conti (1981).
(9) Niemela et al. (1980).
(10) Niemela et al. (1983).
(11) Moffat and Seggewiss (1977).
(12) Davis et al. (1981).
(13) Seggewiss (1974).
(14) Lamontagne et al. (1983b).
(15) Niemela and Sahade (1980).
(16) Massey and Niemela (1981).

Notes to the SB2 systems in Table II:

WR9: M_v(system) and M_v(WR) $= -4.0$ from Turner (1982, private communication). This results in M_v(O) $= -4.5$, corresponding to an O9V star on the scale of Conti *et al.* (1983). Assuming $(b-v)_0 = -0.33$ for the system yields the distance $d = 2.08$ kpc, much closer than the 'HD63077 group' of McCarthy and Miller (1974).

WR11: M_v(system) and M_v(WR) $= -4.9$ from Turner (1982, private communition). Distance from Abt *et al.* (1976). Consequently the system has $(b-v)_0 = -0.32$.

WR21: M_v(WR) from Table IV, M_v(O) $= -5.2$ from scale of Conti *et al.* (1982). Assumption $(b-v)_0 = -0.33$ yields the distance.

WR30: M_v(WR) from Table IV, M_v(O) $= -5.0$ from scale of Conti *et al.* (1983). Assumption $(b-v)_0 = -0.33$ yields the distance.

WR31: M_v(WR) from Table IV, M_v(O) $= -4.9$ from scale of Conti *et al.* (1983). Assumption $(b-v)_0 = -0.33$ yields the distance.

WR42:   Spectral types and $\Delta M_v = 0.2$ from Davis *et al.* (1981). M_v(O) $= -4.9$ from Conti *et al.* (1983) results into M_v(WR) $= -4.7$, in agreement with Table IV.

WR47:   Cluster distance $d = 3.80$, colour excess, and M_v(system) from Lundström and Stenholm (1983). According to Niemela *et al.* (1980) the O star is 3 times fainter than WR star, corresponding to M_v(O) $= -4.5$ and M_v(WR) $= -5.6$. If we take M_v(WR) $= -5.2$ from Table IV, then M_v(O) $= -5.1$, corresponding to O6V, in reasonable agreement with the O-spectral type determined by Niemela *et al.* (1980).

WR48: M_v(WR) from Table IV, M_v(O) from Conti *et al.* (1983). Assumption $(b-v) = -0.33$ yields the distance, somewhat smaller than d(Cen OB1) $= 1.9$ kpc (Humphreys, 1978).

WR79:   Cluster distance, colour excess and M_v(system) from Lundström and Stenholm (1983). M_v(WR) from Table IV implies M_v(O) $= -5.6$, which corresponds with an O5V or an O7III companion according to the scale of Conti *et al.* (1983).

WR97:   Spectral types from Niemela (1982), M_v(WR) from Table IV, M_v(O6V) $= -5.1$ from the scale of Conti *et al.* (1983). Assumption $(b-v)_0 = -0.33$ we find the distance.

WR113:   Distance from Ser OB2 (Humphreys, 1978). $\Delta M_v = 0$ from Massey and Niemela (1981). With M_v(WR) from Table IV, this implies an O8.5V companion on the scale of Conti *et al.* (1983). This yields $(b-v)_0 = -0.39$.

WR127:   Distance Vul OB2 and M_v(system) from Turner (1980). Taking M_v(O) $= -4.0$ from the scale of Conti *et al.* (1983) we find M_v(WR) $= -3.6$ and a ΔM_v not far from $\Delta M_v = 1.3$ found by Massey (1981a). This yields $(b-v)_0 = -0.21$.

WR133:   Distance, reddening and M_v(system) from Lundström and Stenholm (1983). With M_v(O) $= -6.0$ from the scale of Conti *et al.* (1983) this implies M_v(WR) $= -4.3$, in agreement with Table IV.

WR139:   Distance, reddening and M_v(system) from Lundström and Stenholm (1983). With M_v(WR) from Table IV, we find M_v(O) $= -5.1$, consistent with an O6V companion on the scale of Conti *et al.* (1983). Distance in reasonable agreement with d (Cyg OB1) $= 1.82$ (Humphreys, 1978).

WR140:   Spectral type from Lamontagne *et al.* (1983b), M_v(WR) from Table IV, M_v(O4–5V) $= -5.5$ on the scale of Conti *et al.* (1983). This, with the assumption $(b-v)_0 = -0.33$, yields the distance.

WR151:   Spectral types and $\Delta M_v = 0$ from Massey and Conti (1981). M_v(O) from Conti *et al.* (1983) implies M_v(WR) $= -4.6$. Assumption $(b-v)_0 = -0.33$ yields a distance far beyond d(Cep OB2) $= 0.83$ kpc (Humphreys, 1978).

WR153:   Quadrupole system, for which Massey (1981) gives M_v(pair B) $= -4.9$, M_v(O$_A$) $= -4.8$ (i.e., O7.5–8V on the scale of Conti *et al.* (1983)). M_v(WR) from Table IV. This, with the assumption $(b-v)_0 = -0.33$ we obtain a distance, somewhat smaller than d(S132) $= 4.95$ (Crampton, 1971), but beyond Cep OB2 (Humphreys, 1978).

TABLE III

Single-line spectroscopic Wolf–Rayet binaries (SB1) with small and large mass functions (18)

WR	HD/Name	Spectral type	Period (days)	$f(\mathcal{M})$ ($\mathcal{M}_\odot$)	$\mathcal{M}_{WR}$ ($\mathcal{M}_\odot$)	Refs.	Ring nebula	d (kpc)	r (kpc)	z (pc)
128	187 282	WN4	3.85	0.003	28	1	S84	4.90	8.29	−324
6	50 896	WN5	3.763	0.015	12	2	S308	0.91	10.55	−160
43	97 950	WN6+abs	3.7720	0.15	2.6	3		7.0	9.87	−63
71	143 414	WN6	7.690	0.0074	17	4		7.08	6.08	−937
134	191 765	WN6	7.44	0.0055	20	5	S109	2.05	9.62	+55
136	192 163	WN6	4.5	0.00024	104	6	NGC6888	1.57	9.73	+67
138	193 077	WN6+abs	2.3238	0.0009	53	7		1.92	9.69	+37
148	197 406	WN7	4.3174	0.255	1.6	8,9		6.52	11.95	+735
16	86 161	WN8	10.73	0.00024	104	10		2.48	9.83	−110
40	96 548	WN8	4.1584	0.00052	70	11	RCW58	2.48	9.34	−209
123	177 230	WN8	1.7616	0.0019	36	12		6.06	5.68	−502
124	209BAC	WN8	2.3583	0.0005	71	13	S80	3.33	8.27	+192
103	164 270	WC9	1.7556	0.00146	41	14		2.84	7.16	−242
141	193 928	WN6	21.64	4.9	–	15		1.86	9.70	+3
12	CD−45°4482	WN7	23.9	5.5	–	16		5.75	11.95	−198:
22	92 740	WN7+abs	80.35	1.67	–	17,18	NGC3372	2.10	9.59	−31
155	214 419	WN7	1.64 (e)	5.08	–	19		3.77	11.58	−85
145	AS422	WN+WC	22	7.7:	–	20		–	–	–

Notes: (e) = eclipsing system. $\mathcal{M}_{WR}$ calculated by assuming $i = 60°$ and a mass of the unseen companion of $1.6\,\mathcal{M}_\odot$.

References:

(1) Antokhin *et al.* (1982b).
(2) Firmani *et al.* (1980).
(3) Moffat (1982).
(4) Isserstedt *et al.* (1983).
(5) Antokhin *et al.* (1982a).

(6) Koenigsberger *et al.* (1980).
(7) Lamontagne *et al.* (1982).
(8) Bracher (1979).
(9) Moffat and Seggewiss (1980a).
(10) Moffat and Niemela (1982).

(11) Moffat and Isserstedt (1980).
(12) Lamontagne *et al.* (1983a).
(13) Moffat *et al.* (1982).
(14) Isserstedt and Moffat (1981).
(15) Moffat and Seggewiss (1980b).

(16) Niemela (1982).
(17) Moffat and Seggewiss (1978).
(18) Conti *et al.* (1979).
(19) Leung *et al.* (1983).
(20) Pesch *et al.* (1960).

scenario for massive binary systems summarized by van den Heuvel (1976):

$$O_1 + O_2 \underset{w+o}{\rightarrow} O_1 + WR_2 \underset{s.n.}{\rightarrow} O_1 + c_2 \underset{w+o}{\rightarrow} WR_1 + c_2 \underset{s.n.}{\rightarrow} c_1 + c_2$$

(WR SB2)   (WR lmSB1)

(w + o: mass loss and Roche lobe overflow; s.n.: supernova explosion).

The second phase is visible as a WR SB2 system, the fourth phase is visible as a WR lmSB1 system, the last phase is visible as a double pulsar.

Maeder (1982) has pointed out that, next to the binary channel, there are various channels to produce WR stars from single massive stars, depending on mass loss and internal mixing, and that these other channels depend on galactic location, notably on metallicity. It is of great interest to know the relative importance of each of the channels producing WR stars. Therefore it is of importance to determine the exact percentage of WR binaries. As noted above, the percentage of observed binaries is 22%. It should be realized that many of the fainter WR stars listed as single have not yet been investigated for duplicity. In the next chapters we shall look for evidence for more WR binaries by investigating the relative distribution of the known WR binaries in the Galaxy as well as that of the 'single' WR stars.

3. WR Distribution in the Galactic Plane

Recently, Lundström and Stenholm (1983) have reexamined many of the faint WR stars and found new colours and spectral types. In addition Massey and Conti (1983) published some new spectral types. Lundström and Stenholm also re-evaluated the cases of WR stars in open clusters and associations and determined improved intrinsic parameters for these WR stars. We list them in Table IV, with some interpolations and extrapolations. Hidayat et al. (1984) used these values to calculate photometric distances for the 142 of the 159 galactic WR stars for which sufficient data are available.

The distribution of these WR stars projected on the galactic plane is given in Figure 1, where the filled symbols represent the known binaries. The galactic center is indicated at 8.7 kpc from the Sun (Oort and Plaut, 1975), but the heliocentric distances r, given in this paper, are calculated as if the galactic center is at 10 kpc from the Sun. This is because we would like to compare our statistics with other published star counts, e.g. by Maeder (1982), where usually the galactic center is put at 10 kpc.

We confirm that the distribution of the WR stars looks similar to that of the more massive ($\mathcal{M} > 40 \mathcal{M}_\odot$) O-type stars, as found by Conti et al. (1983).

Of the 47 WR stars (21 WN and 26 WC) with $d \leqslant 2.5$ kpc, 16 are known binaries (8 SB2, 2 mSB1 and 6 lmSB1), i.e., 34%, quite similar to the 36% known binaries of the 424 O-type stars in the same volume (Conti et al., 1983). This corresponds to a density projected on the galactic plane of $N(O) = 21.6$ kpc^{-2} and $N(WR) = 2.39$ kpc^{-2} within $d \leqslant 2.5$ kpc, where we can expect that the observations

TABLE IV

Adopted intrinsic parameters for the WR subclasses *

	$(b-v)_0$	M_v
WN2	−0.30	−2.0
WN3	−0.30	−3.6
WN4	−0.25	−4.0
WN4.5	−0.25	−4.3
WN5	−0.20	−4.8
WN6	−0.25	−5.3
WN7	−0.27	−6.4
WN8	−0.30	−5.8
WN9	−0.33	−6.0
WN10	−0.33	−6.0
WC4	−0.25	−2.7
WC5	−0.25	−3.9
WC6	−0.30	−4.4
WC7	−0.35	−4.7
WC8	−0.42	−4.8
WC8.5	−0.42	−5.0
WC9	−0.42	−5.1
WC10	−0.42	−5.0
WO1	−0.33	−2.6
WO2	−0.33	−2.6

* *Note*: These values are based on the recent work of
Lundström and Stenholm (1983) complemented with
assumed values by Conti *et al.* (1983).

are complete. For the binaries the densities are $N(O_{SB}) = 7.8 \, \text{kpc}^{-2}$ and
$N(WR_{SB}) = 0.82 \, \text{kpc}^{-2}$.

Although it seems attractive to compare statistics in a restricted volume around
the Sun, there is a danger here. In Figure 1 it appears immediately that in the
inner region of the Galaxy the density of the WR stars is larger than in the outer
region. This effect is already visible within $d \leqslant 2.5 \, \text{kpc}$: of the 47 WR stars only
6 are outside the solar circle. This galactic star density gradient forces us to do
star counts and statistics as a function of galactocentric distance, and we will
do this in the observable $\pm 90°$ sector of the Galaxy (see Figure 1).

From Figure 1 it also appears that the relative density of WR binaries is smaller
in the inner region, as noted earlier by Maeder (1982). We show this quantitatively
in Table V, where we list the relative distribution of WR subtypes as a function
of galactic distance. Following Maeder we also list the metallicity Z and LMC and
SMC values, and confirm, with our improved data and values, a strong upward
gradient of the WN/WC and WR_{SB}/WR_{total} number ratios with galactocentric distance,
and the correlation with a downward metallicity gradient. It could very well be
that of the various ways of reaching the WR phase, the channels depending on mass

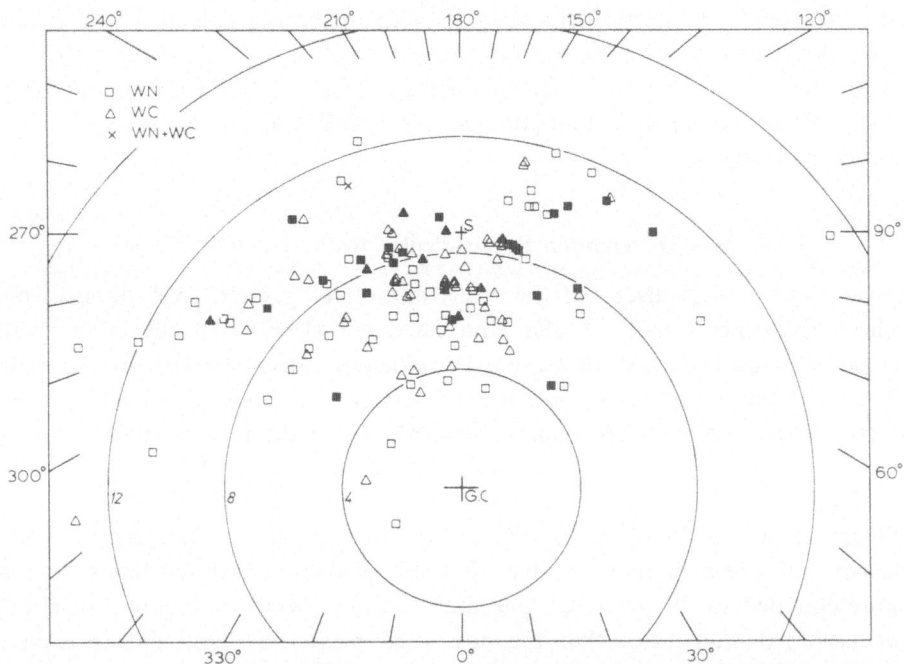

Fig. 1. The (d, l^{II})-distribution of the 141 WR stars for which sufficient photometric and spectroscopic data are available, projected on the galactic plane. The filled-in symbols are spectroscopic binaries.

loss and mixing depend on galactocentric distance (metallicity), while the binary channel is less or not dependent on location. So, if indeed the mass loss/mixing channels operate better in a relative high metallicity environment, this would explain that the relative number of WR binaries between 7 and 9 kpc is smaller (by a factor of 2 in our data) than that between 11 and 13 kpc from the galactic center.

An alternative explanation for more binaries (of all types) in the outer region of the Galaxy is offered by Zinnecker (1982), who argues for a decreasing value

TABLE V

WR heliocentric statistics

r (kpc)	Z	Numbers						Densities (kpc^{-2})	
		WR_{total}	WN	WC	$\dfrac{WN}{WC}$	WR_{SB}	$\dfrac{WR_{SB}}{WR_{total}}$	WR	WR_{SB}
7–9	0.03	49	25	23	1.09	8	0.16	1.95	0.32
9–11	0.02	47	14	23	1.04	18	0.38	1.50	0.57
11–13	0.01	20	15	4	3.75	6	0.30	0.53	0.16
LMC	0.01	100	82	18	4.6	50	0.50	0.15	0.07
SMC	0.002	8	7	1	7	8	1.00	0.05	0.05

of the local mean magnetic field strength with increasing galactocentric distance, in the context of binary formation in general.

In the next section we shall use the observed gradient in the relative number of *known* binaries, to indicate that among *single* WR stars many binaries may be awaiting discovery.

4. WR Distribution Perpendicular to the Galactic Plane

Hidayat *et al.* (1982) discussed the z-distribution of galactic WR stars, in order to check the suggestion by Moffat and Isserstedt (1980) that the large z-values of some WR stars are due to large kick-velocities caused by the first supernova explosion in the evolutionary scenario for massive binaries (given in Section 2), and thus that these stars are binaries (lmSB1). With the now available improved data in Hidayat *et al.* (1984), it is worthwhile to consider the z-distribution in more detail.

Figure 2a shows the (l^{II}, z)-distribution of the 142 galactic WR stars for which sufficient data are available. The fact that this distribution shows less scatter and is more confined to the galactic plane than the one shown in Figure 3 of Hidayat *et al.* (1982) demonstrates that the now used parameters and data are an improvement with respect to earlier values.

We expect Population I objects to be concentrated to the galactic plane within, let us say, $|z| \leqslant 200$ pc. However 28 WR stars (i.e., 20%) have values of $|z| > 200$ pc. In Figure 2b, where we show only the WR binaries, it appears that most of the SB2 system are well confined to the galactic plane, while among the SB1 systems many have large $|z|$-distances. Table VI gives a breakdown of the $|z|$-distribution vs. distance from the Sun. It appears that among the stars with $|z| > 200$ pc 21%

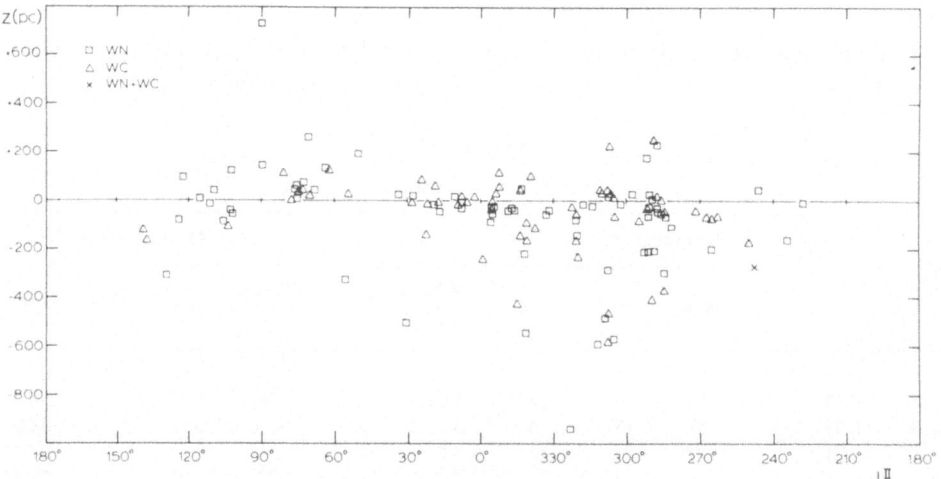

Fig. 2a.   The (z, l^{II})-distribution of WR stars.

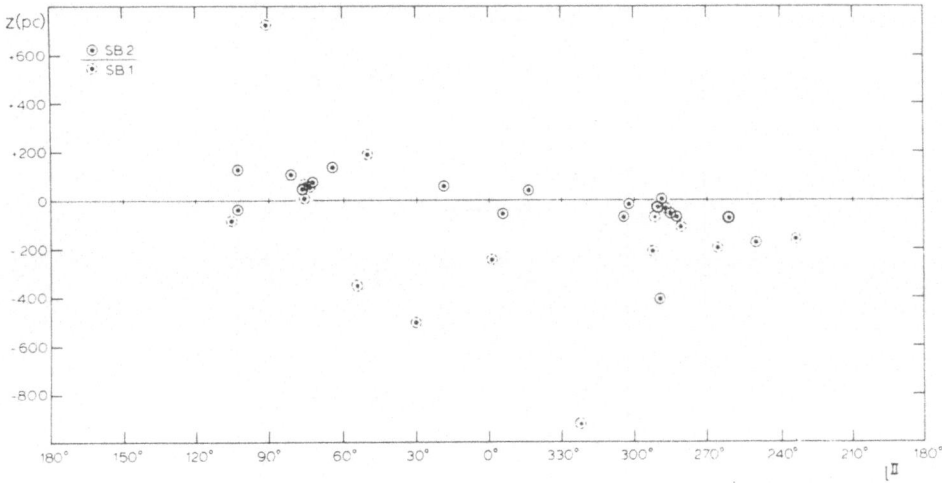

Fig. 2b.   The (z, l^{II})-distribution of known WR binaries.

are lmSB1, while among the stars with $|z| < 200$ pc only 6% are lmSB1. This enforces the probability that the WR stars with large $|z|$-values were ejected out of the plane with supernova-induced kick-velocities.

When we consider the average $\overline{|z|}$-values for single WR stars, for mSB1 and SB2 WR systems and for lmSB1 WR systems, as given in Table VI, we note:

$\overline{|z|}$ (SB2 + mSB1) = 86 pc, quite reasonable for Population I stars;

$\overline{|z|}$ ('single') = 109 pc, about 25% larger than the value for the binaries with massive components; and

$\overline{|z|}$ (lmSB1) = 279 pc, a very large value and explainable by large supernova-induced kick-velocities.

The fact that a difference exists between the $\overline{|z|}$-value for 'single' WR stars and that for WR binaries with massive companions can be explained by assuming that among the 'single' WR stars are many as yet undiscovered binaries (lmBS1's). We shall find evidence for this below. First we shall discuss observational selection effects.

In the breakdown in Table VI in distances from the Sun, we note that only beyond $d > 4$ kpc from the Sun large values of $\overline{|z|}$ (single WR) are reached. This is caused by two observational effects. Firstly, nearby stars have been better investigated for duplicity, so most lmSB1 systems are already known there. Secondly, as correctly pointed out by Garmany *et al.* (1982), at large distances from the Sun, stars in the galactic plane are obscured by interstellar matter, so we will find in general larger $|z|$-values there, as demonstrated in Figure 3. The first observational selection effect reduces $\overline{|z|}$ (single WR) at small distances from the Sun; the second observational selection effect increases $\overline{|z|}$ (single WR) at large distances from the Sun. If we take the average $|z|$ (single WR) over all distances, then these effects may balance out to a certain degree.

TABLE VI

z-distribution WR stars vs distance from the Sun

d (kpc)	Area (kpc²)	Numbers (|z| < 200 pc) (lmSB1)	Numbers (|z| > 200 pc) (lmSB1)	$\bar{|z|}$ (pc) (|z| < 200 pc) (N)	$\bar{|z|}$ (pc) ('single') (N)	(SB2+mSB1) (N)	(lmSB1) (N)	(all) (N)
0–1.8	10	21 (2)	0 (0)	46 (21)	31 (15)	70 (4)	113 (2)	46 (21)
1.8–2.5	10	25 (3)	1 (1)	48 (25)	39 (16)	64 (6)	103 (4)	55 (26)
2.5–3.1	10	14 (0)	3 (1)	36 (14)	60 (15)	58 (1)	242 (1)	71 (17)
3.1–3.6	10	14 (1)	0 (0)	72 (14)	69 (10)	42 (3)	192 (1)	72 (14)
3.6–4.0	10	7 (0)	0 (0)	50 (7)	50 (5)	50 (2)	– (0)	50 (7)
4.0–4.4	10	1 (0)	2 (0)	145 (1)	242 (3)	– (0)	– (0)	242 (3)
4.4–5	18	13 (0)	1 (1)	64 (13)	63 (11)	68 (2)	324 (1)	82 (14)
5–10	236	16 (1)	16 (3)	83 (16)	217 (25)	244 (3)	559 (4)	262 (32)
10–17	594	2 (0)	5 (0)	90 (2)	266 (7)	– (0)	– (0)	266 (7)
0–17	908	113 (7)	28 (6)	58 (113)	109 (107)	86 (21)	279 (13)	121 (141)

Note: mSB1: SB1 with massive unseen companion; lmSB1: SB1 with low mass unseen companion.

In the region $d \leqslant 4$ kpc we have:

$\overline{|z|}$ (single WR) = 48 pc, so these may be the real singles;

$\overline{|z|}$ (SB2+mSB1) = 59 pc; and

$\overline{|z|}$ (lmSB1) = 134 pc.

The latter value is in reasonable agreement with $\overline{|z|}$ (OB runaway stars) = 150 pc (Moffat and Isserstedt, 1980).

Thus, already within $d \leqslant 4$ kpc there is a definite case of difference in $\overline{|z|}$-distances between single WR stars and lmSB1 WR systems, and this effect is persistent even when the data are subdivided by distance within 4 kpc from the Sun as shown in Table VI, contrary to statements by Garmany *et al.* (1982).

Since, as mentioned above, the percentage of known lmSB1's at $|z| > 200$ pc is large (21%) and since the existence of these large $|z|$-values can be best explained by assuming that these stars arrived there after suffering large supernova-induced kick-velocities, it is encouraged to investigate all WR stars with large $|z|$-values for duplicity. In addition, proper motion studies with the HIPPARCOS satellite and radial velocity studies of these stars could give conclusive clues about the supernova dynamics. The stars with $|z| > 200$ pc are listed in Table VII.

Additional evidence can be given for the large probability that among the WR stars with large $|z|$-values more binaries with low-mass companions may be present.

In Section 3, Table V, we have confirmed the finding of Maeder (1982) that known WR binaries are relatively more frequent in the outer than in the inner galactic regions. When we now consider in Table VIII only the 'single' WR stars in their

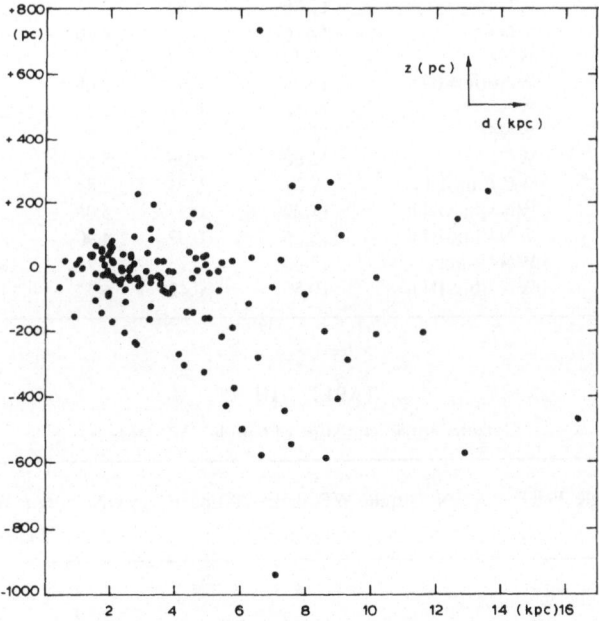

Fig. 3.   The (z, d)-distribution of WR stars.

galactocentric distribution, then we note a similar gradient: the fraction of 'single' WR stars with large $|z|$-values is increasing with galactocentric distance. Therefore, among the 'single' WR stars with large $|z|$-values, *and* thus also among the 'single' WR with smaller $|z|$-values (because supernova-induced kick-velocities will have no directional preference), many WR binaries (lmSB1) may yet be waiting to be discovered.

TABLE VII

WR stars with $|z| > 200$ pc (from Hidayat *et al.*, 1984)

WR	Name	Sp. type	v	$b-v$	d (kpc)	r (kpc)	z (pc)
3	HD9974	WN3+abs	10.79	0.00	4.29	13.14	−309
8	HD62910	WN6+WC4	10.56	0.43	4.12	12.21	−272
17	HD88500	WC5	11.11	0.04	5.81	10.24	−374
20	BS1	WN4.5	14.60	0.74	9.29	11.82	−298
26	MS1	WN5+WC	14.64	0.72	13.60	14.38	+230
29	MS3	WN7	12.65	0.64	11.58	12.66	−204
30	HD94305	WC6+O6–8V	11.73	0.27	8.99	11.00	−410
33	HD95435	WC5	12.34	0.20	7.57	10.45	+250
34	LS5	WN4.5	14.50	0.76	8.55	10.70	−207
35	MS6	WN6	13.83	0.75	10.14	11.56	−209
40	HD96548	WN8(lmSB1)	7.85	0.11	2.48	9.34	−209
49	LSS2979	WN5	13.87	0.56	12.88	10.82	−571
52	HD115473	WC5	9.98	0.15	2.81	8.63	+223
54	LSS3111	WN4	12.99	0.46	6.54	7.97	−285
56	LS8	WC6	13.97	0.26	16.40	13.01	−466
57	HD119078	WC7	10.11	−0.18	6.65	7.91	−582
58	LSS3162	WN4	13.08	0.42	7.36	7.87	−448
61	LSS3208	WN4.5	12.56	0.28	8.66	7.79	−590
69	HD136488	WC9	9.43	0.14	2.80	8.08	−235
71	HD143414	WN6(lmSB1)	10.22	0.06	7.08	6.08	−937
82	LS11	WN8	12.42	0.81	5.42	5.13	−219
83	He3–1344	WN6	12.79	0.65	7.59	3.69	−544
92	HD157451	WC9	10.60	0.06	5.58	4.80	−430
103	HD164270	WC9(lmSB1)	9.01	0.03	2.84	7.16	−242
123	HD177230	WN8(lmSB1)	11.27	0.47	6.06	5.68	−502
128	HD187282	WN4(lmSB1)	10.56	0.02	4.90	8.29	−324
131	IC14–52	WN7+abs	12.40:	0.73:	8.71:	10.77:	+260:
148	HD197406	WN7(lmSB1)	10.50	0.42	6.52	11.95	+735

TABLE VIII

Galactocentric statistics of 'single' WR stars

r (kpc)	N ('single' WR)	N ('single' WR ($|z| > 200$ pc))	N ('single' WR ($|z| > 200$ pc)) / N ('single' WR)
7–9	41	7	0.17
9–11	29	6	0.21
11–13	14	4	0.29

5. Conclusions

Because various channels to produce WR stars may be operational (Maeder, 1982), it is important to assess for what fraction the binary channel is responsible.

The fraction of known WR binaries is 22%. This fraction is 34% in the solar neighbourhood ($d < 2.5$ kpc).

The distribution of WR stars projected on the galactic plane shows that in the inner region of the Galaxy the binary channel is relatively less important. This implies that other channels (mass loss, mixing) which require the conditions of the inner region (larger metallicity) are more active there.

The presence of many lmSB1 WR systems at large distances from the galactic plane indicates that these binaries have received large supernova-induced kick-velocities; this is consistent with the evolutionary scenario summarized by van den Heuvel (1976) for massive binaries.

The presence of many 'single' WR stars at large z-distances indicates that they also may have a binary origin. That these 'single' WR stars with large z-values are in addition relatively more frequent at larger galactocentric distances, like the known WR binaries, improves their chances for duplicity. The total WR binary frequency in the Galaxy could be well above 50%.

Acknowledgement

One of us (KAvdH) greatfully acknowledges financial support for this study from the Dutch Ministery of Education and Science and from the Leids Kerkhoven-Bosscha Fund, as well as warm hospitality by the Institut Teknologi Bandung, to work at the Observatorium Bosscha.

References

Abbott, D. C.: 1982, *Astrophys. J.* **263**, 723.
Abt, H. A., Landolt, A. U., Levy, S. G., and Mochnacki, S.: 1976, *Astron. J.* **81**, 541.
Alter, G., Balász, B. and Ruprecht, J.: 1970, *Catalogue of Star Clusters and Associations*, Prague.
Antokhin, I. I., Aslanov, A. A., and Cherepashchuk, A. M.: 1982a, *Pis'ma Astron. Zh.* **8**, 290 (*Soviet Astron. Letters* **8**, 156).
Antokhin, I. I., Aslanov, A. A., and Cherepashchuk, A. M.: 1982b, *Pis'ma Astron. Zh.* **8**, 734 (*Soviet Astron. Letters* **8**, 395).
Barlow, M. J. and Hummer, D. G.: 1982, in C. de Loore and A. J. Willis (eds.), 'Wolf–Rayet Stars, Observations, Physics, and Evolution', *IAU Symp.* **99**, 387.
Bracher, K.: 1979, *Publ. Astron. Soc. Pacific* **91**, 827.
Conti, P. S.: 1982, in C. de Loore and A. J. Willis (eds.), 'Wolf–Rayet Stars, Observations, Physics, and Evolution', *IAU Symp.* **99**, 3.
Conti, P. S. and Smith, L. J.: 1972, *Astrophys. J.* **172**, 623.
Conti, P. S., Niemela, V. S., and Walborn, N. R.: 1979, *Astrophys. J.* **228**, 206.
Conti, P. S., Garmany, C. D., de Loore, C., and Vanbeveren, D.: 1983, *Astrophys. J.*, (in press).
Crampton, D.: 1971, *Monthly Notices Roy. Astron. Soc.* **153**, 303.
Davis, A. B., Moffat, A. F. J., and Niemela, V. S.: 1981, *Astrophys. J.* **244**, 528.
Firmani, C., Koenigsberger, G., and Bisiacchi, G. F.: 1980, *Astrophys. J.* **239**, 607.
Fraquelli, D. A. and Horn, J.: 1982, in preparation.

Ganesh, V. S., Bappu, M. K. V., and Natajaran, V.: 1967, *Kodaikanal Observatory Bull.*, *Ser.* **A184.**
Garmany, C. D., Conti, P. S., and Chiosi, C.: 1982, *Astrophys. J.* **263,** 777.
Hidayat, B., Supelli, K., and van der Hucht, K. A.: 1982, in C. de Loore and A. J. Willis (eds.), 'Wolf–Rayet Stars, Observations, Physics, and Evolution', *IAU Symp.* **99,** 27.
Hidayat, B., van der Hucht, K. A., Suppelli, K. R., and Admiranto, A. G.: 1984, *Astron. Astrophys.* (to be submitted).
Humphreys, R. M.: 1978, *Astrophys. J. Suppl.* **38,** 309.
Isserstedt, J. and Moffat, A. F. J.: 1981, *Astron. Astrophys.* **96,** 133.
Isserstedt, J., Moffat, A. F. J., and Niemela, V. S.: 1983, in preparation.
Koenigsberger, G., Firmany, C., and Bisiacchi, G. F.: 1980, *Rev. Mexicana Astron. Astrof.* **5,** 45.
Lamontagne, R., Moffat, A. F. J., Koenigsberger, G., and Seggewiss, W.: 1982, *Astrophys. J.* **253,** 230.
Lamontagne, R., Moffat, A. F. J., and Seggewiss, W.: 1983a, *Astrophys. J.* **269,** 596.
Lamontagne, R., Moffat, A. F. J., and Seggewiss, W.: 1983b, *Astrophys. J.*, in press.
Leung, K.-C., Moffat, A. F. J., and Seggewiss, W.: 1983, *Astrophys. J.* **265,** 961.
Lundström, I. and Stenholm, B.: 1983, in preparation.
Maeder, A.: 1982, *Astron. Astrophys.* **105,** 149.
Maeder, A. and Lequeux, J.: 1982, *Astron. Astrophys.* **114,** 409.
Massey, P.: 1981a, *Astrophys. J.* **244,** 157.
Massey, P.: 1981b, *Astrophys. J.* **246,** 153.
Massey, P.: 1982, in C. de Loore and A. J. Willis (eds.), 'Wolf–Rayet Stars, Observations, Physics, and Evolution', *IAU Symp.* **99,** 251.
Massey, P. and Conti, P. S.: 1981, *Astrophys. J.* **244,** 169.
Massey, P. and Conti, P. S.: 1983, *Publ. Astron. Soc. Pacific* **95,** 440.
Massey, P. and Niemela, V. S.: 1981, *Astrophys. J.* **245,** 195.
McCarthy, C. D. and Miller, E. W.: 1974, *Astron. J.* **79,** 1396.
Moffat, A. F. J.: 1974, *Astron. Astrophys.* **34,** 29.
Moffat, A. F. J.: 1982, in C. de Loore and A. J. Willis (eds.), 'Wolf–Rayet Stars, Observations, Physics, and Evolution', *IAU Symp.* **99,** 263.
Moffat, A. F. J. and Isserstedt, J.: 1980, *Astron. Astrophys.* **91,** 147.
Moffat, A. F. J. and Niemela, V. S.: 1982, *Astron. Astrophys.* **108,** 326.
Moffat, A. F. J. and Seggewiss, W.: 1977, *Astron. Astrophys.* **54,** 607.
Moffat, A. F. J. and Seggewiss, W.: 1978, *Astron. Astrophys.* **70,** 69.
Moffat, A. F. J. and Seggewiss, W.: 1980a, in M. J. Plavec, D. M. Popper, and R. K. Ulrich (eds.), 'Close Binary Stars', *IAU Symp.* **88,** 181.
Moffat, A. F. J. and Seggewiss, W.: 1980b, *Astron. Astrophys.* **86,** 87.
Moffat, A. F. J., Lamontagne, R., and Seggewiss, W.: 1982, *Astron. Astrophys.* **114,** 135.
Münch, G.: 1950, *Astrophys. J.* **112,** 266.
Niemela, V. S.: 1980, in M. J. Plavec, D. M. Popper, and R. K. Ulrich (eds.), 'Close Binary Stars', *IAU Symp.* **88,** 177.
Niemela, V. S.: 1982, in C. de Loore and A. J. Willis (eds.), 'Wolf–Rayet Stars, Observations, Physics, and Evolution', *IAU Symp.* **99,** 299.
Niemela, V. S. and Moffat, A. F. J.: 1982, *Astrophys. J.* **259,** 213.
Niemela, V. S. and Sahade, J.: 1980, *Astrophys. J.* **238,** 244.
Niemela, V. S., Conti, P. S., and Massey, P.: 1980, *Astrophys. J.* **241,** 1050.
Niemela, V. S., Méndez, R., and Moffat, A. F. J.: 1983, *Astrophys. J.* **272,** 190.
Oort, J. H. and Plaut, L.: 1975, *Astron. Astrophys.* **41,** 71.
Pesch, P., Hiltner, W. A., and Brandt, J. C.: 1960, *Astrophys. J.* **132,** 513.
Seggewiss, W.: 1974, *Astron. Astrophys.* **31,** 211.
Smith, L. J. and Willis, A. J.: 1982, *Monthly Notices Roy. Astron. Soc.* **201,** 451.
Turner, D. G.: 1980, *Astrophys. J.* **235,** 146.
van den Heuvel, E. P. J.: 1976, in P. Eggleton, S. Mitton, and J. Whelan (eds.), 'Structure and Evolution of Close Binary Systems', *IAU Symp.* **73,** 35.
van der Hucht, K. A.: 1982, in S. D'Odorico, D. Baade, and K. Kjär (eds.), 'The Most Massive Stars', *Proc. ESO Workshop*, 23–25 November, 1981, Garching, p. 157.
van der Hucht, K. A., Conti, P. S., Lundström, I., and Stenholm, B.: 1981, *Space Sci. Rev.* **28,** 227.
Zinnecker, H.: 1982, in Z. Kopal and J. Rahe (eds.), 'Binary and Multiple Stars as Tracers of Stellar Evolution', *IAU Colloq.* **69,** 115.

OBSERVATIONS AND MODELS OF SOME NEGLECTED
SOUTHERN ECLIPSING BINARIES*

DAVID A. H. BUCKLEY

*Mount Stromlo and Siding Spring Observatories, Research School of Physical Sciences,
Australian National University, Australia*

(Received 6 June, 1983)

Abstract. Results of a photometric investigation of some photoelectrically neglected, southern eclipsing binaries, are presented for GW Car, X Car, and RS Sct. Light curve solutions obtained by the Wilson–Devinney and Wood synthetic light curve techniques are described.

1. Introduction

Photometric observations in the BVRI system were made of a selection of photo-electrically neglected, southern eclipsing binaries (Buckley, 1982). Of the six systems observed, reasonably complete light curves were derived for three; GW Carinae, X Carinae, and RS Scuti.

Attempts at determining parameters describing these binaries were made by solving light curves using the traditional approach of Russell and Merrill (1952). Results of these solutions were used to initialize parameters in the synthetic light curve codes of Wilson and Devinney (1971) and Wood (1971, 1972).

An overall consistent solution is presented for the early type (B1) short period system GW Carinae, as derived by both synthesis techniques. Preliminary solutions are also presented for X Car and RS Sct using the Wilson–Devinney program.

This paper summarises results obtained during an MSc thesis project at the University of Canterbury, New Zealand.

2. Observations

Observations were made using a cooled (-30 °C) RCA C31034A photomultiplier on a 61 cm Cassegrain reflector at Mount John University Observatory. Differential measurements were transformed into the standard Kron–Cousins BVRI system (Cousins, 1976). An individual observation consisted of the mean of three to five ten second integrations for each filter. Transformation coefficients were derived for most observing runs, while primary extinction was evaluated nightly. Two comparison stars for each program star gave a check against variations in the principal comparison star.

* Paper presented at the Lembang-Bamberg IAU Colloquium No. 80 on 'Double Stars: Physical Properties and Generic Relations', held at Bandung, Indonesia, 3–7 June, 1983.

Astrophysics and Space Science **99** (1984) 191–197. 0004–640X/84/0991–0191$01.05.

3. Results

The reduced results in terms of H. J. D. vs differential magnitude are given in complete form elsewhere (Buckley, 1982, 1983). Reduced colors for the three complete systems were corrected for interstellar reddening. This was achieved using complementary photoelectric color indices found in the literature.

Times of minimum light were deduced for all the systems using the method of Kwee and van Woerden (1956). The O–C residuals were calculated using ephemerides from Kukarkin *et al.* (1969). These were used to determine periods based upon linear ephemerides and then to phase the observations. The results appear in Table I.

TABLE I

Times of minima and residuals

Star	T (Hel. JD–2 440 000)	Type	E	O–C (days)	P (days)
GW Car	4697.0326 ± 0.0002	I	12 761	– 0.0402	1.128 907 (9)
X Car	2785.1751	I	12 865	– 0.0187	1.082 629 (6)
	4343.0797 ± 0.0010	I	14 304	– 0.0201	1.082 629 (6)
	4612.116 ± 0.001	II	14 552.5	– 0.0176	1.082 629 (8)
RS Sct	4437.1674 ± 0.0010	I	29 542	0.0370	0.664 238 (3)
	4779.9125 ± 0.0005	I	30 058	0.0368	0.664 238 (4)

3.1. GW CARINAE

A photographic light curve was obtained and analysed by O'Connell (1956). His conclusion was that primary eclipse was a total occultation. He further surmised that the system may exhibit apsidal motion by virtue of a displaced secondary eclipse. Published photoelectric photometry includes *UBV* colors by Klare and Nekel (1977) and Garrison *et al.* (1983). Color indices in the *uvbyβ* system have been published by Eggen (1978) for several phase angles, while Wolf's (1982) unpublished measurements are near both eclipses. Good agreement was achieved between results of this study and all other available data (Buckley, 1982, 1983). The values

$$(b - y)_0 = - 0.120 \quad \text{and} \quad (B - V)_0 = - 0.271$$

were derived using Eggen's (1978) color excesses. A B1 spectral type is consistent with these indices and this is in agreement with Garrison *et al.* (1977). An average primary effective temperature of

$$T_{\text{eff}, 1} = 25 300 \pm 100 \text{ K}$$

was adopted from Bessell's (1979) calibration of T_{eff} vs $(b - y)_0$.

3.2. X CARINAE

Published work on X Car includes a determination of the system's parameters by Roberts (1905) from his observations. Sahade's (1952) spectroscopic analysis of rather blended lines resulted in estimates of the system's dimensions. Photoelectric color indices in the $uvby\beta$ system have been obtained by Eggen (1978) and Wolf (1982).

Eggen and Wolf obtain a $(b - y)_0$ value of 0.020 and 0.026 respectively, implying a spectral type between A1 and A2. In this study a $(B - V)_0$ value of 0.06 $\pm$ 0.01 is derived using the same color excess as Eggen. This gives a somewhat later spectral type than A2.

The narrow band indices were adopted for determining temperatures. Eggen's (1978) $(c_1)_0$ value of 1.052 was used in Relyea and Kurucz's (1978) $(c_1)_0$ vs $(b - y)_0$ color-color diagram to estimate both temperature and gravity. These indices apply to primary eclipse and, hence, the secondary component by definition. However, the indices are all essentially equal for both eclipses in this system where the components are of similar temperatures. Bessell's (1979) calibration gave a slightly hotter temperature. The values were, T_{eff}: 9100 $\pm$ 100 K log$g$: 4.1 + 0.1 dex   Relya and Kurucz ,

T_{eff}: 9300 $\pm$ 100 K   Bessell .

Eclipse times are presented in Table I together with Kvíz's (1983) time of minimum.

3.3. RS Scuti

Two visual light curves exist for RS Sct, an early one by Ichinoe (1915) and another by Tsesevich (1954). Numerous visual and photographic times of minima dating back to 1909 have been made. Piotrowski's (1936, 1949) compilation resulted in his conclusion that the scatter in the O–C values was larger than that which could be attributed to observational error. Kwee (1958) found that no linear ephemeris fitted the observations well.

In Table I the times of minima for RS Sct derived in this study are used to calculate the O–C values and, hence, revise the linear period. The initial ephemeris is again taken from Kukarkin et al. (1969). Piotrowski's (1936, 1949) compilation of timings were combined with timings found in IBVS publications and Koch's (1982) compilation to produce a residual plot against epoch. This diagram clearly shows what appears to be a non-linear secular change in period. Furthermore, there is a good deal of scatter, which a to a varying degree must be associated with systematic errors in visual timings.

The $(B - V)_0, (V - R)_0$, and $(V - I)_0$ color indices derived in this study were analysed together with Wolf's (1982) unpublished $uvby\beta$ colors. The β-index at secondary eclipse indicates an F-type star is implied for the primary, as is reported in Wood et al. (1980). A recently obtained spectrum of RS Sct gave calcium H and K and hydrogen line strengths appropriate to an F4 to F5 star (Wolf, 1982).

Reddening in $(b - y)$ and $(B - V)$ was calculated using the method prescribed in Crawford and Barnes (1974) for A–F stars. The following dereddened indices were derived $(b - y)_0 = 0.245, (B - V)_0 = 0.386, (c_1)_0 = 0.569$. Hence, the temperature and gravities were estimated in the same manner as X Car, giving

$T_{eff, 1}$: 6900 $\pm$ 100 K log$g$: 4.0 + 0.1   Relyea and Kurucz ,

$T_{eff, 1}$: 6840 $\pm$ 80 K   Bessell .

4. Synthetic Light Curve Solutions

Both the synthesis programs of Wilson and Devinney (1971) and Wood (1971, 1972, 1978) were used for solving light curves of GW Car. Lack of time precluded a similar analysis of X Car and RS Sct with both codes. Only the Wilson–Devinney (W–D) code was used on these systems. The choice of the synthesis techniques was dictated, in part, by reasons of expediency and utility.

Descriptions of the two models and solutions procedures will not be given here as they are adequately covered in the literature.

Initialization of parameters was achieved using results from Russell–Merrill analyses (Buckley, 1982, 1983).

4.1. GW CARINAE

Simultaneous solutions of the B, V, and R light curves were searched for using modes of detachment from contact, semi-detached to detached in the W–D code. The chief variable parameter set, used in the differential corrector, was inclination (i), secondary polar temperature (T_2), normalized gravitational surface potentials (Ω), mass ratio (q), and luminosities ($L_{1, 2}$).

Runs in all modes resulted in the secondary becoming semi-detached and, hence, further work was conducted in this mode. After initial convergence, some second-order parameters; limb darkening ($x_{1, 2}$) and primary albedo (A_1), were added to the variable set. Improvements to the solution resulted, and this is presented in Table II. Investigation of asynchronous rotation, by systematically changing the ratio of angular velocities (F), did not improve the solution.

Following the derivation of the W–D solution, Wood's (1972, 1978) program WINK was used to find solutions to the B and V light curves. The variable parameter set was taken as i, T_2, ratio of radii (k_v), unperturbed radius of primary (r_1), q, and the scaling factor m_{quad}. Other parameters were initialized from the W–D solution or theoretical values. Since WINK requires gravities for interpolation in the model atmosphere grid, a mass estimate for the system was made from Eggen's (1978) $uvby\beta$ photometry of $15 M_\odot$. This was used with the W–D mass ratio to give radii using Kepler's law, and thence gravities.

Both the W–D and WINK solutions show good consistency. The latter solutions are detached, whereas the W–D solution gives the smaller cooler secondary as filling its Roche-lobe. The WINK solution for V, however, is very close to this model and a 0.06 decrease in q would bring the secondary in contact with its inner critical potential surface.

4.2. X CARINAE

The W–D solution for the B and V light curves was performed in a similar manner to the GW Car analysis. A detached model resulted with a mass ratio of unity, consistent with Sahade's (1952) estimate.

TABLE II

Photometric solutions

	RS Sct	X Car	GW Car			
	W–D solutions				WINK	
	BVR	BV	BVR		B	V
i	87°32 ± 0.56	87°60 ± 0.21	89°93 ± 0.47	i	88°16 ± 0.46	90°0 ± 1.4
k	0.788	0.951	0.852	k_v	0.846 ± 0.004	0.877 + 0.009
r_1	0.381	0.370	0.392	r_1	0.389 ± 0.002	0.392 + 0.002
$T_1(p)$	7000 ± 150 K	9500 ± 100 K	26470 ± 100 K	$T_1(e)$	25300 ± 100 K	25300 ± 100 K
$T_2(p)$	4749 ± 260 K	9500 ± 100 K	21160 ± 170 K	$T_2(e)$	20240 ± 100 K	20230 ± 100 K
Ω_1	3.501 ± 0.011	3.820 ± 0.011	3.214 ± 0.010	$L_1(t)$	0.779	0.767
Ω_2	3.734 ± 0.019	3.944 ± 0.014	3.0101	$L_2(t)$	0.221	0.233
q	0.774 ± 0.007	1.000 ± 0.005	1.0	$L_1(a)$	0.691	0.661
g_1	1.0	1.0	1.0	$L_2(a)$	0.309	0.339
g_2	0.61 ± 0.44	1.0	1.0	q	0.812	0.722
A_1	1.0	1.0	0.6 ± 0.2	a	0.4280	0.4280
A_2	0.5	1.0	1.0	b	0.4000	0.4033
r_1				c	0.3793	0.3831
Pole	0.3605	0.3476	0.3732	a	0.3555	0.3792
Side	0.3768	0.3639	0.3898	b	0.3342	0.3503
Back	0.4005	0.3910	0.4092	c	0.3213	0.3367
Point	0.4323	0.4349	0.4329	u_1	0.319	0.354
r_2				u_2	0.304	0.330
Pole	0.2887	0.3338	0.3101	β_1	0.25	0.25
Side	0.2970	0.3474	0.3237	β_2	0.25	0.25
Back	0.3120	0.3692	0.3563	w_1	0.6 ± 0.2	0.6 ± 0.2
Point	0.3228	0.3984	0.4236	w_2	1.00	1.00
$L_1(B)$	0.933	0.517	0.689	rms	0″0079	0″0077
$L_1(V)$	0.896	0.519	0.677	ε	0.01%	0.01%
$L_1(R)$	0.874	–	0.676			
$x_1(B)$	0.76	0.49	0.319 ± 0.001			
$x_1(V)$	0.60	0.49	0.354 ± 0.033			
$x_1(R)$	0.49	–	0.449 ± 0.030			
$x_2(B)$	0.86	0.49	0.304 ± 0.120			
$x_2(V)$	0.70	0.49	0.330 ± 0.093			
$x_2(R)$	0.60	–	0.443 ± 0.115			
Σ Res²	0.085	0.032	0.017			

Note: The second-order parameters (x, g, A and u, β, w) without errors are theoretical values from Grygar *et al.* (1972), Von Zeipel (1924), and Rucinski (1969).
t = total, a = apparent, p = polar, e = equatorial.

4.3. RS SCUTI

A simultaneous solution of B, V, and R light curves resulted in a detached configuration.

5. Conclusions

The good agreement between GW Car's solutions is encouraging. Clearly the lack of any spectroscopic evidence severely limits how far we can go in building a model. If the radii of the components are well determined, then a limit can be put on the mass ratio, namely

$0.57 < q < 1.15$. Main-Sequence configurations are admitted by both WINK solutions, within uncertainties. For the W–D solution, though, the secondary would appear to be some three times more luminous than a Main-Sequence object. An assumption in this argument is that the primary obeys the upper Main-Sequence mass-luminosity relation. An iterative procedure calculating the radius from Kepler's law and mass from the luminosity resulted in a primary mass of $9 \pm 1 M_\odot$.

If the semi-detached interpretation is correct, then GW Car may join the group of hot semi-detached binaries where the secondary is semi-detached. Similar systems include SX Aur (Chambliss and Leung, 1979) and V Pup (Popper, 1980).

The solutions for RS Sct and X Car are not as yet considered complete. Solutions are likely to be improved and recent work using both WINK and W–D on RS Sct indicates a tendency towards a semi-detached model.

Spectroscopic work is clearly a high priority for these systems. Garrison *et al.* (1977) reports GW Car as a double lined spectroscopic binary, and therefore appears ripe for an independent determination of its dimensions.

Acknowledgements

Thanks are due to Drs R. E. Wilson and P. B. Etzel for providing codes. Discussions with Drs J. B. Hearnshaw and R. H. Koch are gratefully acknowledged. Dr G. W. Wolf kindly provided some of his unpublished data.

References

Bessell, M. S.: 1979, *Publ. Astron. Soc. Pacific* **91**, 589.
Buckley, D. A. H.: 1982, 'Observations and Interpretations of Some Neglected Eclipsing Binaries', MSc Thesis, University of Canterbury, New Zealand.
Buckley, D. A. H.: 1983, (in press).
Chambliss, C. R. and Leung, K.-C.: 1979, *Astrophys. J.* **228**, 828.
Cousins, A. W. J.: 1976, *Monthly Notices Roy. Astron Soc.* **81**, 25.
Crawford, D. L. and Barnes, J. V.: 1974, *Astron. J.* **79**, 687.
Eggen, O. J.: 1978, *Astron. J.* **83**, 288.
Garrison, R. F., Hiltner, W. A., and Schild, R. E.: 1977, *Astrophys. J. Suppl.* **35**, 111.
Garrison, R. F., Hiltner, W. A., and Schild, R. F.: 1983, *Astrophys. J. Suppl.* **52**, 1.
Grygar, J., Cooper, M. L., and Jurkevich, I.: 1972, *Bull. Astron. Inst. Czech.* **23**, 147.
Ichinoe, I.: 1915, *Astron. Nachr.* 4403.
Klare, G. and Nekel, Th.: 1977, *Astron. Astrophys. Suppl.* **27**, 215.
Koch, R. H.: 1982, private communication.
Kukarkin, V. B., Khopolov, P. N., Efremov, Yu. N., Kukarkina, N. P., Kurochkin, N. E., Medvedna, G. I., Perova, N. B., Fedorovich, V. P., and Frolov, M. S.: 1969, *Gen. Cat. Var. Stars*, 3rd ed.
Kvíz, Z.: 1983, private communication.
Kwee, K. K.: 1958, *Bull. Astron. Inst. Neth.* **14**, 485.
Kwee, K. K. and van Woerden, H.: 1956, *Bull. Astron. Inst. Neth.* **12**, 464.
O'Connell, Fr. D. J.: 1956, *Richerche Astronomiche, Specola Astronomica Vaticana* **3**, 313.
Piotrowski, S.: 1936, *Acta Astron.* C3, 26.
Piotrowski, S.: 1949, *Acta Astron.* C4, 120.
Popper, D. M.: 1980, *Ann. Rev. Astron. Astrophys.* **18**, 115.
Relyea, L. J. and Kurucz, R. L.: 1978, *Astrophys. J. Suppl.* **37**, 45.
Roberts: 1905, Report of the British Assoc. for the Advancement of Science, p. 253.

Rucinski, S. M.: 1969, *Acta Astron.* **19**, 125.
Russell, H. N. and Merrill, J. E.: 1952, Contributions from the Princeton University Observatory, No. 26.
Sahade, J.: 1952, *Astrophys. J.* **115**, 134.
Tsesevich, V. P.: 1954, *Odessa Izvestia* **4**, 353.
Von Zeipel, H.: 1924, *Monthly Notices Roy. Astron. Soc.* **84**, 665.
Wilson, R. E. and Devinney, E. J.: 1971, *Astrophys. J.* **166**, 605.
Wolf, G. W.: 1982, private communication.
Wood, D. B.: 1971, *Astron. J.* **76**, 701.
Wood, D. B.: 1972, NASA Publ. X–110–72–473.
Wood, D. B.: 1978, WINK Status Report, No. 9.
Wood, F. B., Oliver, J. P., Florkowski, D. R., and Koch, R. H.: 1980, *A Finding List for Observers of Interacting Binaries*, 5th ed., Publ. University of Pennsylvania, Astronomical Series, Vol. 12.

OBSERVATIONS OF BINARIES AND EVOLUTIONARY IMPLICATIONS*

C. DE LOORE

Astrophysical Institute, Free University of Brussels, VUB, Belgium
and
University of Antwerp, RUCA, Belgium

(Received 1 August, 1983)

Abstract. Comparison of the characteristics of groups of stars in various evolutionary phases and the study of individual systems allow to make estimates of the parameters governing mass loss and mass transfer. Observations enable us in a few cases to determine geometric models for binaries during or after the mass transfer phase (disks, rings, common envelopes, symbiotics, interacting binaries, compact components).

From spectra taken at different phases, radial velocity curves can be derived and masses and radii can be determined. In special cases spectra in different spectral ranges (visual, UV, X-ray) are required for the determination of the radial velocities of the two components (for X-ray binaries, for systems with hot and cool components). Information on parameters related to the mass transfer process enables us to consider non conservative evolution – i.e. the computation of evolutionary sequences with the assumption that mass and angular momentum not only are transferred from one of the components towards the other one, but that also mass and angular momentum can leave the system. Careful and detailed analysis of the observations allows in certain cases to determine the parameters governing this mass and angular momentum loss, and for contact phases, to determine the degree of contact.

1. Introduction

The values of the conservative evolution and the influence of the various parameters was examined by De Grève *et al.* (1978). Non-conservative evolutionary computations were performed by the Brussels group (Vanbeveren *et al.*, 1979) for massive systems, taking into consideration stellar wind mass loss and mass and angular momentum loss during the Roche lobe overflow stages. In all these cases the fractional mass and angular momentum losses were determined by parametrization of general character, and used for groups of systems.

A strong argument against conservative evolution is found in studying total masses M and total angular momenta of binaries as a function of their mass ratios. This has been done by Giannuzzi (1981). For detached systems the regression lines are straight lines parallel to the coordinate axes, since these quantities are independent parameters; for semi-detached systems however the masses (or angular momenta) and the mass ratios are correlated. The mass ratio distribution in detached systems is peaked near 1, hence the semi-detached systems with small mass ratio can be assumed to be in a more advanced evolutionary phase (Giannuzzi, 1981) (see Figures 1 and 2).

Hence, it may be concluded that mass and angular momentum can leave the system. The difficult point for non-conservative evolutionary computations is to

---

* Paper presented at the Lembang-Bamberg IAU Colloquium No. 80 on 'Double Stars: Physical Properties and Generic Relations', held at Bandung, Indonesia, 3–7 June, 1983.

Astrophysics and Space Science **99** (1984) 199–227. 0004–640X/84/0992–0199$04.35.

determine what mass fraction leaves the system as a function of the time during the mass transfer phase, and what happens to the angular momentum. We can try to find values by statistical methods, or by observing individual systems. For a number of binaries the observations reveal that matter expelled by the primary is not immediately accreted by the companion, but is stored in disks, streams, clouds. Observational evidence for the presence of circumstellar matter in close binaries has been reviewed by Batten (1970, 1973).

Disks are usually detected by their emission spectra, which are best seen at primary eclipse, when the star within the disk is hidden behind its companion. Disks are known in systems of different type: short period systems with high temperature components, Algol-type systems, systems with giant and supergiant components like SX Cas or VV Cep. Starting from two ZAMS components we can consider both conservative and non-conservative evolution, and for the latter case we can consider a variety of models and the corresponding possible configurations for solutions in view of the computation of evolutionary sequences. Different evolutionary possibilities are shown in Figure 3.

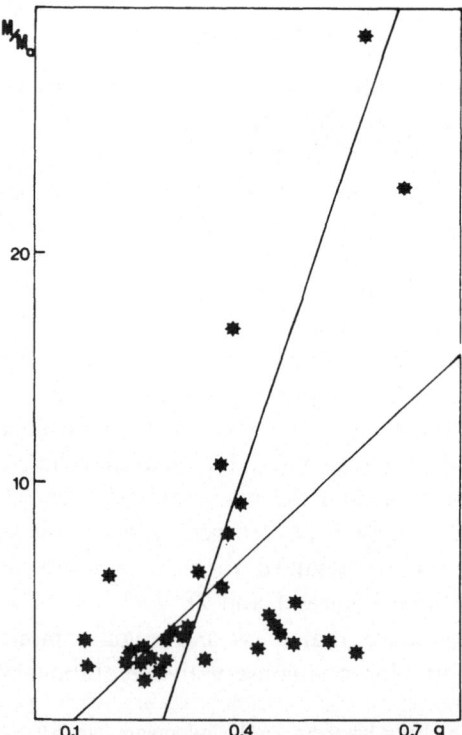

Fig. 1.   Total mass as a function of the mass ratio q $(= M_2/M_1)$ for semidetached systems. These quantities are independent, hence the regression lines should be parallel to the axes. In reality the regression lines show a correlation coefficient of 0.6 (from Gianuzzi, 1981).

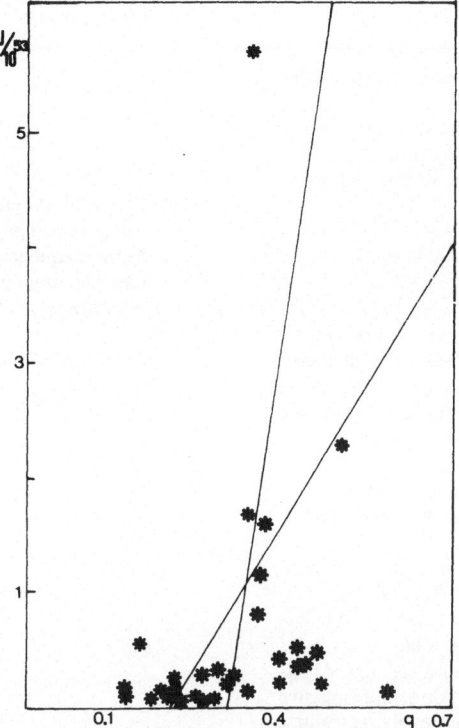

Fig. 2.   Total angular momentum as a function of the mass ratio q.
See further the explanation by Figure 1.

2. Historical Overview

In this overview I will try to indicate the milestones marking the progress in binary research, observations and theory.

Theoretical astrophysicists and observers have mutually helped each order in solving problems, but have also contributed from both sides to pose new problems to the other part of the astronomical community, and this challenge has lead on one hand to a considerable progress in the study of binaries, but on the other hand shows that the problems are far from solved. This will be discussed in the next section.

Successive contributions of theoretical work and observations to our actual knowledge of binary evolution

Theory	Observations
1941: Kuiper mentions the importance of the critical Roche-surface.	
	1942: Joy and Struve.
	1947: discover gaseous rings in binaries.

Theory	Observations
More massive stars evolve faster than lower mass stars.	
1952: Sandage and Schwarzschild show that after core hydrogen burning stars evolve towards the red giant region.	
	1950: Study of Parenagó of 54 Algolsystems: subgiant components less massive than the Main Sequence components, hence less massive star is more evolved = Algolparadox.
1955, 1956: Kopal shows the existence of a group of eclipse-variables, semidetached systems, in which one of the components fills its Roche lobe, for the Algols this is the less massive star. Subdivision of binaries: – detached systems: both components smaller than their Roche-volume; – semidetached systems: one component fills its Roche-lobe; – contact systems: both components fill their Roche-lobe.	
1955: Crawford: solution of the Algol-paradox: dog eats dog; more massive component is fast evolving volume increases and exceeds its Roche volume; mass loss of primary; mass transfer towards secondary, accretion by secondary.	
1960: Computations by Morton.	
1967: Computation of close binary evolution: evolution with mass loss of primary of close binaries. Kippenhahn et al.	Wolf–Rayet stars have abnormal abundances: overabundance of N in WN stars, of C in WC stars. Explanation?
1967: Paczynski.	
1968: Plavec.	
1969: Paczynski explains the Wolf-Rayet abundances for binaries: mass transfer removes the outer layers; the helium core with the products of nuclear burning appears at the suface.	
	1962: Discovery of X-ray binaries.
	1962: Sco X-1.
	1970: Uhuru-catalogue of X-ray binaries: optical identifications; luminous stars + compact object.
1972: Van den Heuvel and Heise: X-ray binaries are an evolutionary phase in the evolution of massive close binaries; the mass transfer from the rejuvenated secondary produces the X-rays.	

Theory	Observations
1977: Evolutionary computions for massive close binaries from ZAMS through two supernova-explosions to two runaway neutron stars. De Loore, De Grève-evolution of accreting stars (= accreting secondaries) Tutukov. Explanation: stellar wind mass loss. 1977: Computation of evolutionary sequences with mass loss for single stars: de Loore, De Grève, Lamers; Chiosi, Nasi, Sreenivasan; Chiosi, Bressan Bertelli.	

Humphreys
1978: Study of luminous stars in nearby galaxies. Lack of evolved supergiants at very high luminosities: region of avoidance in the HRD.

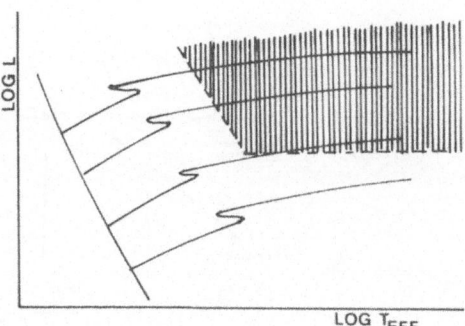

Conflict with evolutionary tracks.

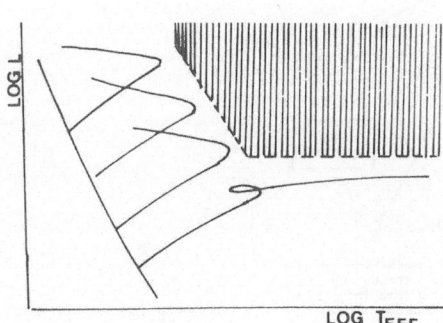

Inclusion of overshooting: larger convective cores
1981: Bressan, Bertelli, Chiosi.
1982: Doom.

Star counts: MS band not wide enough – too many B and A – stars.

Theory	Observations

Study of OB associations (mass and age).
1983: Doom, de Loore, De Grève.
    – Birth of stars in different waves:
    a wave producing stars with
      mass $< 15\,M_\odot$;
    a next wave producing more massive stars.

Conclusion: for the evolution of close binaries, computation programmes taking into account mass loss by stellar wind (massive stars, overshooting by the Roxburgh criterion) and semi-convection for the accreting component give satisfying results.

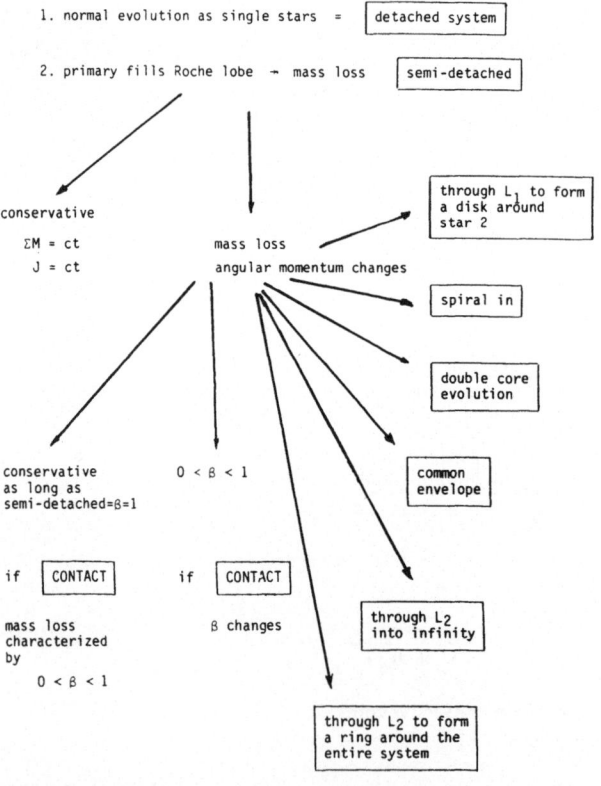

Fig. 3. Possible evolutionary scenarios for close binaries.

3. Observations

(1) From a detailed and careful analysis of spectra, radial velocities can be derived, leading to the determination of mass ratios, and, in special cases, in combination with other observations, to masses, mass luminosity relations, radii.

The spectral range required for the optimum information is determined by the spectral emission range of the components: if both components are of the spectral types A to M, most information can be derived from spectra in the visual. For binaries with hot and cool components, different spectral ranges are needed to collect information on the system, i.e. observations in the visual for the cool component, and in the UV for the hot component. For X-ray binaries the radial velocity curve for the normal star is obtained from optical spectra, while the radial velocity curve for the neutron star component is given by the X-rays.

(2) Recent detailed observations in the visual as well as in the UV (Plavec, 1982) show that a modification of the classical model for semi-detached systems (binary at the end of the mass transfer with little interaction) is necessary. The observations point to large activity during that phase and perhaps to a relationship between Algols and symbiotics.

(3) Structure of the components. – The mass transfer in binaries modifies drastically the structure of both components. In the mass losing star deeper layers, hence with chemical compositions modified by nuclear burning, show up at the surface.

(4) In systems consisting of a normal star and a supergiant, information of the outer layers of the supergiant can be discovered in the UV spectrum of the system in and outside eclipses, by subtraction of the spectra in both phases.

Information concerning the chemical composition by analysis of spectra in various wavelengths is also necessary to compare more advanced model calculations of stellar evolution with observations. As was already mentioned, during the mass transfer phase, more and more nuclear burning products appear at the surface of the mass losing primary, while due to the transferred mass the composition of the outer layers of the accreting secondary changes. In the first phases hydrogen rich matter will be mixed in with the outer layers. Still more complex phases can be imagined when after core helium burning of the remnant of the primary this star fills again its Roche lobe, hence a second phase of mass transfer occurs. This phase has been calculated already (De Grève and de Loore, 1976; Delgado and Thomas, 1981). During this phase pure helium is tranferred towards the secondary. It would be very interesting if observations could give information about these theoretically expected, but not confirmed phases. An upper limit for the primary mass of binaries where two successive stages of mass transfer can occur has been determined by De Grève and de Loore (1976). It would also be very interesting to determine a lower limit for this special case.

During further evolution reversed mass transfer can occur, from the original rejuvenated secondary to the remnant of the primary, i.e. a helium star, or even a white dwarf. Comparison of the theoretical profile of H and He in the two components, calculated with evolution programs, with the observed chemical composition from high resolution spectra, can provide us important clues on mass transfer and mass loss. In Figure 4 is shown how the chemical profile of the mass losing star changes during the mass loss phase, and different possibilities for the behaviour of the chemical profile of the gainer (de Loore and De Grève, 1981) are indicated.

IUE spectra obtained by Plavec and Koch (1980) appear normal and agree with the spectral classification in the optical range.

According to Plavec this is not so surprising since the mass transfer rate in Algols is probably low. However, an abundance analysis of the atmosphere of the primary, mass losing star should reveal an overabundance of helium. During the last phases of the mass transfer, as matter from deeper layers of the primary is expelled, the atmosphere of the secondary should also be enriched in helium if matter is accreted.

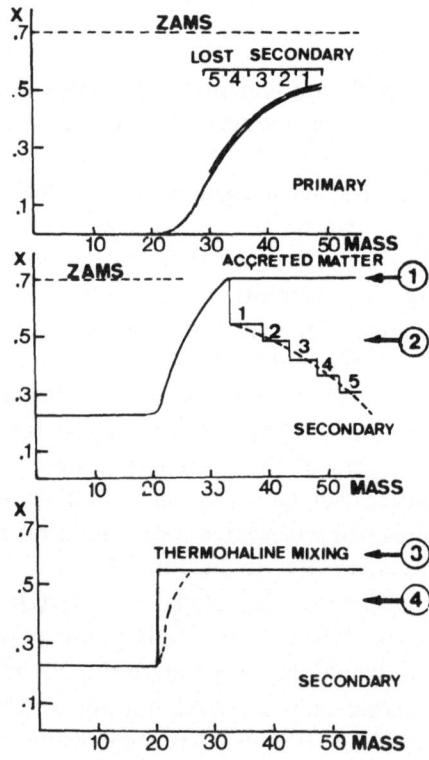

Fig. 4. The chemical profile of the loser (top) and of the gainer (middle and bottom). The figure shows four possible solutions: (1) assuming that the hydrogen profile remains unchanged ($X = 0.7$); (2) stratification and no mixing; (3) thermohaline mixing with a discontinuity; (4) thermohaline mixing with a smooth transition.

4. Differences in Time Scales for Evolutionary Computations and Time Scales for Observations

The aim of the computation of evolutionary sequences for binaries is to explain, starting from an initial ZAMS system i.e., given masses for primary and secondary and an initial period, and applying the exact physical laws for nuclear burning, energy transfer (radiative or convective), the treatment of convection and semiconvection, the treatment of the atmosphere (plan-parallel, spherical, hydrostatic, or dynamic) to calculate successive models for the detached phase, to determine what happens durjng semi-detached and possible contact phases, i.e. again calculating successive evolutionary models, for primary and secondary, but

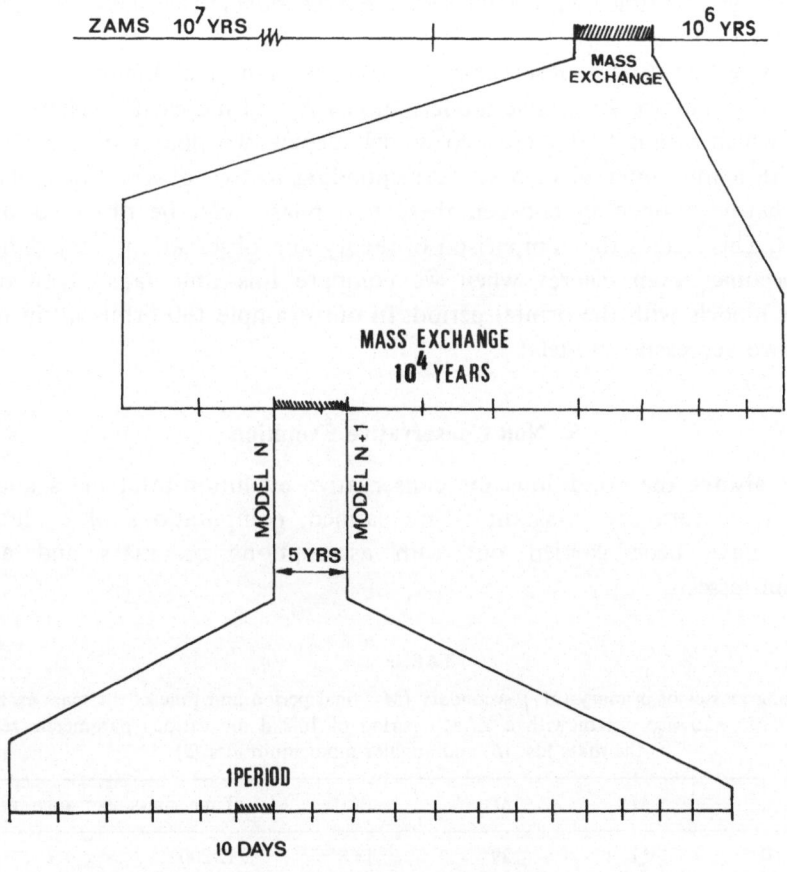

Fig. 5. Time scales for evolutionary computations and observations of binaries.

now with interaction between the material of the two stars, and then to calculate successive evolutionary models for the phase when the contact is interrupted and the system again becomes semidetached and later on detached.

The evolutionary timescale for the various phases, i.e. the time which enters the evolutionary code to calculate two successive models is different.

In Figure 5 are depicted different evolutionary stages for a massive close binary (say a 30 $M_\odot$ + 20 $M_\odot$ system). The core hydrogen burning lifetime is of the order of 10 million years, the subsequent stages are $\sim 10\%$ of this time.

Roche lobe overflow starts during shell hydrogen burning.

The mass transfer occurs on the Kelvin–Helmholtz time scale, which is in this case, of the order of 10^4 yr. While the time step for successive evolutionary models during core hydrogen is of the order of 10^5 yr, and during shell hydrogen of the order of 10^4 yr, the time step for successive models during the mass transfer phase drops to the order of 10 yr.

Let us now turn to the observations, to the lower part of Figure 5. The figure shows clearly that two successive models, model N and model $(N+1)$ are about 5 yr apart, which means that these two models depict two phases in the life of the binary with a time interval of 5 yr, corresponding to two observations, also 5 yr apart. What is happening between these two phases can be observed but not calculated. This makes the comparison of theory and observations very difficult.

This becomes even clearer when we compare this time lapse between two successive models with the orbital period. In our example 180 orbits fill in the gap between two successive models!

5. Non-Conservative Evolution

Since not always the conditions for conservative evolution-total mass and total angular momentum are constant – are satified, computations of evolutionary sequences have been carried out with assumptions on mass and angular momentum losses.

TABLE I

The remaining masses of primary (M_1), secondary (M_2) final period and time of the mass exchange of an initial 40 $M_\odot$ + 20 $M_\odot$ system with a ZAMS period of 10.2 d for various parameters related to the mass loss (β) and angular momentum loss (α)

β	α	M_1	M_2	P(d)	Time mass exchange (in years)
1	0	11	29	18.8	12 200
0.5	1	10.7	23.1	6.8	11 150
	3	10.3	23.2	2.9	7700
0	0	11.2	16.9	62.5	11 200
	1	11.0	16.9	22.3	13 300
	3	10.2	16.9	2.4	14 100

As an example the non-conservative evolution of a $40 + 20\ M_\odot$ system is shown in Figure 6 and Table I (Vanbeveren *et al.*, 1979).

In the computations were included:

(1) Mass loss by stellar wind during core hydrogen burning and shell hydrogen burning (detached phase).

(2) During the Roche lobe overflow stage it was assumed that a fraction β of the mass expelled by the primary is accreted by the companion, and that also a given fraction of the angular momentum is transferred towards this companion, related to α.

6. Survey of Theoretical Conservative Evolutionary Computations

A. COMPUTATIONS

During the last two decades a large number of evolutionary sequences for close binaries have been carried out.

Figures 7 and 8 show the initial masses, mass ratios and periods of systems for which computations exist.

(Cases (A), (B), and (C) are defined according to Kippenhahn; i.e.:

(A) mass transfer starts during core hydrogen burning;

(B) during shell hydrogen burning;

(C) after the start of He-burning.

Figure 7 shows clearly that:

(a) computations with extreme mass ratios ($q < 0.3$) have practically not been carried out;

(b) case (C) is not yet thoroughly investigated;

(c) contact systems (indicated with ($\bigstar$)) have been computed as far as the contact phase, not through the contact phase.

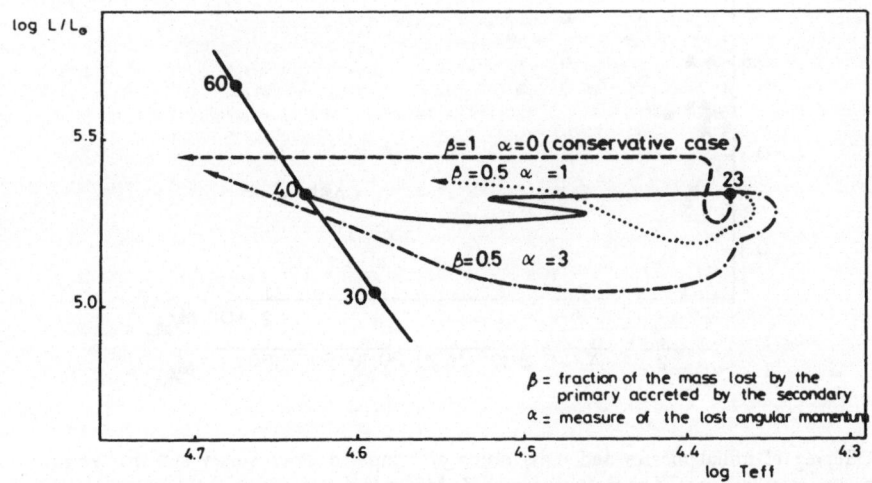

Fig. 6. Non-conservative evolution of a $40\ M_\odot + 20\ M_\odot$ system, for various values of the parameters related to mass loss (β) and angular momentum loss (α).

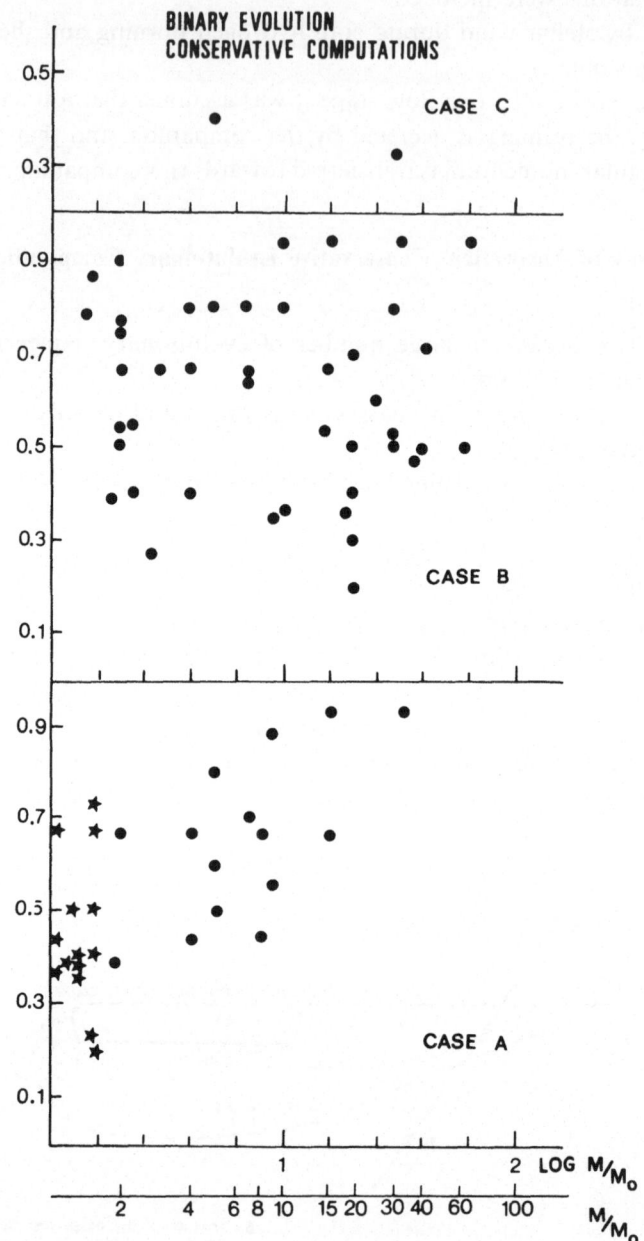

Fig. 7. Survey of initial masses and mass ratios of computed close binary systems. Vertically are displayed the mass ratios q. The dots represent ZAMS-systems for which evolutionary computations were carried out through semi-detached phases; the asterisks represent systems evolving into contact and here the calculations end.

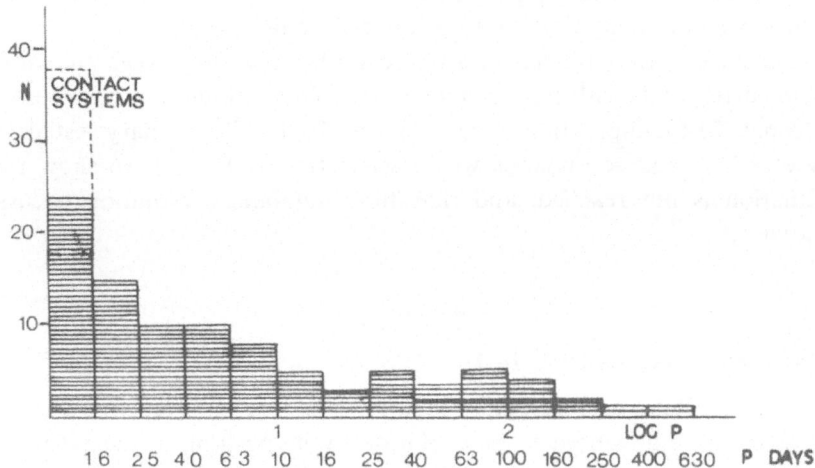

PERIOD DISTRIBUTION OF CONSERVATIVE COMPUTATIONS

Fig. 8. Survey of initial masses and periods of computed close binary systems.

Figure 8, where the period distribution is depicted, reveals clearly that most computations were performed for systems with small periods.

B. TREATMENT OF THE MASS TRANSFER

Successive phases, in the evolution of close binaries are:

(1) *detached phase*;

(2) *semi-detached phase*: i.e. the primary fills its Roche lobe while the secondary lies deeply within its Roche volume;

(3) *contact-phase*: where the two stars are overflowing their Roche lobe, hence filling volumes limited by a common equipotential surface.

In a first approximation the mass exchange was treated in the following way:

(1) *detached phase*: only stellar wind losses;

(2) *semi-detached phase*: conservative mass transfer from primary to secondary;

(3) contact phase: a fraction β of the mass expelled by the primary is accreted by the secondary, determined in such a way that the secondary keeps filling its Roche lobe, the rest leaving the system in a spherically-symmetric way ($\beta < 1$) (Plavec, 1981).

After a certain time the secondary becomes detached from this contact volume, the system becomes again semidetached and the mass transfer occurs again in a conservative way ($\beta = 1$).

The mass transfer comes to an end when the primary shrinks back within its Roche lobe.

Another way to treat the contact phase has been investigated by Packet (1983) in the following sense: the mass transfer rate is determined such that both stars fill a common equipotential surface, outside their Roche lobes. Until now luminosity transfer has been omitted; in the future this will be investigated.

Mass exchange is thus treated in a conservative way, only when the surface of the system corresponds with the equipotential surface through L_2, mass loss has to be considered. Test computations carried out by Packet for a binary system, with a primary of $9\,M_\odot$, and secondaries with mass ratios of 0.9 and 0.6 show that the latter situation is not reached, and that the total binary evolution occurs in a conservative way.

7. Algols

7.1. SHORT-PERIOD SYSTEMS ($P \lesssim 14\,\mathrm{d}$)

Algol binaries are semi-detached close binaries. The less massive component, which appears to be more advanced in its evolution by its position in the HRD (on the giant or subgiant branch of the evolutionary track) is filling its Roche lobe. The more massive component lies in most cases within the Main Sequence band. This component is of an earlier spectral type. Such systems are supposed to be the results of mass transfer and mass loss. The initially most massive one is now the less massive one as a consequence of this mass transfer process. The timescale of the mass transfer is the Kelvin-Helmholtz time scale. Although the Algols are beyond the most important phase of the mass transfer, some mass flow still exists, at rates different from one system to the other. The Algol binaries with subgiant secondaries are very difficult to treat, since as a rule one can only observe the spectrum of the primary component, and in that case the radial velocity curve leads only to the mass function

$$f(m) = \frac{m_2^3 \sin^3 i}{(m_1 + m_2)^2}.$$

For the derivation of the mass ratio complementary information or additional assumptions are needed.

For a number of Algol systems spectroscopic orbits were obtained by Popper (1980a, b). For the systems with periods less than 6 days the orbits found by Popper are without complications. For a number of systems with longer periods the velocities deviate from sinusoidal variations. In the case of U Cep and U Sge these deviations of the velocity curves might be the consequence of absorption in gas streams in the system.

Strong double H emission in the spectra of the longer period systems points to the presence of extra-photospheric matter. The masses and radii of the primaries lie in the range for Main Sequence stars of the same spectral type (except TW And, cooler than predicted by mass and radius).

Table II shows that masses as low as $0.2\,M_\odot$ and $0.3\,M_\odot$ are common for Algols.

TABLE II

Well-known SD systems

Name		M_p	M_s	P
V Pup	HD 65818	$M_1 = 9$	$M_2 = 17$	$P = 1.5$ d
V 356 Sgr	HD 173787	4.7	12.1	8.9 d
u Her	HD 156633	2.7	7.3	2.1 d
Z Vul	HD 181987	2.3	5.4	2.5 d
RZ Cnc	HD 73343	0.5	3.2	21.6 d
AR Mon	HD 37364	0.8	2.7	21.2 d
RV Lib	HD 128171	0.4	2.2	10.7 d
RT Lac	HD 209318	1.5	0.6	5.1 d
RY Per	HD 17043	0.8	5.0	6.9 d
RS Vul	HD 180939	1.4	4.5	4.5 d
U Sge	HD 181182	1.9	5.7	3.4 d
Algol	HD 19356	0.8	3.7	2.9 d
S Cnc	HD 74307	0.2	2.4	9.5 d
RY Gem	HD 58713	0.6	2.6	9.3 d
TT Hya	HD 97528	0.7	2.6	7.0 d
XY Pup	HD 67862	0.3	2.3	13.8 d
AS Eri	HD 21985	0.2	1.9	2.7 d
TW Dra	HD 139319	0.8	1.7	2.8 d
AW Peg	HD 207956	0.3	2.0	10.6 d
RY Aqr	HD 203069	0.3	1.3	2.0 d
TW And	+32 4756	0.4	1.8	4.1 d
U Cep	HD 5679	2.8	4.2	2.49 d
V 701 Sco		9	9	0.76 d

Evolutionary calculations show that it is not easy to explain these systems.

Packet (1980) found that if no mixing occurs the gainer, after the mass transfer, has a position in the HRD hotter than the Main Sequence band, in contradiction with observations. Hence a mixing mechanism has to be involved. Packet suggests differential rotation or pulsational instabilities.

The Interacting Binary U Cephei and the mass losing system HR 2142

According to Plavec (1982) the system consists of a B9V primary and a G8III secondary. The masses are $4.2\,M_\odot$ and $2.8\,M_\odot$ with radii of $2.9\,R_\odot$ and $4.7\,R_\odot$, respectively. The G star is evolved and is losing mass; some returns to the G star.

IUE spectra reveal that the effects of the gas stream are most distinctly visible in the resonance line of Fe II at 2599.395 Å and the resonance M_g doublet at 2795.523 Å and 2802.698 Å. The absorption lines of Si II, III, IV, C II, C IV, Fe III, Al II, III are probably also affected. Most of these absorption lines are broad and shallow and point to rapid rotation of the B star, of ~ 300 km s^{-1}. A geometric model was presented by Kondo et al. (1980) and is shown in Figure 9.

The gas stream leaves the G star from the hemisphere directed towards the B star, circles around the B star; some of the gas probably falls onto this star. The rest leaves the system after orbiting 3/4 of the B star. Some of the matter can fall

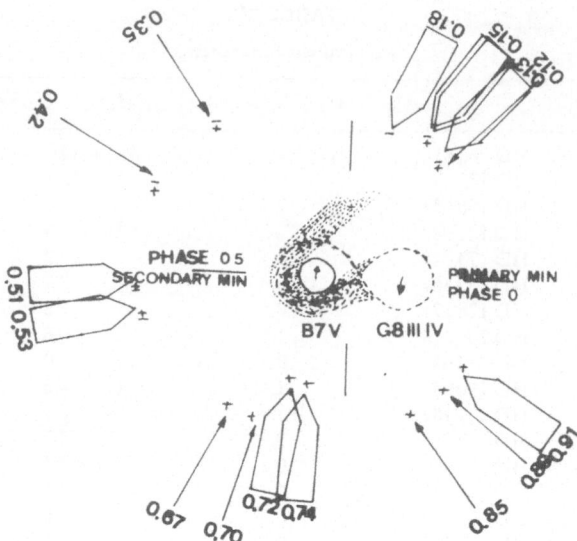

Fig. 9. Schematic representation of the gas stream as observed in the Fe II and Mg II resonance lines. The broad arrows are the data from 1978 June and September and the thin arrows are the data from 1979 March and June: the phases of observation are also given. The + sign indicates that the radial velocity of the gas stream projected against the B-star is away from the observer, the − sign that the radial velocity is toward the observer; the ± sign both away and toward the observer, and the + sign mostly toward, but with a minor component away from, the observer. The + sign at phases between 0.67 and 0.91 is probably due to the combination of the gas streaming around the B-star and the gas falling onto that star creating a hot spot which faces the observer shortly after phase 0.75. The ∓ sign at phases between 0.12 and 0.15 is probably due to the partial occulation of the B-star by the matter leaving the G-star (from Kondo *et al.*, 1980).

back onto the G star. The gas in the stream is hot enough to keep most of the Fe, Si, and Mg atoms ionized.

Observations of HR 2142 (Polidan and Peters, 1980) in the UV show that most of the material that leaves the primary is ejected from the system and not accreted. Evidence of mass ejection from the system can be seen in C II 1335 where rather strong, sharp absorption components are present, not from interstellar origin. As most probable origin for these components a circumsystem cloud or disk is suggested. Also in HR 7084 and λ Tauri gas streams are detected.

7.2. Long-period systems ($13 < P < 600$ d)

A number of Algol systems with longer periods have been studied by Plavec and Koch (1978), in the visual as well as in the UV.

As mentioned before the chemical structure of both components can provide important information on the mass transfer process, on mass loss and mass storage. We know reasonably well how a star losing mass reacts to this mass loss, but how the accretion process works is still not understood. The systems studied by Plavec (1980) display emission lines in the visual (Balmer lines, He I, Fe II); they imply the presence of a much hotter source.

One of these systems, β Lyrae was observed by Hack *et al.* (1975, 1977) with Copernicus. They found that β Lyr has a unique spectrum in the far UV, with strong emission lines.

Koch and Plavec (1978) discovered 6 binaries with the same type of emission line spectra in the far UV. All these W Serpentis stars show emission lines (Balmer lines) in the optical spectra compatible with the optical continuum of a star, too cool (A9) to excite the emission, hence suggesting a hot source. Observations of Plavec and Koch (1978) revealed five of these systems.

The spectra of 'Serpentides' show as common characteristic that:

(a) in the UV we see strong emission lines superposed on;

(b) a hot continuum of 11 500 K.

The optical companions are cooler:

B8II for β Lyr, A6III for SX Cas, F5II for W Ser.

The UV is produced in a region smaller than the observed stellar surface, as shown in Figure 10.

The IUE spectrum of SX Cas out of eclipse, partially and totally eclipsed (Plavec, 1980) is shown in Figure 11.

Optical spectra show that when the eclipse is total, the A6III spectrum disappears completely and the G5III spectrum remains. In the UV during the eclipse also the continuum disappears while the lines remain.

The explanation is that a hot source is present.

This hot source is the hot component.

In these interacting binaries emission lines of C II, C IV, N V, Si II, III, IV, Fe III, Al II, III are present. These emission lines are probably related to the mass flow, and the accretion; the ionization is most likely connected with a hot spot or a hot radiative region in the interior of the thick disk. The matter surrounding the hot component is not completely eclipsed when the star itself disappears behind the companion, and certain components of the shell lines remain visible against the background of continuous hydrogen radiation. The period of SX Cas is variable, and decreases at a rate of $\dot{P}/P = -7.6 \times 10^{-8}\,\mathrm{yr}^{-1}$ (Guinan and Tomczyk, 1979).

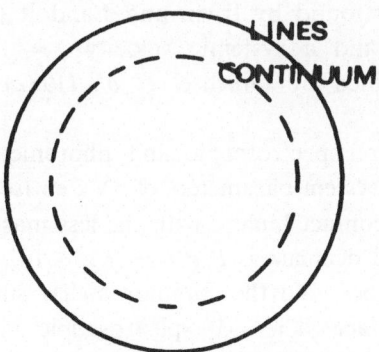

Fig. 10. Explanation in the text.

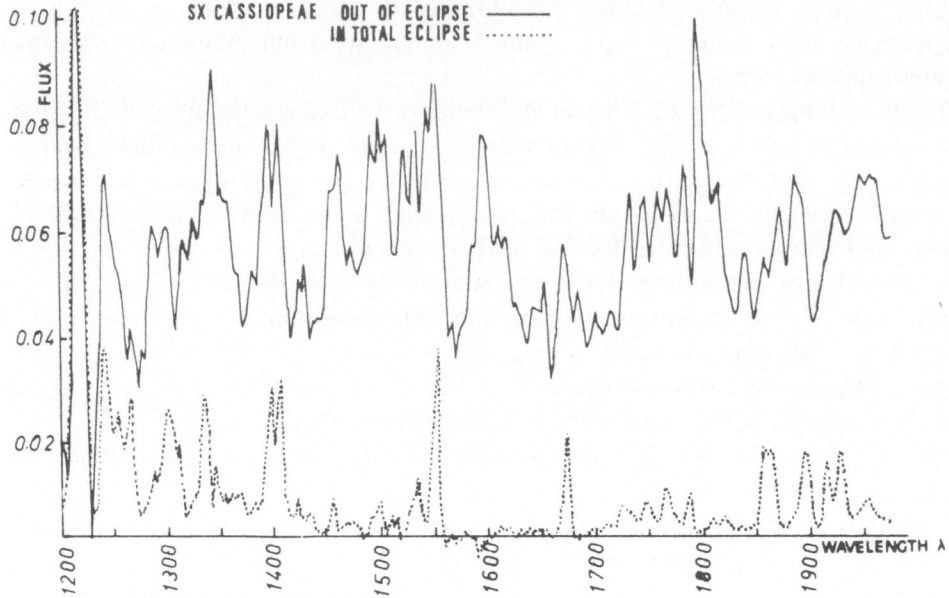

Fig. 11. IUE spectra of SX Cas in- and outside eclipse. During eclipse the continuum disappears, but the UV lines remain.

This period change is comparable to those of β Lyr or W Ser, but the sign is opposite. Simple mass transfer from the less massive component to the other one increases the period. In SX Cas the loser is certainly the less massive one. Probably strong mass loss from the system in the form of a strong stellar wind occurs. A disk model for SX Cas is shown in Figure 12.

8. A Contact System: SV Centuari

SV Centauri has been studied in great detail in the optical and the UV by Drechsel *et al.* (1982), and orbital and absolute dimensions could be derived. A radial velocity curve was found by Irwin and Landolt (1972) leading to a mass ratio of $M_2/M_1 = 1.19$ and a systemic velocity $\gamma = -33.4 \pm 7.5$ km s^{-1}. The improved values determined by Drechsel *et al.* (1982) are $M_2/M_1 = 1.25$ and $\gamma = -27.7 \pm 6.3$ km s^{-1}.

From a combination of spectroscopic and photometric data they found the absolute dimensions and system parameters of SV Cen (see Table III). The analysis reveals that SV Cen is a contact binary, with the less massive component the more luminous star. The period decrease is $\dot{P}/P = -2.15 \times 10^{-5}$ yr^{-1}. Both components are filling their Roche lobes, and the common stellar surface is very close to the outer critical Roche surface. The UV spectroscopic observations point to the presence of an expanding circumbinary envelope, and to a large mass loss from the system. The period decrease can be explained by means of angular momentum

SX CASSIOPEAE

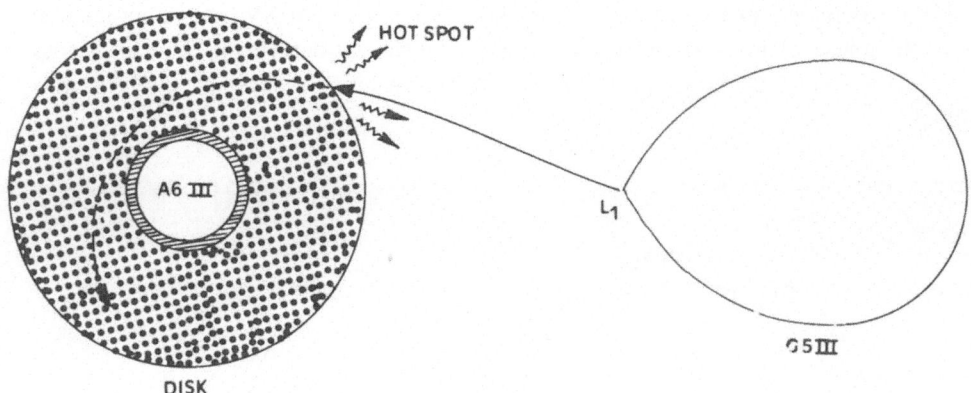

Fig. 12. Model of a close binary during the mass transfer phase, with an accretion disk and a hot spot where the flow of matter, leaving the mass losing star, impinges on the disk. Model for SX Cas by Plavec, Weiland and Koch. According to Plavec (1982) the correct spectral types are probably B7III for the gainer and K3III for the loser.

carried away by matter leaving the system. A geometrical model for the system is shown in Figure 13.

It is suggested that matter is flowing from the less massive component through L_1, near the inner critical equipotential surface. The kinetic overshoot energy of the inflowing material allows a part to escape through L_3. The matter is accelerated by radiation pressure and leaves the system.

Evolutionary computations to reproduce the actual parameters of SV Cen were carried out by Nakamura and Nakamura (1981). They started from ZAMS masses

TABLE III

Absolute dimensions and systemparameters of SV Cen

$a_1 = 8.5 R_\odot$	$M_1 = -5^m4$
$a_2 = 6.8 R_\odot$	$M_2 = -3^m4$
$a = 15.3 R_\odot$	
	$BC_1 = -2^m3$
$m_1 = 7.7 m_\odot$	$BC_2 = -1^m1$
$m_2 = 9.6 m_\odot$	
	$M_1(V) = -3^M1$
$R_1 = 6.8 R_\odot$	$M_2(V) = -2^M3$
$R_2 = 7.4 R_\odot$	$M_{tot}(V) = -3^M5$
$L_1 = 11\,700 L_\odot$	$E(B-V) = 0^m27$
	$A_V = 0^m9$
$L_2 = 1900 L_\odot$	$d = 1800$ pc
	$Sp_1 = $ B1 V
	$Sp_2 = $ B6.5 III

of 13.4 $M_\odot$ and 7 $M_\odot$, with an initial period of 2.21 days. The evolutionary tracks, with calculation of the contact condition according to Robertson and Eggleton (1977) are depicted in Figure 14. The authors conclude that SV Cen is actually in the rapid phase of mass transfer preceding the mass ratio reversal. The mass loss rate in the theoretical computations decreases when the system evolves into contact.

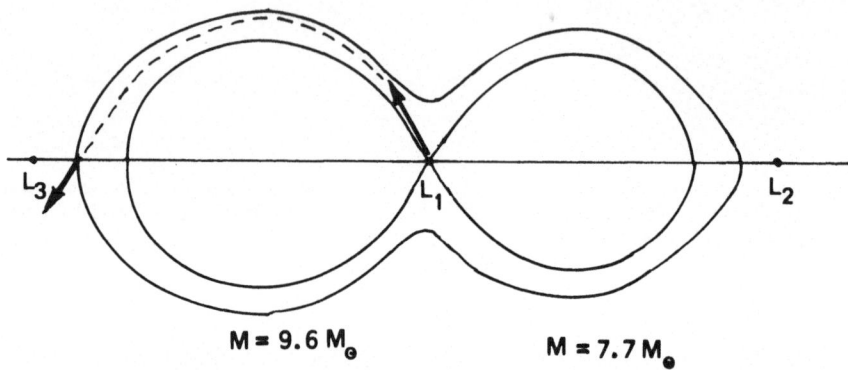

Fig. 13.   Geometrical model for SV Cen. L_1, L_2, L_3: Lagrangian points.

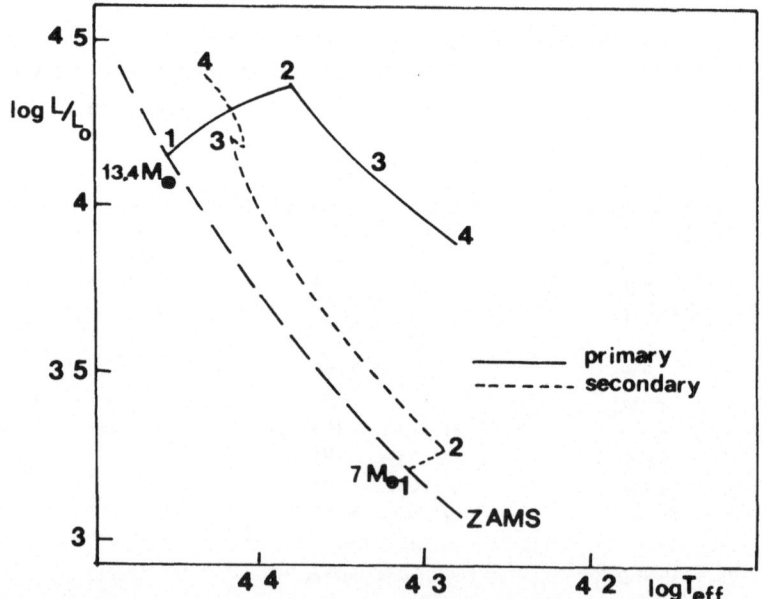

Fig. 14.   Evolutionary tracks leading to the SV Cen system: (1) start of the ZAMS; (2) start of the mass transfer; (3) beginning of the contact phase; (4) actual stage.

9. Supergiant Systems

A few binary systems are known where one of the components is a hot star, much smaller than its companion. Such systems are extremely interesting if they can be observed when the hot star travels behind the outer layers to the supergiant, an atmospheric eclipse. The most famous binaries of this group are listed in Table IV.

TABLE IV

Supergiants

Name	Hot·star	Supergiant	Period	
φ Aurigae		K4 Ib		
31 Cygni	B	K4 Ib		
32 Cygni	B	K4 Ib		
VV Cephei		M2 Ia	20 years	
μ Sagittarii	OV?	B8 Ia		$f(m) = 2.64$
ε Aurigae	B?	F2 Ia	27 years	

The optical is completely dominated by that of the red supergiant component. The hot companion is only marked by the Balmer lines and a couple of He-lines, blended with the late type spectral lines of the supergiants (metals, molecules). The hot component can be studied by ultraviolet during eclipses and outside eclipses. Studies of such systems have been carried out by Stencel *et al.* (1980) for 31 and 32 Cygni. Their observations reveal a hot turbulent region near the *B* component and a cool fast moving wind farther out. μ Sagittarii has been investigated a.o. by Plavec (1980), who, by substraction of the in- and out eclipse spectra, made an estimate of the nature of the hot companion, possible an O Main Sequence star, with $\log T_{eff} \sim 4.60$. Hack and Selvelli (1978) found from IUE spectra indications for a *B*-spectrum for the hot component, with a radius of $2\,R_\odot$. The most astonishing fact is that during the eclipse the supergiant disappears behind the hot star, hence this star should have an envelope of $\sim 850\,R_\odot$. Huang (1965), Morris (1965) suggested that the eclipse is caused by a flat disk, seen edge-on, and formed by material expelled from the F-star. There seem however to remain difficulties, since the F-star is not yet filling its Roche lobe.

10. Cataclysmic Variables

(1) Orbital period $\lesssim 1$ d;
(2) one of the components is a highly evolved star, WD or NS;
(3) the companion has a low mass and low luminosity $M \lesssim 1\,M_\odot$, $L \lesssim 1\,L_\odot$; may also be highly evolved.

The observational characteristics have been reviewed by Robinson (1976) and Warner (1976). Observational and theoretical aspects of novae can be found in Gallagher and Starrfield (1979). One of the components of binaries belonging to this group is a white dwarf surrounded by an accretion disk. The system is in a Roche-lobe overflow phase, and matter flows from the secondary through the inner Lagrangian point L_1 towards the disk. The evolution of cataclysmic variables has been discussed by Paczynski (1981). Observational and theoretical aspects of low mass X-ray sources are discussed by Lewin and Joss (1981) and by van Paradijs (1981). In this group the evolved object is a neutron star instead of a white dwarf. Evolutionary aspects of these objects were reviewed by Ritter (1982a). A list of all cataclysmic binaries and related objects and of the relevant literature is collected by Ritter (1982b).

11. Symbiotic Stars

Symbiotics form a inhomogeneous group of stars, probably representing various evolutionary phases. A number of symbiotics are long-period binaries with one of the components being a hot star. In many cases this hot star is variable, and this is perhaps a common property of symbiotics. An excellent review of symbiotics is given by Plavec (1981) and evolutionary aspects of symbiotics are discussed by Paczynski (1980).

The symbiotics are poorly known. According to Plavec (1981) their evolutionary status could be:

(a) binary systems in the beginning of a second mass transfer stage, with a cool component which is a red giant in expansion but still within its Roche lobe, and a companion situated in the HRD in the same region as the central stars of planetary nebulae (could be a helium star, remnant of an Algol after the first mass transfer stage);

(b) the hot subdwarf can also be a white dwarf where accretion of matter leaving the red giant as stellar wind re-ignites nuclear burning shells (novalike symbiotics);

(c) a system in the first stage of mas transfer in which an accretion disk around a Main Sequence star furnishes the photons required for the ionization of the nebula (Algol-symbiotics).

12. Wolf–Rayet Binaries and OB-Type Binaries

The general characteristics of Wolf–Rayet stars can be summarized as follows:

(1) they are hydrogen poor or they contain no hydrogen at all;

(2) the spectrum is dominated by strong emission lines;

(3) they can be divided into two subgroups, the WC stars, having a strong carbon and oxygen spectrum, and the WN stars, having a nitrogen spectrum;

(4) WR stars are losing mass in a spherically symmetric way (stellar wind); the mass loss rates are of the order of $3 \times 10^{-5} M_\odot \, yr^{-1}$;

(5) single WR stars as well as WR binaries exist.

(For a complete review see Conti (1982).)

Problems connected with the origin of WR stars are:

(1) how are single WR stars formed? By mass loss, during the core hydrogen burning stage, so that the products of nuclear burning appear at the surface (overshooting models, or models with increased mass loss?), or did these stars pass through the red giant stage where a sudden and large mass loss forced them back to the blue part of the HRD or do both scenarios exist?

(2) how are WR binaries formed? Here is the problem simpler, since due to mass transfer during the RLOF stage, the WR characteristics appear in a natural way: enough mass is removed so that the products of nuclear burning appear at the surface:

$$OB1 + OB2 \rightarrow WR1 + OB2$$

$$C.S. + OB2 \rightarrow C.S. + WR1$$

(C.S. means compact object, e.g. neutron star)

$$OB1 \lesssim 30\text{--}40 \, M_\odot;$$

(3) very important is the number of WR stars which are binaries.

Detection of the Binarity

(1) Photometric observations, showing eclipses;
(2) variation of the absorption lines.

In a number of WR stars absorption lines are seen. This has been used until recently as proof for the binarity. There are no systems with companions other than of O-type. In the seventies (Niemela, 1973; Conti, 1976) it was realized that absorption lines could originate in the WR stars as well. Massey (1980), Massey and Conti (1981), Massey *et al.* (1981) have shown that in certain cases the absorption lines could be intrinsic to the Wolf–Rayet star itself, so that the presence of absorption lines is not necessarily a proof for the binarity (e.g. in HD 927407 WR emission lines and absorption lines vary in phase). Hence reliable WR binaries are only those systems for which an orbit has been established. Only 20 of the 159 WR stars in the catalogue of van der Hucht *et al.* (1981) are true WR binaries.

Number of known WR stars (Pop. 1):

159	WR in the Galaxy	(van der Hucht *et al.*, 1981)
100	WR in the LMC	(Breysacher, 1981)
8	WR in the SMC	(Azzopardi and Breysacher, 1979)
40	WR in M33	(Way and Corso, 1972)
20	WR in M31	(Shara and Moffat, 1982)
	WR in NGC 604, M101,	(Boksenberg *et al.*, 1977)
	Tol 3, II Zw 40	(Conti and Massey, 1981).

Hence only 10% of the galactic WR stars are yet established as WR + O systems (SB2) and only 2.5% have optically invisible companions (SB1). This has been examined by Massey (1980, 1981); he concludes that the percentage of close WR + O binaries seems to be maximum 25%, and the overall binary frequence less than 50%.

All WR stars in the SMC show absorption lines. In the LMC the overall fraction of binaries is ∼ 40%.

A list of well defined Wolf–Rayet binaries is given in Table V. For a number of OB-type binaries masses could be determined from binary motion and evolutionary computations. The two components must have the same age and the same sin i; using stellar structure models for OB stars and Of stars allows the determination of consistent t and sin i, hence the masses can be determined for a number of cases (Doom and de Loore, 1984). Masses of OB stars are shown in Table VI.

TABLE V

Wolf–Rayet binaries; well defined systems

Name	P (d)	Type	q	$M_1 \sin^3 i$	$M_2 \sin^3 i$	M_{WR}	M_{OB}
HD 90657	8.2	WN4–O4–6	0.5	6.8	13.6	25	58
HD 152270	8.89	WC7–O6	0.36	1.8	4.9	∼ 20	∼ 60
HD 94546	4.9	WN4–O7	0.34	8	23	> 8	
HD 168206	29.7	WC8–O8–9V	0.37	7	24	∼ 9	25
MR 42	6.3	WN6–O5	0.86	43	50	50	60
HD 186943	9.55	WN4–O9V	0.42	3.36	7.94	13	25
HD 190918	112.8	WN4.5–O9I	0.27	0.7	2.6	9	35
HD 214419	1.6	WN7–O?	1.20	23	19	> 40	> 23
HD 193576	4.21	WN5–O6	0.39	10	25.6	12(17)	45
HD 211853	6.68	WN6–O	> 22	11.5	33	10–25	> 50–60
HD 68273	78.5	WC8–O9I	0.54	17	32	20	35
MR 114	2.13	WN4 + O8	0.43	5.3	12.2	5–11	12–25
HDE 311884	6.34	WN6–O5V	0.84	40		∼ 50	∼ 60
HD 192641		WC7 + abs					
HD 193793		WC7 + abs					
HD 92740	80.34	WN7 + abs					
HD 97152	7.86	WC7–O5–7	0.59	3.6	6.1	20	35
HD 63099	27.63	WC5–O7	0.16	17			
HD 113904	18.34	WC6–O9.5Ia					
HDE 320102	8.83	WN3 + O7	0.33	1.8		11	35
RUNAWAYS							
HD 50896	3.76	WN5					
HD 96548	4.80	WN8					
HD 76536	4.00	WC6					
HD 192163	4.50	WN6					
HD 197406	4.30	WN7					
HD 164270	1.80	WC9					
HD 86161	10.70	WN8					
HD 177230	1.80	WN8					
209 BAC	2.40	WN8					

TABLE VI

Well defined O-type binaries

Name	Type	P (d)	$M_1 \sin^3 i$	$M_2 \sin^3 i$	$t/10^6$ yr	M_{1i}	M_{2i}	M_1	M_2	Remarks
AO Cas	O9III + O9III	3.5	10.1	12.9	6.5	25	33	23	29	3% overcontact
	O9.5III + O9III	3.5	10.1	12.9	6.7	19	26	19	24	
HD 19820	O9IV + O9IV	3.4	18.9	9.18						
ι Ori	O9III + O9III	29.1	15.9	9.4	6.1	42	24	35	21	
29 CMa	O8.5If + O7	4.4	20	24						23% overcontact
HD 93205	O3V + O8	6.1	39	15	0.0	63	24	63	24	
HD 93403	O5f + O7.5	15.1	5.2	3.4	5.4	57	33	45	29	
HD 159176	O7V + O7V	3.4	10.8	11.4	5.7	35	38	32	34	
HD 191201	B0III + B0III	8.3	13.8	12.9	11.1	19	17	17	16	
HDE 228766	O5. 5If + O7.5I	10.7	16.3	15.7	6.0	47	44	38	37	
HD 228854	O6.5V + O7.5	1.88	37.3	32.7	5.7	44	37	37	33	both fill RL
HD 206267	O6.5V + O9	3.7	18.3	6.37						
14 Cep	O9V + O9V	3.1	6.16	2.91						
HD 215835	O6 + O6	2.1	23.4	19.1	5.0	51	39	43	35	both fill RL
HD 166734	O7If + O9I	34.5	29	31	6.7	41	40	32	34	
HD 149404	O8.5I + O7IIIf	9.8	1.58	2.66	5.7	24	45	23	39	
HD 165052	O6.5V + O6.5V	6.1	2.5	2.2	5.4	45	38	38	33	
HD 167771	O7IIIf + O9III	4.0	2.7	2.3	6.4	43	33	32	29	

13. Massive X-Ray Binaries

13.1. RADIAL VELOCITIES AND MASS DETERMINATIONS

From a collection of 92 blue spectrograms of HD 77581 taken with the 152 cm spectrographic telescope of ESO, heliocentric radial velocities were determined. Using the observed pulse arrival times of Rappaport et al. (1976) the orbital elements for the compact object were derived. Combination leads to the masses of the components of the Vela X-1 system:

$$M_x \sin^3 i = 1.67 \pm 0.12 \, M_\odot,$$

$$M_{\mathrm{opt}} \sin^3 i = 20.5 \pm 0.9 \, M_\odot.$$

If we adopt an average value of $\sin^3 i = 0.96$, as 'best value' for the masses, $M_x = 1.74 \, M_\odot$ and $M_{\mathrm{opt}} = 21.3$ were obtained (van Paradijs et al., 1977).

13.2. VARIATIONS IN THE PROFILES OF Hβ AND He II 4686

The observed periodic variation of the Hα emission profile is commonly interpreted as due to gaseous streams in the system. Also the profile of Hβ is phase dependent and consists of two absorption components, superimposed on one another. One of these components is the steady photospheric profile, the other is variable in strength and velocity (Zuiderwijk, thesis, 1979). The profile of He II 4686 is of the P Cygni type around phase 0.7 and in absorption with variable strength between $\phi = 0.9$ and 0.5. These observations point to a gaseous stream in the system, and an asymmetrically expanding atmosphere.

13.3. DETAILED ANALYSIS OF THE Hα LINE

The wide undisplaced emission component of Hα is clearly produced far from the stellar surface in an extended envelope outflowing radially. The absorption component is formed in the part of the envelope between observer and stellar surface, hence the absorption line velocity reflects the outflow velocity of the envelope at the mean Hα forming level into the direction of the observer. An analysis of the Hα profile as a function of phase was carried out by Zuiderwijk *et al.* (1974). Figure 15 shows the variation of the Hα profile.

Already in the case of a circular orbit the outflow pattern is expected to be asymmetric as a consequence on the asymmetric gravitational field of the neutron star. The eccentricity ($e = 0.096$) makes this still worse. The figure shows that the

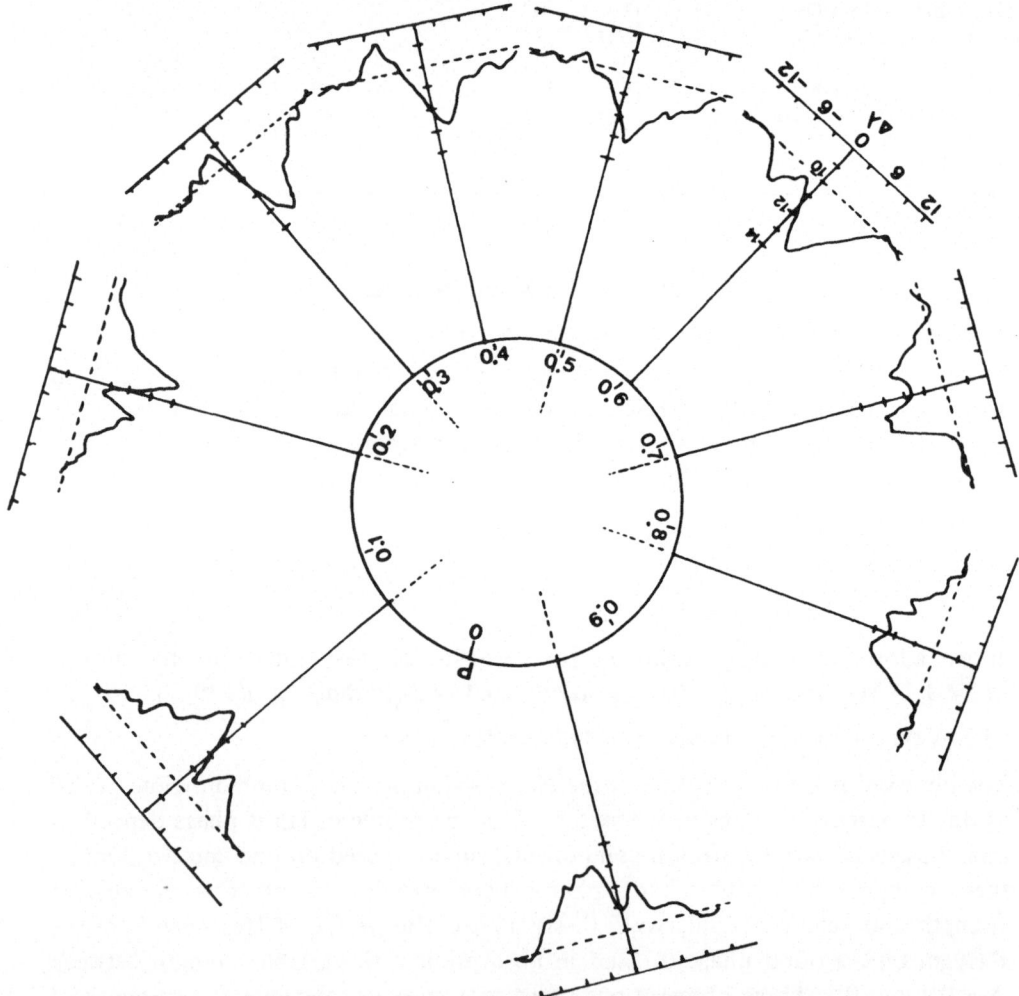

Fig. 15.   The variable Hα profile of HD 77581 as a function of phase (from Zuiderwijk, 1974).

largest outflow occurs between the phases 0.25 and 0.50. The figure suggests that the outflow is not very strongly dependent on the size of the Roche lobe as the Hα profiles before and after apastron passage are different although the sizes of the lobes are the same. Probably the B0.51a star is well within the Roche lobe, so that the mass loss process is probably due to stellar wind, not to Roche lobe overflow.

14. Barium Stars

According to McClure *et al.* (1980) all stars with a Ba ii anomaly vary in radial velocity and could be binaries. In a couple of cases they could even go farther and could derive a mass function, indicating companions of 1 to 2 solar masses. The low masses, low luminosities and small radial velocity ranges suggest that the systems are reasonably wide and that the companions probabiy are degenerate (Plavec, 1982).

The Ba ii star ζ Cap (G511) has such a component (Böhm–Vitense, 1980). From the observed flux distribution is derived that $T_{eff} \sim 22\,000$ K, and the mass is $\sim 1\ M_\odot$. These observations strongly suggest that the Ba ii anomaly may be due to mass transfer.

15. Conclusions

Many problems concerning the evolution of close binary systems remain unsolved. More detailed and accurate observations are needed, as well as more evolutionary computations, allowing comparison between theory and observations. For more systems should be derived the masses of the components, the rates of mass flows, a geometric picture of the mass flow, the evolutionary status.

We have a considerable amount of information on the final stages of binary evolution, massive X-ray binaries, binary pulsars, cataclysmic variables, X-ray bursters and on advanced phases of the evolution of massive stars, e.g. Wolf–Rayet binaries, and we have good ideas about the progenitors of such systems. Observations in different wavelength regions, optical, UV, X-ray, IR, radio, in many cases simultaneously, have especially for these cases provided relevant information. For more normal cases, the mass transfer phase itself which is very short, especially the rapid phase, but also for the phases following the mass exchange, more information is required.

Very important for the determination of the evolutionary status is the chemical composition of the atmosphere. The hydrogen to helium ratio of the atmosphere of the two components of a close binary can give us information on the mass loss and accretion processes as explained in Section 3. A contaminated atmosphere for the gainer is a proof for accretion; the hydrogen to helium ratio, combined with a theory on accretion and mixing of accreted material and existing material tells us how much is accreted. The same analysis of H/He gives information on the internal structures of the loser and the amount of expelled matter.

Systems with well defined masses, periods and information on the abundances for primary and secondary, compared with evolutionary computations can provide important contributions to our ideas on the evolution of close binaries.

References

Azzopardi, M. and Breysacher, J.: 1979, *Astron Astrophys.* **75**, 120.
Batten, A. H.: 1970, *Publ. Astron. Soc. Pacific* **82**, 574
Böhm-Vitense, E.: 1980, *Astrophys. J.* **239**, L79.
Boksenberg, A., Willis, A. J., and Searle, L.: 1977, *Monthly Notices Roy. Astron. Soc.* **180**, 158.
Breysacher, J.: 1981, *Astron. Astrophys. Suppl.* **43**, 203.
Conti, P. S.: 1976, *Mém. Soc. Roy. Sci. Liège, 6° série* **9**, 193.
Conti, P. S.: 1982, in C. de Loore and A. J. Willis (eds.), 'Wolf–Rayet Stars: Observations, Physics, Evolution', *IAU Symp.* **99**, 3.
Conti, P. S. and Massey, P.: 1981, *Astrophys. J.* **249**, 471
Crawford, J. A.: 1955, *Astrophys. J.* **121**, 71.
De Grève, J. P. and de Loore, C.: 1976, *Astrophys. Space. Sci.* **43**, 35.
De Grève, J. P., de Loore, C., and van Dessel, E. L.: 1978, *Astrophys. Space Sci.* **53**, 105.
De Loore, C. and De Grève, J. P.: 1982, in S. D'Odorico (ed.), *The Most Massive Stars*, ESO Workshop.
Delgado, A. J. and Thomas, H. C.: 1981, *Astron. Astrophys.* **96**, 142.
Doom, C. and de Loore C.: 1984, *Astrophys. J.*, in press.
Drechsel, H., Rahe, J., Wargau, W., and Wolf, B.: 1982, *Astron. Astrophys.* **110**, 246.
Gallagher, J. S. and Starrfield, S.: 1978, *Ann. Rev. Astron. Astrophys.* **16**, 117.
Giannuzzi, M. A.: 1981, *Astron. Astrophys.* **103**, 111.
Guinan, E. F. and Tomczyck, S.: 1979, *Inf. Bull. Var. Stars*, No. 1623.
Hack, M., Hutchings, J. B., Kondo, Y., McCluskey, G. E., Plavec, M., and Polidan, R. S.: 1975, *Astrophys. J.* **198**, 453
Hack, M., Hutchings, J. B., Kondo, Y., and McCluskey, G. E.: 1977, *Astrophys. J. Suppl.* **34**, 565.
Huang, S. S.: 1965, *Astrophys. J.* **141**, 976.
Irwin, J. B. and Landolt, A. U.: 1972, *Publ. Astron. Soc. Pacific* **84**, 686.
Kondo, Y., McCluskey, G. E., and Stencel, R. E., 1980, in M. J. Plavec, D. M. Popper, R. K. Ulrich (eds.), 'Close Binaries: Observations and Interpretation', *IAU Symp.* **88**, 237.
Kopal, Z.: 1955a, *Mem. Roy. Soc. Liège (4)* **15**, 684.
Kopal, Z.: 1955b, *Ann. Astrophys.* **18**, 379.
Kopal, Z.: 1956, *Ann. Astrophys.* **19**, 298.
Lewin, W. H. G. and Joss. P.C.: 1981, *Space Sci. Rev.* **28**, 3.
Massey, P.: 1980, *Astrophys. J.* **236**, 526.
Massey, P. and Conti, P. S.: 1981, *Astrophys. J.* **244**, 173.
Massey, P., Conti, P. S., and Niemela, V.: 1981, *Astrophys. J.* **246**, 145.
Morris, S. C.: 1965, *Astron. J.* **70**, 685.
Morton, D. C.: 1960, *Astrophys. J.* **132**, 146.
Nakamura, M. and Nakamura, Y.: 1982, *Astrophys. Space Sci.* **83**, 163.
Niemela, V. S.: 1973, *Publ. Astron. Soc. Pacific* **85**, 220.
Paczynski, B.: 1980, *Astron. Astrophys.* **82**, 349.
Paczynski, B.: 1981, *Acta Astron.* **31**, 1.
Packet, W.: 1980, in M. J. Plavec, D. M. Popper, R. K. Ulrich (eds.), 'Close Binaries: Observations and Interpretation', *IAU Symp.* **88**, 211.
Packet, W.: 1983, personal communication.
Plavec, M. J.: 1973, in A. H. Batten (ed.), 'Extended Atmospheres and Circumstellar Matter in Spectroscopic Binary Systems', *IAU Symp.* **51**, 216.
Plavec, M. J.: 1980, in M. J. Plavec, D. M. Popper, and R. K. Ulrich (eds.), 'Close Binaries: Observations and Interpretation', *IAU Symp.* **88**, 251.
Plavec, M. J.: 1981, in C. Chiosi and R. Stalio (eds.), 'Effects of Mass Loss on Stellar Evolution', *IAU Colloq.* **59**, 431.

Plavec, M. J. 1982, in Z. Kopal and J. Rahe (eds.), 'Binary and Multiple Stars as Tracers of Stellar Evolution', *IAU Colloq.* **69**, 159.

Plavec, M. J. and Koch, R. H.: 1978, *Inf. Bull. Var. Stars*, No. 1482.

Plavec, M. J., Ulrich, R. K., and Polidan, R. S.: 1973, *Publ. Astron. Soc. Pacific* **85**, 508.

Popper, D. M.: 1980a, in M. J. Plavec, D. M. Popper, R. K. Ulrich (eds.), 'Close Binaries: Observations and Interpretations', *IAU Symp.* **88**, 203.

Popper, D. M.: 1980b, *Ann. Rev. Astron. Astrophys.* **18**, 115.

Rappaport, S., Joss, P. C., and Mc Clintock, J. E.: 1976, *Astrophys. J.* **206**, L105.

Ritter, H.: 1982a, *High-Energy Astrophysics*, Proc. Workshop, Nanking, 1982.

Ritter, H.: 1982b, 'Catalogue of Cataclysmic Binaries, Low Mass X-Ray Binaries and Related Objects', preprint.

Robinson, E. L.: 1976, *Ann. Rev. Astron. Astrophys.* **14**, 119.

Shara, M. M. and Moffat, A. F. J.: 1982, in C. de Loore and A. J. Willis (eds.), 'Wolf–Rayet Stars: Observations, Physics and Evolution', *IAU Symp.* **99**, 3.

Stencel, R. E., Kondo, Y., Bernat, A. P., and Mc Cluskey, G.: 1980, in M. J. Plavec, D. M. Popper, R. K. Ulrich (eds.), 'Close Binary Stars: Observations and Interpretation', *IAU Symp.* **88**, 555.

Vanbeveren, D., De Grève, J. P., van Dessel, E. L., and de Loore, C.: 1979, *Astron. Astrophys.* **73**, 19.

van der Hucht, K., Conti, P. S., Lundström, I., and Stenholm, B.: 1981, *Space Sci. Rev.* **28**, 227.

van Paradijs, J.: 1981, *Astron. Astrophys.* **103**, 140.

Zuiderwijk, E. J., van den Heuvel, E. P. J., and Hensberge, G.: 1974, *Astron. Astrophys.* **35**, 353.

J., and de Loore, C.: 1977, *Astron. Astrophys. Suppl.* **30**, 195.

Wray, J. D. and Corso, C. J.: 1972, *Astrophys. J.* **172**, 577.

Zuiderwijk, E. J., van den Heuvel, E. P. J., and Hensberge, G.: 1974, *Astron. Astrophys.* **35**, 353.

Zuiderwijk, E. J.: 1979, Ph. D. Thesis, University of Amsterdam.

PHOTOMETRIC AND SPECTROSCOPIC STUDY OF R CM.a*

K. R. RADHAKRISHNAN, M. B. K. SARMA, and K. D. ABHYANKAR

Centre of Advances Study in Astronomy, Osmania University, Hyderabad, India

(Received 19 July, 1983)

Abstract. *UBV* light curves and spectrograms of R CMa obtained with the 48-inch telescope of Japal-Rangapur Observatory during 1980–82 have been used for deriving the eclipse and orbital elements as well as the absolute dimensions of the components. The primary is found to be a Main-Sequence F2V star of mass 1.52 $M_\odot$ and the secondary a subgiant star of spectral type G8 and mass 0.20 $M_\odot$ which fills its Roche lobe, in agreement with Kopal and Shapley (1956) results, Kopal (1959), or Sahade's (1963) results. From a consideration of the possible evolution of this system it is concluded that a large fraction of the original mass of the secondary is lost from the system. A study of the period changes indicates the possible presence of a third component of mass of about 0.5 $M_\odot$ which is most likely to be an *M* dwarf.

1. Introduction

R Canis Majoris is the prototype of a group of Algol systems having very low mass function. It is known to exhibit peculiarities in its light curve that change from time to time and is thought to be surrounded by gas. It also shows changes in period which are not yet fully understood. Hence, this star was included in the photometric and spectroscopic observing programme of Japal-Rangapur Observatory.

R CMa was observed photometrically with the 48-inch telescope in the standard *UBV* colors on 34 nights during 1980–82 covering nine minima. The individual observations and times of minima have already been published by Radhakrishnan and Sarma (1982). Further 55 spectra of the star with a dispersion of 33 A mm^{-1} at Hγ were obtained with the Meinel spectrograph attached to the 48-inch telescope during 1980–81. The results from this data are summarised in this short paper, the details will be published elsewhere.

2. Analysis of Period Changes

Dugan (1924)‚found an abrupt shortening in the period of R CMa in 1914. This was corroborated by Dugan and Wright (1939), Wood (1946), and Guinan (1977). Koch (1960) felt, however, that the (O–C) curve could also be represented by a sine curve of semi-amplitude 0.032 day. We have combined our 9 times of minima

---

* Paper presented at the Lembang-Bamberg IAU Colloquium No. 80 on 'Double Stars: Physical Properties and Generic Relations', held at Bandung, Indonesia, 3–7 June, 1983.

Astrophysics and Space Science **99** (1984) 229–236. 0004–640X/84/0992–0229$01.20.

with the 120 other published visual and photoelectric times of minima for making
a fresh analysis of the period changes in R CMa.

On plotting the (O–C) curve with the second ephemeris of Guinan (1977):

$$\text{Primary minimum} = \text{HJD } 2\,420\,213.1347 + 1^{d}135\,938\,72E, \tag{1}$$

an almost sinusoidal variation was noted. After an upward revision of Guinan's
period the following revised ephemeris was obtained:

$$\text{Primary minimum} = \text{HJD } 2\,430\,436.5832 + 1.135\,941\,97E. \tag{2}$$

The (O–C) diagram based on this ephemeris is shown in Figure 1. Its sinusoidal
nature indicates apparent change of period either due to apsidal motion or due to
the presence of a third body. Since the eccentricity of the binary orbit is almost
zero and the secondary minima have always been observed at phase 0.5 the period
variation is most likely to be due to the presence of a third body. On applying
Irwin's (1952) method and correcting the semi-amplitude by the method of least-
squares we obtained the following elements:

$$P = 29\,000 \text{ binary cycles} = 91.44 \text{ yr}, e = 0.45, \omega = 25°, K = 0^{d}029. \tag{2}$$

The theoretical (O–C) curve based on these elements is shown by the solid line in
Figure 1. From these elements we obtain

$$a_{12} \sin i = 8.22 \times 10^{8} \text{ km}, \quad f(m) = 0.02 \, M_{\odot}. \tag{3}$$

If we assume a total mass of 1.72 $M_{\odot}$ for the binary system the mass of the third

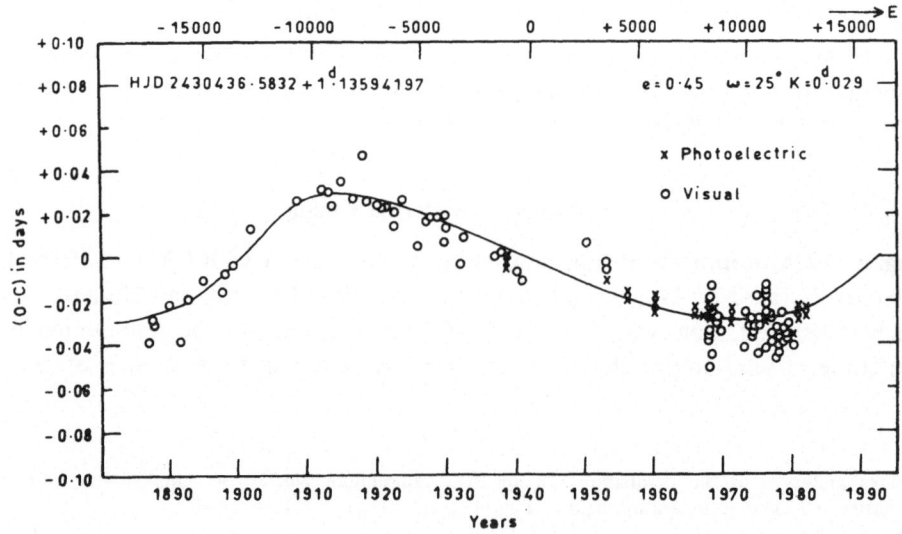

Fig. 1.   (O–C) diagram for R CMa.

body comes out to be 0.46 $M_\odot$ for $i = 90°$ and 0.54 $M_\odot$ for $i = 60°$. This low mass suggests that it may be a white dwarf or an M dwarf. Recent findings by Needham *et al.* (1980) of infrared excess in R CMa may be an indication of the presence of a cool M-type third body.

3. Spectroscopic Orbit

Fifteen stellar lines were measured for deriving the radial velocities of R CMa. The same lines were also measured on 13 spectra of the radial velocity standard α Lep of spectral type F0. The mean heliocentric velocity of α Lep was found to be 24.2 ± 0.8 km^{-1} s which is in good agreement with the catalogue value of 24.7 ± 0.2 km s^{-1}.

Preliminary elements of the spectroscopic orbit were calculated both by the method of Russell and Wilsing (Russell, 1902; Wolfe *et al.*, 1967) and by Sterne's (1941) method for a nearly circular orbit. They were corrected by the least squares differential correction method for small eccentricities. Our spectroscopic elements were similar to those derived by the previous authors: Jordan (1916), Sitterly (1940), and Struve and Smith (1950). Hence, all observations were combined to get a common solution by adopting the following procedure:

(i) Phases were calculated by applying the light-time correction due to the motion of the third body.

(ii) The observed radial velocities for each set were corrected for third body motion appropriate for that epoch.

(iii) Differences in the individual systemic velocities for different sets were found to be larger than the corrections for the third body motion; attributing them to the differences in the radial velocity systems all radial velocities were converted to our system.

The final elements are given in Table I and all observations are compared with the theoretical radial velocity curve in Figure 2. The characteristic low mass functions $f(m) = 0.0025$ $M_\odot$ is confirmed. The scatter of about 5 km s^{-1} in the observed radial velocities seems to be intrinsic and it may be due to the rotationally broadened and weak lines in the spectrum which limit the accuracy of measurement.

TABLE I

Spectroscopic elements of R CMa

e	$= 0.049 \pm 0.019$
V_0	$= 37.19 \pm 0.78$ km s^{-1}
K_1	$= 27.78 \pm 0.48$ km s^{-1}
$a_1 \sin i$	$= 4.33 \pm 0.08 \times 10^5$ km
ω	$= 173°.9 \pm 21°.2$
$f(m)$	$= 0.002\,51 \pm 0.000\,14\ M_\odot$

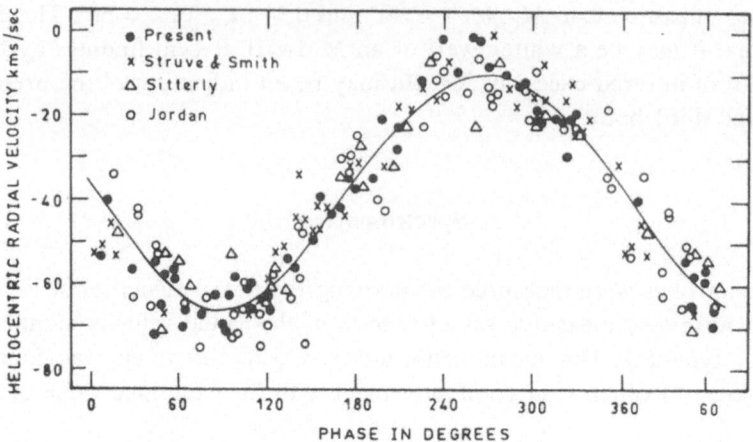

Fig. 2.   Radial velocity curve of R CMa.

4. Photometric Solution and Absolute Dimensions

Figure 3 shows the V light curve of R CMa based on 874 observations made on 34 nights; the B and U light curves are similar. Like Wood's (1946) observations our light curves are free from peculiarities of various kinds found by other observers: Pickering (1904), Wendell (1909), Koch (1960), Kitamura and Takahashi (1962), and Sato (1971). Apparently there is no evidence of circumstellar gas at the present epoch.

The UBV light curves were analysed by the standard Russell–Merrill method; the primary eclipse was found to be a transit. The derived inclination and radii of the two components are given in the first row of Table II while the spectral types obtained from the derived colors are given in the fourth row. Now, in order to determine the absolute dimensions of the system we need the mass ratio m_2/m_1 which was estimated in three ways: (i) by assuming that the primary is a Main-Sequence star of spectral type F2V; (ii) by assuming that the secondary fills its Roche lobe; and (iii) by comparing the observed Fourier coefficients of the light curves with those calculated theoretically from Merrill's (1970) formulae. All of them gave a consistent value of $m_2/m_1 = 0.13 \pm 0.01$. The masses and radii obtained by combining it with the spectroscopic elements are given in the fifth and sixth rows of Table II while the absolute bolometric luminosities are given in the last row. Our results confirm the conclusion of Sahade (1963) that R CMa systems are normal semi-detached Algol type binaries. It is found that the secondary, which fills its Roche lobe, is overluminous by about 5 magnitudes and much hotter for its mass. The position of the two components of R CMa in the HR-diagram is shown in Figure 4.

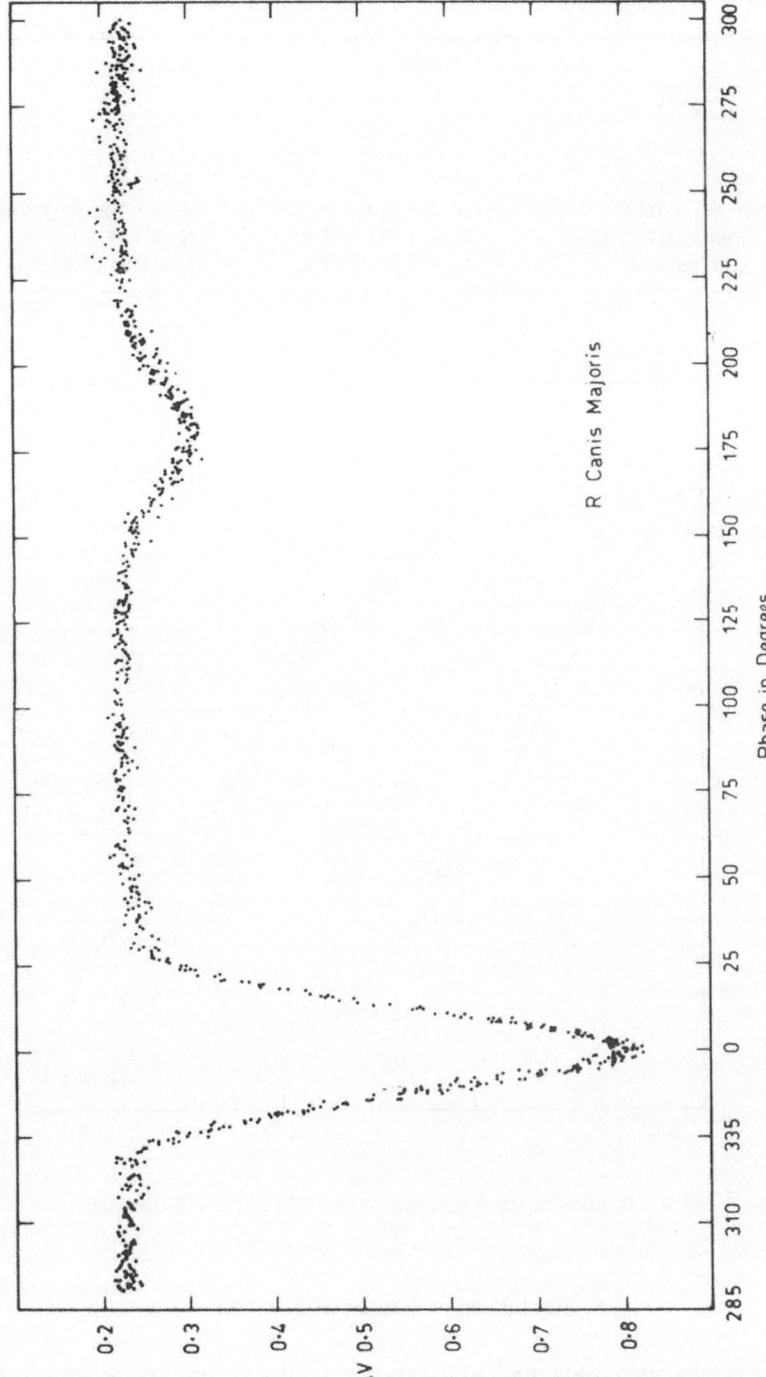

Fig. 3. *V* light curve of R CMa.

TABLE II

Photometric elements and absolute dimensions of R CMa

	Primary	Secondary
$i = 79°.93$	$r_h = 0.319$	$r_c = 0.217$
$B - V$	$+0.31 + 0.01$	$+0.77 \pm 0.01$
$U - B$	$+0.04 \pm 0.02$	$+0.60 \pm 0.02$
Spectral type	F2V	G8IV-V
$m_2/m_1 = 0.13$	$m_h = 1.52\ M_\odot$	$m_c = 0.199 \pm 0.004\ M_\odot$
$f(m) = 0.0025\ M_\odot$	$R_h = 1.73 \pm 0.04\ R_\odot$	$R_c = 1.18 \pm 0.02\ R_\odot$
L(bolometric)	$L_h = 6.64 \pm 0.29\ L_\odot$	$L_c = 0.86 \pm 0.03\ L_\odot$

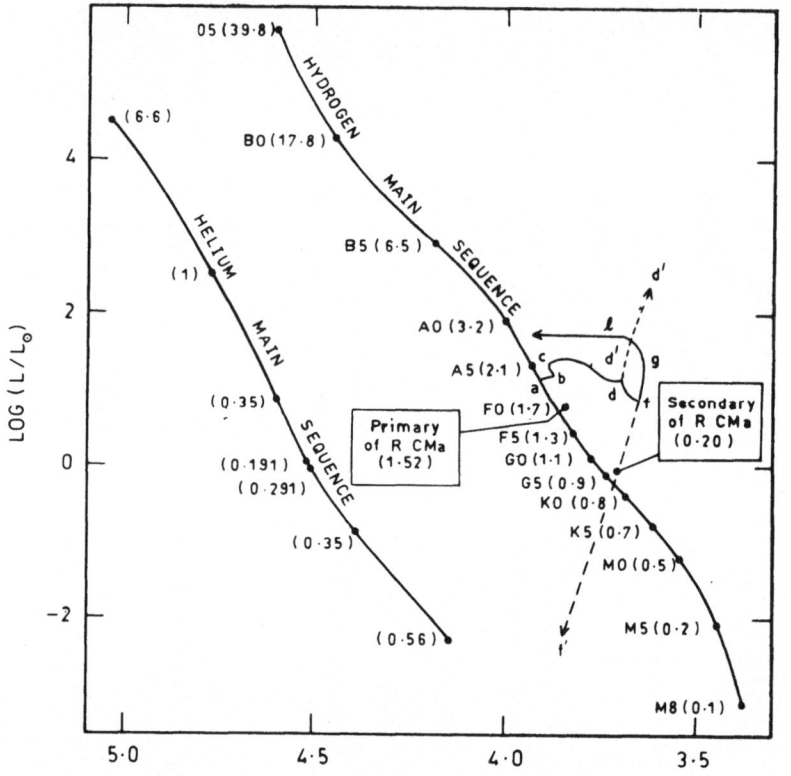

Fig. 4.   Position of the components of R CMa in the HR-diagram.

5. Evolutionary Status of R CMa

R CMa is a typical semi-detached Algol system with a secondary of very low mass.
Evolution of such binaries is found to be due to case B mass transfer (Refsdal and

Weigert 1969; Plavec, 1973). If we assume the mass transfer to be conservative one can start with a system having $M_1 = 1.0$ to $1.2\ M_\odot$ and $M_2 = 0.7$ to $0.5\ M_\odot$. Then there are two difficulties: (i) the more massive star would take about 10^{10} yr to reach the stage of case B mass transfer while the system is much younger; according to Guinan and Ianna (1983) it is a high-velocity star of age between 2 to 6 billion years, and (ii) the less massive star cannot absorb a large mass of 0.8 to 1.0 $M_\odot$ as shows by Neo *et al.* (1977). An assumption of a complete reversal of mass by taking $M_1 = 1.5\ M_\odot$ and $M_2 = 0.2\ M_\odot$ will also not be tenable, because then M_2 would still be in the gravitational contraction phase and so would not be in a position to receive the matter lost by M_1. We have, therefore, to conclude that we are dealing with non-conservative mass transfer.

We suggest that the present primary of R CMa would have started with a slightly less mass of 1.3 $M_\odot$ and the secondary was much more massive with an initial mass of 1.8 $M_\odot$. The evolutionary track of such a secondary as calculated by Refsdal and Weigert (1969) is shown in Figure 4. The star begins to lose mass at point '*d*'. Only a small fraction of the lost mass is absorbed by the companion which has increased its mass to its present value of 1.52 $M_\odot$. The remainder is lost from the system through the outer Lagrangian points. However the fact that the star has now apparently moved down from point f instead of going up indicates that it has lost its hydrogen envelope very rapidly and is already on its way to become a helium white dwarf by contraction along the Hayashi track. Similar conclusion about the final stage of evolution of this component is arrived at by Guinan and Ianna (1983).

References

Dugan, R. S.: 1924, *Contr. Princeton Univ. Obs.* **6**, 49.

Dugan, R. S. and Wright, F. W.: 1939, *Contr. Princeton Univ. Obs.* **19**, 34.

Guinan, E. F.: 1977, *Astron. J.* **82**, 51.

Guinan, E. F. and Ianna, P. A.: 1983, *Astron. J.* **88**, 126.

Irwin, J. B.: 1952, *Astrophys. J.* **116**, 211.

Jordan, F. C.: 1916, *Alleghany Publ.* **3**, 49.

Kitamura, M. and Takahashi, C.: 1962, *Publ. Astron. Soc. Japan* **14**, 44.

Koch, R. H.: 1960, *Astron. J.* **65**, 326.

Kopal, Z.: 1959, in *Close Binary Systems*, Chapman-Hall and John Wiley, London and New York; Chapter VII.

Kopal, Z. and Shapley, M. B.: 1956, *Jodrell Bank Annals* **1**, 143.

Merril, J. E.: 1970, in *Vistas Astron.* **12**, 43.

Needham, J. D., Phillips, J. P., Selby, M. J., and Magro, C. S.: 1980, *Astron. Astrophys.* **83**, 370.

Neo, S., Miyatti, S., Nomoto, K., and Sugimoto, D.: 1977, *Publ. Astron. Soc. Japan* **29**, 249.

Pickering, E. C.: 1904, *Harvard Annals* **46**, 184.

Plavec, M.: 1973, in A. H. Batten (ed.), *Extended Atmospheres and Circumstellar Matter in Spectroscopic Binary Systems*, D. Reidel Publ. Co., Dordrecht, Holland, p. 216.

Radhakrishnan, K. R. and Sarma, M. B. K.: 1982, *Nizamiah and Japal-Rangapur Obs. Contr.*, No. 16.

Refsdal, S. and Weigert, A.: 1969, *Astron. Astrophys.* **1**, 167.

Russell, H. N.: 1902, *Astrophys. J.* **15**, 252.

Sahade, J.: 1963, *Ann. Astrophys.* **26**, 80.

Sato, K.: 1971, *Publ. Astron. Soc. Japan* **23**, 335.
Sitterly, B. W.: 1940, *Astron. J.* **48**, 190.
Sterne, T. E.: 1941, *Proc. U.S. Nat. Acad. Sci.* **27**, 108.
Struve, O. and Smith, B.: 1950, *Astrophys. J.* **111**, 27.
Wendell O. C.: 1909, *Harvard Annals* **69**, 66.
Wolfe, R. H., Horak, H. G., and Storer, N. W.: 1967, in M. Hack (ed.), *Modern Astrophysics*, Gauthier-Villars, Paris, p. 251.
Wood, F. B.: 1946, *Contr. Princeton Univ. Obs.* **21**, 31.

A REVIEW OF EARLY-TYPE CLOSE BINARY SYSTEMS*

K. C. LEUNG

Behlen Observatory, University of Nebraska, Lincoln, Nebr., U.S.A.

(Received 30 August, 1983)

Abstract. This is a review of close binary systems with very early spectral types (B, O, Of, and WR stars). We limit our selection to systems with Roche model photometric solution only. There are 10 contact systems known at present. Three of them (V701 Sco, BH Cen, and RZ Pyx) are located in the vicinity of the zero-age contact (ZC) according to a Spectral Type – Period diagram. Only the first two systems have absolute dimensions available. Both of these fall into the ZC zone in the $\log P - \log m$ diagram and the $\log m - \log R$ diagram. The system TU Mus was thought to be a ZC system is located in the evolved contact zone (EC) in the above diagrams. Both V701 Sco and BH Cen were thought to be having mass ratios about unity. With the much improved light curves of BH Cen and new analysis the mass ratio has been revised to 0.84! This result may favor Shu's model for early-type zero-age contact systems. The evolved systems might be undergone Case A mass exchange except for two systems (V729 Cyg and RY Sct) which might be from Case B. The systems V367 Cyg was classified by Plavec as a W Ser system may be a evolved contact system.

It is interesting to note that V729 Cyg (O7f + OfIa), UW CMa(O7f + O) and CQ Cep(WN7 + O) are all highly evolved contact systems. The Max II of UW CMa and CQ Cep are displaced to 0.78 and 0.80 phase, respectively. The displacement for V729 Cyg is not known due to poor coverage at this phase. The most distinct feature going from Of to Wolf-Rayet stars is the increasing domination of emission lines. It is suspected that this phaseshifts reflect the increasing activity of mass-flow in the common envelope.

There are 8 semidetached systems with reliable absolute dimensions. It is believed that 6 of them are derived from Case A while the remaining 2 are from Case B mass exchange.

Acknowledgement

This work is supported by NSF INT8120404 grant.

* Paper presented at the Lembang-Bamberg IAU Colloquium No. 80, on 'Double Stars: Physical Properties and Generic Relations', held at Bandung, Indonesia, 3–7 June, 1983.

PHOTOMETRY OF THE RS CVn TYPE ECLIPSING
BINARY UV PISCIUM *

P. VIVEKANANDA RAO and M. B. K. SARMA

Centre of Advanced Study in Astronomy, Osmania University, Hyderabad, India

(Received 12 August, 1983)

Abstract. The analysis of the UBV photoelectric study of the short period RS CVn eclipsing binary, UV Psc, has suggested that the primary is a transit with $\theta_e = 27°$, $i = 88°5$, and $k = 0.75$. The spectral type and luminosity of the hotter component is estimated to be G4–6V and that of the cooler component to be K0–2V. Absolute dimensions for the components of UV Psc were derived by combining the present analysis with that of the spectroscopic analysis given by Popper.

The out-of-eclipse observations have showed large amount of scatter and an investigation of this showed that hotter component could be an intrinsic variable. No periodicity for this variation has been fixed due to lack of sufficient data.

1. Introduction

The light variability of UV Piscium (BD + 6°189 = BV 149) was noticed from the photographic patrol plates and photographic light curves for this system were published by Huth (1959) and Strohmeier and Knigge (1960). Hall (1976) classified UV Psc as a member of the short period RS CVn group. These binaries exhibit various peculiarities such as H and K of Ca II in emission indicating the presence of active chromospheres and coronae and also radio and soft X-ray emission, which strongly suggest the presence of a large scale solar-type activity such as the spots on the surface of one or both the components. As a result, these binaries often show a wave like distortion (Hall, 1981; Rodono, 1981) in their out-of-eclipse light curves.

The presence of this photometric wave in the RS CVn binaries distorts the light curve such that the times of primary and secondary minima will be displaced and the shape of the light curve during the eclipses will also be distorted. It is very essential to correct the observations for the effect of this wave before one uses them for determining the times of minima or for solving orbital elements from the light curves. Before undertaking the analysis for the computation of elements, it is essential to know which component of the system (hotter or cooler) is responsible for the wave nature so that suitable corrections can be applied to the observations of the primary and secondary eclipses. Photometric analysis of UV Psc by Carr (1969), Oliver (1974), and Sadik (1979) were concentrated only on the solution of the light curves without taking into consideration the properties of the distortion

---

* Paper presented at the Lembang-Bamberg IAU Colloquium No. 80 on 'Double Stars: Physical Properties and Generic Relations', held at Bandung, Indonesia, 3–7 June, 1983.

wave, the source responsible for it, its effect on the period determination and the nature of the intrinsic variation.

However, by combining our present observations with those of Carr (1969), Oliver (1974), and Sadik (1979), the authors (Vivekananda Rao and Sarma, 1983b; hereafter referred to as Paper II) have showed that UV Psc exhibits a double-peaked distortion wave – suggesting the presence of two cool dark regions located on the surface of the hotter G4–6 component. By use of the constants derived for the wave, the observations during the primary minima were corrected for distortion and the times of minima were determined. With these corrected minima and other times of minima available in the literature, a period study was undertaken by the authors (Vivekananda Rao and Sarma, 1983a; hereafter referred to as Paper I). From this study it was found that no reliable period changes took place in this system during the interval 1966–81.

The present communication mainly deals with the nature and source of the intrinsic variation, the light curve solution after correcting for the distortion wave effect and the evolutionary nature of the system.

2. Observations

UV Psc was observed for 47 nights in U passband and 54 nights in B and V passbands on the standard UBV system during 1976–77, 1977–78, and 1978–79 observing seasons using the 1.22 m reflecting telescope of the Japal–Rangapur Observatory. The details of the photometric equipment and the reduction techniques were described in an earlier paper (Vivekananda Rao and Sarma, 1981b). The UBV light curves obtained during these three observing seasons were already published (Vivekananda Rao and Sarma, 1981c).

3. Rectification

In Paper II it was shown that the hotter component is responsible for the distortion wave in UV Psc. In Section 5 of this paper it is also shown that the primary eclipse is a transit where the cooler component is transiting over the hotter component. By virtue of these properties, all the observations obtained during October 1976–December 1978 were corrected for the distortion wave effects with the aid of the equation

$$l^{cor} = l^{obs} - (1 - f^{tr})[A_1 (\text{wave})\cos \theta + B_1 \sin \theta] -$$
$$- (1 - f^{tr})[A_2 (\text{wave})\cos 2\theta + B_2 \sin 2\theta], \tag{1}$$

where

$$f^{tr}(x, k, p) = \tau(x, k)\alpha_0^{tr}(x, k, p)n \tag{2}$$

is the fractional loss of light of the hotter and spotted star during the primary eclipse. A detailed derivation of A_1 and A_2 (wave-constants) for each year of observations and the procedure adopted for calculating f's during the primary minimum were given in Papers I and II, respectively. The value of $(1-f)$ is equal to unity for points outside the eclipse and also during the secondary minimum, which is the eclipse of the unspotted cool star. All the observed points corrected according to the above equation were converted into magnitudes (by taking the magnitude corresponding to unit light intensity for the respective years) to obtain a unified light curve for each passband during Oct. 1976–Dec. 1978. The observations of Carr (1969), Oliver (1974), and Sadik (1979) were not included in the unified plot in order to avoid the uncertainties involved in transforming their observations into our system because of the differences in the photometric equipment, the comparison and check stars used. The light curves obtained during our period of observation should be identical within the limits of the observational errors as we had removed the effects of the wave from the observations. Figure 1a shows the combined plot of actual observations (not corrected for the wave) whereas Figure 1b shows the combined plot of observations that are corrected for the wave for the V-passband. From these two figures (Figures 1a and 1b) it is evident that the observations of all three years could be merged satisfactorily in all phases except that of the primary eclipse portion. Even here, the differences in the depths observed during the interval had reduced considerably. These conclusions

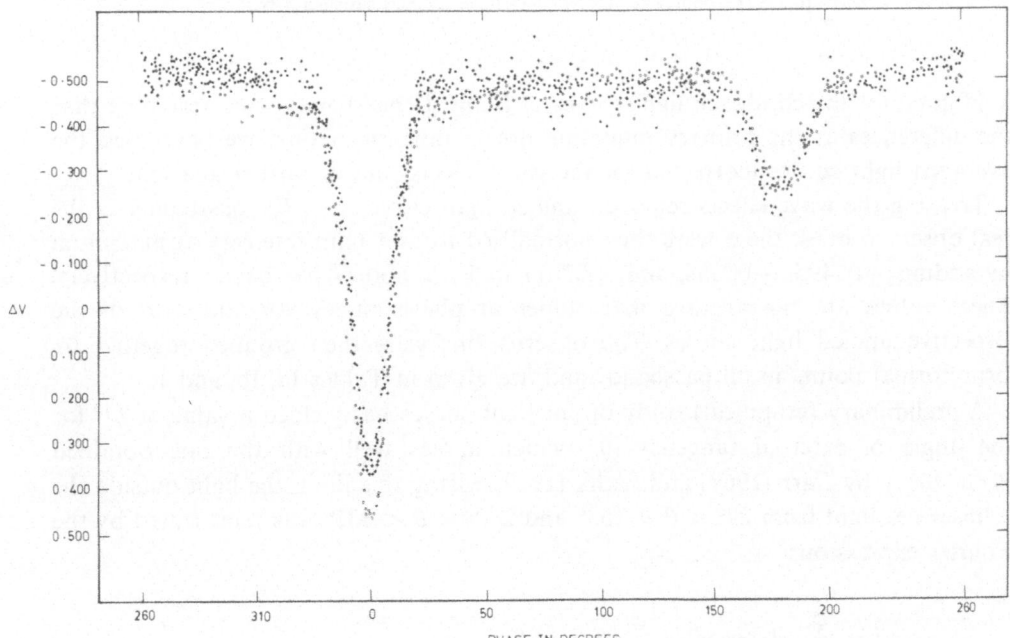

Fig. 1a. UV Psc: Plot of observed ΔV versus phase in degrees for the combined light curve. ○ represents 1976–77 data. + represents 1977–78 data. ● represents 1978–79 data. The effect of the distortion wave can be seen in both primary and secondary minima.

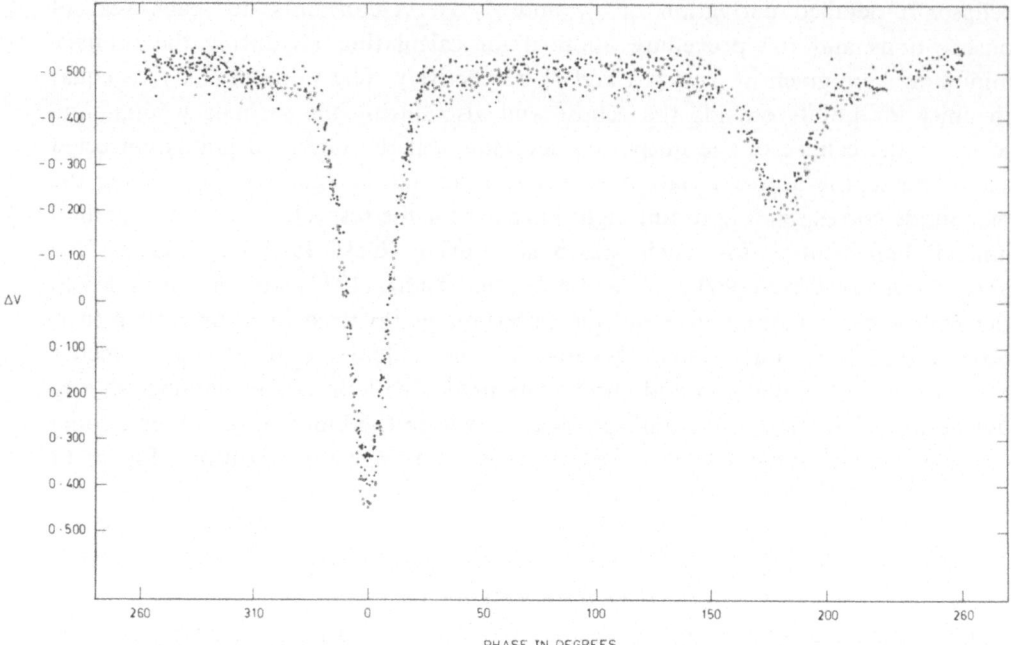

Fig. 1b. UV Psc: Plot of observed ΔV versus phase in degrees for the combined light curve after the removal of the distortion wave. Symbols same as in Figure 1a. It can be seen that except at the zero phase the points at other phases have merged satisfactorily.

hold good for the combined light curves of B and U passbands also. Assuming that the differences at the primary minimum are of intrinsic nature, we have used the averaged light curves (corrected for the wave effects) in our further analysis.

Treating the wave effects removed unified light curves in UBV passbands as the real observed ones, these were then normalised to unit light intensity at maximum by adding $-0^{m}.495$, $-0^{m}.258$, and $+0^{m}.011$ in V, B, and U passbands, respectively. These values are the average magnitudes at phase angles 90° and 270° of the respective unified light curves. The observations were then grouped together to form normal points in all passbands and are given in Tables Ia, Ib, and Ic.

A preliminary (graphical) study of our light curves has yielded a value of 27° for the angle of external tangency, θ_e, which agrees well with the one obtained ($\theta_e = 26^\circ.5$) by Carr (1969) and Sadik (1979). Using this limit, the light outside the eclipses i.e., light from 27° $< \theta <$ 153° and 270° $< \theta <$ 333° was represented by the Fourier expression:

$$l = \sum_{n=0}^{4} A_n \cos n\theta + \sum_{n=1}^{4} B_n \sin n\theta. \tag{3}$$

The values of A_n and B_n along with their probable errors were determined by the

TABLE Ia

UV Psc: normal points in yellow

Phase (in degrees)	ΔV (Var-Comp)	No. of points	Phase (in degrees)	ΔV (Var-Comp)	No. of points
0.010	+0.379	5	55.668	−0.485	12
0.742	+0.388	5	58.724	−0.492	9
1.450	+0.368	4	61.149	−0.514	8
2.587	+0.334	4	64.669	−0.504	10
3.276	+0.357	4	68.665	−0.509	11
3.650	+0.378	4	71.572	−0.521	10
4.334	+0.266	4	74.851	−0.515	11
5.367	+0.280	4	78.699	−0.498	8
5.764	+0.221	5	81.694	−0.489	8
6.390	+0.171	5	84.353	−0.491	10
7.295	+0.144	3	87.335	−0.508	8
7.748	+0.122	4	89.882	−0.491	4
8.012	+0.117	4	·92.275	−0.505	7
8.557	+0.065	5	95.561	−0.487	7
9.329	+0.016	4	98.427	−0.504	5
9.883	−0.044	5	101.163	−0.495	5
10.511	−0.029	5	104.790	−0.489	7
11.496	−0.090	5	107.606	−0.492	9
12.096	−0.126	5	110.270	−0.493	10
12.740	−0.150	5	113.353	−0.486	8
13.785	−0.224	5	116.507	−0.497	11
14.502	−0.219	4	119.739	−0.488	9
15.122	−0.251	5	122.602	−0.509	9
15.628	−0.250	4	125.844	−0.492	9
16.573	−0.288	4	128.447	−0.498	9
17.104	−0.329	4	131.461	−0.511	9
17.669	−0.334	4	134.659	−0.505	10
18.222	−0.336	4	137.529	−0.496	8
18.682	−0.379	4	140.446	−0.494	11
19.264	−0.378	4	143.755	−0.496	9
20.551	−0.404	4	146.784	−0.475	9
21.275	−0.427	4	149.619	−0.472	9
21.808	−0.430	4	152.084	−0.474	4
22.418	−0.446	4	153.398	−0.493	3
23.120	−0.472	5	154.967	−0.452	3
24.395	−0.472	4	156.558	−0.451	3
25.796	−0.479	3	158.884	−0.429	3
26.747	−0.458	3	161.684	−0.406	4
27.707	−0.490	2	163.985	−0.397	3
28.936	−0.458	7	164.664	−0.398	3
31.574	−0.469	4	165.689	−0.375	3
34.653	−0.461	9	166.296	−0.379	4
36.742	−0.493	7	167.922	−0.363	4
40.522	−0.481	7	169.500	−0.326	4
43.377	−0.498	5	170.265	−0.324	4
46.337	−0.483	8	171.557	−0.272	4
49.259	−0.492	9	172.217	−0.273	4
52.657	−0.481	9	173.337	−0.240	3

(continued)

Table Ia (continued)

Phase (in degrees)	ΔV (Var-Comp)	No. of points	Phase (in degrees)	ΔV (Var-Comp)	No. of points
174.517	−0.253	3	278.648	−0.505	9
175.800	−0.226	4	281.714	−0.512	13
177.341	−0.242	3	284.511	−0.510	9
178.279	−0.217	4	287.519	−0.517	11
180.002	−0.210	4	290.566	−0.502	12
181.900	−0.246	3	293.569	−0.522	9
182.673	−0.226	3	296.566	−0.502	14
184.234	−0.251	3	299.688	−0.502	9
185.304	−0.208	3	302.837	−0.495	9
186.303	−0.242	4	306.292	−0.481	10
187.438	−0.244	4	309.408	−0.484	10
188.518	−0.284	3	312.175	−0.480	9
190.167	−0.302	3	315.342	−0.483	7
191.053	−0.308	4	318.516	−0.462	7
191.576	−0.328	3	321.350	−0.477	9
192.655	−0.329	4	324.093	−0.477	5
193.454	−0.350	4	327.760	−0.464	6
194.930	−0.368	4	330.667	−0.479	4
196.020	−0.355	4	333.178	−0.474	3
197.407	−0.403	4	334.181	−0.461	4
198.424	−0.425	3	335.306	−0.429	4
199.681	−0.395	4	336.675	−0.438	5
200.714	−0.440	4	337.623	−0.402	5
202.213	−0.466	4	338.640	−0.427	5
204.298	−0.467	3	339.950	−0.405	4
204.980	−0.440	4	341.455	−0.379	3
206.104	−0.465	4	341.731	−0.351	4
207.725	−0.466	4	342.664	−0.282	4
210.325	−0.458	6	343.301	−0.331	5
213.575	−0.460	6	343.524	−0.290	4
216.468	−0.472	3	344.502	−0.234	3
218.839	−0.482	3	345.311	−0.226	4
221.435	−0.478	7	345.799	−0.236	3
224.464	−0.485	6	346.498	−0.210	4
227.328	−0.495	3	347.536	−0.175	4
231.618	−0.505	5	347.973	−0.115	5
235.443	−0.496	4	348.792	−0.094	4
238.502	−0.501	4	349.832	−0.070	4
241.440	−0.508	7	350.468	−0.004	4
244.654	−0.521	4	351.633	+0.004	5
247.913	−0.514	5	352.257	+0.029	3
251.806	−0.525	4	352.804	+0.095	3
254.471	−0.513	9	353.396	+0.152	5
257.553	−0.516	9	354.459	+0.211	4
260.169	−0.524	9	354.844	+0.210	3
263.424	−0.501	10	355.655	+0.318	4
266.703	−0.510	9	356.204	+0.277	4
268.652	−0.520	4	356.754	+0.310	4
269.963	−0.500	3	357.564	+0.348	3
272.779	−0.515	8	358.197	+0.400	5
275.693	−0.518	12	358.831	+0.360	5

TABLE Ib

UV Psc: normal points in blue

Phase (in degrees)	ΔB (Var-Comp)	No. of points	Phase (in degrees)	ΔB (Var-Comp)	No. of points
0.778	+0.735	4	56.831	−0.214	11
1.388	+0.740	4	60.471	−0.235	10
2.585	+0.721	4	63.382	−0.228	9
3.072	+0.720	4	66.408	−0.239	7
3.648	+0.682	4	69.708	−0.242	11
4.445	+0.644	4	71.851	−0.237	8
4.834	+0.587	3	75.322	−0.255	9
5.795	+0.585	5	78.641	−0.251	6
6.027	+0.583	5	81.147	−0.232	9
6.416	+0.507	5	84.306	−0.233	12
7.064	+0.494	4	87.795	−0.244	5
7.901	+0.441	4	90.099	−0.252	7
8.569	+0.357	5	92.708	−0.246	5
9.074	+0.352	4	96.469	−0.242	9
9.690	+0.305	4	99.080	−0.231	4
10.294	+0.233	4	102.386	−0.240	6
11.128	+0.254	5	105.606	−0.239	6
11.564	+0.176	4	108.454	−0.214	12
11.922	+0.200	4	111.910	−0.221	12
12.548	+0.150	4	115.781	−0.228	10
13.400	+0.088	4	118.981	−0.222	10
14.259	+0.047	4	122.530	−0.234	10
14.728	+0.042	5	125.662	−0.214	9
15.676	0.000	4	128.196	−0.239	9
16.174	+0.012	5	131.818	−0.254	8
16.708	−0.034	4	134.802	−0.234	11
17.545	−0.072	4	138.309	−0.230	9
18.021	−0.073	5	140.913	-0.218	9
18.704	−0.095	4	144.479	−0.212	11
19.530	−0.131	5	147.379	−0.206	9
20.306	−0.132	4	149.827	−0.212	5
21.566	−0.127	4	151.949	−0.220	7
22.128	−0.189	5	154.014	−0.234	3
22.808	−0.181	4	155.691	−0.202	3
23.429	−0.203	5	157.186	−0.165	3
24.830	−0.216	4	160.166	−0.174	4
26.301	−0.221	3	162.728	−0.186	4
27.611	−0.203	3	164.750	−0.135	3
28.534	−0.200	4	165.738	−0.120	3
30.214	−0.199	11	166.377	−0.102	3
33.183	−0.215	11	167.321	−0.091	4
36.262	−0.210	10	169.107	−0.072	3
38.738	−0.211	3	170.372	−0.053	4
41.640	−0.231	7	171.035	−0.054	4
44.631	−0.229	5	172.028	−0.042	4
47.632	−0.221	12	173.590	−0.020	4
51.832	−0.218	10	174.847	−0.017	4
54.314	−0.201	6	175.755	+0.002	4

(continued)

Table Ib (continued)

Phase (in degrees)	ΔV (Var-Comp)	No. of points	Phase (in degrees)	ΔV (Var-Comp)	No. of points
176.860	+0.026	4	285.205	−0.246	8
178.443	+0.015	4	288.230	−0.239	12
179.972	+0.024	4	291.520	−0.252	12
181.872	−0.016	4	294.836	−0.244	10
183.338	−0.007	3	297.282	−0.244	12
184.819	−0.016	3	300.632	−0.221	10
185.760	−0.001	3	303.596	−0.227	6
186.772	−0.019	3	306.753	−0.216	11
187.594	−0.016	4	309.704	−0.214	9
188.754	−0.048	4	312.627	−0.198	8
190.276	−0.063	3	315.260	−0.209	7
191.788	−0.068	4	318.431	−0.212	7
192.461	−0.082	4	321.288	−0.200	10
193.237	−0.104	4	325.637	−0.202	8
194.512	−0.098	3	329.132	−0.206	2
195.506	−0.115	3	330.583	−0.197	3
196.463	−0.135	4	332.314	−0.238	2
197.478	−0.136	4	333.990	−0.197	4
198.801	−0.148	4	335.108	−0.170	5
200.425	−0.171	4	336.354	−0.182	5
201.624	−0.177	4	337.593	−0.163	5
202.802	−0.171	3	338.529	−0.127	5
204.176	−0.186	4	339.963	−0.139	5
205.335	−0.195	3	341.392	−0.112	4
206.184	−0.197	3	342.003	−0.051	5
208.135	−0.187	5	342.803	−0.029	5
211.207	−0.177	7	343.731	−0.018	5
213.980	−0.199	4	344.169	−0.002	4
217.327	−0.204	5	345.200	+0.021	5
221.366	−0.198	8	346.464	+0.090	5
224.745	−0.221	7	347.423	+0.140	5
229.454	−0.252	4	348.137	+0.138	5
233.974	−0.247	7	348.634	+0.161	5
238.462	−0.223	6	349.507	+0.185	3
242.196	−0.232	7	350.344	+0.265	4
245.667	−0.234	5	351.365	+0.318	3
248.379	−0.224	3	352.447	+0.418	4
252.556	−0.262	7	352.870	+0.414	4
255.050	−0.222	9	353.750	+0.485	5
258.872	−0.251	13	354.256	+0.547	4
262.577	−0.261	6	354.978	+0.593	5
264.216	−0.255	9	355.837	+0.656	4
267.602	−0.255	11	356.659	+0.691	5
270.099	−0.267	5	357.244	+0.684	4
273.761	−0.267	11	357.924	+0.698	4
276.378	−0.255	8	358.426	+0.714	4
279.118	−0.245	8	359.165	+0.724	4
282.559	−0.257	16	359.984	+0.754	4

TABLE Ic

UV Psc: normal points in ultraviolet

Phase (in degrees)	ΔU (Var-Comp)	No. of points	Phase (in degrees)	ΔU (Var-Comp)	No. of points
0.066	1.110	4	75.165	−0.011	6
1.324	1.102	4	77.390	0.010	5
2.247	1.132	4	80.727	0.021	11
2.874	1.076	4	83.913	0.047	10
4.444	1.011	4	86.911	0.029	8
5.196	0.951	4	89.884	0.037	6
5.603	0.936	4	92.118	0.014	5
6.444	0.845	4	95.988	0.041	6
6.852	0.850	4	97.723	0.048	7
7.180	0.850	4	102.548	−0.012	8
8.034	0.750	4	106.149	0.039	5
9.020	0.657	4	108.365	0.039	11
9.631	0.628	4	111.421	0.056	10
10.437	0.556	4	114.750	0.046	9
10.926	0.543	4	118.665	0.036	13
11.772	0.494	3	122.676	0.027	12
12.285	0.451	4	125.779	0.055	6
13.402	0.391	4	127.790	0.022	9
14.479	0.315	4	131.686	0.031	6
14.905	0.257	3	134.458	0.064	8
15.662	0.328	3	137.356	0.035	10
16.164	0.298	4	141.244	0.042	10
16.980	0.293	4	144.893	0.048	7
17.673	0.181	3	147.761	0.050	7
18.512	0.166	4	150.174	0.040	5
19.694	0.133	4	152.162	0.053	8
20.028	0.118	4	154.819	0.039	3
21.387	0.091	3	157.328	0.107	3
22.207	0.115	3	160.586	0.105	4
22.577	0.082	4	165.036	0.100	4
23.408	0.091	3	166.706	0.089	3
23.898	0.038	4	167.139	0.152	3
25.523	0.048	4	170.427	0.150	3
27.262	0.060	4	171.211	0.152	3
29.483	0.069	11	171.771	0.160	2
32.628	0.046	10	174.251	0.168	3
35.479	0.074	6	175.393	0.202	3
37.456	0.030	6	176.286	0.163	2
41.759	0.046	9	178.535	0.196	3
44.801	0.018	2	179.985	0.206	3
47.327	0.058	13	181.934	0.182	2
51.591	0.048	5	183.970	0.190	3
54.524	0.055	6	185.612	0.196	4
57.109	0.047	6	187.723	0.180	4
61.427	0.007	8	189.045	0.153	3
64.384	0.019	5	189.910	0.144	2
67.149	−0.008	7	191.675	0.199	3
70.980	0.012	10	192.484	0.145	4

(continued)

Table Ic (continued)

Phase (in degrees)	ΔU (Var-Comp)	No. of points	Phase (in degrees)	ΔU (Var-Comp)	No. of points
193.722	0.121	4	308.090	0.045	10
196.232	0.098	4	311.227	0.060	7
197.242	0.118	4	314.112	0.061	8
199.820	0.038	2	318.513	0.053	5
200.907	0.063	2	320.986	0.068	7
201.511	0.072	3	325.875	0.083	8
203.177	0.064	3	330.102	0.066	6
205.426	0.056	3	333.995	0.082	4
206.041	0.061	3	335.919	0.086	5
208.340	0.068	4	337.307	0.089	4
211.006	0.082	5	338.154	0.112	4
213.901	0.063	5	339.882	0.134	4
217.483	0.064	5	340.380	0.152	4
221.512	0.066	8	340.955	0.116	3
225.575	0.045	9	342.534	0.224	3
233.642	0.017	10	342.817	0.201	3
239.205	0.021	6	343.729	0.266	4
243.193	0.023	8	344.178	0.242	5
248.524	0.011	8	344.618	0.248	4
254.073	0.021	9	344.998	0.075	4
257.624	0.013	11	346.489	0.391	4
260.492	0.006	11	347.042	0.416	4
264.450	0.003	11	347.498	0.400	3
267.771	−0.001	9	348.466	0.452	4
270.206	−0.020	5	348.793	0.429	4
273.773	−0.013	9	349.241	0.465	4
276.323	0.005	11	351.399	0.656	4
280.075	0.000	10	352.593	0.729	5
283.243	0.016	15	353.066	0.768	3
286.607	−0.007	10	353.606	0.800	4
289.406	0.010	10	354.463	0.852	3
292.372	0.009	12	355.603	1.000	4
296.023	0.025	13	356.307	1.071	3
298.154	0.034	6	356.978	1.062	4
301.165	0.052	11	357.656	1.083	4
305.266	0.040	9	358.262	1.097	3
			359.106	1.142	4

method of least squares and the derived coefficients are given in Table II. For rectification of light, the $\cos 3\theta$, $\cos 4\theta$, and all the sine terms representing the light were substracted from the wave corrected light.

The reflection coefficients C_0, C_1, and C_2 (given in Table II) were obtained using the following equations (Russell and Merrill, 1952):

$$C_0 = -(0.75 - 0.25 \cos^2 i) \frac{G_c + G_h}{G_c - G_h} A_1 \operatorname{cosec} i, \tag{4}$$

$$C_1 = -A_1, \tag{5}$$

$$C_2 = -0.25 \frac{G_c + G_h}{G_c - G_h} A_1 \sin i. \tag{6}$$

The G_c and G_h values used in the above equations were obtained from Cester's (1969) tables for the spectral types (G5V, K1V) of the individual components and an inclination (i) equal to $88°.5$ was used. The coefficients thus derived were used in the equation

$$Nz = \frac{-4(A_2 - C_2)}{(A_0 - C_0) - (A_2 - C_2)} \tag{7}$$

to compute the values of z. The values of N used in the above equation and given in Table III were taken from Princeton contribution No. 26 except for the value which represents the cooler component in the ultraviolet passband. In this case the value of N was estimated using the following equation given by Russell and Merrill

TABLE II

UV Psc: Fourier and rectification coefficients

	V	B	U
A_0	+0.9874	+0.9629	+0.9671
	±0.0001	±0.0001	±0.0001
A_1	−0.0042	−0.0027	−0.0069
	±0.0001	±0.0001	±0.0001
A_2	−0.0292	−0.0263	−0.0306
	±0.0001	±0.0001	±0.0001
A_3	−0.0008	−0.0012	−0.0050
	±0.0001	±0.0001	±0.0001
A_4	−0.0056	+0.0012	+0.0013
	±0.0001	±0.0001	±0.0001
B_1	−0.0017	−0.0009	−0.0022
	±0.0001	±0.0001	±0.0001
B_2	−0.0010	−0.0012	+0.0015
	±0.0001	±0.0001	±0.0001
B_3	+0.0064	+0.0079	+0.0100
	±0.0001	±0.0001	±0.0001
B_4	−0.0046	−0.0022	−0.0036
	±0.0001	±0.0001	±0.0001
C_0	+0.0116	+0.0102	+0.0462
C_1	+0.0042	+0.0027	+0.0069
C_2	+0.0039	+0.0034	+0.0154
z	0.045	0.038	0.049

(1952):

$$N = \frac{(15+x)(1+y)}{15-5x},$$

where x (limb darkening coefficient) = 1.0 and y (gravity darkening coefficient) = 2.012 were used (Kopal, 1959), Table 4–5).

TABLE III

UV Psc: adopted values of 'x' and 'N' for UBV colours

Colour	Star	x	N
V	hotter component	0.6	2.6
	cooler component	0.8	3.2
B	hotter component	0.8	3.2
	cooler component	0.8	3.2
U	hotter component	0.8	3.2
	cooler component	1.0	4.8

The phase angle θ was rectified using the equation

$$\sin^2 \Theta = \frac{\sin^2 \theta}{1 - z \cos^2 \theta}, \tag{9}$$

where $z = 0.044$ (an average value for all UBV passbands). The light curves were then rectified using the formula given below and the coefficients given in Table II.

$$l_{\mathrm{rec}}^{\mathrm{cor}} = [l^{\mathrm{cor}} + C_0 + C_1 \cos\theta + C_2 \cos 2\theta - \sum_{n=1}^{4} B_n \sin n\theta -$$

$$- \sum_{n=3}^{4} A_n \cos n\theta] / [(A_0 + C_0) + (A_2 + C_2)\cos 2\theta]. \tag{10}$$

A plot of $l_{\mathrm{rec}}^{\mathrm{cor}}$ versus rectified phase Θ for UBV passbands is shown in Figure 2.

4. Intrinsic Variation

4.1. SOURCE OF THE INTRINSIC VARIATION

A close inspection of Figures 1a-1i (Vivekananda Rao and Sarma, 1981b) show that outside of the eclipses, the observations have a scatter of $\pm 0^{m}\!.05$ which is larger than the estimated internal probable error ($\pm 0^{m}\!.02$). Such a large scatter was also found in some other RS CVn type binaries like SS Cam (Arnold *et al.*, 1979), TY Pyx (Vivekananda Rao and Sarma, 1981a), and AR Lac (Chambliss, 1976).

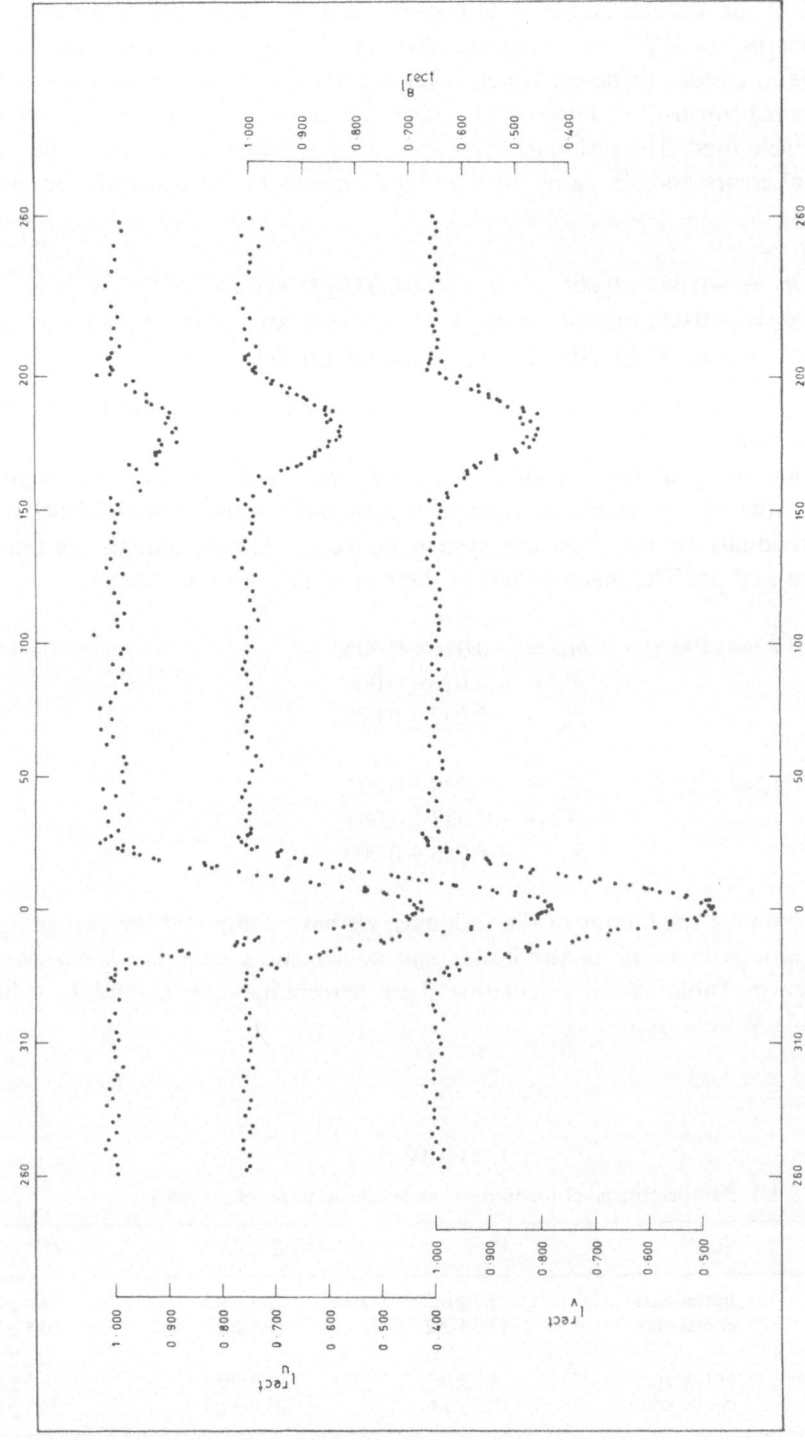

Fig. 2. UV Psc: Plot of l_{rec}^{cor} versus rectified phase in U, B, and V colours for normal points.

This variation has been attributed to the intrinsic variability of one or both components of the system. Hence, it is possible that the large scatter observed in the light curves of UV Psc could be due to the intrinsic variability of its component(s). In order to detect which component(s) of the system is responsible for the observed intrinsic variation all the rectified individual observations outside the eclipses were used. The residuals $(1 - l_{\text{rec}}^{\text{cor}}) = \delta l_i$ were found to be larger than the observational errors and the range of the nightly means of the residuals are given below.

$$\delta l_V = -0.042 \pm 0.005 \quad \text{to} \quad +0.033 \pm 0.003,$$
$$\delta l_B = -0.032 \pm 0.004 \quad \text{to} \quad +0.046 \pm 0.007,$$
$$\delta l_U = -0.047 \pm 0.004 \quad \text{to} \quad +0.058 \pm 0.004.$$

The following methods were used to detect the component responsible for this intrinsic variation.

(i) The largest and best defined residuals were noticed on two nights: JD 2443490.5 (where the residuals showed a systematic trend) and JD 2443861.5 (where the residuals did not show any systematic trend). These residuals are shown in Figures 3a and 3b. The mean values of these residuals are given below:

$$\text{JD } 2443490.5: \quad \delta l_V = -0.026 \pm 0.004,$$
$$\delta l_B = -0.026 \pm 0.004,$$
$$\delta l_U = -0.032 \pm 0.008,$$

$$\text{JD } 2443861.5: \quad \delta l_V = +0.038 \pm 0.003,$$
$$\delta l_B = +0.037 \pm 0.003,$$
$$\delta l_U = +0.058 \pm 0.003.$$

In order to estimate the colour of the residuals, we have computed the percentages of these variations in terms of the hotter and cooler star's light in all passbands and are given in Table IV. In calculating these percentages the L_h and L_c values given in Table V were used.

TABLE IV

UV Psc: percentage of the intrinsic variations in terms of L_h and L_c

Date	Star	V	B	U
JD 2443490.5	hotter star	3.2 ± 0.5	3.1 ± 0.5	3.6 ± 0.7
	cooler star	14.2 ± 2.2	16.1 ± 2.5	30.5 ± 7.6
JD 2443861.5	hotter star	4.0 ± 0.4	4.4 ± 0.4	6.5 ± 0.4
	cooler star	18.1 ± 1.6	23.0 ± 1.8	55.2 ± 2.8

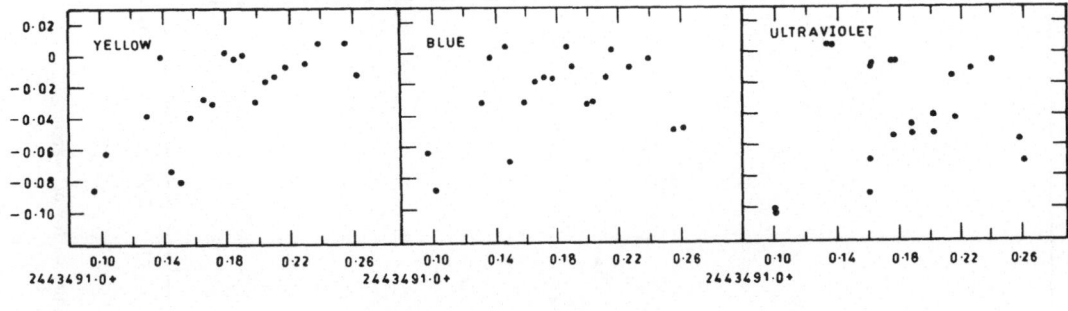

Fig. 3a.   UV Psc: Plot of (O-C) in light versus heliocentric Julian Day for *UBV* colours. The (O-C) represents the residual between the rectified observed points outside the eclipse and the computed light (equal to unity). A systematic trend can be seen in this plot.

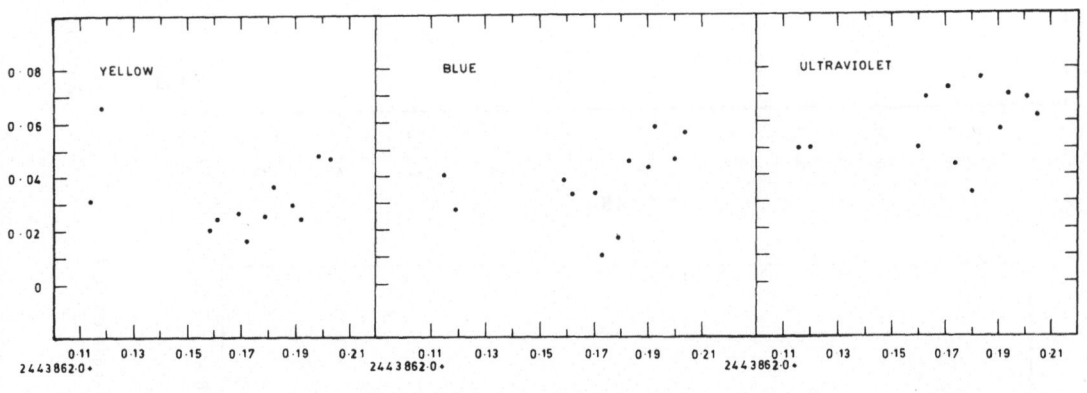

Fig. 3b.   UV Psc: Same as Figure 3a for another day where no systematic trend is found.

It is very clear from Table IV that these residuals have nearly the same percentage of the hotter star's light in all the three passbands and on both the nights.

(ii) Taking mean residuals on all the nights except those in eclipses, we have pplotted δl_V versus δl_B and also δl_U versus δl_U in Figures 4a and 4b. The first figure gave a slope of $(\delta l_B/\delta l_V) = 1.15$ and it is expected to have a slope of 1.02 if the hotter (G5) component is varying whereas a value of 0.78 is expected if the cooler (K1) star is varying. The second figure gave a slope of $\delta l_U/\delta l_V = 1.20$ and a value of 0.80 is expected if the hotter component is variable and a value of 0.54 if the cooler component is variable. Thus, the slopes obtained here are in agreement with the values expected if the hotter component is varying and hence we conclude that the hotter star is responsible for the intrinsic variation.

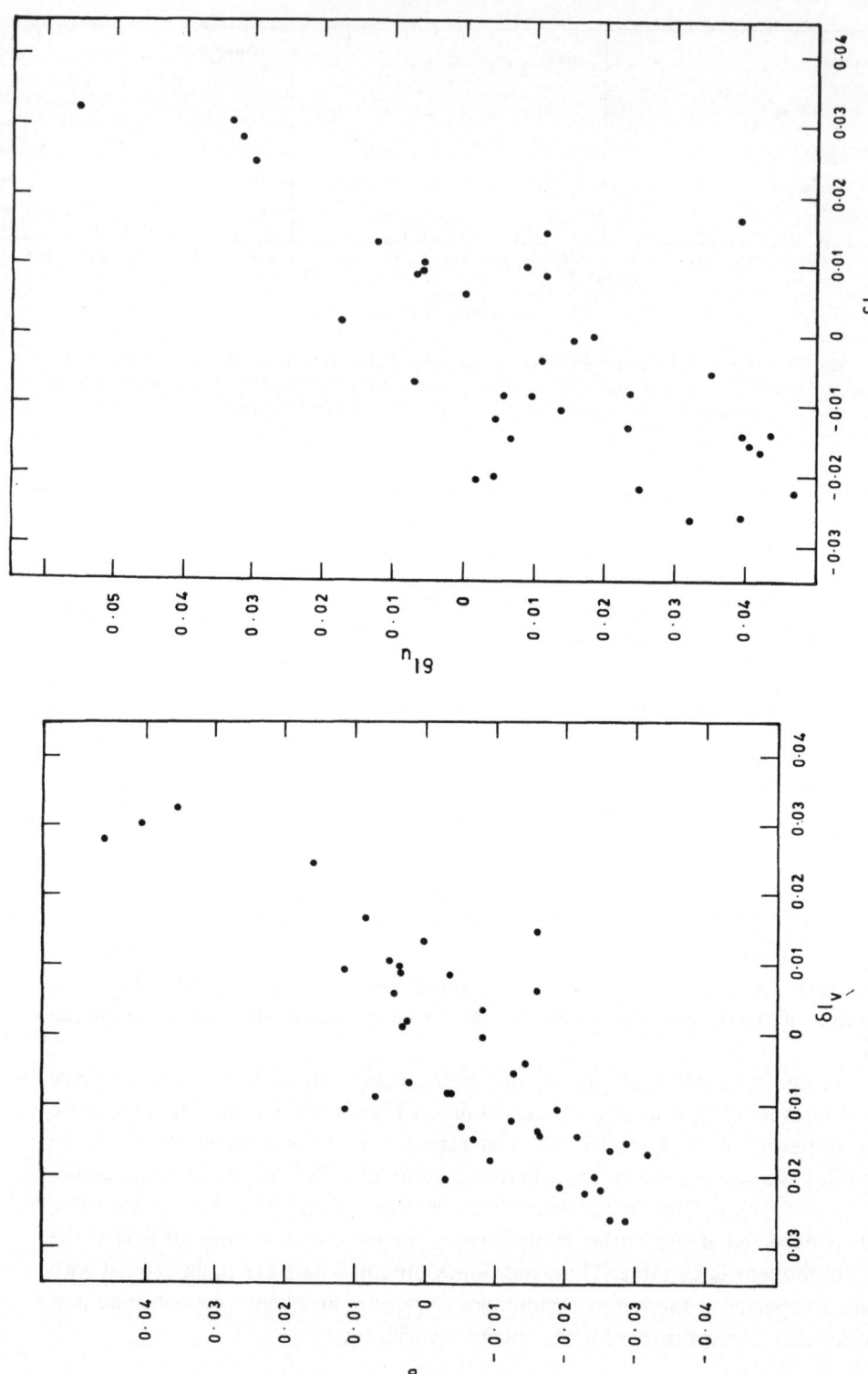

Fig. 4b. UV Psc: Same as Figure 4a for δl_V versus δl_U.

Fig. 4a. UV Psc: Plot of δl_V versus δl_B. Here δl represents the mean residual of each night for outside eclipse observations.

4.2. PERIODICITY OF THE INTRINSIC VARIATION

Individual residuals within each night were examined carefully to determine periodic nature of the variability. The residuals were found to be more or less constant during a night without any significant trend except the night of JD 2443490.5. Even an examination of the residuals over several nights did not yield any trend or periodicity except for a random distribution of the residuals. Since no periodicity could be established for these variations, it is not possible to remove its effect from the eclipses data. This will not have any significant effect on the solution of the primary and secondary eclipses for computation of elements as the percentage of contribution of these variations is only about 5% of the total light and in addition, due to averaging of data from large number of nights, the net effect is expected to be negligible.

5. Solution

A plot of l_{rec}^{cor} versus $\sin^2 \Theta$ (for normal points) for all passbands is made and the appearance of the light curves suggests that the eclipses are total. Further as $\chi_{0.8}^{sec} > \chi_{0.8}^{pr}$ the primary eclipse is found to be a transit and secondary an occultation. This conclusion is in agreement with that of Carr (1969) and Sadik (1979). Modified Wellmann's method was used to solve the light curves in all passbands and the details of this method were given by Vivekananda Rao and Sarma (1981a). Tsesevich (1940) tables were used to read the 'p' values for the assumed limb darkening coefficients of $x_h = 0.6$ and $x_c = 0.8$ in V passband; $x_h = x_c = 0.8$ in B passband; and $x_h = 0.8$ and $x_c = 1.0$ in U passband. As the solution of primary eclipse alone and primary plus secondary eclipse combined together yielded about the same value for $\sin^2 \Theta_e$, the data for both primary and secondary eclipses together were used to derive $\sin^2 \Theta_e$ for different 'k' values. But in the case of U light curve, due to very small depth of the secondary eclipse, only primary eclipse was used to derive the value of $\sin^2 \Theta_e$. Figures 5a, 5b, and 5c show the plot of $\sum w(l_0 - l_c)^2$ against k for VBU passbands. A weighted average of all the three passbands gave $k = 0.75 \pm 0.01$. This value of k agrees quite well with the value of $k = 0.75$ determined by Carr (1969) and Oliver (1974) and $k = 0.762$ determined by Sadik (1979). Corrections for the depths (primary and secondary) of the hand drawn curve were obtained according to the modified Wellmann's method. These corrections were small (~ 0.003 in luminosity units) in B and V passbands and gave the same value for the geometric depth. But in the case of ultraviolet light curve, the correction amounted to a large value (~ 0.04 in luminosity units) which the observations did not permit. Hence, no depth correction could be applied to the U light curve. Since both B and V light curves gave an unique solution, elements were computed using these curves for $k = 0.75$. The averaged elements for the system are given in Table V. Using these elements, theoretical curves were computed for VBU passbands and Figures 6a–6f show the

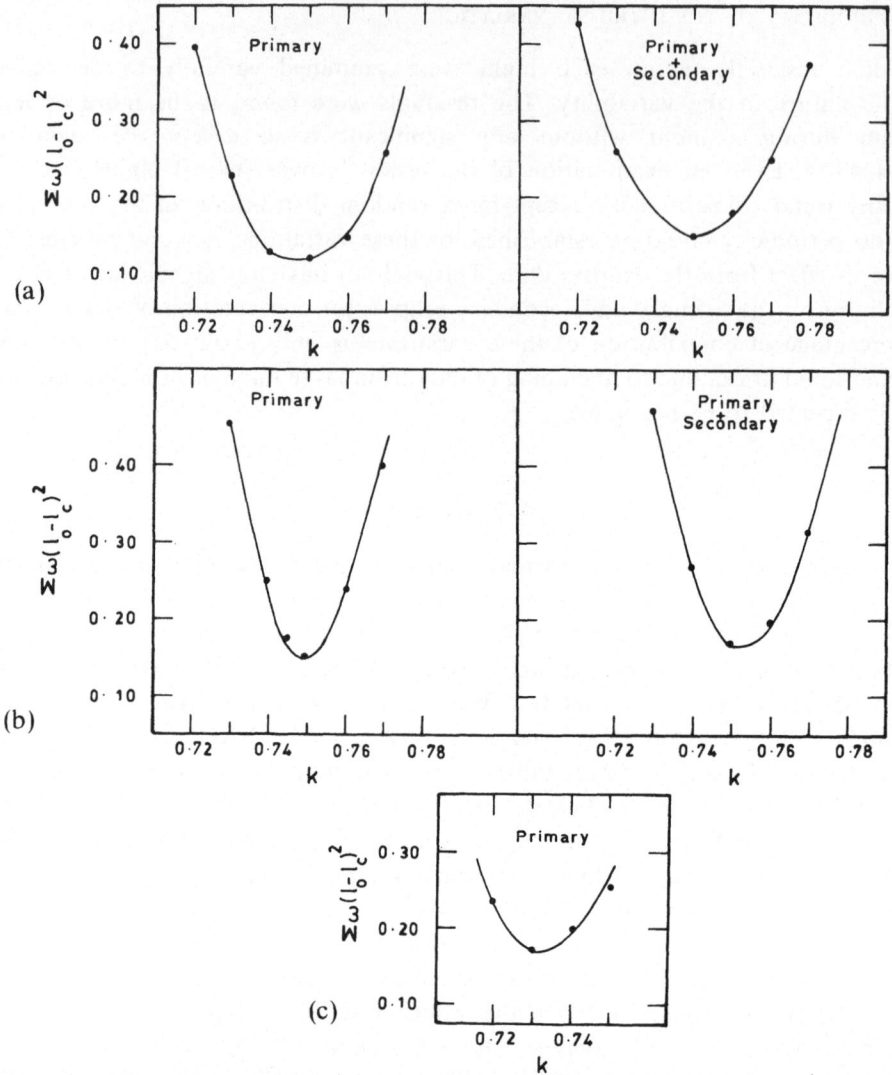

Fig. 5. UV Psc: Plot $\sum w(l_0 - l_c)^2$ versus k for primary and secondary minima. (a) in yellow colour
(b) in blue colour and (c) ultraviolet colour (primary only).

plot of the observed normals and the theoretical fit (solid line). The fit of the
theoretical curves to the observed normal points in all passbands (in primary and
secondary) is found to be quite satisfactory.

The absolute radii $R_{h,c}$ for the two components were estimated from the
relationship

$$m_h + m_c = \frac{1}{74.55} \frac{1}{P^2} \left(\frac{R_{h,c}}{r_{h,c}} \right)^3, \tag{11}$$

where P is the orbital period in days, $r_{h,c}$ are the fractional radii of individual

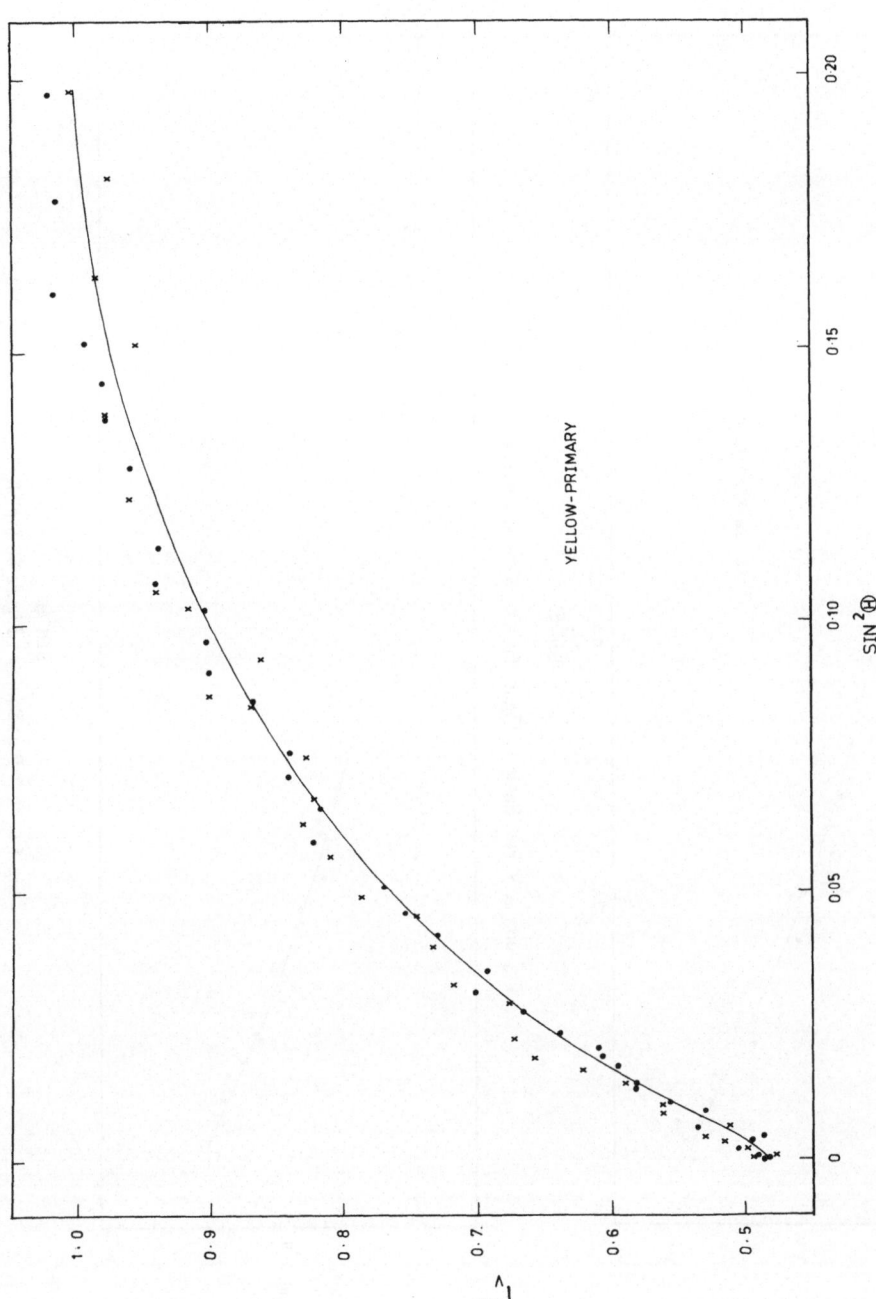

Fig. 6a. UV Psc: Plot of I_{rec}^{cor} versus $\sin^2 \Theta$ for yellow primary minimum. Dots represent the observed normal points on the ascending branch and crosses the observed normal points on the descending branch of the light curve. The theoretical light curve is shown as continuous line.

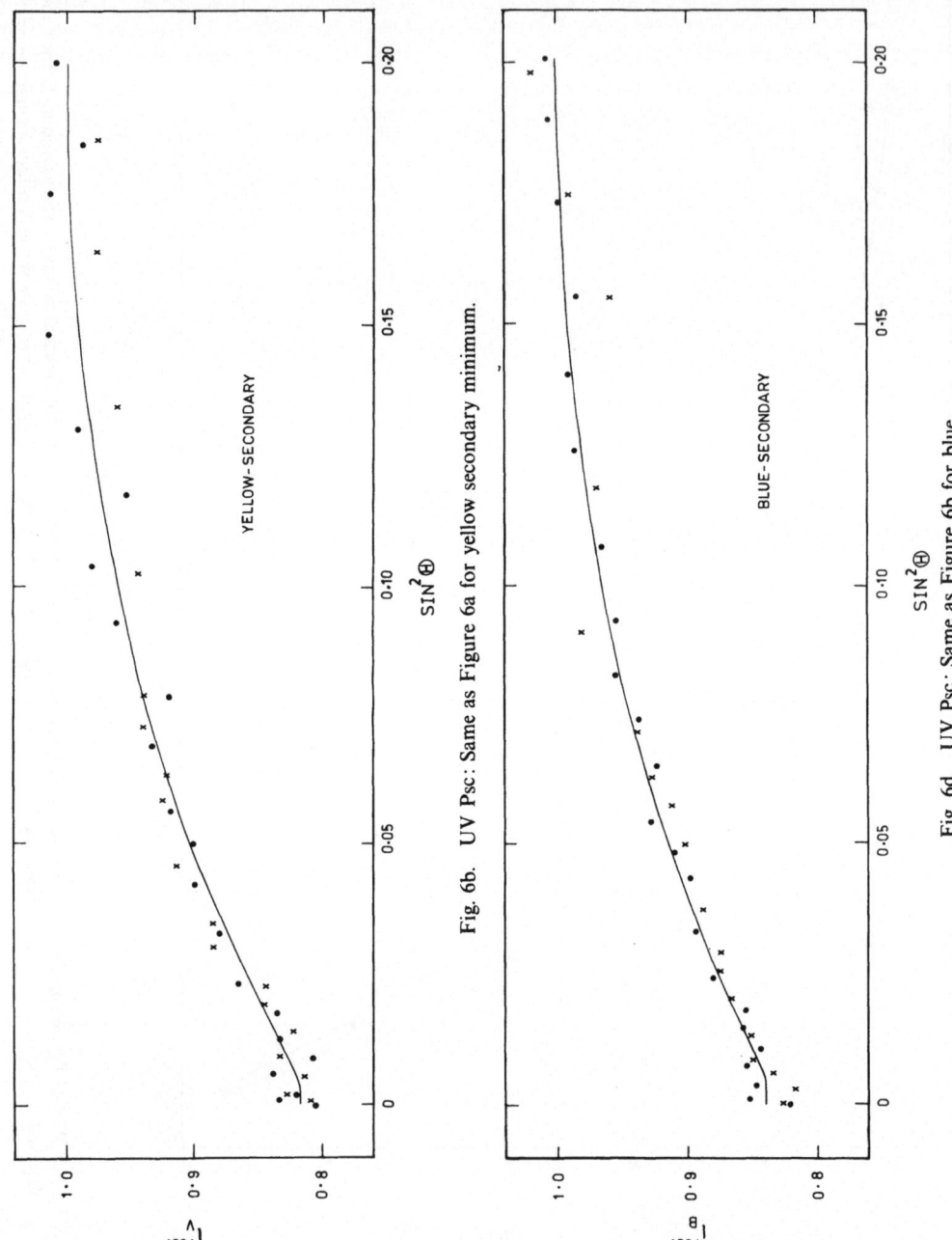

Fig. 6b.   UV Psc: Same as Figure 6a for yellow secondary minimum.

Fig. 6d.   UV Psc: Same as Figure 6b for blue.

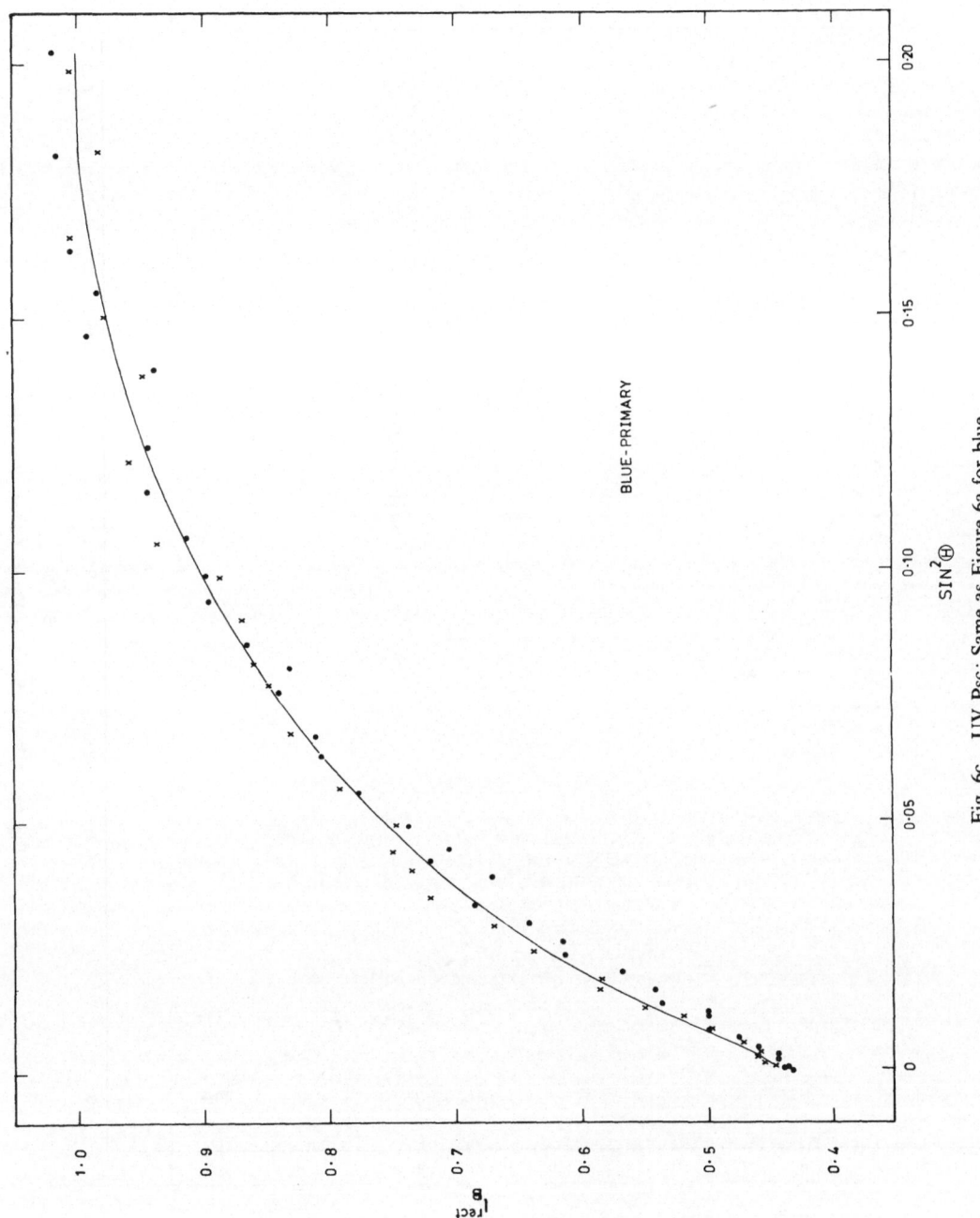

Fig. 6c. UV Psc: Same as Figure 6a for blue.

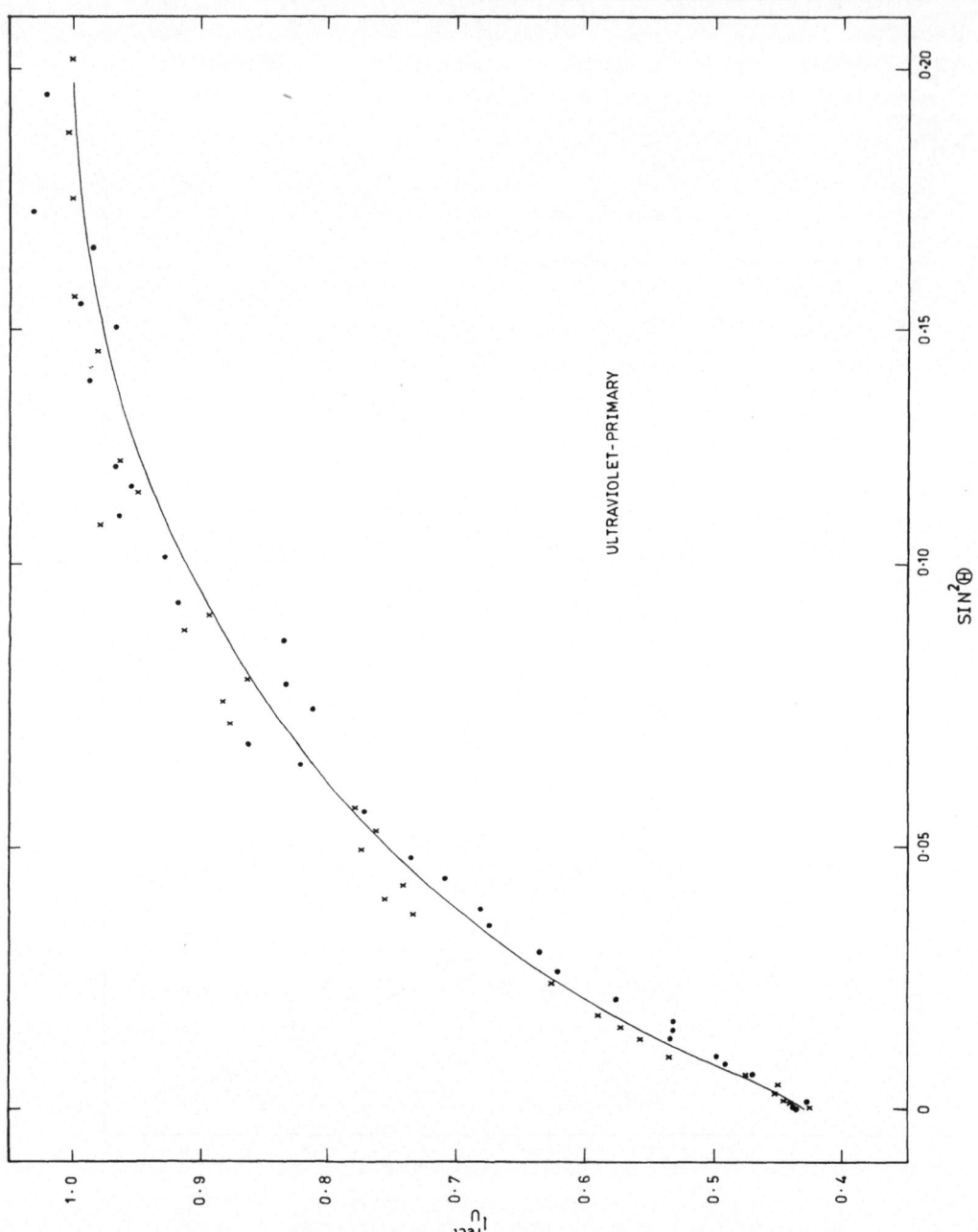

Fig. 6e.   UV Psc: Same as Figure 6a for ultraviolet.

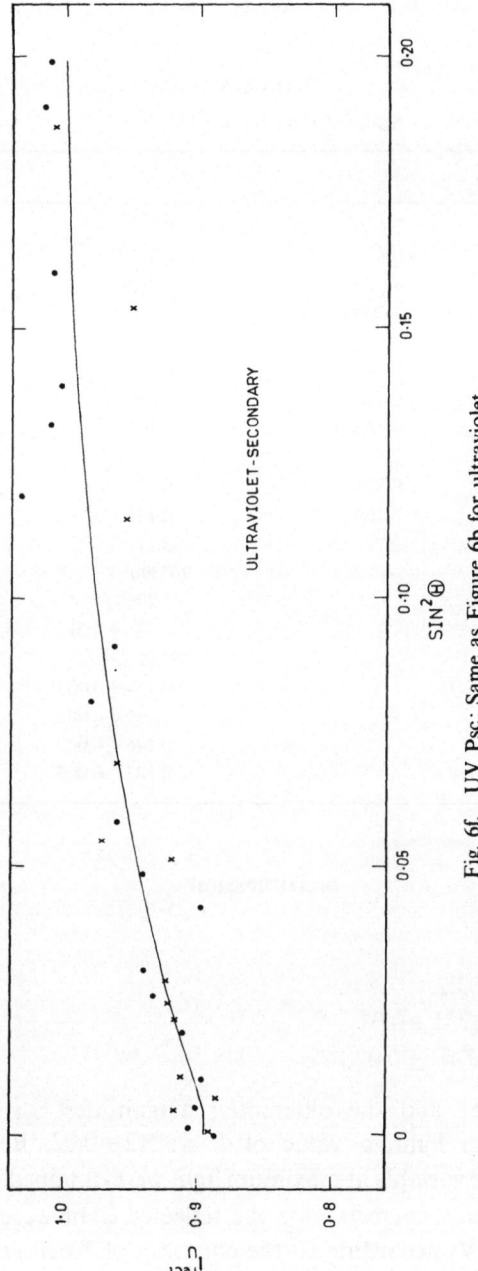

Fig. 6f. UV Psc: Same as Figure 6b for ultraviolet.

components, and m_h and m_c are the individual masses of the components. Popper (1976) had estimated the masses as 1.20 $m_\odot$ and 0.90 $m_\odot$ for the hotter and cooler components, respectively. Thus, the estimated radius of the hotter and cooler components of UV Psc will be 1.24 ± 0.01 $R_\odot$ and 0.93 ± 0.01 $R_\odot$, respectively.

TABLE V

Adopted elements for UV Psc

Element	V	B	U
x_h	0.6	0.8	0.8
x_c	0.8	0.8	1.0
α_0^{tr}	1.030	1.043	1.043
α_0^{oc}	1.000	1.000	1.000
$1 - l_0^{tr}$	0.518	0.559	0.570
$1 - l_0^{oc}$	0.182	0.161	0.105
L_h	0.818	0.839	0.895
L_c	0.182	0.161	0.105
J_h/J_c	2.54	2.95	4.59
L_h''	0.831	0.851	0.905
L_c''	0.169	0.149	0.095
J_h''/J_c''	2.77	3.21	5.36
Θ_e		$26°390$	
p_0		-1.20	
k		0.75 ± 0.01	
j		$88°53 \pm 0.5$	
$r_h = a_h$		0.254 ± 0.002	
$r_c = a_c$		0.191 ± 0.002	
b_h		0.248 ± 0.002	
b_c		0.187 ± 0.002	

6. Discussion

By use of the values

$$V = 9^m\!.62 \pm 0.02,$$
$$B - V = +0^m\!.51 \pm 0.01,$$
$$U - B = -0^m\!.01 \pm 0.02,$$

for the comparison star and the differential magnitudes corresponding to unit luminosity at maximum light, a value of $V = 9^m\!.12 \pm 0.02$, $B = 9^m\!.87 \pm 0.02$, and $U = 10^m\!.13 \pm 0.02$ for the variable at maximum light were obtained. The luminosities of the individual components corrected for the reflected light were obtained from L_h and L_c (given in Table V) according to the equation of Koch et al. (1970)

$$\left. \begin{aligned} L_h' &= L_h - 0.8 L_c (ab)_h \, E_h/E_c, \\ L_c' &= L_c - 0.8 L_h (ab)_c \, E_c/E_h, \end{aligned} \right\} \tag{12}$$

where the values of E_h and E_c were taken from Cester's (1969) tables. The results of this calculation applied to the luminosity in V, B, and U are given in Table V as L_h' and L_c' after normalization. From these values the magnitudes and colours for the two components were obtained.

	hotter	cooler
V	$9^m32 \pm 0.02$	$11^m05 \pm 0.02$
$B-V$	$0^m72 \pm 0.01$	$0^m89 \pm 0.01$
$U-B$	$0^m20 \pm 0.02$	$0^m74 \pm 0.02$

Assuming that the interstellar reddening is negligible, the colours obtained above suggest a spectral type of G4–6 ($T_e = 5520 \pm 100$ K) for the hotter component and K0–2 ($T_e = 4740 \pm 100$ K) for the cooler component (Allen, 1976). These spectral types are in agreement with those derived by Carr (1969), Oliver (1974), and Sadik (1979). The spectral type of the cooler component can also be estimated by assuming that the radiation of both the hotter and cooler components may be approximated by the Planck's function over the passbands of U, B, and V filters (Wood, 1971). Assuming a temperature of $T_e = 5520$ K for the hotter component of spectral type G5, we got a temperature $T_e = 4654$ K in V, $T_e = 4700$ K in B, and $T_e = 4525$ K in U for the cooler component with an uncertainty of ± 100 K. The average temperature $T_e = 4626 \pm 100$ K for the cooler component corresponds to a spectral type of K0-2. Thus as the spectral types of the components derived from the above method and also estimated from the colours closely agree with one another, a spectral type of G4–6 ($T_e = 5520 \pm 100$ K) for the hotter component and K0–2 ($T_e = 4740 \pm 100$ K) for the cooler component is justified. This conclusion is in agreement with the results of Sadik (1979) who has estimated a temperature of $T_e = 5740$ K and $T_e = 4750$ K for the hotter and cooler components, respectively. From the values of temperature and radii given above for each component the bolometric luminosities estimated from Stefan–Boltzmann's law are

$$\left.\begin{array}{l} \log (L_h/L_\odot) = +0.10 \pm 0.10, \\ \log (L_c/L_\odot) = -0.41 \pm 0.11, \end{array}\right\} \tag{13}$$

for the hotter and cooler components, respectively. From these luminosities, the absolute bolometric magnitudes are determined using

$$M_{bol} = 4.75 - 2.5 \log (L/L_\odot) \tag{14}$$

and these values are

$$\left.\begin{array}{l} M_{bol}(\text{hotter}) = +4^m50 \pm 0.25, \\ M_{bol}(\text{cooler}) = +5^m83 \pm 0.28. \end{array}\right\} \tag{15}$$

Applying standard bolometric correction of $-0\overset{m}{.}07$ and $-0\overset{m}{.}27$ for the Main Sequence stars of spectral type G5 and K1 the absolute visual magnitudes determined for the two components are

$$\left. \begin{array}{l} M_V(\text{hotter}) = +4\overset{m}{.}57\pm0.25, \\ M_V(\text{cooler}) = +6\overset{m}{.}10\pm0.28. \end{array} \right\} \tag{16}$$

These absolute visual magnitudes are in good agreement with the values of M_V determined from the photometric distance of $0\overset{''}{.}012\pm0.006$ $[M_V = (\text{hotter}) = = +4\overset{m}{.}72\pm0.86$ M_V (cooler) $= +6\overset{m}{.}45\pm0.86]$ given by Dworak (1973). The estimated distance obtained from these magnitudes is of the order of 85 parsec and hence our assumption of no interstellar reddening is quite justified. The absolute visual magnitudes given by Allen (1976) (1976) for the corresponding spectral types of G5V and K1V are $M_V(\text{hotter}) = +5\overset{m}{.}1$ and M_V (cooler) $= 6\overset{m}{.}2$, respectively.

Assuming the hotter and cooler components to be of spectral type and luminosity class as G4–6V and K0–2V, a mass ratio of $m_2/m_1 = 0.80\pm0.04$ was obtained from Allen (1976). This value of mass ratio is in close agreement with the value of $m_2/m_1 = 0.75$, given by Popper (1976) from his spectroscopic studies. From Plavec and Kratochvíl's (1964) tables, for a mass ratio of 0.80, the sizes of the Roche lobes are found to be $r_h^* = 0.395$ and $r_c^* = 0.354$. Comparing these values with $b_h = 0.248$ and $b_c = 0.187$ obtained in the present investigation (Table V), we conclude that both the components of UV Psc are well within their Roche lobes suggesting that UV Psc is a detached system like other members of the RS CVn group.

The Main-Sequence luminosity class derived for both the components of UV Psc in the present investigation agrees with that of Carr (1969) but differs from that of Oliver (1974) and Sadik (1979). Oliver (1974) had obtained the primary eclipse to be an occultation and the secondary to be transit and had determined the spectral types and luminosity classes as G2V or G2IV with ultraviolet excess for the hotter component and K0IV for the cooler component. But, the colours obtained from our investigation suggest that the hotter component must be a normal Main Sequence star of spectral type G5V rather than a subgiant with UV-excess as suggested by Oliver (1974). Next, the nature of the primary eclipse as occultation rather than a transit made Oliver (1974) to conclude that the cooler component should be a subgiant. Sadik (1979) concluded the primary eclipse to be a transit and the secondary an occultation and suggested that the spectral type and the luminosity class for the hotter and cooler components as G2V and K0IV, respectively. In such a case, the cooler K0IV star will have a radius larger than the hotter G2V component and hence, the primary eclipse will be an occulation rather than a transit which is contrary to that of his own solution. If the primary eclipse is a transit as obtained by Carr (1969), Sadik (1979) and the present investigation, the cooler component must have a radius smaller than that of the hotter star. This can happen only if the cooler component of UV Psc is a Main Sequence star rather than a subgiant. Hence, we suggest that both the components of UV Psc have to be of Main Sequence stars of spectral types G5 and K1 in order to explain the nature of the primary and secondary eclipses. Further, if the cooler component of UV Psc is a subgaint (K0–IV) as suggested by both Oliver (1974) and Sadik (1979) and has to share the properties of other RS CVn candidates like Z Her (Popper, 1956) RS CVn (Popper, 1961), AR Lac (Chambliss, 1976), SZ Psc (Jakate *et al.*, 1976), snd WW Dra (Mardirossian *et al.*, 1980), then its radius should be in the order of around 2.5 $R_\odot$. However, a radius (R_c) of 0.93 $R_\odot$ obtained in the present investigation, and also of 0.929 $R_\odot$ by Sadik (1979) does not support the subgaint classification criterion at all. Hence, we conclude that the system UV Psc should consist of Main Sequence stars of spectral types G4–6V and K0–2V in order to have an agreement with the nature of the eclipses, colours and radii of the components.

7. Evolution

By use of the derived masses, radii, temperatures, and luminosities of the components of UV Psc, their position on the HR-diagram is shown in Figures 7a and 7b. These figures clearly show that both the components lie very close to the Main Sequence. Hence, we can conclude that the components of UV Psc belong to the Main Sequence. This conclusion agrees with that of the other short period group of RS CVn binaries like CG Cyg (Milone and Naftilan, 1979), RT And (Mancuso *et al.*, 1979), SV Cam (Hilditch *et al.*, 1979), ER Vul (Al-Naimiy, 1981), and WY Cnc (Awadalla and Budding, 1979), where both the components occupy the Main Sequence domain. Hence, we suggest tentatively that the short period group of RS CVn binary components belong to the Main Sequence only.

8. Conclusions

Like several other RS CVn-systems, UV Psc is found to show a large intrinsic variation ($\pm 0^m05$) which cannot be accounted for by the presence of the distortion wave alone. The source of the intrinsic variation is found to be the hotter star as in the case of SS Cam and AR Lac. No periodicity of the intrinsic variation could be established with the present data. Further observations are needed to solve this problem. As neither component fills its Roche lobe, we conclude that UV Psc is a detached binary, a property it shares with other members of the RS CVn group. From the present work, the spectral types of the two components are found to be: G4–6 and K0–2 for the hotter and cooler components, respectively. No ultraviolet excess has been detected for either of the components. The derived colours, temperatures, absolute dimensions and nature of the eclipses strongly suggest that the components of UV Psc belong to the main sequence. This system has the same evolutionary status as other short period group of RS CVn systems like CG Cyg, RT And, SV Cam, ER Vul, and WY Cnc.

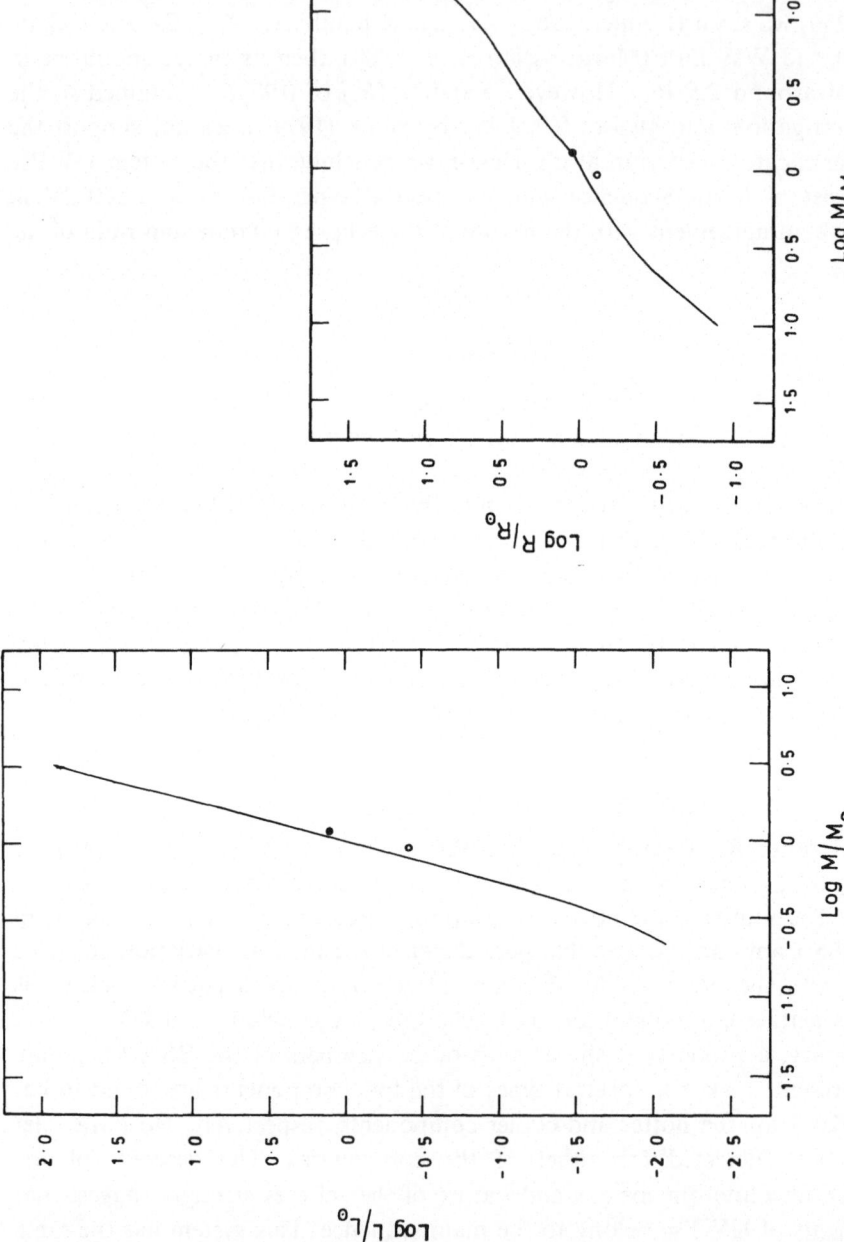

Fig. 7b.   UV Psc: Same as Figure 7a.

Fig. 7a.   UV Psc: Position of the components of UV Psc on the HR diagram of the Main Sequence. ● represents hotter component. ○ represents cooler component.

Additional spectroscopic studies of this system are needed to get a better picture regarding its spectral types, masses and evolutionary status. RI observations are needed to know more about the K0–2 secondary.

Acknowledgements

The authors thank Drs K. D. Abhyankar and G. C. Kilambi for useful discussions. PVR would like to thank the University Grants Commission, New Delhi, for providing financial assistance during the tenure of this work.

References

Allen, C. W.: 1976, *Astrophysical Quantities*, Univ. of London, The Athlone Press.
Al-Naimiy, H. M. K.: 1981, *Astron. Astrophys. Suppl.* **43**, 85.
Arnold, C. N., Hall, D. S., Montle, R. E., and Stuhlinger, T. W.: 1979, *Acta Astron.* **29**, 243.
Awadalla, N. S. and Budding, E.: 1979, *Astrophys. Space Sci.* **63**, 479.
Carr, R. B.: 1969, Ph. D. Thesis, Univ. of Florida (unpublished).
Chambliss, C. R.: 1976, *Publ. Astron. Soc. Pacific* **88**, 762.
Cester, B.: 1969, *Mem. Soc. Astron. Ital.* **40**, 169.
Dworak, T. Z.: 1973, *Inf. Bull. Var. Stars*, No. 846.
Hall, D. S.: 1976, in W. S. Fitch (ed.), 'Multiple Periodic Variable Stars, Part I', *IAU Colloq.* **29**, 287.
Hall, D. S.: 1981, in A. K. Dupree and M. Bonnet (eds.), *Solar Phenomena in Stars and Stellar Systems*, Reidel Publ. Co., Dordrecht, Holland, p. 431
Hilditch, R. W., Harland, D. M., and Mclean, B. J.: 1979, *Monthly Notices Roy. Astron. Soc.* **187**, 797.
Huth, H.: 1959, *Mitteilungen über veränderliche Sterne*, Nr. 424.
Jakate, S., Bakos, G. A., Fernie, J. D., and Heard, J. F.: 1976, *Astron. J.* **81**, 250.
Koch, R. H., Plavec, M., and Wood, F. B.: 1970, *A Catalogue of Graded Photometric Studies of Close Binaries*, Publication of the Univ. of Pennsylvania, Astronomical Series, Vol. XI.
Kopal, Z.: 1959, *Close Binary Systems*, Chapman and Hall, London and Wiley, New York.
Mancuso, S., Milano, L., and Russo, G.: 1979, *Astron. Astrophys. Suppl* **3**, 415
Mardirossian, F., Mezzetti, M., Cester, B., and Giuricin, G.: 1980, *Astron. Astrophys. Suppl.* **39**, 73.
Milone, E. F. and Naftilan, S. A.: 1979, in M. J. Plavec, D. M. Popper, and R. K. Ulrich (eds.), 'Close Binary Stars: Observations and Interpretation', *IAU Symp.* **88**, 419.
Oliver, J. P.: Ph. D. Thesis, Univ. of California, Los Angeles (unpublished).
Plavec, M. and Kratochvil, P.: 1964, *Bull. Astron. Inst. Czech.* **15**, 165.
Popper, D. M.: 1956, *Astrophys. J.* **124**, 196.
Popper, D. M.: 1961, *Astrophys. J.* **133**, 148.
Popper, D. M.: 1976, *Inf. Bull. Var. Stars*, No. 1083.
Rodono, M.: 1981, in E. B. Carling and Z. Kopal (eds.), *Photometric and Spectroscopic Binary Systems*, Reidel Publ. Co., Dordrecht, Holland, p. 285.
Russell, H. N. and Merrill, J. E.: 1952, *Contr. Princeton Univ. Obs.*, No. 26.
Sadik, A. R.: 1979, *Astrophys. Space Sci.* **63**, 319.
Strohmeier, W. and Knigge, R.: 1960, *Veröffentlichungen der Remeis Sternwarte ZU*, Bamberg, VCS.
Tsesevich, V. P.: 1940, *Bull. Astron. Inst. USSR Acad. Sci*, No. 45.
Vivekananda Rao, P. and Sarma, M. B. K.: 1981a, *Acta Astron.* **31**, 107.
Vivekananda Rao, P. and Sarma, M. B. K.: 1981b, *Contributions form Nizamiah and Japal-Rangapur Observatories*, No. 14.
Vivekananda Rao, P. and Sarma, M. B. K.: 1981c, in E. B. Carling and Z. Kopal (eds.), *Photometric and Spectroscopic Binary Systems*, D. Reidel Publ. Co., Dordrecht, Holland, p. 305.
Vivekananda Rao, P. and Sarma, M. B. K.: 1983a, *Bull. Astr. Soc. India* **11**, 75.
Vivekananda Rao, P. and Sarma, M. B. K.: 1983b, *J. Astrophys. Astron.* (in press).
Wood, D. B.: 1971, *Publ. Astron. Soc. Pacific* **83**, 286.

A SPECTROSCOPIC STUDY OF EPSILON AURIGAE*

MAMORU SAITŌ

Department of Astronomy, Kyoto University, Kyoto, Japan

SHUSAKU KAWABATA

Kyoto Gakuen University, Kameoka, Kyoto, Japan

KEIICH SAIJO

National Science Museum, Ueno Park, Tokyo, Japan

and

HIDEO SATO

Tokyo Astronomical Observatory, Mitaka, Tokyo, Japan

(Received 2 August, 1983)

Abstract. Epsilon Aurigae has been observed during ingress and totality between 1982 and 1983 at Okayama. Analyses of profiles of Hα line and of radial velocities of neutral hydrogen and metals show that the secondary component consists of at least three parts in structure.

1. Introduction

An eclipsing binary Epsilon Aurigae has a period of 27.1 yr and the eclipse is occurring between 1982 and 1984. For the previous eclipses many observations were made in optical wavelength regions. The observed results have derived various models of the structure and physical state of the invisible secondary component (Kuiper *et al.*, 1937; Gaposchkin, 1954; Hack, 1959; Huang, 1965; Kopal, 1954, 1971; Wilson, 1971). Campaign Newsletters of Epsilon Aurigae eclipse being published by Drs Hopkins and Stencel have announced that the present eclipse continues to progress on schedule and that many astronomers have been observing the eclipse on ultraviolet and infrared wavelength regions as well as optical region. Observations of polarization are also being done. We can expect that nature of the secondary is unveiled by these observations.

This report is preliminary results obtained by the 188 cm reflector of Okayama Astrophysical Observatory for variations of Hα profile and radial velocities of atoms with phase around the second contact. The results may give a constraint for models of the secondary.

2. Profile of Hα Line

Figure 1 shows profiles of the Hα line on spectrograms with a dispersion of 8.3 Å per mm. The profile obtained outside eclipse, Figure 1a, is characterized by a relatively

---

* Paper presented at the Lembang-Bamberg IAU Colloquium No. 80 on 'Double Stars: Physical Properties and Generic Relations', held at Bandung, Indonesia, 3–7 June, 1983.

Astrophysics and Space Science **99** (1984) 269–272. 0004–640X/84/0992–0269$00.60.
© 1984 *by D. Reidel Publishing Company.*

narrow absorption line with emissions at both sides. We can see that the central absorption increases and progresses towards the red side with phase, and the central reversal emission appears in Figures 1c and 1d. In totality just after the second contact the emission at the red-side disappears, as shown in Figure 1d.

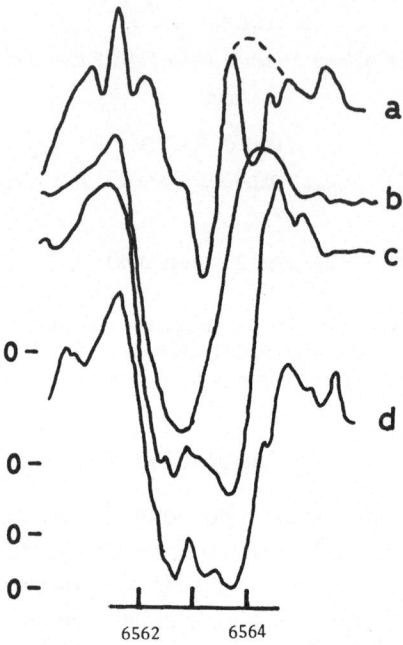

Fig. 1.   Profiles of Epsilon Aurigae around the Hα line. Plate No. and the date observed are: (a) C10–3636, 21 November, 1978, (b) C10–3970, 10 December, 1981, (c) C10–4082, 9 December, 1982, and (d) C10–4104, 27 January, 1983. There is a small flaw on plate (a). A corrected profile is drawn by a dashed curve in profile (a). The abscissas are arbitrary scale.

The disappearance of the red-side emission in totality was reported by Guinan (1983) and Boehm and Ferluga (1983). In the last eclipse, Wright and Kushwaha (1957) found the same phenomenon.

Figure 2 shows intensities at three phases against outside eclipse as functions of wavelength around Hα. The decrease of continuum radiation has been estimated from the V-magnitude light curve of Ingvarsson (1983) at each phase. We can see from Figure 2 that (1) strong absorption of Hα line has appeared with radial velocity of -5 km s^{-1} even at 10 December, 1981, at seven months before the first contact, 29 July, 1982 (Gyldenkerne, 1970), and the absorption gradually increases with phase, (2) at ingress and totality, absorption has been rapidly increasing at red side with radial velocity of 40 km s^{-1}, and (3) the eclipse at the Hα line becomes almost complete by the two absorption components, although half the continuum radiation is appearing during totality.

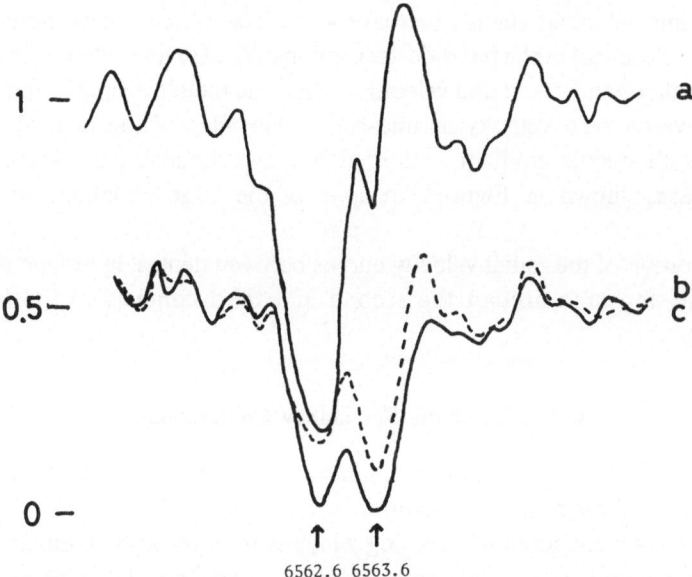

6562.6 6563.6

Fig. 2. Intensities around the Hα line at eclipsing phases relative to intensity of the plate C10–3636 obtained outside eclipse. Plate No. and the date observed are: (a) C10–3970, 10 December, 1981, (b) C10–4082, 9 December, 1982, and (c) C10–4102, 27 January, 1983.

3. Radial Velocities of Absorption Lines

Figure 3 shows radial velocity curves of absorption lines of neutral hydrogen and metals around the second contact. The center of gravity of the binary system moves -2.5 km s^{-1}. Our measurements have been made for seven plates of blue and ultraviolet regions with a dispersion of 4.1 Å per mm.

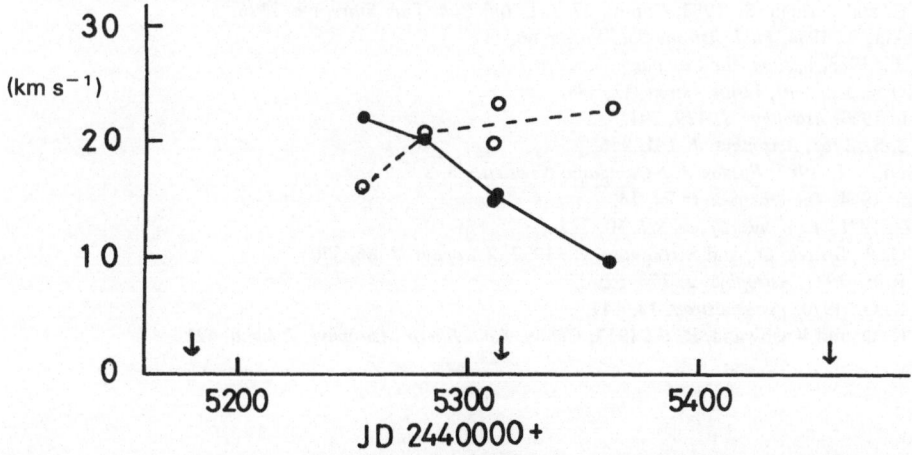

Fig. 3. Radial velocities of neutral hydrogen, denoted by open circles, and neutral metals, denoted by filled circles. Three arrows show first and second contacts and mid-eclipse predicted by Gyldenkerne (1970).

The Balmer line velocities slightly increase with phase. The velocity increase corresponds to the development of the red-side absorption of the Hα line with phase mentioned above. On the other hand, the radial velocities of neutral metals decrease almost linearly with phase towards zero velocity at mid-eclipse. Profiles of the neutral metals are asymmetrical with steeper gradient at the red side and the intensities scarcely change during the phases shown in Figure 3 in spite of the large variations of the radial velocities.

Such a separation of the radial velocity curves between neutral hydrogen and neutral metals has also appeared around the second and third contacts of the last eclipse (Wright, 1970).

4. Model of the Secondary Component

We may consider from the results obtained in the previous sections that the secondary consists of at least three parts in structure:

(1) Neutral metals are confined in a ring which is rotating with 20 km s^{-1} or more, because of the linearly decreasing radial velocity curve and of the almost constant intensities of the absorption lines.

(2) Neutral hydrogen with radial velocities of 40 km s^{-1} distributes in the ring of metals and also inside the ring and it eclipses all over the photosphere of the primary. The observations of Wright and Kushwaha (1957) for the last eclipse show that neutral hydrogen layer seems to be rotating with velocities increasing towards the center.

(3) A low-density neutral hydrogen envelope extends at least twice the radius of the ring of metals. The radial velocity of the envelope is almost equal to that of the binary system.

References

Boehm, C. and Ferluga, S.: 1983, Comm. 27, *IAU Inf. Bull. Var. Stars*, No. 2326.
Gaposchkin, S.: 1954, *Publ. Astron. Soc. Pacific* **66**, 112.
Guinan, E.: 1983, *Epsilon Aur Campaign Newsletter* **7**, 7.
Gyldenkerne, K.: 1970, *Vistas Astron.* **12**, 199.
Hack, M.: 1959, *Astrophys. J.* **129**, 291.
Huang, S.-S.: 1965, *Astrophys. J.* **141**, 976.
Ingvarsson, S. I.: 1983, *Epsilon Aur Campaign Newsletter* **6**, 8.
Kopal, Z.: 1954, *The Observatory* **74**, 14.
Kopal, Z.: 1971, *Astrophys. Space Sci.* **10**, 332.
Kuiper, G. P., Struve, O., and Strömgren, B.: 1937, *Astrophys. J.* **86**, 570.
Wilson, R. E.: 1971, *Astrophys. J.* **170**, 529.
Wright, K. O.: 1970, *Vistas Astron.* **12**, 147.
Wright, K. O. and Kushwaha, R. S.: 1957, *Comm. Coll. Intern. Astrophys. Liège* **8**, 421.

SPECTROSCOPIC SEGREGATION IN BINARY SYSTEMS*

JON DARIUS

Science Museum and University College, London, U.K.

(Received 4 August, 1983)

Abstract. Binary systems displaying spectroscopic segregation, whereby line spectra of the two components overlap little or not at all in wavelength, account for a small subset of LUV objects (late-type ultraviolet). Optical and ultraviolet (IUE) spectra for one such system, HD 15351 + UV0225 + 13 (F4–5 V + sdO), are presented and a few related systems described. The role of LUV objects in establishing an absolute luminosity calibration for hot subdwarfs is emphasized, along with their significance for stellar evolution. A search strategy for additional spectroscopically segregated LUV objects is outlined.

Spectroscopic and spectrophotometric observations in the near and far ultraviolet have revealed a small but intriguing new class of binary stars. Ultraviolet surveys such as those conducted by Celescope on OAO-2 (Davis *et al.*, 1973), the S2/68 experiment on TD-1A (Thompson *et al.*, 1978) and the UV experiment on ANS (Wesselius *et al.*, 1982) turn up occasional contradictory objects, apparently single, whose UV colours are unacceptably blue for their nominal spectral type as determined from observations in the visual region. For convenience these can be called LUV objects (standing for late-type ultraviolet – properly speaking a contradiction in terms). For the most part the denomination of LUV objects is factitious since many stars in this category will disappear on closer scrutiny.

Sky surveys can generate spurious stellar LUV objects in the following ways: (1) erroneous fluxes – rarely a problem except for faint sources covered by few satellite passes or else observations contaminated by high, variable background radiation such as in the South Atlantic Anomaly; (2) misidentification through the presence of a rich star field or conceivably through an error in the attitude solution; (3) blending with an extended source or by two sources of comparable magnitude with a strong UV component – less of a problem on ANS (aperture $2\rlap{.}{'}5 \times 2\rlap{.}{'}5$) than on S2/68 (aperture $11' \times 17'$); and (4) misclassification of the assumed late-type component. In practice the vast majority of spurious LUV objects arise through errors of the last type (see, for example, Barbier *et al.*, 1978; Berger and Fringant, 1980; Jaschek and Jaschek, 1980) although the difficulties are increasingly attenuated by publication of revised spectral catalogues such as those by Houk and Cowley (1975) and Houk (1978) among others. It should be emphasized that on first inspection the sources we are considering appear to be single field objects, known binary systems being *de facto* excluded. Naturally enough, systems of eclipsing binaries of β Lyrae type (such as BF Aur) and late-type

---

* Paper presented at the Lembang-Bamberg IAU Colloquium No. 80 on 'Double Stars: Physical Properties and Generic Relations', held at Bandung, Indonesia, 3–7 June, 1983.

primaries with recognized hot secondaries (such as α Sco, comprised of M1 Ib + B4 V) in part mimic the characteristics of LUV objects.

What then do we make of a star of confirmed spectral type late A or cooler which nevertheless displays a steep UV slope, say m (1550 Å) – m (2750 Å) < – 1.3, on the basis of uncontaminated UV data? Following Doyle's dictum (Doyle, 1890) that after the elimination of the impossible whatever remains must be the truth, we infer that non-spurious LUV objects must be examples of binary systems displaying an extreme case of spectroscopic segregation. That is to say, they consist of late-type primaries with subluminous hot secondaries where the former is completely dominated by the latter in the ultraviolet, and the latter by the former in the optical region.

An outstanding instance of spectroscopic segregation is furnished by HD 17576, considered by Darius and Whitelock (1978). Spectra obtained at a dispersion of 30 Å mm^{-1} clearly vindicated its classification as G0V; yet S2/68 observations of UV0246–37 (precisely identified with HD 17576 in position) indicated a temperature estimated to be in excess of 35 000 K. The identification of UV0246–37 as a hot subdwarf secondary to the late primary HD 17576 has since been corroborated by studies with the International Ultraviolet Explorer (IUE). The UV spectrum (SWP 4104) displays Lα λ1215, Si II λ1260, O I λ1302, and the Si IV doublet $\lambda\lambda$1392/1402 prominently, along with weaker evidence for C IV λ1550 and N IV λ1718.

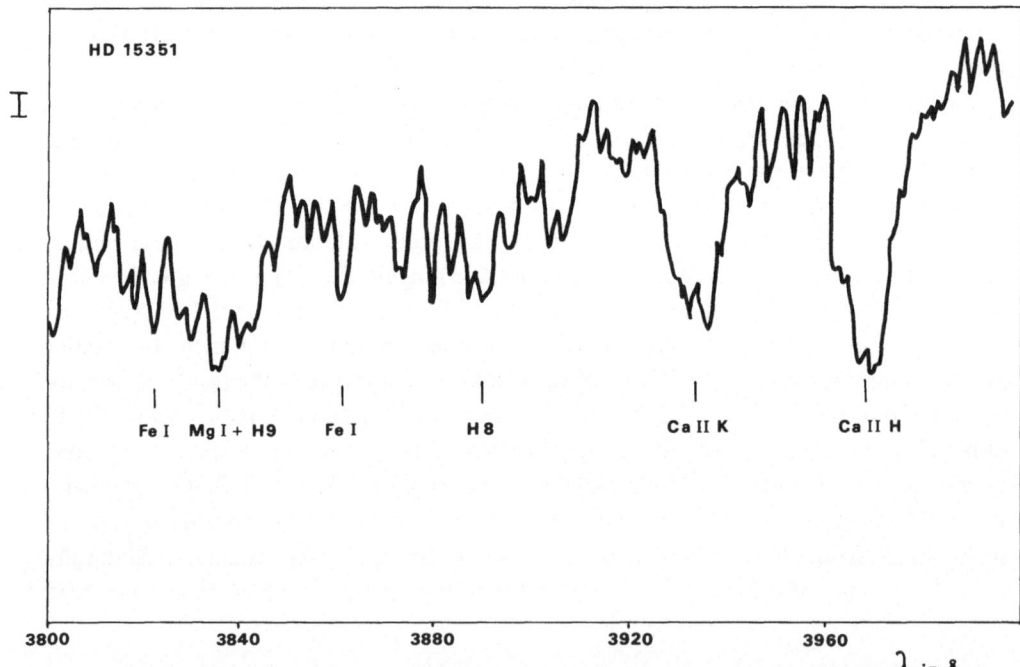

Fig. 1. Portion of optical spectrum of HD 15351, taken with f/1.4 EMI image-tube spectrograph at 30 Å mm^{-1} (second-order blue) on 1.9 m telescope at South African Astronomical Observatory. Salient features are identified.

The presence of the hot companion, almost indistinguishable spectroscopically in the optical region, can nevertheless be inferred from Strömgren photometry.

I should like to describe another remarkable system in this paper, the LUV object UV0225 + 13 identified with HD 15351. It was classified as F5 in the HD catalogue, yet registered as a strong UV source according to S2/68 observations, its colours best approximated by B1. Barbier *et al.* (1978) estimated F2 from an objective-prism spectrum. A high-quality spectrum at 30 Å mm^{-1} was obtained with the Image-Tube Spectrograph on the 1.9 m telescope at SAAO by Whitelock (1977), and it was found that the mid-F classification is vindicated. From the relative strengths of the hydrogen lines and metallic-line luminosity-class discriminants, coupled with a photometric measurement at Hβ (Whitelock, 1977), we can confidently claim HD 15351 to be F4–5 V. The only slight indication of a binary component is a marginal lowering in the Ca II K : H ratio (see Figure 1) through infilling of the K line.

When HD 15351 was re-observed in the ultraviolet, this time with IUE, the results were unequivocal (Figure 2). Those familiar with the appearance of late-type stars in the short-wavelength SWP camera (1150 to 1950 Å) know how abruptly the short-wavelength tail peters out at one end of the channel. Again, spectroscopic segregation explains the seeming contradiction between the visible late-type object and the ultraviolet early-type. Although it may not be immediately clear from Figure 3, comparison of the large- and small-aperture spectra of SWP 4083 confirms the presence of several features

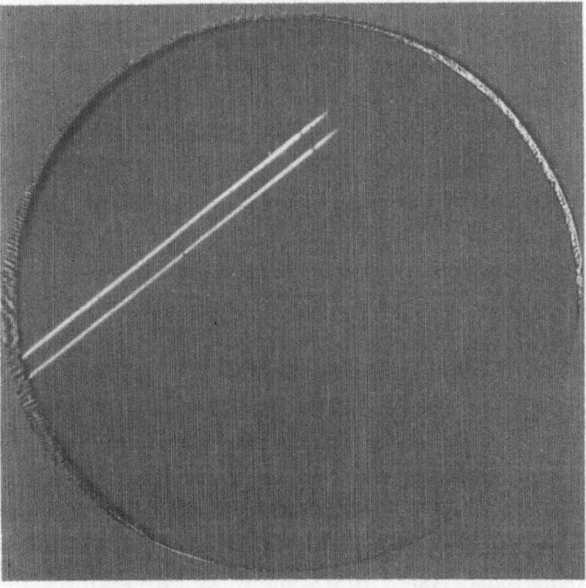

Fig. 2.   GPHOT image (geometrically and photometrically corrected) showing short-wavelength, low-resolution IUE spectra of UV0225 + 13 (HD 15351) in large and small apertures, to left and right respectively. Wavelength decreases upward, the strong feature toward upper right being Lα.

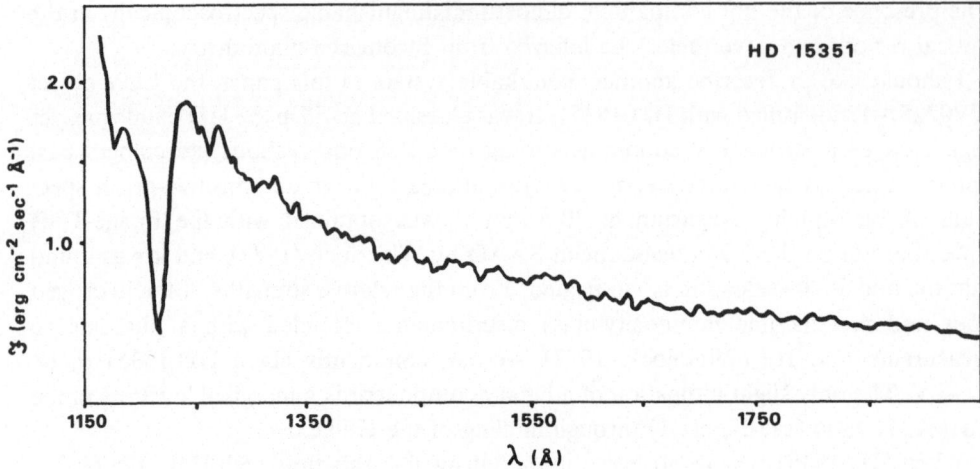

Fig. 3.   Low-resolution, large-aperture IUE spectrum SWP 4083 of UV0225 + 13 (HD 15351). Réseau
marks have been deleted and the adjacent continuum smoothed.

besides Lα in the SWP channel: C III λ 1175, Si II blended at λ 1260, possibly C II λ 1334,
Fe II + Si I at λ 1427 and He II λ 1640. Fe II λ 1690 and N IV λ 1718 are marginally
possible.

TABLE I

Low-resolution IUE observations discussed in text

Target	Camera	Image number	Aperture	Exposure time (min)
UV0225 + 13 (HD 15351)	LWR	7044	Large	2.1
			Small	3.4
	SWP	4083	Large	1.6
			Small	2.5
UV0246 − 37 (HD 17576)	LWR	3635	Large	1.3
			Small	2.0
	SWP	4104	Large	1.2
			Small	1.6
UV2158 − 02 (BD-3°5357)	LWR	1855	Large	10.0
			Small	14.6
	SWP	2055	Large	6.0

In the long-wavelength channel (Table I) it might at first appear that there are strong
features (Figure 4), but again comparison of large- and small-aperture data is invaluable
– albeit in the opposite sense. Aside from some weak Fe II features, the LWR spectrum
appears nearly line-free at low resolution. Note too the absence of any λ 2200 interstellar
absorption feature.

To infer the absolute magnitude of the hot secondary, we use the distance modulus

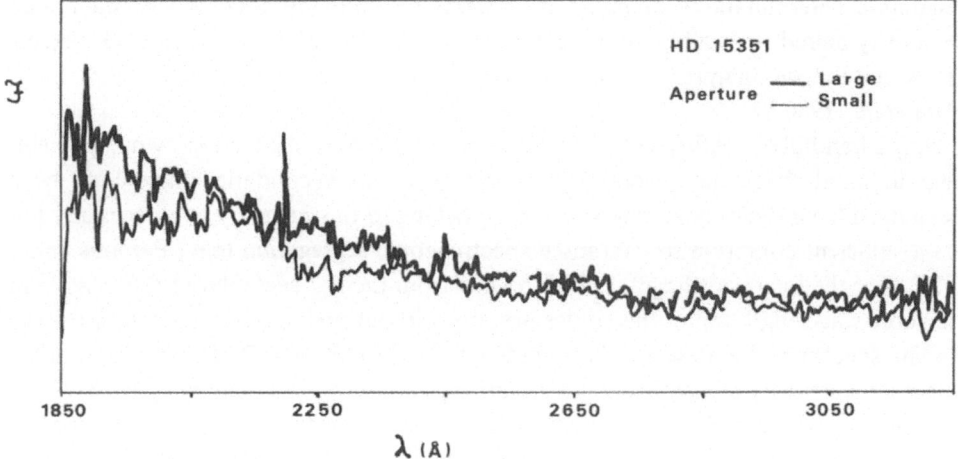

Fig. 4.   Low-resolution IUE spectra LWR 7044, both apertures, of UV0225 + 13 (HD 15351). Gaps mark réseaux. Flux units are arbitrary.

provided by the cool primary, coupled with an extrapolation from the UV slope past the A1 datum at $\lambda 2740$ provided by S2/68 (m_{A1} = 8.2) to yield an apparent magnitude of 10.2. Through Allen (1973) we find the absolute magnitude to be 5.1 – though it could be as low as 4.7 on a different luminosity calibration (Mikami, 1978) – and a new hot subdwarf has declared itself. The unreddened Strömgren indices $[m_1]$ = 0.153 and $[c_1]$ = 0.209 found from photometry by Whitelock (1977) are, if anything, more anomalous than those of HD 17576.

Another, better-known instance of a LUV object is UV2158–02, or BD − 3°5357. A spectrum at 42 Å mm^{-1} taken at Mount Wilson points to a G8 III or G5 III–IV classification, whilst the broad-band UV colours imply B2. In this case, however, unexpected evidence of the system's peculiarity was found in its variability and in the presence of Hα (as well as Ca II H and K components) in emission. In the event, UBV photometry reveals an eclipsing binary system with a 13^h eclipse every 9^d2 (ΔU = 1^m2). The radius of the hot component is $\sim 2\%$ that of the cool subgiant. Dworetsky et al. (1977) suggest that the secondary has M_v = 4.7 and $\log g$ = 6.1 – an evident hot subdwarf which at 300 pc has m_v = 12.2, 3 mag. fainter than the subgiant. IUE long-wavelength spectra at low resolution (Table I) manifest no certian features, but SWP spectra show well-defined C III $\lambda 1175$, Lα $\lambda 1215$, Si IV doublet $\sim \lambda 1400$, C IV doublet $\sim \lambda 1550$. The temperature is perhaps 5000 K hotter than that of HD 17576.

To allow us to infer absolute magnitudes for these hot subdwarfs is perhaps the greatest service that spectroscopically segregated LUV objects can render us. Subdwarf luminosities are very poorly known (see Greenstein and Sargent, 1974) save for such examples as the two in the galactic plane reported by Walker (1981) with $M_v \geq 3.2$ for one and $M_v \geq 5.6$ for the other. There is also the famous example of probable hot subdwarfs found by Kilkenny et al. (1978), LB 3459, whose primary they considered to be a $0.5 \mathfrak{M}_\odot$ star with M_v = 7.0 ± 0.8 (Kilkenny et al., 1981). But undoubtedly the best magnitude determinations originate in systems of cool primaries of well calibrated luminosity paired with subluminous hot companions. (In addition to the LUV objects above, note those analyzed by Goy, 1977; Stickland and Harmer, 1978; Goy, 1980; Gilra *et al.*, 1980.)

Only a handful of confirmed LUV objects of the type described are known at present. Bear in mind that other evolved primaries with hot secondaries have long been recognized from their composite spectra; to belong to this category is a necessary but not a sufficient condition to guarantee spectroscopic segregation (e.g., Parsons *et al.*, 1976). For the long-wavelength tail of the flux from the hot secondary to overlap with the short-wavelength tail of the cooler primary without making its presence evident in the line spectrum, the prescription is evidently F, G, or K + sdO. The question arises whether there are many more such sdO's lurking in unrecognized binary systems. While it is true that this possibility is unlikely for Main-Sequence stars down to $\sim 10^m$, the low intrinsic luminosity of hot subdwarfs would tend to conceal them if the primary were either much fainter or much farther. If the primary were a supergiant, the sheer magnitude difference would militate against detection of the subdwarf companion except in an interacting system.

As shown by Darius and Whitelock (1978) for HD 17576 and again for HD 15351 above, *ubvy* and Hβ photometry proves quite effective at distinguishing candidate LUV objects from mid-F to late K. So does the seven-colour photometric system of Geneva, as shown by Goy (1977, 1980) for HD 128220 and HD 113001. Primaries of earlier type impede discrimination by standard photometric means, but R. Barbier and the author are working on an alternative approach. Walraven *VBLUW* photometry affords better UV passbands ($V - B$, $U - W$) for testing stars of earlier type for possible duplicity, and we are now reducing ESO observations of likely candidates taken on the Dutch 90 cm telescope at La Silla. There are strong reasons for resorting to Walraven photometry: among the LUV objects scanned by S2/68 there was a very large population of A0 (and some A1) stars rather too blue for their spectral type. It is the fate of most LUV objects to evaporate on closer scrutiny, but this approach stands a good chance of turning up a few more genuine binaries with spectroscopic segregation.

Double-star specialists whose main concern lies in mass transfer during evolution may consider these non-interacting systems uninteresting. However, the fact that no evidence of mass exchange is found in most systems of the type described here does not imply that it has not taken place. The canonical position, one infers, must be that evolution of the original primary is very nearly complete and that the next phase must see the Main-Sequence companion begin to expand, fill its Roche lobe and transfer mass and angular momentum to the compact 'primary' provided that their separation not be too great. HD 15351 and its ilk may be seen as precursors of V471 Tauri-type systems containing a white dwarf plus Main-Sequence star (except that their separation is known to be small). It will be argued that the reason so few LUV objects are known is the rapidity of evolution through the subdwarf phase according to some models, but equally the observational selection effects mentioned earlier hinder their recognition. Let us hope that deeper surveys in the satellite ultraviolet and the successful application of

Walraven photometry will increase the number of these systems and allow us (1) to obtain the first secure absolute luminosity calibration for hot subdwarfs, and (2) to account for their evolutionary position.

Acknowledgements

The British Council is warmly thanked for a travel grant and the Bandung Institute of Technology through the kind offices of Bambang Hidayat for its hospitality. Provision of spectroscopic and photometric observations by Patricia Whitelock is gratefully acknowledged.

References

Allen, C. W.: 1973, *Astrophysical Quantities*, 3rd ed., Athlone Press, London.
Barbier, R., Dossin, F., Jaschek, C., Jaschek, M., Klutz, M., Swings, J. P., and Vreux, J. M.: 1978, *Astron. Astrophys.* **66**, L9.
Berger, J. and Fringant, A.-M.: 1980, *Astron. Astrophys.* **85**, 367.
Darius, J. and Whitelock, P. A.: 1978, *Nature* **275**, 428.
Davis, R. J., Deutschman, W. A., and Haramundanis, K.: 1973, *Celescope Catalog of Ultraviolet Stellar Observations*, Smithsonian Institution Press, Washington, D.C.
Doyle, A. C.: 1890, *The Sign of Four*, Spencer Blackett, London.
Dworetsky, M. M., Lanning, H. H., Etzel, P. B., and Patenaude, D. J.: 1977, *Monthly Notices Roy. Astron. Soc.* **181**, 13P.
Gilra, D. P., Wesselius, P. R., and Rao, N. K.: 1980, *Proceedings of Second European IUE Conference*, ESA SP-157, p. 227.
Goy, G.: 1977, *Astron. Astrophys.* **57**, 449.
Goy, G.: 1980, *Astron. Astrophys.* **88**, 370.
Greenstein, J. L. and Sargent, A. I.: 1974, *Astrophys. J. Suppl.* **28**, 157.
Houk, N.: 1978, *University of Michigan Catalogue of Two-Dimensional Spectral Types for the HD Stars*, Vol. 2, Department of Astronomy, University of Michigan, Ann Arbor.
Houk, N. and Cowley, A. P.: 1975, *University of Michigan Catalogue of Two-Dimensional Spectral Types for the HD Stars*, Vol. 1, Department of Astronomy, University of Michigan, Ann Arbor.
Jaschek, M. and Jaschek, C.: 1980, *Astron. Astrophys. Suppl.* **42**, 115.
Kilkenny, D., Hilditch, R. W., and Penfold, J. E.: 1978, *Monthly Notices Roy. Astron. Soc.* **183**, 523.
Kilkenny, D., Hill, P. W., and Penfold, J. E.: 1981, *Monthly Notices Roy. Astron. Soc.* **194**, 429.
Mikami, T.: 1978, *Publ. Astron. Soc. Japan* **30**, 207.
Parsons, S. B., Wray, J. D., Kondo, Y., Henize, K. G., and Benedict, G. F.: 1976, *Astrophys. J.* **203**, 435.
Stickland, D. J. and Harmer, D. L.: 1978, *Astron. Astrophys.* **70**, L53.
Thompson, G. I., Nandy, K., Jamar, C., Monfils, A., Houziaux, L., Carnochan, D. J., and Wilson, R.: 1978, *Catalogue of Stellar Ultraviolet Fluxes*, Science Research Council, U.K.
Walker, A. R.: 1981, *Monthly Notices Roy. Astron. Soc.* **197**, 241.
Wesselius, P. R., van Duinen, R. J., de Jonge, A. R. W., Aalders, J. W. G., Luinge, W., and Wildeman, K. J.: 1982, *Astron. Astrophys. Suppl.* **49**, 427.
Whitelock, P. A.: 1977, unpublished data.

VARIABLE, OPTICALLY THICK, HOT PLASMA OBSERVED IN INTERACTING BINARIES*

Y. KONDO

Laboratory for Astronomy and Solar Physics, NASA-Goddard Space Flight Center, Greenbelt, Md., U.S.A.

G. E. McCLUSKEY

Division of Astronomy, Department of Mathematics, Lehigh University, Bethlehem, Pa., U.S.A.

and

S. B. PARSONS

Space Telescope Science Institute, Johns Hopkins University, Baltimore, Md., U.S.A.

(Received 5 August, 1983)

Abstract. We report recent International Ultraviolet Explorer (IUE) observations of two interacting binaries, R Arae and HD 207739. The ultraviolet spectra indicate the presence of optically-thick, variable hot plasma in those binary systems. These two binaries may belong to a class of binaries that are currently undergoing a rarely observed and probably short-lived phase in their evolution. Their properties are compared with those of two other interacting binaries, U Cephei and β Lyrae.

1. Introduction

One of the important characteristics of interacting binary systems is the presence of gas, often a high temperature plasma within and/or surrounding the entire system and in the form of streams, disks, circumstellar rings, circumbinary shells, winds, hot spots, etc. The physical properties of this gas are affected by the binary nature of the system in which it originates. The signature of this gas may be observed photometrically or spectroscopically.

Here we report on two binaries which were recently observed with the International Ultraviolet Explorer (IUE) satellite observatory. The data indicate that these systems, R Arae (McClusky and Kondo, 1983) and HD 207739 (Parsons *et al.*, 1983), may belong to a class of binaries that are currently undergoing a rarely observed and probably short-lived phase in their evolution. Their properties will be compared with two other binaries, which have also been studied in the ultraviolet, U Cephei and β Lyrae.

2. Discussion

The orbital period of R Arae (B9Vp + ?) is about 4.4 days (Sahade, 1952) and that of HD 207739 (F8II + ?) is about 18.6 days. Additional IUE high-resolution spectra

---

* Paper presented at the Lembang-Bamberg IAU Colloquium No. 80 on 'Double Stars: Physical Properties and Generic Relations', held at Bandung, Indonesia, 3–7 June, 1983.

obtained since the completion of the two papers referred to above show strong evidence of the presence of a variable, optically thick, hot plasma which gives rise to non-monotonic changes in the spectral energy distribution.

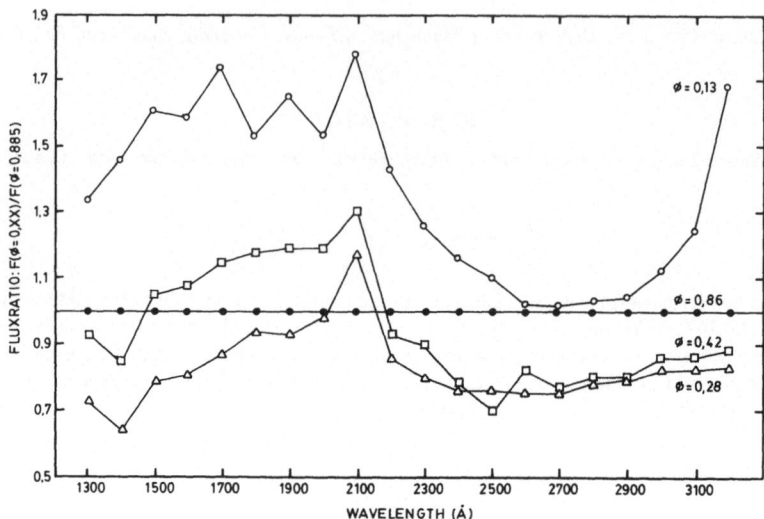

Fig. 1.   IUE continuum variations of R Arae.

Figure 1 shows the spectral energy distribution of R Arae at several phases. These curves are normalized to the distribution at phase 0.86 since this particular normalization yields curves of roughly the same shape. Figure 2 shows the spectral energy curves of HD 207739 at two phases, one phase arbitrarily normalized to the other. A similar spectral energy variation of R Arae was first observed with the Astronomical Netherlands Satellite (ANS) (Kondo *et al.*, 1981).

If these spectral energy variations were simply the result of a temperature change of the plasma involved, the result would be a monotonic increase or decrease of the spectral gradient toward short wavelengths. Instead we observe an irregular increase in the ultraviolet flux at some wavelengths and a decrease at other wavelengths. These non-monotonic variations in the spectral energy curve may occur if two or more regions of the plasma at different temperatures are of variable volume and/or temperature. The most striking variations occur in the continuum levels but the equivalent widths and profiles of a number of spectral lines also vary; in a few cases lines are not detectable at some phases and quite prominent at others. In order for the continuum to change so dramatically, the optical thickness of the plasma clouds involved must be substantially greater than unity. If the mass flow rate is near the conventionally quoted supercritical

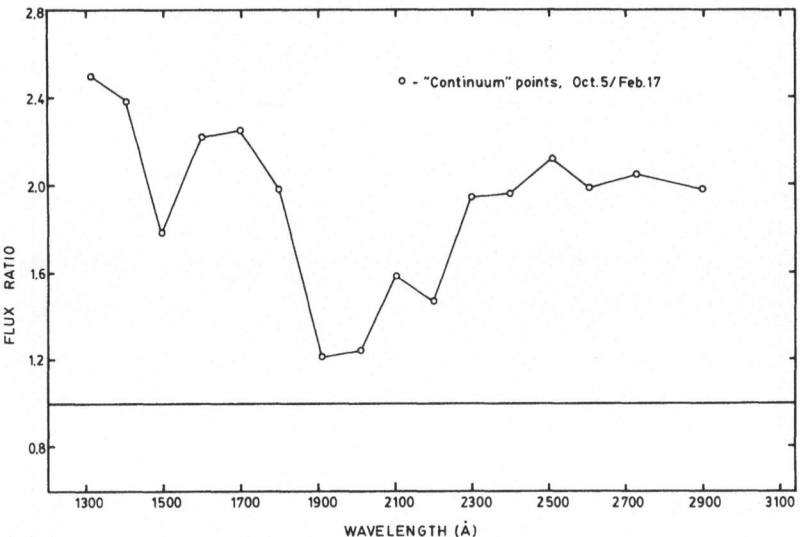

Fig. 2.   Ultraviolet continuum variations of HD 207739.

transfer rate of some 10^{-6} solar mass per year, it can probably replenish, more or less annually, a layer of plasma of several optical depths (in the ultraviolet) surrounding one of the components or the entire system. On the other hand, rather than a relatively complete replenishment of the emitting plasma, what is occurring may be the injection of sufficient gas to significantly modify the temperature and volume of some regions of the circumstellar or circumbinary plasma envelope. The plasma could be heated by the conversion of kinetic energy to heat through such processes as accretion, shocks or colliding gas streams. The optical thickness of this plasma is significantly larger than the shells, streams, winds, or other gas flows commonly found in interacting binaries. Until a more complete understanding of the processes involved is achieved, we will tentatively call this emitting plasma 'surface' the 'pseudo-photosphere'. This does not imply that it must comprise an extension of the true photosphere or of one of its components.

R Arae was observed with IUE in 1982 over nearly one complete orbital period. Light curves were obtained from IUE's Fine Error Sensor (FES) reading, which is essentially in the blue, and from the continuum flux levels (at various ultraviolet wavelengths) at different phases over one orbital period. The continuum flux levels have been normalized to a standard exposure time. The light curves thus obtained, though relatively crude, exhibit two remarkable characteristics. Figure 3 shows the FES light curve, which is the longest wavelength observed, while Figure 4 shows the 1540 and 1295 Å light curves, the latter being the shortest wavelength at which accurate measurements can be made.

First, there is a secondary minimum near phase 0.6. Second, the depth of this secondary minimum increases toward short wavelengths (as does the primary minimum as may be expected). The appearance of a secondary minimum at phase 0.6 is puzzling.

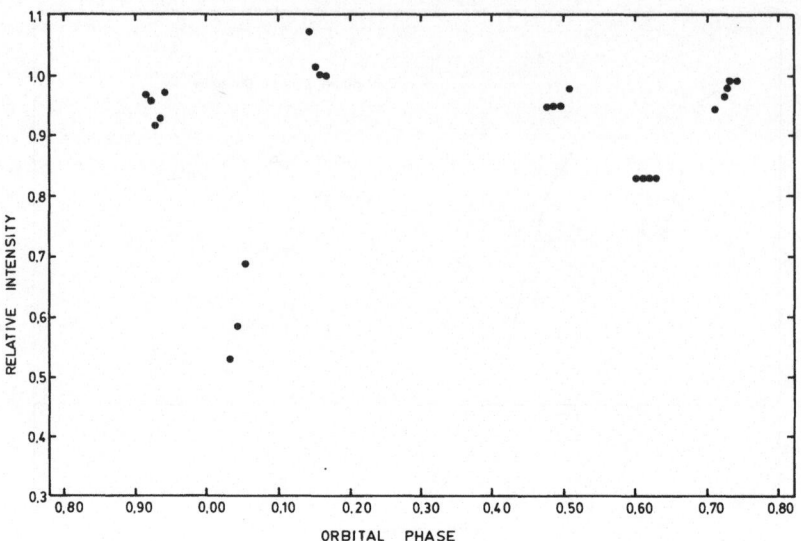

Fig. 3.   IUE fine error sensor light curve of R Arae.

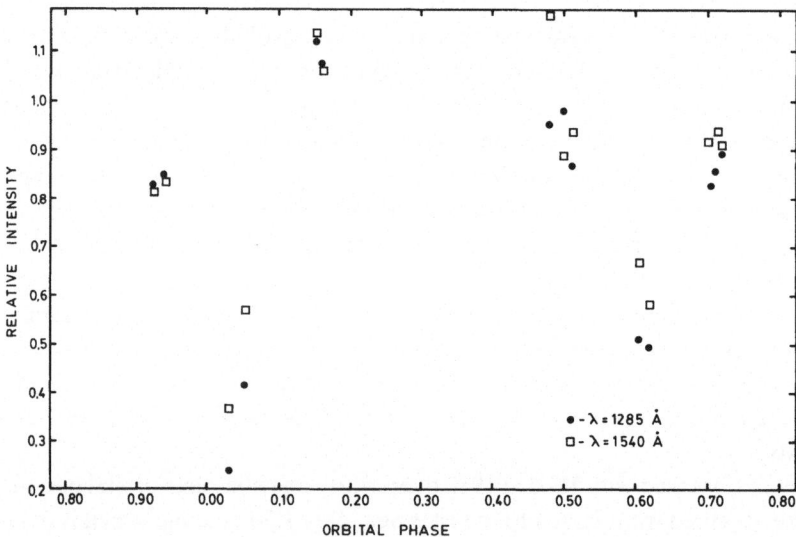

Fig. 4.   IUE light curves of R Arae.

Sahade (1952) did not derive the eccentricity for this binary but his velocity curves do not preclude the possibility of some eccentricity in the orbit. In order for the secondary minimum (due to an eclipse) to be displaced to phase 0.6, the minimum eccentricity required (when the major axis of the ellipse is in the plane of the sky) is about 0.16. It is, however, not certain that the secondary minimum is simply due to the eclipse of the undetected star by the B8p star.

The interacting system U Cephei (B7V + G8III–IV, P = 2.5 days – Batten, 1974) is a much more extensively studied binary. It has a circular orbit with a shallow secondary eclipse occurring at phase 0.5. However, during an active mass transfer event observed in 1974–75, a secondary minimum was observed at phase 0.6 in the ultraviolet light curves with ANS (Kondo *et al.*, 1978). Perhaps, during an active mass transfer event physical conditions in U Cephei resemble those observed in R Arae (in 1982), which seems to be currently in a continuous state of active mass transfer. If so, the secondary minimum observed at phase 0.6 in R Arae may not have been an eclipse of the secondary component. In fact, for reasons discussed below, the secondary minimum observed in the ultraviolet is quite unlikely to be an eclipse of the secondary star.

An indication of this is the depth of this eclipse and the dependence of this depth on wavelength. The secondary minimum in R Arae becomes deeper toward short wavelengths. Blackbody-emitting stars cannot produce such an effect. The eclipse of the cooler star must become shallower at shorter wavelengths. This phenomenon has been observed in two other binaries. It was first reported in β Lyrae (B9p + ?, P = 12.9 days) from photometric data obtained with the Orbiting Astronomical Observatory (OAO-2) by Kondo *et al.* (1976). The ANS observations of the secondary minimum at phase 0.6 in U Cephei during its active mass transfer event also show that feature to be deepening toward shorter wavelengths; in fact, the entire ANS ultraviolet light curves of U Cephei look very peculiar. As stated earlier, U Cephei has a circular orbit and the secondary minimum observed could not have been the result of an eclipse of the G-type secondary component in that system. In the case of β Lyrae, the secondary eclipse, as discussed in the references listed above, shows a very anomalous wavelength dependent behavior but has not been observed to deviate from phase 0.5.

We suggest that the displaced secondary minimum observed in the ultraviolet for R Arae and U Cephei is due to a cooler region in the optically thick plasma which was facing us at phase 0.6. Contrasted to the hotter regions surrounding it, the minimum would become deeper toward short wavelengths. With only two examples, we cannot be certain at the moment whether or not the appearance of a minimum at the same phase, i.e., 0.6, is indicative of some common process occurring in the mass flow patterns of these systems.

We speculate that R Arae and HD 207739 (and U Cephei during one of its active events) are perhaps in an evolutionary stage just preceding (or following) the short-lived supercritical mass transfer in which β Lyrae appears currently to find itself.

Acknowledgement

One of us (GEM) was partially supported in this research by NASA Grant NSG 5386.

References

Batten, A. H.: 1974, *Publ. Dom. Astrophys. Obs.* **11**, 191.
Kondo, Y., McCluskey, G. E., and Eaton, J. A.: 1977, *Astrophys. Space Sci.* **41**, 121.

Kondo, Y., McCluskey, G. E., and Houck, T. E.: 1971, *IAU Colloq.* **15**, 308.
Kondo, Y., McCluskey, G. E., and Wu, C.-C.: 1978, *Astrophys. J.* **222**, 635.
Kondo, Y., McCluskey, G. E., and Wu, C.-C.: 1981, *Astrophys. J. Suppl.* **47**, 333.
McCluskey, G. E. and Kondo, Y.: 1983, *Astrophys. J.* **266**, 755.
Parsons, S. B., Holm, A. V., and Kondo, Y.: 1983, *Astrophys. J.* **264**, L19.
Sahade, J.: 1952, *Astrophys. J.* **116**, 27.

MAGNETIC BRAKING IN CATACLYSMIC AND LOW-MASS
X-RAY BINARIES*

O. VILHU**

Nordita, Copenhagen, Denmark

(Received 22 July, 1983)

Abstract. The chromospheric-coronal emission of lower Main-Sequence single and binary stars can be correlated with an activity parameter of type $R = g(B-V)P^{-1}$, where P is the rotation or orbital period and $g(B-V)$ a function of the color resembling the convective turnover time. Observations indicate that the active region area coverage filling factor grows as R^2, and the whole stellar surface becomes filled with closed loop structures at $R \approx 3$. A braking formula is proposed (Equation 4) to include all periods ($0\overset{d}{.}1 \lesssim R \lesssim 30^d$) and spectral types F–M. On the basis of this equation, the mass transfer rates in compact binaries (driven by the gradual loss of orbital angular momentum) are discussed. It is concluded that the magnetic braking has good chances of being that mechanism which drives the mass transfer in cataclysmic variables and galactic bulge X-ray sources.

1. Introduction

In cataclysmic variables (CV) and galactic bulge X-ray sources (bursters), the mass exchange is often much higher than predicted by gravitational radiation alone. This applies to models where a low mass Main-Sequence star fills its Roche lobe and is forced to lose mass onto the compact companion by gradual loss of angular momentum.

Rappaport *et al.* (1982) have shown that the mass exchange rate due to radiation of gravitational waves is constrained to a value lower than 2×10^{-10} $M_{\odot}$ yr^{-1}, while the rate for many CV's exceeds this limit by one hundred times (Tutukov and Yungelson, 1979). Verbunt and Zwaan (1981) give a similar number to power the brightest galactic bulge X-ray sources, if the model by Joss and Rappaport (1979) is accepted where a neutron star is accompanied by a low-mass ($< 0.8\,M_{\odot}$) Roche-lobe filling Main-Sequence star. For this reason, other mechanisms besides the gravitational radiation, have been searched for.

Verbunt and Zwaan (1981) and Tutukov (1983) extrapolated the empirical braking law by Skumanich (1972): $V_{eq} = f\,10^{14}\,t^{-0.5}$ cm s^{-1}, where f is a constant of the order of unity. This leads to the angular momentum loss rate of the form $dJ/dt \sim f^{-2}\,P^{-3}$, where P is the binary period. Applying this law for the synchronously co-rotating cool secondary in a compact binary, they found mass transfer rates of the order 10^{-8}–10^{-9} $M_{\odot}$ yr^{-1}.

Spruit and Ritter (1983) used implicitly the same idea, and in particular

* Paper presented at the Lembang-Bamberg IAU Colloquium No. 80 on 'Double Stars: Physical Properties and Generic' Relations, held at Indonesia, 3–7 June, 1983.
** On leave from Observatory and Astrophysics Laboratory, University of Helsinki.

explained the $2^h - 3^h$ gap in CV's by a sudden decrease in the braking when then secondary becomes fully convective at around 0.3 $M_\odot$ (having some observational support).

Strictly speaking, Skumanich's law applies for relatively *slowly rotating solar-type stars* only ($P > 3^d$), and its extrapolation to over 10 times shorter periods and much later spectral types may be dangerous. In particular, Vilhu (1982) showed that the best explanation for the close binary period statistics and for contact binary production is achieved if the exponent of P in the angular momentum loss formula changes from -3 to -1.5 when $P < 3$ days. Rucinski (1983) reached a similar conclusion by studying the period distribution of a sample of close detached binaries.

Further, although the time dependence ($t^{-0.5}$) may be independent of the spectral type (see Skumanich and Eddy, 1981), the constant f may well have a strong dependence on it.

In this colloquium, I will present some recent results on chromospheric-coronal activity of Main-Sequence stars (Section 2) and discuss the braking law in this context (Section 3).

2. The Activity Parameter

Noyes (1982) suggested the use of an activity parameter of type

$$R = g(B-V)P^{-1} \tag{1}$$

for Main-Sequence stars of spectral types G–K. Vilhu (1983) used a similar parameter and extended the function $g(B-V)$ to also include F-stars. This function, shown in Figure 1, was constructed to obtain the best correlations between R and fractional fluxes f/f_{bol} from the chromospheric-coronal plasma.

If the function $g(B-V)$ is interpreted as the *convective turnover time*, then R is *equal to the inverse Rossby number or to the square root of the dynamo number* (Durney and Latour, 1978).

Figure 2 shows fractional Ca K + H fluxes for single Main-Sequence stars vs the activity parameter, as reproduced from Figures 4 and 5 of Noyes (1982), and supplemented with binary star observations taken from Middelkoop (1981) and then reduced to the fractional fluxes f/f_{bol}. Figures 3–5 show similar correlations for hotter plasma levels, as observed with IUE and Einstein satellites (for details and references see Vilhu, 1983).

Striking features of these diagrams are the increasing slopes when the emitting plasma temperature increases (CHR $\rightarrow$ TR $\rightarrow$ COR) and the 'saturation' observed at $R \equiv R_c \approx 3$, corresponding roughly to the period $P = 3$ days for G2-stars. Binary and single stars seem to follow the same trend, and for $R < 3$, the binary effects (apart from the synchronism) seem to be small. The lowest point in Figure 2 is a binary (HD 92168, $P = 7.^d8$, F8V). It deviates mostly from the general trend,

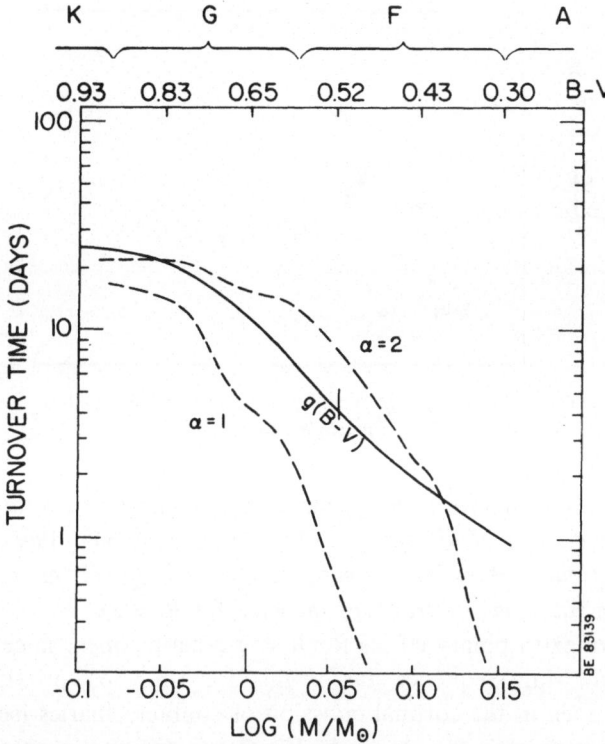

Fig. 1. The function $g(B-V)$ used to compute the activity parameter R. For comparison, the convective turnover time curves for mixing lengths $\alpha = 1$ and $\alpha = 2$ are shown as taken from Gilman (1980). The stellar masses and spectral types are marked. To the left from the small vertical bar, $g(B-V)$ is the same as in Noyes (1982).

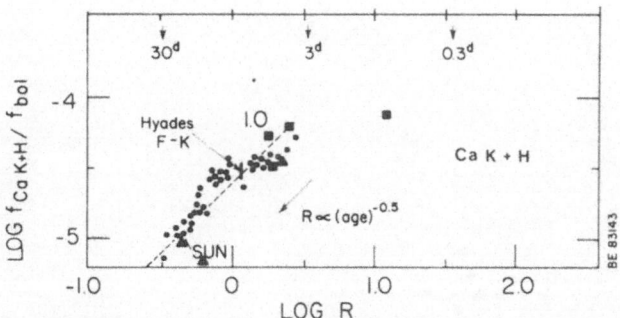

Fig. 2. Calcium K + H-line fluxes (scaled with the bolometric flux) versus the activity parameter R for single Main-Sequence stars (dots) as taken from Noyes (1982). Binary star observations (filled squares and triangles) are from Middelkoop (1981) and reduced further to fractional fluxes f/f_{bol}. The vertical bar shows the range for Hyades stars between $0.5 < B - V < 1.0$ (taken from Noyes, 1982 and reduced to f/f_{bol}). The arrow shows the direction of the time-evolution for a single star. The rotation period for a G2 star (in days) is marked. For the meaning of the symbols see also Figures 4 and 5.

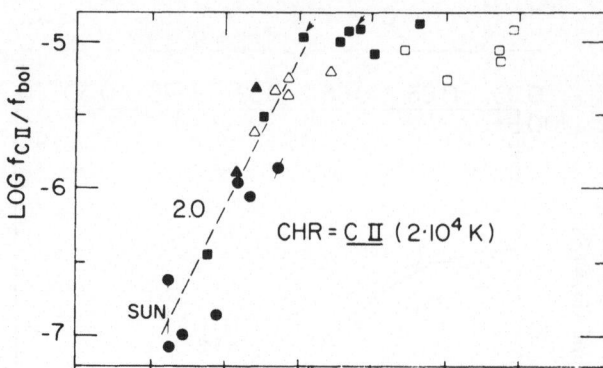

Fig. 3.   The fractional C II 1335 fluxes vs the activity parameter R. The symbols used are explained in Figures 4 and 5.

indicating possible departure from synchronism which generally may be assumed to be valid at periods $P \leqq 9^d$ (see, e.g., Middelkoop, 1981). *Thus a synchronized binary seems to produce the same chromospheric-coronal fluxes as a single star with the same rotation rate and spectral type* (at least for $R < 3$).

The question of extra binary effects for $R > 3$ remains open, since there does not exist known single Main-Sequence stars in this domain. Close proximity effects are, however, clearly seen in the coronal emission of contact binaries (see Figure 5).

Vilhu (1983) interpreted the slopes in Figures 2–5 assuming a two-component

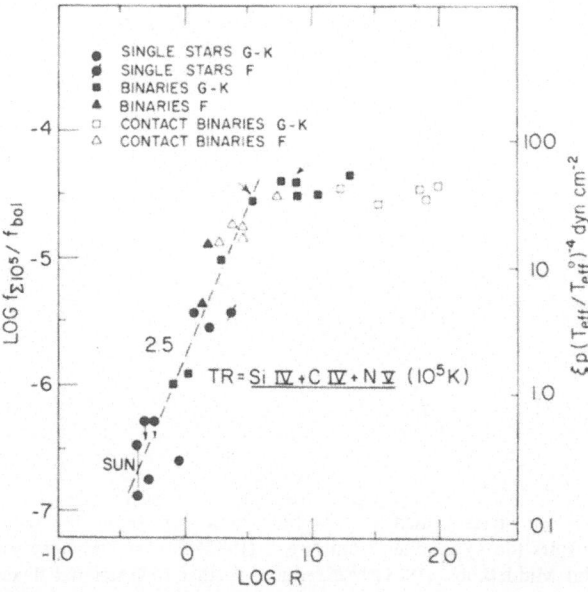

Fig. 4.   The same as Figure 3, but for N v 1240 + C iv 1549 + Si iv 1400. The right vertical scale gives the values for mean loop pressures (for details, see Vilhu, 1983).

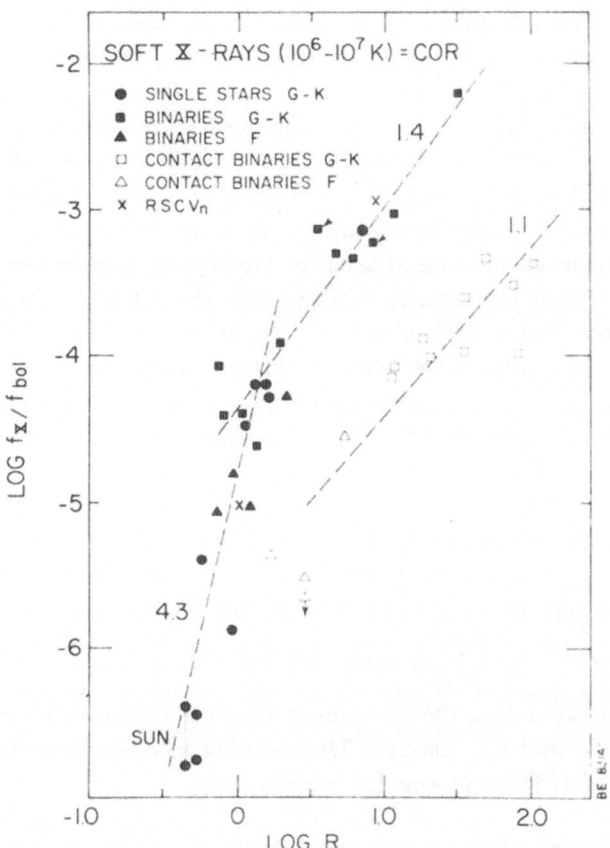

Fig. 5. The same as Figure 3, but for soft X-rays (0.2–4 keV). The two crosses show the domain for RS CVn stars if the activity parameter R is further multiplied by (gravity)$^{-0.25}$ (for details see Vilhu, 1983). The numbers in Figures 3–5 give the slopes of $\log f/f_{bol}$ vs $\log R$.

stellar surface (quiet plus active regions) where the active region filling factor grows as $\propto R^2$ (for $R < 3$). *Since the dynamo number is also proportional to R^2 (if $g(B-V)$ is interpreted as the convective turnover time) this means that the filling factor is proportional to the dynamo number.* The steeper slopes in TR- and COR-fluxes can be explained by increasing coronal loop pressures and sizes with growing R. Lower X-rays from contact binaries is probably due to their smaller photospheric twisting velocities.

3. Magnetic Braking

The relatively good correlations of Figures 2–5 suggest that single Main–Sequence stars with $R \lesssim 3$ ($P > 3$d for G2-stars) move down along these diagrams when aging. Skumanich and Eddy (1981) suggest that the common braking law for spectral types $F - K$ is of the form $\omega = h(B-V)t^{-0.5}$, where ω is the angular velocity

(P^{-1}) and $h(B-V)$ an unknown function of the colour. We give here some evidence that $h(B-V) \sim g(B-V)^{-1}$, where $g(B-V)$ is the 'turnover time' function in Equation (1) and Figure 1.

If indeed $h(B-V) \propto g(B-V)^{-1}$, then stars at a fixed age should have the same R-value (independent of the spectral type), since then $R \equiv g(B-V)\omega \propto t^{-0.5}$, independently of $B-V$. In this way, each stellar cluster should have its own characteristic R-value, the larger one the younger the cluster is.

That this seems to be the case at least for the Hyades cluster (age 5×10^8 yr) for $0.5 < B-V < 1.0$ is demonstrated in Figure 2, where Ca K+H observations of Hyades stars (the thick bar) given by Noyes (1982) are shown after being converted into f_{CaK+H}/f_{bol}. The bar is placed at a mean Hyades R-value ($R_{Hya} \simeq 1.25$). If the Hyades stars follow $f_{CaK+H}/f_{bol} \propto R$ (as can be reasonably assumed), this result means that the color dependence of R is very small for a fixed age. Catalano and Marilli (1983) find that in Hyades and Pleiades clusters the calcium-luminosity depends on mass roughly as $L_K \propto M^5$. This is not very different from the mass-luminosity relation in lower Main-Sequence, from which it follows that f_{CaK}/f_{bol} is rather insensitive on stellar mass in Hyades and Pleiades.

This evidence suggests the use of a braking law of the form

$$R \propto t^{-0.5} \quad \text{or} \quad V_{eq} \propto rg(B-V)^{-1}t^{-0.5}, \tag{2}$$

where r is the stellar radius, the function $g(B-V)$ is shown in Figure 1, V_{eq} is the equatorial velocity, and t is the age. This braking law allows us to write for the momentum loss (J is the spin angular momentum)

$$\frac{1}{J}\frac{dJ}{dt} \propto R^2. \tag{3}$$

This means (in our interpretation) that $1/J(dJ/dt)$ is proportional to the dynamo number or to the active region filling factor.

Comparing this braking law with that used by Verbunt and Zwaan (1981), one finds easily that the mass transfer rates should be multiplied by a factor of $\dot{M}/\dot{M}_{VZ} = (g(B-V)/g_\odot r_\odot/r^2$, where, $\dot{M}_{VZ}$ is the rate computed by Verbunt and Zwaan. For a CV with a $0.3\ M_\odot$ secondary (corresponding roughly to a period of 3 hr) this would mean nearly one hundred times larger mass-transfer rates than predicted with $V_{eq} = f10^{14}\ t^{-0.5}$, where $f = $ const.

On the other hand, this extra gain in the mass-transfer rate may be lost by a possible change in the exponent of Equation (3) after the saturation point $R \simeq 3$. Hence, we propose to use a generalized braking law of the type

$$\frac{1}{J}\frac{dJ}{dt} \propto (R/R_c)^\alpha, \tag{4}$$

where $\alpha = 2$ for $R \lesssim R_c \simeq 3$ and α is still unknown for $R > R_c$, but the results by Vilhu (1982) and Rucinski (1983) suggest that $\alpha \simeq 0.5$ for $R > R_c$. The constant of proportionality in this relation is the *same* for all spectral types F–M and rotational periods $P \lesssim 30^d$.

The simplest magnetic braking theory with a co-rotating wind inside the Alfvén radius predicts $J^{-1} dJ/dt \propto B_0^2$, where B_0 is the surface magnetic field (see e.g., Durney and Latour, 1978). This law is the same as Equation (4) for $R \lesssim 3$ provided that the mean surface magnetic field is proportional to R, as Figure 2 and the linear connection between local solar magnetic fields and CaK emission suggest (Frazier, 1970).

For $R > R_c$ (≈ 3), however, something qualitatively new takes place in the chromospheric-coronal radiative losses. It is clear, however, that at present very little can be predicted of the behaviour of *open field lines* (responsible for the braking) on the basis of the *closed* ones (responsible for the upper chromosphere-coronal emission).

A simple and straightforward interpretation of the saturation in Figures 2–5 might lead to the assumption that $\alpha \to 0$ [in Equation (4)] when $R \to 3$. However, although the stellar surface may be totally covered with closed field regions at $R > 3$ (allowing, seemingly, no place for open field lines), some closed field lines may well open up and allow the braking to increase ($\alpha > 0$) even at $R > 3$.

The effect of α on the numerical computations by Verbunt and Zwaan (1981) can be estimated. Assuming a braking law [Equation (4)] with $\alpha = 2$ for $R < 3$ and $\alpha = 0.5$ for $R > 3$, one easily computes how much smaller mass transfer rate a CV with a 3 hr orbital period would have as compared with the case where $\alpha = 2$ everywhere. The result is roughly 1/100, and this would then completely cancel the gain due to the lowering of the secondary's mass as discussed above. This estimate shows that the results by Verbunt and Zwaan (1981) and Tutukov (1983) have good chances of being correct.

As mentioned previously, the value 0.5 for α is exactly what can be deduced from close binary statistics (Vilhu, 1982; Rucinski, 1983), and so far represents the best guess.

One observational test for the parameter α might be accurate long-term (O–C)-observations of very close (but detached) binaries like ER Vul (G0V + G5V, $P = 0.7$ days) and UV Leo (G0V + G1V, $P = 0.6$ days). For these systems we predict a systematic decrease of the orbital period by amounts $|P/\dot{P} \simeq -10^9$ yr and $P/\dot{P} \approx -10^8$ yr for $\alpha = 0.5$ and 2.0, respectively.

Observations of stars of different spectral types and ages (in clusters) would also be important to check whether each cluster indeed has its own characteristic R-value, as the Hyades data suggest (Figure 2).

References

Durney, B. R. and Latour, J.: 1978, *Geophys. Astrophys. Fluid Dyn.* **9**, 241.
Catalano, S. and Marilli, E.: 1983, *Astron. Astrophys.* **121**, 190.
Frazier, E. N.: 1970, *Solar Phys.* **14**, 89.

Gilman, P. A.: 1980, in D. Gray and J. Linsky (eds.), *Stellar Turbulence*, Springer, Berlin, p. 19.

Joss, P. C. and Rappaport, S.: 1979, *Astron. Astrophys.* **71**, 217.

Middelkoop, F.: 1981, *Astron. Astrophys.* **101**, 295.

Noyes, R. W.: 1983, in J. O. Stenflo (ed.), 'Solar and Stellar Magnetic Fields: Origins and Coronal Effects', *IAU Symp.* **102**, 133.

Rappaport, S., Joss, P. C., and Webbink, R.: 1982, *Astrophys. J.* **254**, 616.

Rucinski, S. M.: 1983 (preprint).

Skumanich, A.: 1972, *Astrophys. J.* **171**, 565.

Skumanich, A. and Eddy, J.: 1981, in R. M. Bonnet and A. K. Dupree (eds.), *Solar Phenomena in Stars and Stellar Systems*, D. Reidel Publ. Co., Dordrecht, Holland, p. 349.

Spruit, H. C. and Ritter, H.: 1983, *Astron. Astrophys.* **142**, 267.

Tutukov, A. V.: 1983, (preprint).

Tutukov, A. V. and Yungelson, L. R.: 1979, *Acta Astron.* **25**, 665.

Verbunt, F. and Zwaan, C.: 1981, *Astron. Astrophys.* **100**, L7.

Vilhu, O.: 1982, *Astron. Astrophys.* **109**, 17.

Vilhu, O.: 1983, NORDITA preprint 83/21 (*Astron. Astrophys.*, in press).

STEADY MASS LOSS ASSOCIATED WITH NOVA OUTBURSTS*

MARIKO KATO

Department of Astronomy, Faculty of Science, University of Tokyo, Bunkyo-ku, Tokyo, Japan

(Received 6 June, 1983)

Abstract. The structure of optically thick mass-losing envelopes of white dwarfs are studied in relation to nova outbursts. A sequence of steady mass-loss solutions is constructed for a nova outburst from the maximum photospheric radius to the disappearance. Much of mass of the envelope will be blown out.

1. Nova Outbursts

Nova is thought to be an outburst phenomena triggered by the nuclear shell flash in the hydrogen-rich envelope of an accreting white dwarf in a close binary system. The onset of shell flash and the initial phase of expansion of envelope in the nova outburst have been studied by various authors (see, e.g., Truran, 1981; Nariai *et al.*, 1980, and references therein). Such time-dependent calculations show that the envelope greatly expands and the velocity of outermost mesh point exceeds the escape velocity. However, the subsequent stages of great expansion have not been calculated yet. Such a extended stage can be treated by steady mass-loss (e.g., Ruggles and Bath, 1979). The present paper concerns a sequence of steady mass-loss solutions for a nova outburst from the maximum expansion to the termination of the mass loss (for details, see Kato, 1983b).

2. Steady Mass-Loss Solution

The basic equations governing the structure of the envelopes are the equations of motion, continuity, energy conservation, and energy transfer by diffusion process (Kato, 1983a). The chemical abundances of hydrogen, helium, and heavy elements are assumed to be $X = 0.73$, $Y = 0.25$, and $Z = 0.02$ by weight, respectively. The mass and the radius of the white dwarf are taken to be $M = 1.3 M_\odot$ and $R = 0.0041 R_\odot$.

Some of the solutions obtained are shown in Table I. Here $\dot{M}$ is the mass loss rate. The subscript ph denotes the values at the photosphere. Figure 1 shows the variations in the diffusive luminosity L_r and the Eddington luminosity $L_{Ed} = 4\pi c GM/\kappa$. The short vertical lines and the crosses denote the photosphere and the critical point, respectively. The Eddington luminosity is a local variable and of which dips correspond to the ionization zones of helium and hydrogen. The luminosity becomes super-Eddington there.

---

* Paper presented at the Lembang-Bamberg IAU Colloquium No. 80 on 'Double Stars: Physical Properties and Generic Relations', held at Bandung, Indonesia, 3–7 June, 1983.

Astrophysics and Space Science **99** (1984) 295–298. 0004–640X/84/0992–0295$00.60.

TABLE I

Properties of the thermal equilibrium envelopes

Model	$\dot{M}$ (g s^{-1})	$\log r_{\mathrm{ph}}$ (cm)	$\log T_{\mathrm{ph}}$ (K)	$\log v_{\mathrm{ph}}$ (cm s^{-1})	$\log v_{\mathrm{esc}}$ (cm s^{-1})	L_{ph} (10^{38} erg s^{-1})	$\log \Delta M$ ($M_\odot$)
E1	8.20 (23)	13.18	3.78	7.41	6.53	2.13	-3.59
E2	2.05 (22)	12.98	3.84	6.88	6.63	1.48	-4.70
E3	1.12 (22)	12.96	3.85	6.83	6.64	1.51	-4.87
E4	2.85 (21)	12.84	3.92	6.64	6.70	1.58	-5.22
E5	2.08 (21)	12.80	3.94	6.55	6.72	1.60	-5.29
E6	3.25 (20)	11.74	4.47	6.53	7.25	1.69	-5.86
E7	7.22 (19)	11.47	4.61	6.39	7.38	1.75	-6.18

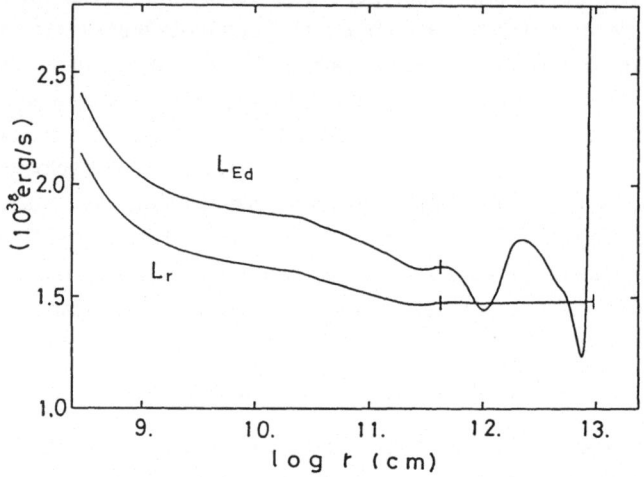

Fig. 1.   Distributions of the energy flux by diffusion and the Eddington luminosity $L_{\mathrm{Ed}} = 4\pi cGM/\kappa$.

3. The Condition of Mass Loss to Occur

We obtain the solutions only when the diffusive luminosity at the bottom of the envelope L_c is larger than the minimum value of the Eddington luminosity – i.e.,

$$L_c > L_{\mathrm{Ed, min}} .$$

This suggests that the mass loss occurs when the diffusive flux L_c is greater than a critical value which is determined by the opacity not at the bottom of the envelope but in the surface region.

When a shell is static and mass loss does not occur, the diffusive luminosity is constant ($L_r = L_c$) except in the nuclear burning region. On the other hand, the Eddington luminosity is a local variable and generally decreases outward. If L_c is less than the minimum value of L_{Ed}, mass loss may not occur. But if L_c becomes as large as $L_{\mathrm{Ed, min}}$, mass loss will begin. The luminosity then decreases outward in order to raise the matter against the gravity of the white dwarf. Consequently, the luminosity becomes everywhere sub-Eddington in models where mass loss just sets in. In the model where L_c is much larger than $L_{Ed, min}$, regions of super-Eddington luminosity appear near the surface of the greatly extended envelopes, where the opacity steeply increases in the partial ionization zones of hydrogen and helium.

4. Nova Outbursts and Steady Mass Loss

Fiure 2 gives a schematic diagram for the existence of the solutions with mass loss. Because the acceleration of matter occurs in the envelope ($R_{WD} < r_{cr} < r_{ph}$), the solutions exist in the region enclosed by two curves $r_{cr} = r_{ph}$ and $r_{cr} = R$. In the lower region to the line $r_{cr} = r_{ph}$, static solutions may exist instead (for detail, see Kato, 1983b).

Progress of a nova outburst is schematically explained by using Figure 2. According to the theory of shell flash, the thermal run-away commences with hydrogen ignition, when the mass accumulated on the white dwarf exceeds some critical value. Just after the ignition of the hydrogen burning the envelope begins to expand because the energy generation by nuclear burning is much larger than the outward energy flux. When the envelope expands greatly, the early explosive expansion will settle down to a milder mass-loss phase in a steady state. Such a state is not only dynamically steady but also thermally steady. The locus denoted as thermal equilibrium in Figure 2 describes the thermally steady-state models in which the energy loss is balanced by the energy generation. The envelope will stay on this line for a long time and then it will slowly move along leftward as the mass ΔM is decreased by mass loss. After the envelope reaches point B, the mass-loss will stop and the envelope may settle down to a static structure. It will shrink according to the decrease of ΔM due to the hydrogen burning and eventually the nuclear burning will be extinguished.

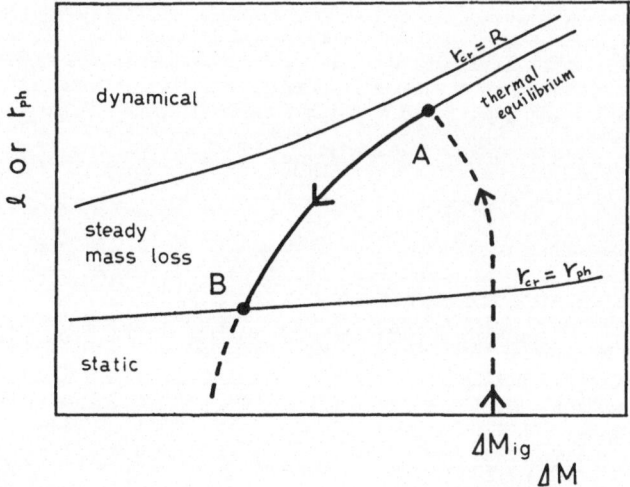

Fig. 2. Schematic diagram for the existence of steady solutions and the evolution of a nova outburst.

Table I shows a sequence of models corresponds to the line E. In the theory of shell flash, the ignition mass ΔM_{ig} is estimated to be $2 \times 10^{-6} \sim 10^{-4} M_\odot$ for $M = 1.3 M_0$ when the accretion rate is $10^{-7} \sim 10^{-11} M_\odot$ yr^{-1} (Nariai and Nomoto, 1979). The time scale to move leftward on the line E, which is estimated by $\Delta M/\dot{M}$, is about 6×10^6 s for Model E1 and about 2×10^7 s for Model E7.

References

Kato, M.: 1983a, *Publ. Astron. Soc. Japan* **35**, 33.

Kato, M.: 1983b, *Publ. Astron. Soc. Japan* **35**, in press.

Nariai, K. and Nomoto, K.: 1979, Proceedings of Fourth Annual Workshop on *Dwarf Novae and Other Cataclysmic Variables*, Rochester, p. 525.

Nariai, K., Nomoto, K., and Sugimoto, D.: 1980, *Publ. Astron. Soc. Japan* **32**, 473.

Ruggles, C. L. N. and Bath, G. T.: 1979, *Astron. Astrophys.* **80**, 97.

Truran, J. W.: 1981, in D. Wilkinson (ed.), *Progress in Particle and Nuclear Physics*, Vol. 6, Pergamon Press, Oxford.

THE q-LOG P DISTRIBUTION OF CLASSICAL ALGOLS*

EDWIN BUDDING

Carter Observatory, Wellington, New Zealand

(Received 11 July, 1983)

Abstract. Data on a large sample (~ 400) of candidates for the classical Algol-type eclipsing binary configurations have been collected. The mass-ratio q has been calculated on the assumption of their semi-detached nature, and then compared with corresponding period for some subsets of the sample. The tendency to slight overluminosity (compared with corresponding MS stars) of the primaries in classical Algols is confirmed.

   The role of the loss of angular momentum in classical Algol evolution is considered, together with a method for determining representative parameters to characterize such effects using the sample data. The possible existence of a class of post-contact classical Algols is noted.

1. Catalogue of Classical Algol Candidate Stars

A list of some 400 candidate objects has been compiled to provide the basis for a catalogue of stars, which have been suspected at some time or other, for more or less well established reasons, as falling into that category of semi-detached binary exemplified by β Per (Algol). One broad purpose of this undertaking is to provide a more extensive range of material which might be useful in providing statistical tests on some theories of binary evolution, rather in the way considered by Giuricin and Mardirossian (1981a), the receipt from whom of published and unpublished work the author is glad to acknowledge.

   Also of basic usefulness to the present work has been the catalogue of Brancewicz and Dworak (1980, hereafter referred to as BD) whose uniform procedure, even if carrying through systematic effects associated with uncertainties attaching to some adopted relations, provides a potentially important basis for statistical comparisons.

   A good proportion of the candidate systems (311) were previously compiled by the present author in a rather preliminary survey of their properties (Budding, 1981). These systems were taken from the listing of those eclipsing binaries described as EA in the well-known *General Catalogue* of Kukarkin *et al.* (1969), and the designation EA2 was applied to the binaries in question to distinguish them from another group – EA1 – showing 'Algol' type light curves in the traditional sense, but being made up of pairs of essentially unevolved stars. The EA2 designation implies the classical semi-detached evolved star containing configuration as typified by Algol itself.

* Paper presented at the Lembang-Bamberg IAU Colloquium No. 80 on 'Double Stars: Physical Properties and Generic Relations', held at Bandung, Indonesia, 3–7 June, 1983.

Astrophysics and Space Science **99** (1984) 299–311. 0004–640X/84/0992–0299$01.95.

Algol data

	Name	Period	M_1	Spectrum	q_{BD}	r_{L_2}	q_{SD}	Δm	r_1	SD status	Remarks
1	TT AND	$2^d.7652$	3.25	A0:+F7:III	0.60	74	0.25	2.5v	0.17	0.7	Well documented case.
			1.8				0.22				
2	TW AND	$4^d.1228$	2.11	A8+K2	0.20	96	0.20	2.3v	0.14	0.9	Lowish mass.
3	UU AND	$1^d.4863$	1.84	F5+K5IV:	0.25	-148	0.40	3.0v	0.32	0.7	
4	WW AND	$23^d.2853$	2.92	A5+(F6IV)	0.30	37	0.05	1.1v	0.08	0.5	Solution errors if s/d. Perhaps EA1 solution possible.
			3.2				0.4				
5	XZ AND	$1^d.3573$	3.26	A0+G8:III:	0.69	91	0.51	3.0p	0.30	0.9	
6	BO AND	$5^d.7973$	5.42	B8	0.50	80	0.24	1.0p	0.19	0.7	
7	CO AND	$1^d.8277$	1.45	F8	0.69	86	0.42	1.0p	0.15	0.7	EA1 solution possible? If s/d must be one of coolest.
8	CP AND	$3^d.6089$	2.14	A7	0.62	68	0.16	1.5p	0.17	0.7	
9	DW APS	$2^d.3130$	2.93	A0	0.71	86	0.42	1.3p	0.18	0.7	
			1.3				0.23				
10	RY AQR	$1^d.9666$	3.98	A3+G8:IV:	0.30	83	0.18	1.3p	0.20	0.9	Popper's mass seems low; his q seems high.
11	CX AQR	$0^d.5560$	1.47	F2p+(G5)	0.70	106	0.95	1.1p	0.23	0.3	EA1 solution likely.
12	CZ AQR	$0^d.8628$	2.96	A5p	0.50	100	0.52	1.5p	0.30	0.7	Early stage of s/d phase?
13	XZ AQL	$2^d.1392$	2.73	A2	0.48	67	0.13	1.3p	0.23	0.5	Perhaps occultation possibility should betried.
14	YZ AQL	$4^d.6723$	3.13	A3	0.82	87	0.50	3.7v	0.15	0.7	Solution errors.
15	FK AQL	$2^d.6509$	6.29	B9+G5III:	0.35	119	0.75	2.4p	0.20	0.7	
			1.48				0.26oc				
16	KO AQL	$2^d.8640$	2.90	A0V+(F8IV)	0.20	63	0.05tr	1.0p	0.20	0.9	Transit solution probably wrong.
17	KP AQL	$3^d.3675$	2.68	A3V	0.87	33	0.02?	0.8p	0.12	0.3	
18	QS AQL	$2^d.5133$	6.82	B5V+(G8IV)	0.17	161	>1	0.15p	0.17	0.3	Solution ambiguous. Eclipse depths shallow and secondary depth ~ one third that of primary!
19	QY AQL	$7^d.2296$	2.69	F0+(K3IV)	0.28	83	0.17	3.2p	0.19	0.9	

Fig. 1. First page of provisional form of catalogue of classical Algol (EA2) binaries.

The first page of a provisional form of the catalogue is presented as Figure 1. A few remarks may be made about this. The name and period of each binary is straightforward enough, though a good many of the candidates show period variations affecting higher decimal digits of the period than those included. Each entry contains (where possible) the mass of the primary star as given by BD, (to two significant decimal figures). Sometimes additional estimates of this mass are included for comparison. These generally well-known examples will be remarked on subsequently. The spectra can be traced to the sources listed by BD. Their mass ratio q_{BD} is also listed. Later, reasons will be given why it is thought that there has been a tendency to systematically overestimate this quantity (at least for EA2 systems). Also of interest is the quantity tabulated by BD as RL2 – since it provides a clue to the likelihood of a semi-detached binary.

The quantity q_{SD} corresponds to the mass-ratio which would be implied by the quoted relative radius of the subgiant component, were this to be in contact with its surrounding Roche lobe. The value comes from an interpolation on the tabulated data of Kopal (1959) (rather than any use of formulae). Also supplied is the depth in magnitudes of the primary eclipse. This usually comes directly from Kukarkin *et al.* (1969), who supply the suffix p, v, etc., depending on the wavelength of observation. The key point here is the fact that EA binaries with deep primary (and shallow secondary) minima are very likely to conform to the EA2 type characteristics in other respects (i.e., to be semi-detached). The relative radius of the primary star r_1 is useful since, taken together with the other items, it can allow calculation of an overall picture of the salient physical quantities characterizing the binary system.

The quantity described as SD status is meant to indicate the likelihood of a genuine classical evolved Algol binary based on the information available. It takes the 5 odd values from 0.1 up to 0.9 in steps of 0.2.

The weight 0.9 refers to well-known cases, whose semi-detached state has usually been attested by a number of authorities. It is this group for which additional comparison data is often presented. 0.7 attaches to those binaries which show similar properties to the 0.9 examples, but have received relatively little attention. 0.5 corresponds to candidates which might possibly be EA2 systems, but for which an alternative EA1 or other configuration is about equally likely. 0.3 is given to those candidates for which a semi-detached configuration may just be possible but the supporting evidence seems poor, or indeed an alternative arrangement seems more likely. 0.1 is given to those cases which have somehow appeared as candidates in an initial search, but further examination has made appear very unlikely to be EA2 binaries. An additional column contains a few brief words on possible peculiarities of the system in question. The author would be glad to receive early comments on the style of, or possible improvements which could be made to, this catalogue.

Figure 2 shows the plot of q_{BD} against q_{SD} from which the general excess of q_{BD} values can easily be ascertained. An explanation of this is offered as follows: the

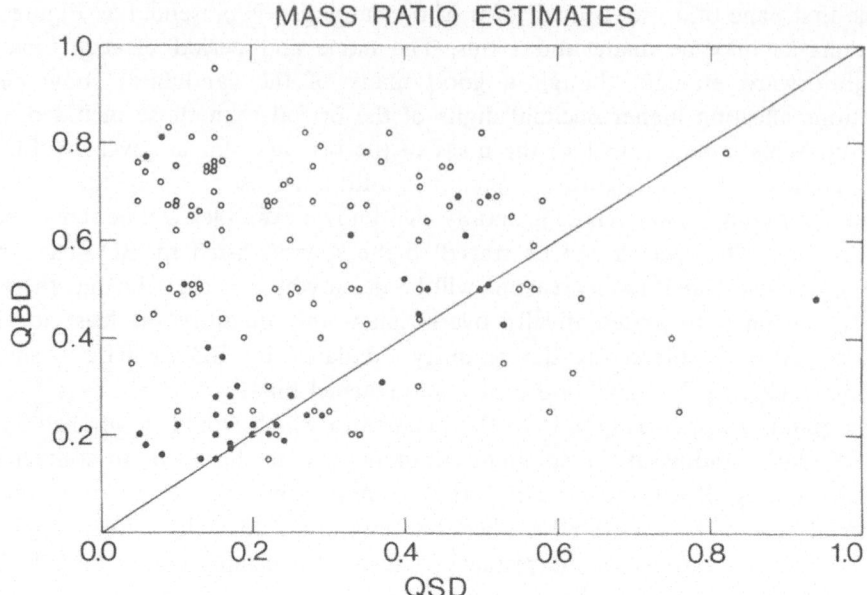

Fig. 2. Plot of q_{BD} against q_{SD} for 147 EA2 candidate binaries with primaries in the mass range $2 < M_{1(BD)} < 4 M_\odot$. The trend of $q_{BD} > q_{SD}$ is evident even for the s.d. status 0.9 systems (full circles). Open circles denote s.d. status 0.7 systems.

five basic relations put together in BD – essentially the same as those considered, for example, by Hall (1974), and including a mass-luminosity relation which might be applicable, but only for the primary component in the case of EA2 systems – refer to eight separate quantities, some of which can be regarded as known. The final reduced expression, obtained from elimination between these five relations, can be cast in the form

$$\log M_1 = f(\log P, \log r_1, \log T_1, \log(1+q), \alpha_i), \qquad (1.1)$$

where M_1 is the primary mass, P the period, r_1 the primary's radius in units of the binary mean separation, T_1 its surface temperature, q the mass ratio, and α_i are some known constants.

The binaries considered all have known primary spectral type from which it is assumed T_1 can be derived. P is, of course, known to high accuracy, while r_1 obtains from the solution of the light curves. We are left with M_1 and q, the former turning out to have a relatively small dependence on the assumed value of the latter. Subgiant components in semi-detached binaries are known to show luminosity excesses over the normal mass: luminosity law which, in some cases may be very considerable (see, e.g., Giannone and Gianuzzi, 1972). BD do not make clear what their procedure would be in the case of semi-detached suspects, though in those well-known cases where secondary masses have been more confidently specified in source material (status 0.9 systems) the general agreement between q_{SD} and q_{BD} is better.

If we concentrate just on the primary, however, and keep in mind the iterative improvement solution method of BD which starts with a q value of 0.75 (quite higher than average for a semi-detached system); the adopted Main-Sequence mass luminosity relation would tend to impart a Main-Sequence character to this primary star. A reasonable form for (1.1) can be shown to be

$$\log (1+q) = 4.76 \log M_1 - 5.90 \log T_1 - 1.95 \log P - 2.95 \log r_1 + 20.24,$$
$$(1.2)$$

where T_1, P, and r_1 are taken to be known for a particular case. If a value of q satisfying the above equation is perturbed downwards in value, a corresponding reduction of M_1 is clearly entailed – i.e., if M_1 were already in keeping with its derived R_1 and T_1 as a Main-Sequence star – a lower value would imply the star to be somewhat overluminous for its mass. This is indeed what has been found to be the case for certain well studies semi-detached binaries (Hall, 1974). The implication is that a more generally self consistent solution is found when the q-values are reduced from those of BD to those of the trend of q_{SD}, on the basis that the primaries in these systems tend, possibly as a result of processes connected with the binary evolution, to be rather over-luminous for their mass, compared with normal Main Sequence stars.

2. Existence of Trends in the q-log P Diagram

In pursuing the aforementioned aim of providing material to test binary evolution theory the distribution of the systems in the $q : \log P$ plane was examined. These two quantities can be readily related to various schemes of binary evolution, and, for genuine EA2 stars with known light curve solutions (i.e., known r_2) q_{SD} ($= q(r_2)$), as well as P, is directly accessible. This distribution has been considered with such purposes in mind by such authors as Svechnikov (1969), Ziółkowski (1976), and Giuricin and Mardirossian (1981a).

At first data was taken from the compilation of Giuricin and Mardirossian (1981a) for those 0.9 status systems for which a primary mass of less than $6 M_\odot$ had been calculated. The distribution of 76 such relatively low-mass classical Algol candidates in the $q : \log P$ plane is shown in Figure 3. A negative correlation of q with $\log P$ can be eye-judged from this diagram ($r = -0.45$), and though the scatter in the plane is quite wide it can be observed that there are relatively few points with very small q and period, or relatively large q ($\gtrsim 0.5$) and large period.

The existence of this general trend is in broad agreement with the Roche-lobe overflow (RLOF) theories of binary evolution, and even on fairly general grounds we might expect that if the contact component does tend to lose mass then lower mass ratios will tend to go with longer orbital periods. Further insight into Figure 3 may thus be afforded by considering the form of how q would vary with $\log P$ according to some of the discussed evolutionary schemes.

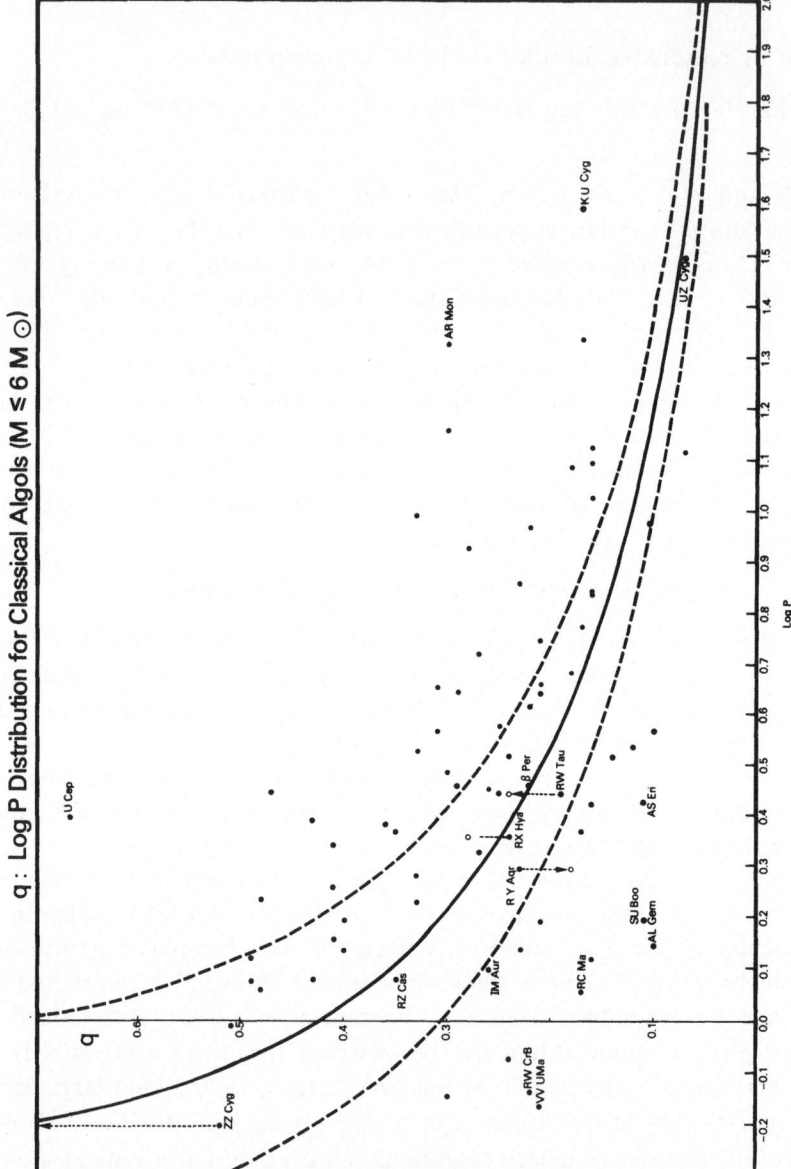

Fig. 3.   Distribution of a selection of low mass EA2 systems in the $q:\log P$ plane, taken from the compilation of Giuricin and Mardirossian (1981a).

Perhaps the simplest case to consider is the 'conservative' trend (which corresponds to the curves actually shown in Figure 3) which is given by

$$\log P = \log P_0 + 6 \log (1+q) - 3 \log q - \log 64, \tag{2.1}$$

where P_0 corresponds to the minimum period, obtaining when $q = 1$. The curve shown as a continuous line corresponds to $P_0 = 0.587$ days, which might be appropriate for a conservative RLOF process for a pair of stars of total (constant) mass M about $5 M_\odot$, which would just be in contact as an equal mass pair. P_0 (contact) is actually specified by

$$P_0 \text{ (contact)} = 0.082\alpha^{3/2} (M/2)^{(3n-1)/2}, \tag{2.2}$$

where n is a mass : radius relationship index, which might be expected to be ~ 0.7 for the stars in question; while α expresses the ratio of separation of the centres of equal mass to the nominal ('volumetric') radii of the corresponding critical lobe surfaces (typically, $\alpha \sim 2.6$).

From (2.2) we may deduce that any appropriate abscissal shifts of the continuous curve shown in Figure 3, in dependence on the expected range of total mass of the 76 systems should be relatively slight (or order 0.1 either way). Hence, while the majority $(\sim \frac{2}{3})$ of the systems considered lying to the right of the continuous curve would not, on this evidence, be in conflict with a conservative RLOF scheme, a substantial number of systems exist which, if projected backwards along a conservative path in the $q : \log P$ plane, would, at some point, no longer be able to satisfy the initial premise of being a semi-detached system. This result is not new. The likelihood of non-conservative evolutionary trends in the properties of classical Algol systems has been considered by a number of authors, at least since the work of Paczyński and Ziółkowski (1967), and the compilation of Svechnikov (1969), if not before. However, the accumulation of systems on the right of the conservative limit in Figure 3 is notable. The distribution of these EA2 systems over a range of possible values of $\log P_0$ is shown as a histogram in Figure 4. In what follows, features and possible interpretations of such diagrams as Figures 3 and 4 will be considered.

3. Possible Interpretations and Checks

Apart from the majority of systems clustering somewhat above the continuous curve in Figure 3, there exists a group of systems rather more spread out, below and somewhat to the left of the curve. Some care may be required to establish facts in the case of such systems. For example, KO Aql would have appeared among this group if earlier solutions for q had been adhered to. Such solutions were calculated on the premise that the primary eclipse is caused by a transit of the smaller star across the disk of the larger one. It was shown, essentially already by Russell (1912), that occultation primary solutions producing a quite similar form of light curve can generally be found as an alternative to transit primary solutions (though

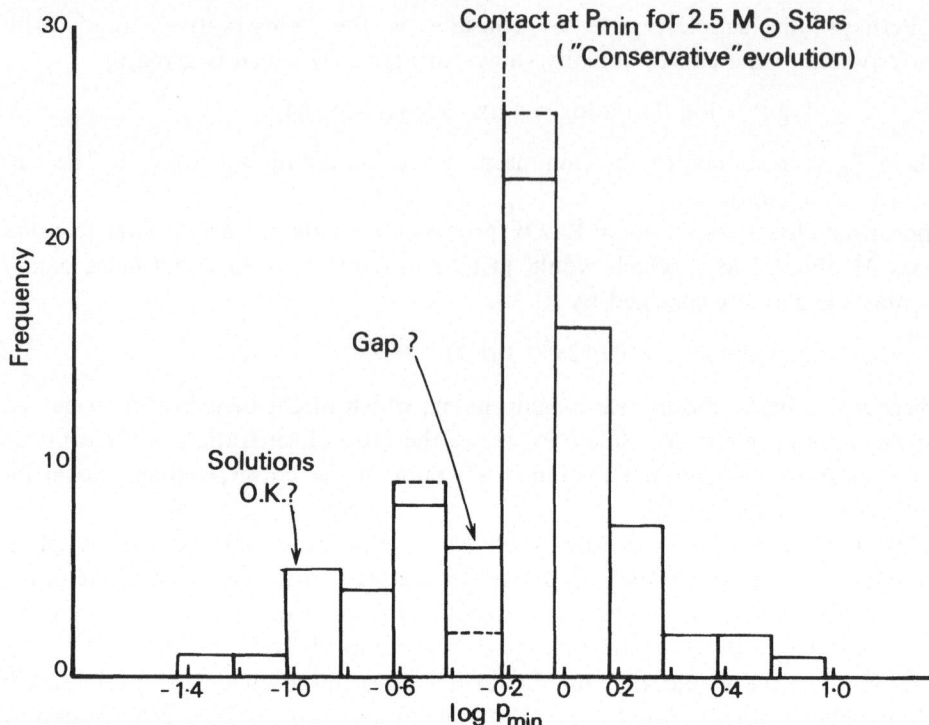

Fig. 4.   Numbers of Algols falling into successive intervals of 0.2 in $\log P_{min}$ (days) on the assumption of 'conservative' evolution (Equation (2.1)). A sizeable number of systems (to the left of the contact limit) violate an initial premise of the conservative RLOF mechanism since their Roche lobes, at minimum period, are not large enough to contain both stars. Some of these systems may result from faulty photometric solutions producing too low q-values. On the other hand, the suggestion of a 'gap' may indicate a qualitative difference in the evolution progress for such systems.

the reverse is not always the case). The more recent discussions of KO Aql (see, e.g., Blanco and Cristaldi, 1974; Giuricin and Mardirossian, 1981b) have shown, in a reasonable way, that the occultation alternative provides a much more satisfactory explanation of the evidence. Perhaps such an explanation might fit some members of the low q, $\log P$ group – transit primary hypotheses account for some 46 % of the solutions in this group, though only 30 % of the entire set are solved in this way. An occultation alternative for a given model, since it raises the secondary relative radius, will clearly increase the corresponding value of q, perhaps to allow a point to move into the majority domain.

This point can be carried a little further by considering what radii of primaries might be reasonably expected for the mass of systems involved, and following through the implications on the relative dimensions of components. In this way, it would be very difficult to expect any real EA2 system, of any normally encountered mass ratio (i.e., $q \lesssim 0.6$), of total mass several $M_\odot$ with a period less than one day, for example, to have a secondary star actually larger than the primary. Similarly, cases like

AS Eri or RV Oph, for which occultation primary solutions have already been advanced, presumably cannot allow larger q values, despite their relatively short periods. Systems which might, perhaps, be profitably re-examined to check the assumed eclipse type, however, include SU Boo, XX Cep, IM Aur, UX Her, V338 Her, and AL Gem.

Prominent among those stars well to the lower right of Figure 3 is the well-known system R CMa, once regarded as the prototype for a subgroup of peculiar overluminous Algol systems (Kopal, 1959). The particular problems posed by the original members of the 'R CMa group' appear to have been largely disposed of with careful reobservation and analysis (Sahade and Ringuelet, 1970; Okazaki, 1977; Hall and Neff, 1979). In the case of R CMa itself, its overluminosity problem was made less acute by reducing the adopted mass ratio; however, let us note here that this operation would accentuate its evolution problem in the sense previously mentioned. The careful consideration which has been given to R CMa, as well as some other members of the low $q : \log P$ group, such as AS Eri (cf. Popper, 1973; Refsdal *et al.*, 1974), do allow us to have confidence that at least some of these points really do lie below the main congregation.

Keeping in mind the matter of confident positioning of points in the $q : \log P$ plane, Figure 4, nevertheless, directs attention to the possibility of some definite gap between the main aggregation and the low $q : \log P$ group. This possibility is enhanced by the fact that of the 6 binaries in the range $0.372 < P_0 < 0.587$, 3 (RY Aqr, RX Hya, and RW Tau) are re-positioned outside this region by the q-values of the catalogue, while the case of ZZ Cyg as a *bona fide* EA2 system is not so well established (Hall and Cannon, 1974). The possible existence of a gap in the $q : \log P$ distribution at the minimum period-contact limit, raises intriguing notions about the possible role of a common envelope in providing angular momentum loss in binary evolution. Derivation of specific information on possible mass or angular momentum loss was perhaps the immediate stimulus to the compilation of the catalogue, so that further testing of any particular features of the distribution might be checked from a larger data-base of systems confined within a smaller range of masses.

Let us also note in this context, however, the third body in the R CMa-system (cf. Radhakrishnan *et al.*, 1983). Total angular momentum, in the evolution of some Algol systems, might not be so much lost as redistributed, if there happened to be some third orbit in which it could be deposited.

The distribution of 147 points corresponding to BD primary masses in the range 2 to 4 $M_\odot$ is shown in Figure 5, together with some possible interactive evolution tracks generalized to mass and angular momentum loss by the f and g parameters, which have become widely referred to (see, e.g., Paczyński and Ziółkowski, 1967; Thomas, 1977), and corresponding histograms to Figure 4, can be drawn up for a range of values of f and g. Some examples are shown in Figure 6.

There appears to have been some confusion in the literature over the precise meaning of f and g. Here we shall adopt f to mean the fraction of matter lost by the loser which is subsequently lost entirely from the system, (with the implication that

$1 - f$ is the fraction transferred to the gainer). The orbital angular momentum of the system J $(= (PG^2/2\pi M)^{1/3} M_1 M_2)$ is taken to decrease with a power law dependence on the total mass $(J/J_0) = (M/M_0)^g$. The formulae (5) and (6) of Giuricin and Mardirossian (1981a), attributed to Vanbeveren *et al.* (1979), are then straightforward to derive, and are general enough to encompass a range of possibilities.

In what follows, it will be convenient to write p in place of $\log P$. An interpretation may be put on the observed frequencies h_i in histograms like those of Figure 6, by referring to the observed density of points in the p, q-plane, $\sigma \equiv \sigma(p, q ; \chi(p_0), f_1, g_1, \alpha_j)$, where p and q are regarded as the basic independent variables. $\chi(p_0)$ represents a range of initial condition, for given mass, in terms of a distribution of the logarithm of the minimum period p_0, and f and g are the parameters already referred to. The additional quantities α_j are meant to refer to other matters influencing the observed density distribution, or possible selection effects; we shall neglect the role of such quantities in the vicinity of the highest density of points. An estimate of the particular values f_1 and g_1 which best characterize the observed distribution might be empirically arrived at in the following way.

Curves $\sigma \equiv$ const. represent contours of constant density in the p, q-plane. One such curve $\sigma = \sigma_{max}$, say, refers to an elongated region of minimum diameter $\sim \varDelta p_0$, wherein the point density attains a maximum. Now the curve ϕ_{ev} $(p, q ; f, g) = $ const. (p_0)

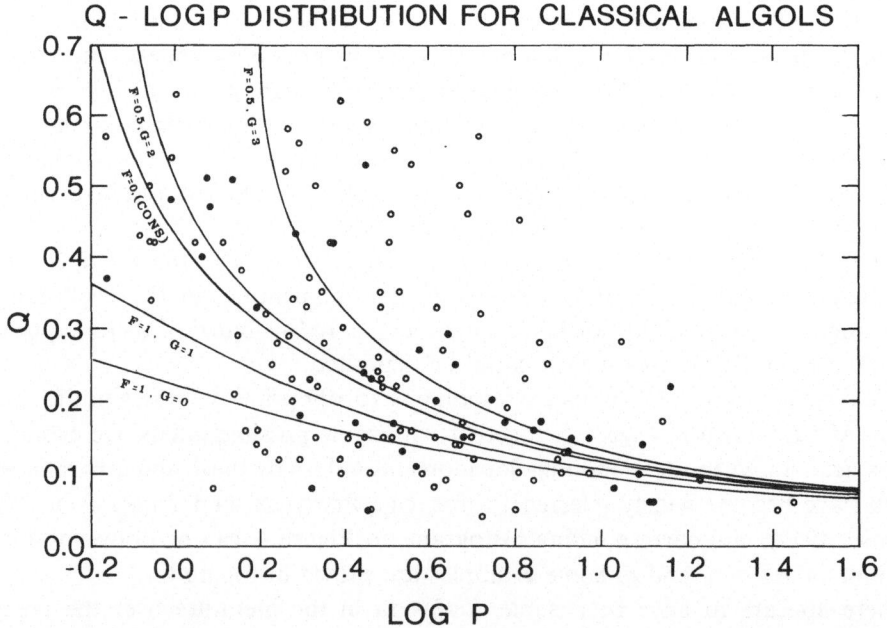

Fig. 5. Similar to Figure 3, but with the data corresponding to the 147 system from the catalogue, referred to also in Figure 2.

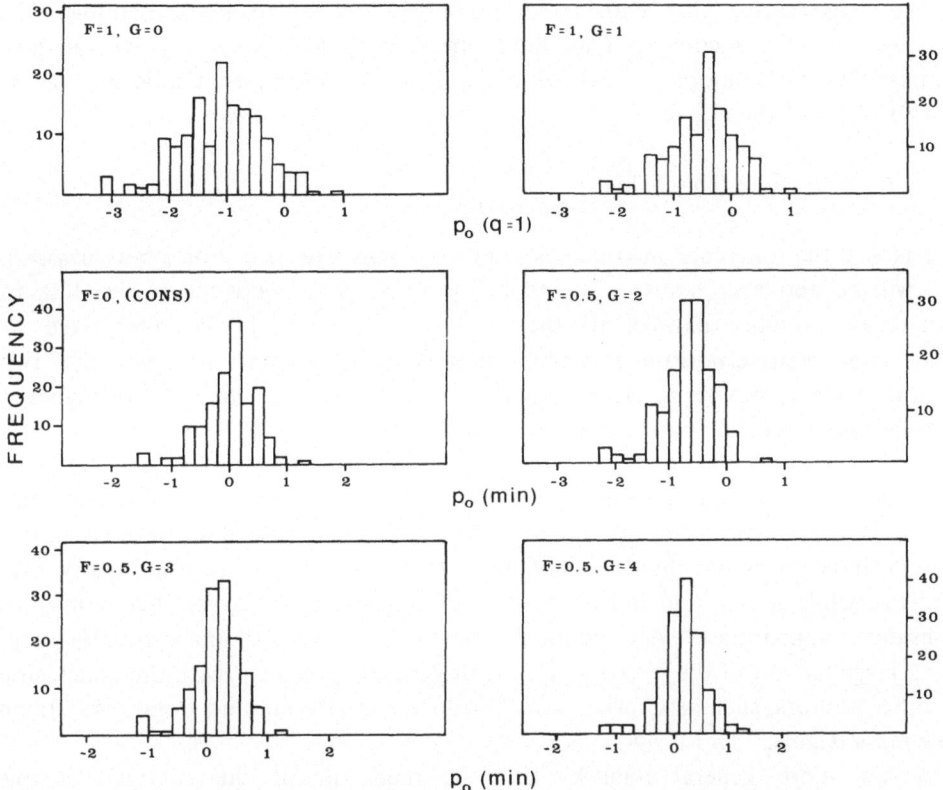

Fig. 6. Similar to Figure 4, but with the inclusion of non-conservative evolution trends such as those shown in Figure 5. The low g histograms clearly involve unacceptably low initial periods for the protomorph systems, as well as indicating non-parallelism between the densest region of the $p-q$ plane and the evolution tracks. The matching ameliorates with higher g and lower f. Thus the $f = 0.5$, $g = 4$ combination clearly aligns well with the density maximum as well as allowing the great majority of evolution tracks to be free from the 'over-contact' problem of Figure 4. With all considered combinations of f and g, however, there is a persistent knot of low p, q-systems whose explanation must remain puzzling.

represents a trial evolution track dependent on the parameters f and g, and specified for given p_0. In the most straightforward approach, we attribute the observed density distribution, for given mass, in the p, q-plane, only to different values of p_0 as expressed by the function $\chi(p_0)$. The existence of some maximum density region $\sigma = \sigma_{max}$ would imply, neglecting any observational selection effects, some largest value in the frequency $\chi(p_0)$ of EA2 binaries, at, say $P_{0\,max}$. The number of points $h_i(f, g)$ between $\phi_{ev} = c(p_0 - (\Delta p_0)/2)$ and $\phi_{ev} = c(p_0 + (\Delta p_0)/2)$, will be maximal when $\phi_{ev} = c(p_0)$ coincides with the major axis of the elongated region $\sigma = \sigma_{max}$. This condition occurs when $f = f_1, g = g_1$ (and $p_0 = p_{0\,max}$); i.e., maximizing the histogram maximum $h_{i\,max}$ in terms of f and g is equivalent to determining their best representative values, and will also lead to an estimate for $p_{0\,max}$ (which might be compared to the 'contact' p_0, for example).

The conservative case considered previously proves to be something of a discriminant in this context. Thus $\log P$ specified by (2.1), $\equiv p_c$ say, we can now compare with the more generalized formula of (for example) Vanbeveren *et al.* (1979). We will then have

$$p - p_c = (3g - 5) \log \frac{1+q}{1+q(1-f)}.$$

If f is ~ 0 the difference in ϕ curves is not too sensitive to g. For $f > 0$, if $g > \frac{5}{3}$, ϕ_{ev} will lie above $\phi_{ev}(\text{cons}) = c$, and below if $g < \frac{5}{3}$. A tendency of the data to congregate roughly parallel to the $\phi_{ev}(\text{cons})$ curve might be interpreted as evidencing relatively little systemic mass loss in general, however the two parameters may not be so clearly resolved in this way – i.e. the effect of higher f may be simulated, to some extent, by higher g.

In view of this, the numerous uncertainties surrounding this rather simplistic approach and inherent noise in the histogram distributions which is not negligible, it would be dangerous to place too much weight on any of its results so far; though indications are that f is not likely to be so far from zero for many EA2 systems, while $g > \frac{5}{3}$, and indeed a value of g significantly larger than $\frac{5}{3}$ may be considered appropriate*. A value for $f \sim 1$ with $g < \frac{5}{3}$ leads to unacceptably low p_0 for a large faction of the binaries. This is in general agreement with the conclusion of other authors, such as Giuricin and Mardirossian (1081a), and Popov (1970), on low-mass Algols.

A few more general remarks may be made about the enlarged sample distribution shown in Figure 5. Again there are a number of cases of possible incorrect eclipse type assumption, and, even if the basic type of eclipse is right, the accuracy of the solutions for r_2 may be doubtful, particularly in the case of the status 0.7 systems. Also, the gap of Figures 3 and 4 seems to have disappeared on Figures 5 and 6. It is still, perhaps, too preliminary to rule out a possibly qualitative change in the evolution pattern which may occur with over-contact at p_0, though we have found no clear corroboration of this possibility by taking into account a larger number of classical Algols whose primaries fall into the mass range $3 \pm 1 \ M_\odot$.

Acknowledgements

In addition to some support generously made available through the Local Organizing Committee, the author wishes to acknowledge a travel grant from the Kingdon–Tomlinson Trust (provided through the *Roy. Astron. Soc. of New Zealand*) and generous funding from the Carter Observatory Board which enabled me to take part in this conference.

---

* A 'good' pair of values, in the previously considered optimization sense, was found to be $f = 0.5$, $g = 4$ (see Figure 6).

References

Blanco, C. and Cristaldi, S.: 1974, *Publ. Astron. Soc. Pacific* **86**, 187.

Brancewicz, H. K. and Dworak, T. S.: 1980, *Acta Astron.* **30**, 501.

Budding, E.: 1981, in F. D. Kahn (ed.), *Investigating the Universe* D. Reidel Publ. Co., Dordrecht, Holland, p. 271.

Giannone, P. and Gianuzzi, M. A.: 1972, *Astron. Astrophys.* **6**, 309.

Giuricin, G. and Mardirossian, F.: 1981a, *Astrophys. J. Suppl.* **46**, 1.

Giuricin, G. and Mardirossian, F.: 1981b, *Astron. Astrophys. Suppl.* **5**, 85.

Hall, D. S.: 1974, *Acta Astron.* **24**, 215.

Hall, D. S. and Cannon, R. O.: 1974, *Acta Astron.* **24**, p79.

Hall, D. S. and Neff, S. G.: 1979, *Acta Astron.* **29**, 641.

Kopal, Z.: 1959, *Close Binary Systems*, Chapman and Hall, London.

Kukarkin, B. V., Kholopov, P. N., Efremov, Yu. N., Kukarkina, N. O., Kurochkin, N. E., Medvedeva, G. I., Perova, N. B., Fedorovich, V. P., and Frolov, M. S.: 1969, *General Catalogue of Variable Stars* (with supplements), Nauka, Moscow.

Okazaki, A.: 1977, *Publ. Astron. Soc. Japan* **29**, 289.

Paczyński, B. and Ziółkowski, J.: 1967, *Acta Astron.* **17**, 7.

Popov, M. V.: 1970, *Perem. Zvezdy* **17**, 412.

Popper, D. M.: 1973, *Astrophys. J.* **185**, 265.

Radhakrishnan, K. R., Sarma, M. B. K., and Abhyankar, K. D.: 1983, *Astrophys. Space Sci.* **99**, 229 (this volume).

Refsdal, S., Roth, M. L., and Weigert, A.: 1974, *Astron. Astrophys.* **36**, 113.

Russell, H. N.: 1912, *Astrophys. J.* **36**, 54.

Sahade, J. and Ringuelet, A.: 1970, *Vistas Astron.* **12**, 143.

Svechnikov, M. A.: 1969, *Katalog: Orbitalnyich Elementov, Mass i Svetimostei Tesnyich Dvoinyich Zvezd* A. M. Gorky Univ. (Sverdlovsk).

Thomas, H. C.: 1977, *Ann. Rev. Astron. Astrophys.* **15**, 127.

Vanbeveren, D., De Grève, J. P., van Dessel, E. L., and de Loore, C.: 1979, *Astron. Astrophys.* **73**, 19.

Ziółkowski, J.: 1976, in P. Eggleton, S. Mitton, and J. Whelan (eds.), 'Structure and Evolution of Close Binary Systems', *IAU Symp.* **73**, 321.

SEMI-DETACHED BINARIES: THE ORIGIN AND PRESENT STATUS OF TU MON, SX CAS, AND DM PER*

J. P. DE GRÈVE and W. PACKET

Astrophysical Institute, Vrije Universiteit Brussel, Belgium

(Received 16 July, 1983)

Abstract. An attempt is made to trace back the possible progenitor systems of the Algol-type binaries TU Mon, SX Cas, and DM Per. The present characteristics are compared to the result of the evolution of $9\,M_0 + 5.4\,M_0$. The position of the hot components in the HRD is discussed with regard to the theoretical models.

1. Introduction

Before coming to the issue of this paper we should like to begin with a synthesis of Kam Leung's paper, because we think that it reflected beautifully some small questions concerning generic relations, that still remain at present (see Figure 1).

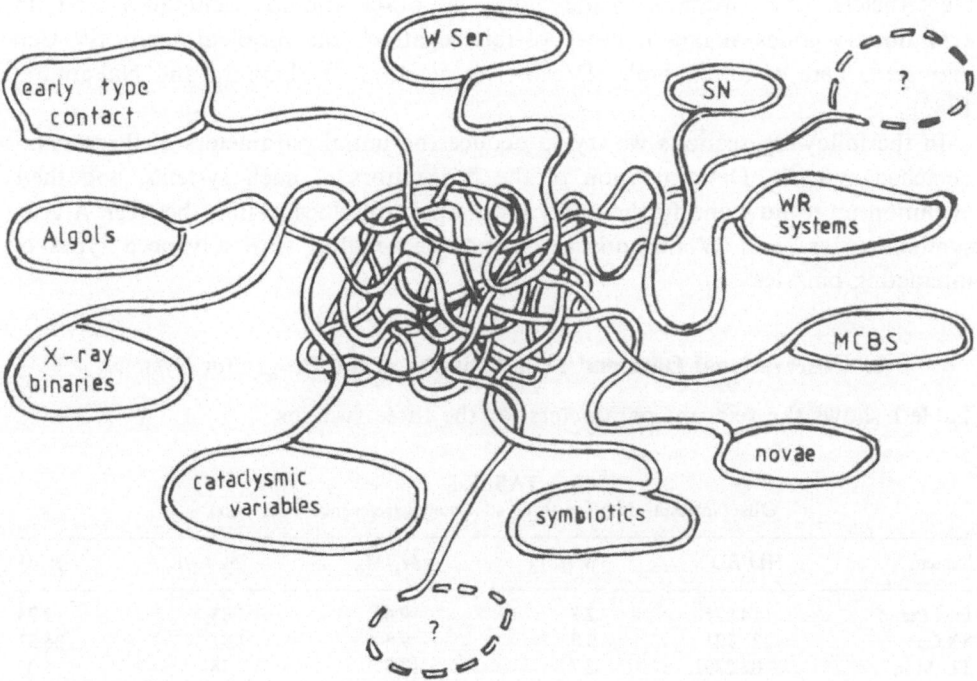

Fig. 1. The knot: the most recent version of the evolution for close binary systems.

* Paper presented at the Lembang-Bamberg IAU Colloquium No. 80 on 'Double Stars: Physical Properties and Generic Relations', held at Bandung, Indonesia 3–7 June, 1983.

Astrophysics and Space Science **99** (1984) 313–317. 0004–640X/84/0992–0313$00.75.

One word on the choice of the systems. At the time the paper was prepared only one (new) evolutionary computation (a test sequence which turned out to be succesfull!) was available. We therefore looked for sd-systems with approximately the same total mass as the theoretical model (14.4 M_0). The present three systems resulted, although we admit that their dimensions are much less well known than those discussed by Popper (1980). We intend to account for that in the future sequences.

In this talk we comment on one aspect of the 'Knot': the relation between semi-detached (sd) binaries and the (unknown) parental systems. Furthermore, we will discuss here only the conservative approach.

From the various review papers on the evolution of close binaries (see Paczynski, 1971, 1980) it follows that the characteristics of the mass transfer as well as the resulting system depend heavily on the initial parameters, i.e., mass of the loser M_{1i}, mass ratio $q_i = M_{2i}/M_{1i}$ and period P_i. But the relationship between these parameters and different kinds of interacting or post-mass-transfer binaries is still unsettled. Plavec (1973) already showed the inconsistencies arising when one tries to match individual systems to individual theoretical computations. Nevertheless, the progress made both in observational techniques and in evolutionary codes, makes it now possible to attack the problem again. SV Cen may serve here as an example (Drechsel *et al.*, 1982; Nakamura and Nakamura, 1982).

In the following sections we try to deduce the initial parameters of three semi-detached systems. Determination of the progenitors of such systems, and their evolution up to now and further, may give important clues to links between Algols, symbiotic stars and W Sepentis stars and the relation with advanced types of interacting binaries.

2. Observational Data and First Estimate of the Progenitor Systems

Table I shows the relevant parameters for the three systems.

TABLE I

Observational data for three sd-systems (references: see text)

System	HD/BD	M_1/M_0	M_g/M_0	$\log L_1/L_0$	P (d)
DM Per	14 871	2.3	9.4	1.85	2.73
SX Cas	232 121	3.8	9.5	1.91	36.57
TU Mon	-02.2331	2.7	12.7	2.26	5.05

They are derived from the paper of Giuricin *et al.* (1981) for DM Per, from Giuricin and Mardirossian (1981) for the W Serpentis system SX Cas, with spectral types and radius of the cool component taken from Plavec *et al.* (1982), and from Cester *et al.* (1977) for TU Mon. Using the initial mass ratio q_i as a free parameter (between 0 and 1) limits are set on the initial mass M_{1i} of the loser through the relations:

$$M_{1i} \geq 0.5 \times (M_1 + M_h)$$

and

$$M_{1i} \leq M_i^{\max}.$$

$M_i^{\max}$ is given by $M_i^{\max} = f^{-1}(M_1 = M_f)$, where

$$M_f = f(M_{1i}) = M_{1i}/(9.645 - 0.342\, M_{1i}),$$

represents the remnant mass at the end of a case B of RLOF for $3 \leq M_{1i} \leq 12$ (De Grève, 1980). With $q_i \geq 0$ this leads to the boundary limits shown in Figure 2. The values of the maximum mass are 12.4 M_0, 16.8 M_0, and 13.5 M_0, respectively, for DM Per, SX Cas, and TU Mon. The corresponding initial periods (assuming conservative mass transfer) are shown in the same figure, with distinction between case A and case B of RLOF. It follows that the progenitor systems of TU Mon are restricted both in mass range (7.5 M_0 − 13.5 M_0) and in period (0.7–12 d).

Tracks of single stars in the H–R-diagram are more or less horizontal after core hydrogen burning (and for $\log T_{\mathrm{eff}} \geq 4.0$). The mass luminosity relation in that phase, as derived from theoretical models, is given by

$$\log L/L_0 = 4.04\, (\log M/M_0)^{0.81} \quad \text{for} \quad 1.5 \leq M/M_0 \leq 15.$$

With the condition that the radius R_1 equals the critical radius R_R at the onset of

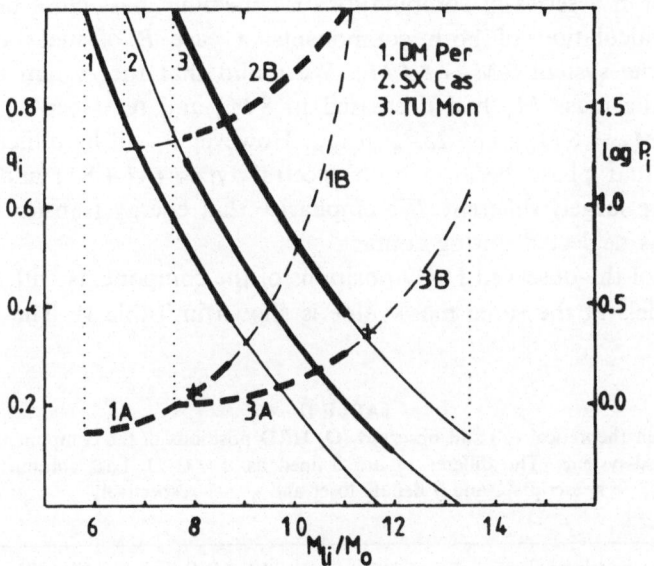

Fig. 2. Possible initial mass ratios (full curves) and initial periods (dashed curves) as a function of initial mass of the loser for 3 sd-systems, assuming conservative mass exchange. The labels of the period curves refer to the systems (1, 2, 3) as well as to the case of mass transfer (A, B), the boundary between A and B (if existing) is indicated by X. The thick parts of the curves represent the values allowed by the mass-luminosity position of the loser (see Section 3).

the mass transfer, the effective temperature at that specific moment can be calculated. So for all the combinations (M_{1i}, q_i, P_i) the location of the onset of mass transfer can easily be determined.

3. The Mass-Luminosity Position

Restraints can be set on the values of q_i and M_{1i} from the position in the $\log M_1 - \log L/L_0$ diagram. In the present study conservative computations are used. The position of the minimum luminosity during early case B mass transfer is given by (Devos, 1983)

$$\log L_{min} = 0.495\, M_{1i}^{0.634} - 19.581\, q_i^{-0.05} + 19.983,$$

$$M_{min} = 0.257\, M_{1i}^{1.256} + 10.746\, q_i^{-0.05} - 10.496,$$

in solar units. For early case B the influence of the period is negligible. Comparison of the present values of L_1 and M_1 with families of simplified tracks from the L_{min} positions to the final position in the mass-luminosity diagram, for different initial masses, leads to the results shown in Figure 2 (thick part of the curves).

4. Comparison with the Evolution of $9\, M_0 + 5.4\, M_0$

As the first one in a series of computations for medium mass close binaries, with simultaneous calculation of both components, a case B of mass transfer was computed for the system $9\, M_0 + 5.4\, M_0$. We found that the system entered into contact when the mass M_1 had decreased to $8\, M_0$ and remained in contact for 8000 yr (until $M_1 = 6\, M_0$), i.e., 25 % of t_{KH}. However it will be difficult to detect the system in that phase because both spectral types ($A1 + B1$) and luminosity ($\Delta M_V = 1.7$) are largely different. We emphasize that energy transfer between the components was neglected during contact.

Comparison of the observed HRD-positions of the components with those of the theoretical models at the same mass-value is shown in Table II. The last column

TABLE II

Comparison between theoretical (C) and observed (O) HRD positions of the components (at the same mass value) of 3 sd-systems. The differences are defined as $\Delta = C-O$. Last column: see text. The subscripts 1 and g denote loser and gainer respectively

System	$\Delta(\log L/L_0)_1$	$\Delta(\log T_e)_1$	$\Delta(\log L/L_0)_g$	$\Delta(\log T_e)_g$	M_{MS}/M_0
DM Per	1.60	0.13	1.23	0.24	5.5
SX Cas	−0.06	0.14	0.85	0.31	6.7
TU Mon	0.74	0.05	0.42	0.14	8.5

gives the mass of a normal single star corresponding with the present HRD-position of the gainer.

The differences for the loser may mainly be due to the differences in period between theoretical and observed system (as this component is in contact with its critical surface). The observed gainers however, are systematically underluminous, compared to the models. The latter may result from the poor quality of the observations of the three systems (cf. the comment of Giuricin *et al.*, 1981, for DM Per).

On the other hand these results indicate that for modeling observed sd-systems, one has to start with very well known systems. This in turn means that computations must be performed for lower initial masses, as very few high mass systems have reliable absolute dimensions (Popper, 1980).

References

Cester, B., Fedel, B., Giuricin, G., Mardirossian, F., and Pucillo, M.: 1977, *Astron. Astrophys.* **61**, 469.
De Grève, J. P.: 1980, *Astrophys. Space Sci.* **72**, 411.
Devos, Y.: 1983, Second Cycle Thesis, V.U.B., Brussels.
Drechsel, H., Rahe, J., Wargau, W., and Wolf, B.: 1982, *Astron. Astrophys.* **110**, 246.
Giuricin, G. and Mardirossian, F.: 1981, *Astrophys. J. Suppl.* **46**, 1.
Giuricin, G., Mardirossian, F., and Predolin, F.: 1981, *Astron. Astrophys. Suppl.* **43**, 251.
Nakamura, M. and Nakamura, Y.: 1982, *Astrophys. Space Sci.* **83**, 163.
Paczynski, B.: 1971, *Ann. Rev. Astron. Astrophys.* **9**, 183.
Paczynski, B.: 1980, *Highlights of Astronomy* **5**, 27.
Plavec, M.: 1973, in A. H. Batten (ed.), *Extended Atmospheres and Circumstellar Matter in Spectroscopic Binary Systems*, D. Reidel Publ. Co., Dordrecht, Holland, p. 216.
Plavec, M., Weiland, J. L., and Koch, R. H.: 1982, *Astrophys. J.* **256**, 206.
Popper, D.: 1980, *Ann. Rev. Astron. Astrophys.* **18**, 115.

NUTATIONAL EFFECTS IN SS 433*

REMO RUFFINI and DOO JONG SONG

Dipartimento di Fisica 'G. Marconi', Universita' di Roma, Italia

(Received 2 November, 1983)

Abstract. The nutation effects of an accreting disk around SS 433 are analyzed within the framework of the fully relativistic model of Fang and Ruffini.

1. Introduction

The theoretical interpretation of the shifts in the moving lines of SS 433 ((Mammano *et al.*, 1980; Margon *et al.*, 1980), presenting large asymmetries between the blue (λ_-) and the redshifted lines (λ_+) with respect to the lines at rest give the SS 433 is a relativistic system which has a periodic change in moving lines.

In order to explain the observed shifts in Hα, Hβ, etc. lines, at least two basic geometries have advanced (Milgram, 1979): in one (the 'twin jet' model), the emission comes from two highly collimated beams moving at a speed $v/c \simeq 0.25$ away from a central source, in the other (the 'ring' model), the emission occurs from two opposite points of a ring orbiting a gravitationally collapsed object at a radius of $\sim 45M$ (Fang and Ruffini, 1979; Ruffini and Stella, 1980a). And the possible processes of stimulated emission in the ring model have been considered at work in SS 433 (Ruffini and Stella, 1980b).

In the ring model the modulations in the amplitude of the shifted lines with a period of $\sim 164^{\text{d}}$ is assumed to a precessional effect of the ring around the compact object and the observational data (Frasca *et al.*, 83) is well fitted by the general relativistic line shift equation (Ruffini and Stella, 1980a), a combination of gravitational and Doppler shift

$$1 + Z^{\pm} = (1 - 3M_1/r)^{-1/2} \left\{ 1 \pm (M_1/r)^{1/2} \cos \delta [\tan^2 \xi + (1 - 2M_1/r)^{-1/2}] \right\}, \tag{1}$$

where the $+$ $(-)$ sign refers to the red (blue) shifts, and

$$\cos \delta = (\cos i \sin \alpha - \sin i \cos \alpha \cos \varphi) (1 - \sin^2 i \sin^2 \varphi)^{-1/2}, \tag{2}$$

$$\sin \xi = \sin i \sin \varphi, \tag{3}$$

with the following values of parameters in Equations (1) and (2):

$$r \simeq 45M_1, \quad (i, \alpha_0) = (65°7, 54°0) \quad \text{or} \quad (36°0, 24°3).$$

Further analysis of the spectroscopical data of the shifted lines has shown significant statistical deviation from a simple sinusoidal behaviour in the modulations of the shifted

* Paper presented at the Lembang–Bamberg IAU Colloquium No. 80 on 'Double Stars: Physical Properties and Generic Relations', held at Bandung, Indonesia, 3–7 June, 1983.

Astrophysics and Space Science **99** (1984) 319–327. 0004–640X/84/0992–0319$01.35.

TABLE I

The relevant short period modulations (Frasca *et al.*, 1983)

Period	Amplitude		Difference of phase
	Z_+	Z_-	
$6\overset{d}{.}29$	0.0080	0.0075	$\sim 180°$
$5\overset{d}{.}84$	0.0049	0.0041	$\sim 180°$

lines with a period of $\sim 164^d$ (Newsom and Collins, 1980, a, b, 1981; Wagner *et al.*, 1981; Katz *et al.*, 1982; Mammano *et al.*, 1983). According to the observational data analysis of Frasca *et al.* (1983), the relevant modulations in the shifted lines among many periodic modulations are given in Table I. The square module of these amplitudes corresponding to the $\sim 6^d$ periodicities gives a relation (Mammano, 1982)

$$|A_{6.28}|^2 \simeq 3 |A_{5.84}|^2. \tag{4}$$

The aim of this talk is to show these $\sim 6^d$ periodicities can be accounted for by a nutational effect induced by the companion star on the accretion disk, initially introduced in the ring model, in the binary system model for SS 433. If it is confirmed, this effect will establish unequivocally the binary nature of the system.

2. Model

From the analysis of spectroscopical data (Crampton *et al.*, 1980, 1981) and the photometric data (Margon *et al.*, 1980, 1983), it has been inferred that the SS 433 is very probably a member of a binary system with a period $P \sim 13\overset{d}{.}08$. Therefore we make a binary system simply introducing a companion star into ring model (see Fang and Ruffini, 1979; or Fang *et al.*, 1981).

In our model the accretion disk in the binary system are formed by a succession of precessional rings inclined at the same angle α with respect to the orbital plane. The perturbational modes generated at the border of the accretion disk are then assumed to propagate down to the inner emitting ring by viscosity (see Meritt and Petterson, 1980; Hatchett, 1981).

Each ring will generally experience a torque due to the coupling of its quadrupole moment with the tidal field of the companion star, M_2. This torque which can be calculated from the transport law of the spin angular momentum of a ring in the gravitational field of the companion star is the origin of the nutation type perturbations of a ring and determines the quadrupole angular velocity of a ring. In the Newtonian limit, the torque can be expressed by the equation

$$\frac{dL^i}{dt} = -\varepsilon_{ijk} R_{kolo} f_{jl} \equiv N^i, \tag{5}$$

where L^i is the spin angular momentum of the ring,

$$f_{jl} = \int \rho(x^j x^l - \tfrac{1}{3} r^2 \delta_{jl}) \, \mathrm{d}^3 x$$

is the reduced quadrupole moment tensor, and the Riemann curvature tensor R_{kolo} is given by

$$R_{\mathrm{kolo}} = \partial^2 \Phi / \partial x^k \, \partial x^l, \tag{6}$$

where Φ is the Newtonian potential of the companion star defined by

$$\Phi(x, y, z) = -M_2 / [(x - x_R)^2 + (y - y_R)^2 + (z - z_R)^2]^{1/2} .$$

The (X_R, Y_R, Z_R) is the coordinate of the companion star.

In order to describe a ring in the binary and evaluate the perturbations on a ring due to the torque defined in Equation (5), we introduce three systems of coordinates (see Figure 1):

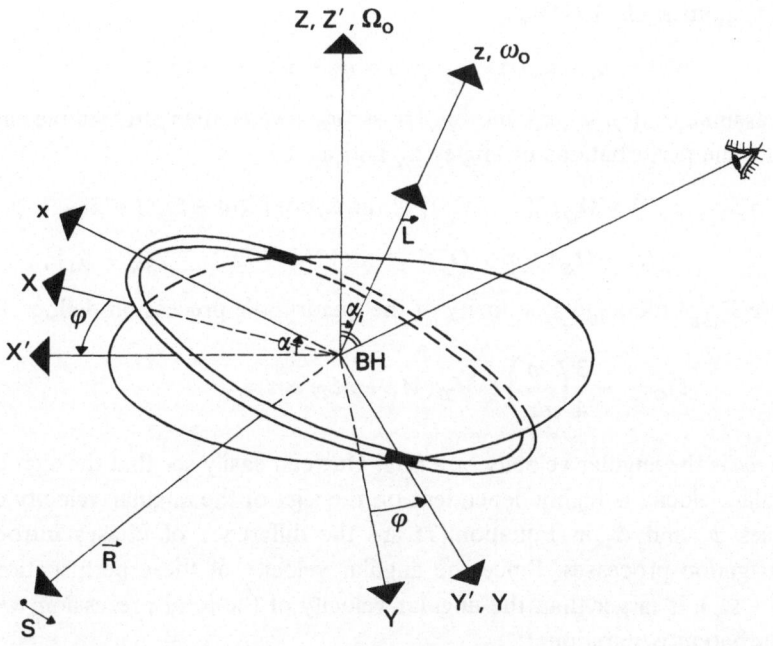

Fig. 1. A coordinate system (X', Y', Z') is defined with its origin at the centre of the ring, with the Z'-axis normal to the binary orbital plane and the Y'-axis along the line of the nodes, defines as the intersection between the binary orbital plane and the plane of the ring. The ring's local plane is also defined with the y-axis coincident with Y'-axis and the x-axis in the plane of the ring. The plane of the ring is inclined by an angle α with respect to the binary orbital plane. The centre of the companion star rotates in the $X - Y$ plane with an angular velocity ω and is identified by the position R. Finally we define a non-rotating coordinate (X, Y, Z), still with its origin at the centre of the ring, with $Z = Z'$ and the Y-axis, initially intersecting the nodes, forming an angle with respect to the Y'-axis. The line-of-sight from the Earth to the centre of the ring substent the Z-axis by an angle i.

(i) The fixed system (X, Y, Z) with its origin at the center of the ring, the Y-axis initially along the line of nodes defines as the intersection between the binary orbit and the plane of the ring, and the X-axis in the plane of the binary.

(ii) The ring's local frame (x, y, z) has the y-axis coincident with the nodal line, the x-axis in the plane of the ring and α is the inclination angle between the plane of the binary orbit and the plane of the ring.

(iii) The moving (rotating) system (X', Y', Z') still has its origin at the center of the ring with the axis $Z' = Z$ and the Y'-axis which is always along the line of node, displaced from the Y-axis by the angle φ given by the precession and nutation of the ring in the binary system. The (X, Y, Z) coordinate system is fixed with respect to the asymptotic inertial frame at rest at infinity. The center of companion star lies in the $X - Y$ plane.

We now assume that the ring is precessing at a rate $\Omega_p \sim 2\pi/164^d$ and the angular velocity of the binary system is $\omega \simeq 2\pi/13^d08$. Using Equation (5) in (X, Y, Z) coordinate system in which we are doing the observations, we can evaluate the perturbations induced by the companion star on the angles α and φ. If we write the change of angles (see Cogotti *et al.*, 1983) as

$$\alpha(t) = \alpha_0 + \alpha_N(t), \qquad \varphi(t) = \varphi_0 + \Omega_p t + \varphi_N(t);$$

and assume $|\alpha_N(t)| \ll |\alpha_0|$ and $|\varphi_N(t)| \ll |\varphi_0 + \Omega_p t|$, then after simple calculations, we obtain the perturbations of angles α_N and φ_N as

$$\alpha_N(t) = \Omega_Q [2(\omega - \Omega_p)]^{-1} \tan\alpha_0 \cos[2(\omega - \Omega_p)t - 2\varphi_0 + \psi_1],$$

$$\varphi_N(t) = \Omega_Q [2(\omega - \Omega_p)]^{-1} \cos[2(\omega - \Omega_p)t - 2\varphi_0 + \psi_2], \qquad (7)$$

where Ω_Q is the angular velocity of the quadrupole precession defined by

$$\Omega_Q = -\frac{3}{4}\left(\frac{\omega}{\omega_0}\right) \omega M_2/(M_1 + M_2) \cos\alpha_0 . \qquad (8)$$

The ω_0 is the angular velocity of a ring. One can easily see that the sign of quadrupole angular velocity is highly dependent on the sign of the angular velocity of a ring. The phases ψ_1 and ψ_2 in Equation (7) are the difference of phases introduced by the propagation processes. Since the angular velocity of these perturbations on angles, $2(\omega - \Omega_p)$, is larger than the angular velocity of the total precession, we can say the perturbation is nutation.

3. Applications to the Foruma of the Line Shift

According to the observational data of Ciatti *et al.* (1983), the sum of the red- and blue-shift, $Z^+ + Z^-$, is time independent up to the relevant Fourier component (see also Frasca *et al.*, 1983), we may assume that the radius of the emitting ring is constant. In order to demonstrate the effect of nutational perturbation on the shifted lines, we expand the relativistic formula given in Equation (1) in terms of $(M_1/r)^{1/2}$ and α_N and

φ_N given in Equation (7). Keeping the expansion terms up to the order of $(M_1/r)^{3/2}$ and choosing linear terms of α_N and φ_N only, we have for the expressions of the shifted lines $1 + Z^\pm = 1 + Z_0^\pm + \Delta Z^\pm$, where

$$1 + Z_0^\pm = (1 - 3M_1/r)^{-1/2}[1 \pm (M_1/r)^{1/2} \times$$

$$\times (\cos i \sin \alpha_0 - \sin i \cos \alpha_0 \cos(\Omega_p t + \varphi_0))] \tag{9}$$

is the usual formula for the line shift with long term precessional rate $\Omega_p = 2\pi/163\overset{d}{.}34$ and, according to Cogotti *et al.* (1983), the $\Delta Z^\pm$ terms are

$$\Delta Z^\pm = \pm \{ +A_1(2\Omega_p)\cos(2\Omega_p t + 2\varphi_0) - A_2\{3\Omega_p\}\cos(3\Omega_p t + 3\varphi_0) +$$

$$+ A_3(2\omega - 2\Omega_p)\cos[2(\omega - \Omega_p)t + \Theta_3] +$$

$$+ A_4(2\omega - \Omega_p)\cos[(2\omega - \Omega_p)t - \Theta_4] -$$

$$- A_5(2\omega - 2\Omega_p)\cos[(2\omega - 3\Omega_p)t - \Theta_5] -$$

$$- A_6(2\omega)\cos(2\omega t + \Theta_6) - A_7(2\omega - 4\Omega_p)\cos[2\omega - 4\Omega_p)t + \Theta_7] -$$

$$- A_8(2\omega + \Omega_p)\cos[(2\omega + \Omega_p)t - \Theta_8] +$$

$$+ A_9(2\omega - 5\Omega_p)\cos[(2\omega - 5\Omega_p)t - \Theta_9]\}, \tag{10}$$

where A_i, $i = 1, \ldots, 9$, are the amplitudes given by

$$A_1 = \tfrac{1}{2}\Gamma(M_1/r)^{3/2}\sin^3 i \cos i \sin \alpha_0,$$

$$A_2 = \tfrac{1}{4}\Gamma(M_1/r)^{32}\sin^3 i \cos \alpha_0,$$

$$A_3 = \Gamma(M_1/r)^{1/2}\Omega_Q[2(\omega - \Omega_p)]^{-1}\cos i \sin \alpha_0[1 + \tfrac{1}{2}M_1/r(1 + \cos^2 i)] \times,$$

$$\times [(1 + \tan^4 \alpha_0) - 2\cos\Delta\psi\tan^2\alpha_0]^{1/2},$$

$$A_5 = \tfrac{1}{2}\Gamma(M_1/r)^{1/2}\Omega_0[2(\omega - \Omega_p)]^{-1}\sin i \cos \alpha_0[1 + \tfrac{3}{4}(M_1/r) \times$$

$$\times (1 + \tfrac{1}{3}\cos i)][(1 + \tan \alpha_0) + 2\cos\Delta\psi\tan^2\alpha_0]^{1/2},$$

$$A_6 = \tfrac{1}{4}\Gamma(M_1/r)^{3/2}\Omega_Q[2(\omega - \Omega_p)]^{-1}\sin^2 i \cos i \sin \alpha_0(5 + 4\cos\Delta\psi)^{1/2},$$

$$A_7 = \tfrac{1}{4}\Gamma(M_1/r)^{3/2}\Omega_Q[2(\omega - \Omega_p)]^{-1}\sin^2 i \cos i \sin \alpha_0(5 - 4\cos\Delta\psi)^{1/2},$$

$$A_8 = \tfrac{1}{8}\Gamma(M_1/r)^{3/2}\Omega_Q[2(\omega - \Omega_p)]^{-1} Am^{-3}i \cos \alpha_0[(9 + \tan^4 \alpha_0) -$$

$$- 6\cos\Delta\psi\tan^2\alpha_0]^{1/2},$$

$$A_9 = \tfrac{1}{8}\Gamma(M_1/r)^{3/2}\Omega_Q[2(\omega - \Omega_p)]^{-1}\sin i \cos \alpha_0[(9 + \tan^4 \alpha_0) +$$

$$+ 6\cos\psi\tan^2\alpha_0]^{1/2}; \tag{11}$$

where the symbol $\Gamma = (1 - 3M_1/r)^{-1/2}$ and $\Delta\psi = \psi_1 - \psi_2$. The first two amplitudes A_1 and A_2 are the Fourier harmonics of Ω_p and all the rest are the combinations between $(2\omega - 2\Omega_p)$, the nutational rate, and Ω_p. The phases Θ_i, $i = 3, \ldots, 9$, are given by

$$\Theta_3 = \psi_1 - 2\varphi_0,$$

$$\Theta_4 = \arctan\{[\tan^2\alpha_0\cos(\psi_1 - \varphi_0) - \cos(\psi_2 - \varphi_0)] \times$$
$$\times [\tan^2\alpha_0\sin(\psi_1 - \psi_0) - \sin(\psi_2 - \varphi_0)]^{-1}\},$$

$$\Theta_5 = \arctan\{[\tan^2\alpha_0\cos(\psi_1 - 3\varphi_0) - \cos(\psi_2 - 3\varphi_0)] \times$$
$$\times [\tan^2\alpha_0\sin(\psi_1 - 3\varphi_0) - \sin(\psi_2 - 3\varphi_0)]^{-1}\},$$

$$\Theta_6 = \arctan[(\sin\psi_1 + 2\sin\psi_2)(\cos\psi_1 + 2\cos\psi_2)^{-1}],$$

$$\Theta_7 = \arctan\{[\sin(\psi_1 + 4\varphi_0) - 2\sin(\psi_2 + 4\varphi_0)] \times$$
$$\times [\cos(\psi_1 - 4\varphi_0) - 2\cos(\psi_2 - 4\varphi_0)]^{-1}\},$$

$$\Theta_8 = \arctan\{[\tan^2\alpha_0\cos(\psi_1 + \varphi_0) - 3\cos(\psi_2 + \varphi_0)] \times$$
$$\times [\tan^2\alpha_0\sin(\psi_1 + \varphi_0) - 3\sin(\psi_2 + \varphi_0)]^{-1}\},$$

$$\Theta_9 = \arctan\{[\tan^2\alpha_0\cos(\psi_1 - 5\varphi_0) + 3\cos(\psi_2 - 5\varphi_0)] \times$$
$$\times [\tan^2\alpha_0\sin(\psi_1 - 5\varphi_0) + 3\sin(\psi_2 - 5\varphi_0)]^{-1}\}. \tag{12}$$

We can then conclude from the above relations that the combination of the precession of the disk and the nutation due to the tidal torque exerted by the companion star leads to the modulations in the shifted lines.

4. Applications to the SS 433

4.1. THE FIT OF THE OBSERVATIONAL DATA

In order to evaluate the amplitudes of the frequency shifts $\Delta Z^{\pm}$ given by Equation (10), we have determine the parameters r, i, α_0, $\Delta\psi$, and Ω_Q. We use the assumed angular velocity of the disk by $\Omega_p \simeq 2\pi/163^d34$ (Mammano *et al.*, 1983) and binary period as $p \simeq 13^d08$ (Crampton *et al.*, 1980). According to the recent data analysis of observations

TABLE II

The nine amplitudes of modulations, given in Equation (11), and their corresponding periods are given

H	T (days)	$|A_i|$
$2\Omega_p$	81.7	0.0043
$3\Omega_p$	54.4	0.0004
$2\omega - \Omega_p$	6.29	0.0078
$2\omega - 3\Omega_p$	5.84	0.0045
$2\omega - 2\Omega_p$	6.06	0.0036
2ω	6.54	2×10^{-5}
$2\omega - 4\Omega_p$	5.64	4×10^{-5}
$2\omega + \Omega_p$	6.81	6×10^{-5}
$2\omega - 5\Omega_p$	5.45	3×10^{-5}

(Mammano et al., 1982; Frasca et al., 1983) and the relation given in Equation (4), we obtain the only viable values of parameters as $(i, \alpha_0) \simeq (65°7, 54°0)$, $\Delta\psi \simeq 127°3$ at $r \simeq 45M_1$. It is interesting that the set $(i, \alpha_0) \simeq (65°7, 54°0)$ is obtained both for the $\sim 164^d$ periodicity and for the nutation of $\sim 6^d$ periods. Finally we can evaluate the value of Ω_Q in Equation (11), using the experimental value for the amplitude of the 6^d28, by $\Omega_Q \simeq 2\pi/83^d8$. In Table II we give the amplitudes and the corresponding periods of the modulation in Equations (11) and (10).

4.2. A DISCUSSION ON THE BINARY PERIOD

The value of the total precessional angular velocity has been determined quite accurately by a variety of observations. Although the binary period of 13^d08 which is first reported by Crampton et al. (1980) is widely circulated, it is interesting to analyze strictly the value of the binary period.

According to the data analysis of the observations (Frasca et al., 1983) the significant amplitudes occur at periods of 6^d29 and 5^d84. In Table III and Table IV we give the values of the expected periodicities in ΔZ corresponding to the above major modulations. The co-precessional case (Table III) corresponds to the results of Giles et al. (1980) that they obtained the binary period as 11^d8, and the counter-precessional case (Table IV) is similar to the results of Crampton et al. (1980).

TABLE III

The periodicities of the three principal harmonics given in Equation (10) are given as a function of selected values of the binary period for a corotating precessional period of 164^d

H \ P	$\Omega_p > 0$		
	12^d11	11^d68	11^d27
$2\omega - \Omega_p$	6.28	6.06	5.84
$2\omega - 2\Omega_p$	6.84	6.28	6.05
$2\omega - 3\Omega_p$	6.81	6.84	6.28

TABLE IV

Same as Table III, for a counter-rotating precessional period

H \ P	13^d08	13^d62	14^d22
$2\omega - \Omega_p$	6.28	6.54	6.81
$2\omega - 2\Omega_p$	6.06	6.28	6.54
$2\omega - 3\Omega_p$	5.84	6.05	6.28

We assume that the total precessional angular velocity of the disk mainly from the quadruple precessional velocity given in Equation (8), i.e. $\Omega_p \sim \Omega_Q = = -\frac{3}{4}(\omega/\omega_0)\omega M_2/(M_1 + M_2)|\cos\alpha_0|$. We used $|\cos\alpha_0|$ instead of $\cos\alpha_0$ to give sign of rotational disk to ω_0. According to this relation we know the following thing that the sign of Ω_p is directly related to the sign of ω_0, the rotation of the disk. If ω_0 and ω have the same sign, the precession is now counter-precession, but ω_0 has opposite sign to ω, the Ω_p has co-precession with ω. If the widely used $\sim 13^d$ period which is confirmed by Katz et al. (1982) is correct, ω_0 has the same sign of ω and binary system has counter-precessional angular velocity. It is worth noting that, since $\Omega_Q \sim 2\pi/83^d$ and $\Omega_p \sim 2\pi/164^d$, the required additional angular velocity may be $\Omega_{add} \sim 2\pi/164^d$ and the sign of Ω_{add} may be opposite to Ω_Q, if Ω_p is obtained by additional combination of many precessional mode.

4.3. THE MASSES OF THE BINARY SYSTEM

From the amplitude of the nutational period we have determined the numerical value of $\Omega_Q \simeq 2\pi/83^d8$. By the equation for Ω_Q in Equation (8) with known values of parameters we obtain the relation

$$\frac{M_1}{M_\odot} \simeq K(r)/[1 - K(r)(M_2/M_\odot)^{-1}], \tag{13}$$

where $K(r) \simeq 3.4 \times 10^7 (r/45M_1)^{-3/2}$ and, as usual, r is the radius of the disk. Combining the above equation with the mass function of Crampton and Hutchings (1981)

$$\frac{M_2^3}{(M_1 + M_2)^2} \simeq 13M_\odot. \tag{12}$$

We can infer constraints on the masses of the system (see Table V).

TABLE V

Here are shown for selected values of the radius of the accretion disk around a compact object, the masses of binary system (M_1 for the compact object and M_2 for the companion star), the interstellar distance R, and velocities (v_1 for the velocity of the compact object and v_2 for the companion star). Reading from the top of the table, all the data are obtained by the assumption $v_1 \sim v_{orb} \simeq 195$ km s^{-1} (see Crampton and Hutchings, 1981) and, reading from the bottom, we have assumed $v_2 \sim v_{orb}$

M_1/M	M_2/M	R (cm)	r (cm)	v_2 (km s^{-1})	
3.1×10^8	1.0787×10^6	1.10×10^{15}	4.84×10^{15}	5.60×10^4	7.16×10^{13}
1.76×10^6	3.467×10^4	1.98×10^{14}	1.17×10^{14}	9.88×10^3	2.30×10^{13}
3.04×10^4	1.088×10^3	5.13×10^{13}	2.02×10^{14}	5.44×10^3	7.23×10^{12}
1.79×10^3	3.96×10^2	2.11×10^{13}	2.64×10^{13}	8.82×10^2	5.84×10^{12}
74.56	62.31	8.39×10^{12}	4.97×10^{12}	2.33×10^2	4.13×10^{12}
52	52	7.65×10^{12}	4.13×10^{12}	1.95×10^2	4.13×10^{12}
15.59	30.01	5.81×10^{12}	2.3×10^{12}	1.01×10^2	4.43×10^{12}
1.16	15.07	4.1×10^{12}	7.7×10^{11}	15	1.00×10^{13}
M_2/M	M_1/M	R (cm)		V_1 (km s^{-1})	r (cm)

We see from the table that the possibility that SS 433 is a very massive black hole (see, e.g., Terlevich and Pringle, 1979;. Amitai-Milchgrub *et al.*, 1979; and Shaham, 1980) can be discarded on the ground of the above formulae. We are probably dealing with stars of mass $\lesssim 50M_\odot$ both in the case of collapsed object and that of the companion star. It is also worth stressing that the disk should have dimensions comparable to the interstellar distance.

5. Conclusions

In this article we have considered as a model of SS 433 a binary system which consists of companion star, compact object and accretion disk around a compact object, precessing with a period of $\sim 164^d$. We have considered the nutational effects on the disk induced by a companion star with a binary period of $\sim 13^d$. Within the ring model, the theoretical expressions for the $\sim 6^d$ modulations in the shifted lines are obtained.

In the background of our model we have reached general conclusions:

(1) Strong and independent evidence for the binary nature of the system obtained from the different Fourier components of the 6^d periodicities.

(2) The sense of the precessional angular velocity of the disk is opposite to that of binary angular velocity which has the same sign of angular velocity of the disk.

(3) The presence of an extended disk of dimensions comparable to the separation distance of the system.

(4) Evaluation of the masses of the system.

References

Amitai-Milchgrub, A., Piran, T., and Shabam, J.: 1979, *Nature* **279**, 505.

Ciatti, F., Mammano, A., Iijima, T., and Vittone, A.: 1983, *Astron. Astrophys.*, in press.

Cogotti, R., Ruffini, R., and Song, D. J.: 1983, *Astron. Astrophys.*, in press.

Crampton, D. and Hutchings, J. B.: 1981, *Astrophys. J.* **251**, 604.

Crampton, D., Cowley, A. R., and Hutchings, J. B.: 1980, *Astrophys. J.* **235**, 431.

Fang Li Zhi and Ruffini, R.: 1979, *Phys. Letters* **86B**, 193.

Fang Li Zhi, Ruffini, R., and Stella, L.: 1981, *Vistas Astron.* **25**, 185.

Frasca, S., Ciatti, F., and Mammano, A.: 1983, preprint.

Giles, A. D., King, A. R., Jameson, R. F., Sherrington, M. R., Hough, G. H., Bailey, J. A., and Cunningham, E. C.: 1980, *Nature* **286**, 289.

Hatchett, S. P., Begelman, M. C., and Sarazinn, C. L.: 1981, *Astrophys. J.* **247**, 677.

Katz, J. I., Anderson, S. F., Margon, B., and Grandi, S. A.: 1982, *Astrophys. J.* **260**, 780.

Mammano, A., Ciatti, F., and Vittone, A.: 1980, *Astron. Astrophys.* **85**, 14.

Mammano, A., Margoni, R., Ciatti, F., and Cristiani, S.: 1983, *Astron. Astrophys.*

Margon, B., Anderson, S. F., and Grandi, S. A.: 1983, *Astrophys. J.* (submitted).

Margon, B., Ford, H. C., Grandi, S. A., and Stone, R. P. S.: 1983, *Astrophys. J.* **233**, L63.

Margon, B., Grandi, S. A., and Downes, D. A.: 1980, *Astrophys. J.* **241**, 306.

Milgram, M.: 1979, *Astron. Astrophys.* **76**, L3.

Meritt, D. and Petterson, J. A.: 1980, *Astrophys. J.* **236**, 255.

Newsom, G. H. and Collins, II, G. W.: 1980, *IAU Circ.*, No. 3459.

Newsom, G. H. and Collins, II, G. W.: 1981a, *Astron. J.* **86**, 1250.

Newsom, G. H. and Collins, II, G. W.: 1981b, *Astrophys. J.* **262**, 714.

Ruffini, R. and Stella, L.: 1980a, *Nuovo Cimento Letters* **27**, 529.

Ruffini, R. and Stella, L.: 1980b, *Phys. Letters* **93B**, 107.

Shahman, J.: 1980, *Astrophys. Letters* **20**, 115.

Terlevich, R. J. and Pringle, J. E.: 1979, *Nature* **278**, 719.

Wagner, R. M., Newsom, G. H., Foltz, C. B., and Byard, P. L.: 1981, *Astron J.* **86**, 1671.

PERIODS RANGING FROM 5 TO 1500 DAYS
IN THE ANTICORRELATED MOVING LINES
OF SS 433*

S. FRASCA

Department of Physics, University of Rome

and

F. CIATTI and A. MAMMANO**

Asiago Astrophysical Observatory, University of Padova

(Received 6 October, 1983)

Abstract. Some tens of harmonic components are found in the moving lines of SS 433, leaving a standard deviation of 0.002 in Z units, that is comparable to the observational error.

Ten harmonics appear anticorrelated in the two branches, among which those at 80, 155, and 1500 days are also strong. Their behaviour indicates changing structure in the ejecting mechanism.

No secular change is occurring in the precessional 163-day period beyond the estimated error of $5 \times 10^{-5} \, d^{-1}$. No periodicity occurs in beam velocity. One quarter of cases of $(Z_1 + Z_2)$ are found outside their mean value by more than 5σ. This implies either variable beam velocity or absence of strict anti-parallelism.

A harmonic analysis of 216 anticorrelated wavelengths of the relativistically displaced lines in SS 433 = V 1343 Agl (Ciatti *et al.*, 1983) has been performed, modelizing the results by a simulation program which avoids aliases due to spectral windows.

The red branch (Z_1) can be represented with a standard deviation of 0.0026 (in $\Delta\lambda/\lambda = Z$ units) if 41 harmonic components, with periods ranging from 1 to 1500 days, are included. The blue branch (Z_2) is still better represented by 63 harmonic components to ± 0.0013, i.e. at the observational error level (see Figures 1 and 3).

The most significant harmonics are listed in Table I. It appears that the phase differences of harmonics with about the same period in the two branches are nearly $180°$, i.e. they result anticorrelated. A confirmation of their anticorrelation follows from harmonics found in the combination of $(Z_1 - Z_2)$ data, which moreover supply nearly double amplitudes, while the $(Z_1 + Z_2)$ data supply null results, as expected from anticorrelated harmonics.

Anticorrelated wavelength displacements are predicted in both disc and jet models for SS 433 (Cogotti *et al.*, 1983), where periods of about 165, 6.28, 5.84, and 80 days agree with observations, given in Table I.

* Paper presented at the Lembang-Bamberg IAU Colloquium No. 80 on 'Double Stars: Physical Properties and Generic Relations', held at Bandung, Indonesia, 3–7 June, 1983.
** Also at Department of Mathematics, University of Messina.

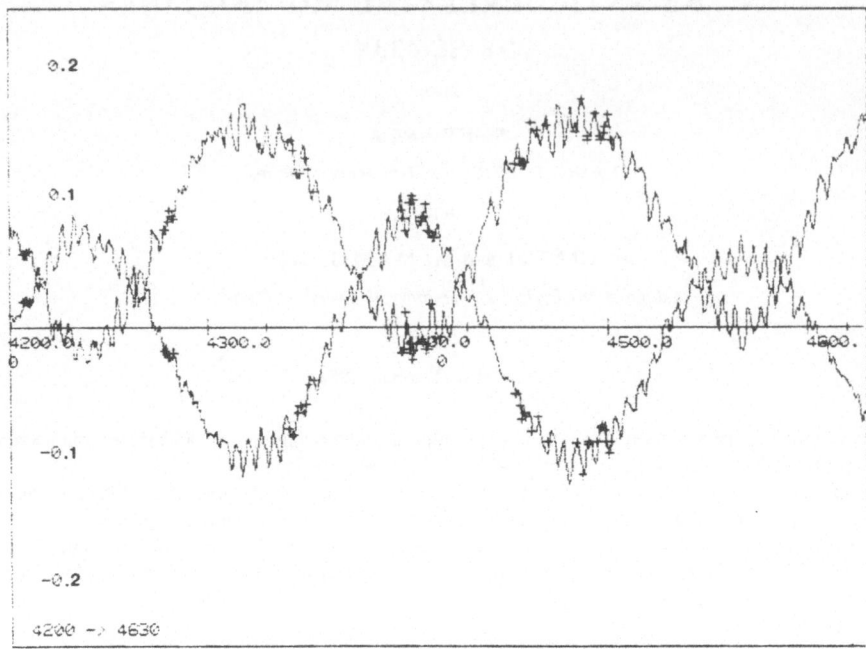

Fig. 1. Fit of SS 433 moving lines by means of 41 red and 63 blue harmonic components in the time interval 4200–4600 JD. Ordinates are in Z units. Crosses represent observations.

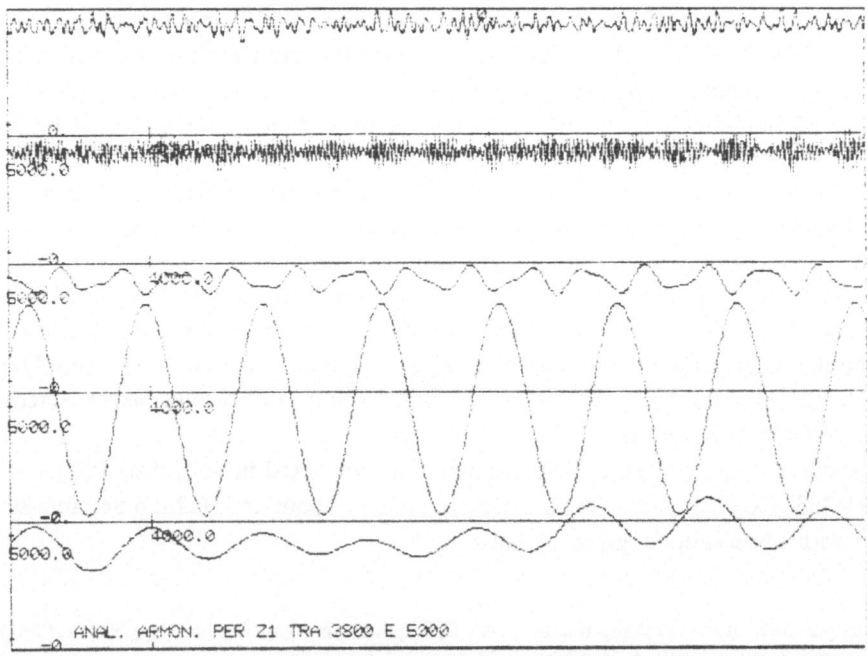

Fig. 2. Grouping of harmonic components of the red branch in SS 433 for all known observations from 3800 to 5000 JD. Ordinates are in Z units. Note the variable behaviour of the 80-day grouping and of the 150–180 day grouping.

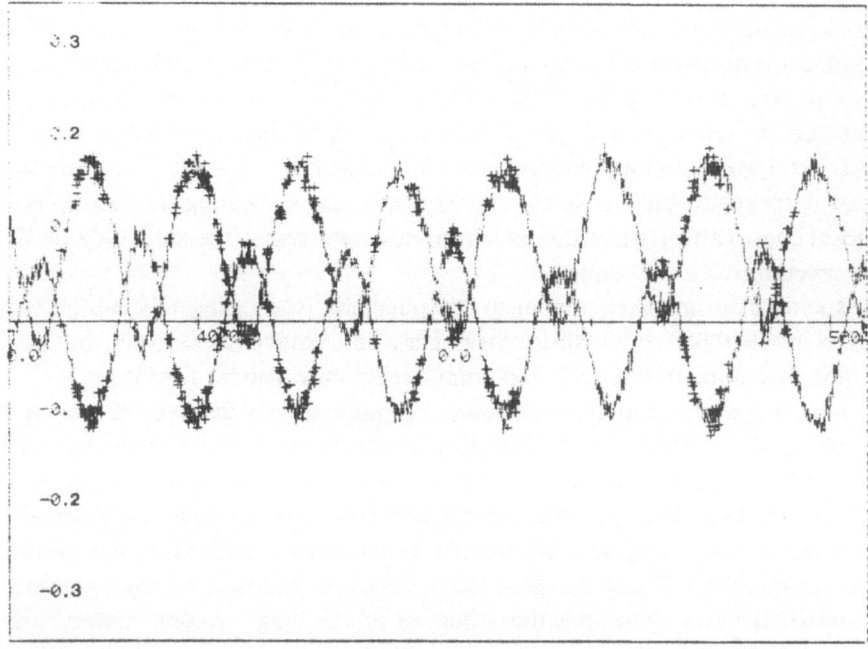

Fig. 3.   Fit of the blue and red branches as in Figure 1, but in the time interval 3600–5000 JD, shows the strong influence of variable amplitude of some harmonic components. We can no more speak of a simple sinusoidal behaviour.

TABLE I

Most significant harmonic components in SS 433

ψ period (amplitude)		Z_1 period (amplitude)			Z_2 period (amplitude)		
164.8	(18°7)	164.8	± 0.2	(0.082)	164.8	± 0.2	(0.083)
1311	(2.7)	1491	± 79	(0.009)	1584	± 121	(0.005)
153.8	(2.0)	156.6	± 0.9	(0.011)	155.7	± 0.9	(0.011)
6.29	(2.0)	6.283	± 0.001	(0.008)	6.289	± 0.001	(0.008)
81.1	(1.8)	81.6	± 0.3	(0.009)			
77.9	(1.3)		–		70.5	± 0.3	(0.002)
–	(*)	186.6	± 1.4	(0.006)	172.6	± 1.2	(0.009)
					192.2	± 1.6	(0.006)
5.84	(1.1)	5.833	± 0.002	(0.005)	5.839	± 0.002	(0.004)
11.28	(0.9)		–		11.28	± 0.01	(0.003)
54.3	(0.9)	48.0	± 1.3	(0.004)		–	
6.54	(0.9)	6.793	± 0.004	(0.002)	6.541	± 0.003	(0.002)
3.05	(0.4)		–		3.086	± 0.001	(0.002)
–		18.10	± 0.03	(0.002)	18.50	± 0.03	(0.002)
4.21	(0.4)		–		4.215	± 0.001	(0.001)

Standard deviation:

0°9	0.0026	0.0013
when using 26 periods	with 41 periods	with 63 periods

* Periodicity of 182.5 ± 1.3 is present in $(Z_1 - Z_2)$ data.

We searched for periodicities in the direction of jets (ψ, according to Ciatti *et al.*, 1981) by assuming them antiparallel, and found 26 harmonic components whose periods again span from about 1 to 1300 days. The resulting standard deviation of $0°9$ is close to the expected error inferred from the observations. Some of them are listed in Table I. We suggest that those harmonic components common at least to two of the parameters Z_1, Z_2, and ψ are really anticorrelated and therefore may be among the most physically significant ones. Table I lists just such harmonic components. The amplitudes in Z_1 and Z_2 are given in $\Delta\lambda/\lambda = Z$ units.

The standard deviations pertaining to the ephemeris, when using only the components in Table I, rise to 0.008 and 0.009 for the red and blue branch respectively, thus possibly indicating that some of the neglected components may also be significant.

We remark that the amplitudes of some components now discovered, namely those with periods about 80, 155, and 1500 days, are comparable to or larger than those of the nutation-like components at 6.28 and 5.84 days (Ciatti *et al.*, 1983). Analysis of time subsets of the data show variable contributions of such individual long-period components versus time. One given period is being replaced in amplitude by one of the other two, as shown also in Figure 2, where the influence of groups of nearby components is represented. It thus seems that the structure of the body ejecting matter is slowly changing with time. In this connection we note that the time subsets show the components around 80 days to behave opposite to those of 165 days, thus excluding connections with its second harmonic.

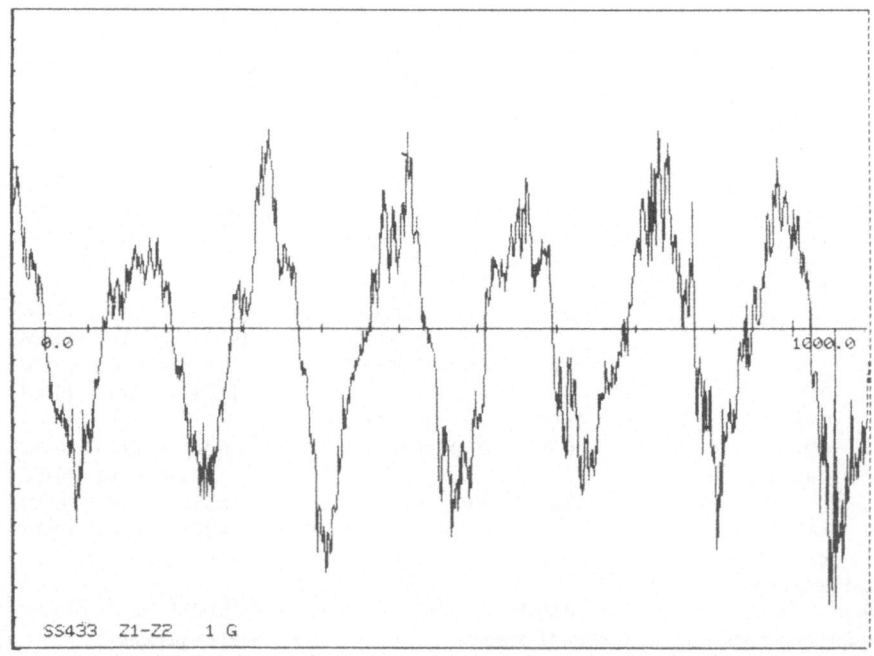

Fig. 4. Autocorrelation function of $(Z_1 - Z_2)$ for 1000 days, with one day interval, showing no damping.

However no secular trend is inferable from our data. While the 6.28-day component is stable in one part over 10^4, analysis of the autocorrelation functions indicates that the precessional period has not changed beyond the estimated error of 0.8% (Figure 4).

Another important point is related to possible variability of beam velocity. Neglecting gravitational redshifts, the beam velocity would be a function of the combination $(Z_1 + Z_2)$, if the jets were strictly antiparallel (Ciatti *et al.*, 1981). We find in 25% of the cases the absolute values of the difference from the average are larger than 5 standard deviations. Therefore, if we assume constant beam velocity, the antiparallelism is not strictly true and viceversa.

Harmonic analysis in the antiparallelism hypothesis seems to show apparent periodicities in beam velocity from 5 to 238 days. However all their amplitudes are smaller than the standard deviation, so that their meaning remains doubtful. Moreover, by developing the formulae given by Milgrom (1980) in the assumption of a constant misalignment angle, the interferencies with the precessional period should appear in the red and blue branches, contrary to the observations. In this case we can therefore rule out periodicities in the beam velocity.

Acknowledgements

We thank Prof. M. Anile, A. Donato, G. Pizzella, L. Rosino, and R. Ruffini for useful discussions and for hospitality. This research has received financial support by the National Research Council, and by Ministry of Education under contract 3177 with University of Messina.

References

Ciatti, F., Mammano, A., and Vittone, A.: 1981, *Astron. Astrophys.* **94**, 251.
Ciatti, F., Mammano, A., Iijima, T., and Vittone, A.: 1983, *Astron. Astrophys. Suppl. Ser.* **52**, 443.
Cogotti, R., Doo Jong, S., and Ruffini, R.: 1983, preprint.
Milgrom, M.: 1980, *Astron. Astrophys.* **87**, L15.

LATE STAGES OF THE EVOLUTION OF CLOSE BINARIES*

C. DE LOORE

Astrophysical Institute, Vrije Universiteit Brussel, België

and

W. SUTANTYO

Bosscha Observatory, Bandung, Indonesia

(Received 21 July, 1983)

Abstract. Close binaries can evolve through various ways of interaction into compact objects (white dwarfs, neutron stars, black holes). Massive binary systems (mass of the primary M_1 larger than 14 to 15 M_0) are expected to leave, after the first stage of mass transfer a compact component orbiting a massive star. These systems evolve during subsequent stages into massive X-ray binaries. Systems with initial large periode evolve into Be X-ray binaries.

Low mass X-ray sources are probably descendants of lower mass stars, and various channels for their production are indicated. The evolution of massive close binaries is examined in detail and different X-ray stages are discussed. It is argued that a first X-ray stage is followed by a reverse extensive mass transfer, leading to systems like SS 433, Cir X1. During further evolution these systems would become Wolf–Rayet runaways. Due to spiral in these system would then further evolve into ultra short X-ray binaries like Cyg X-3.

Finally the explosion of the secondary will in most cases disrupt the system. In an exceptional case the system remains bound, leading to binary pulsars like PSR 1913 + 16. In such systems the orbit will shrink due to gravitational radiation and finally the two neutron stars will coalesce. It is argued that the millisecond pulsar PSR 1937 + 214 could be formed in this way.

A complete scheme starting from two massive ZAMS stars, ending with a millisecond pulsar is presented.

1. Introduction

At this moment hundreds of strong galactic X-ray sources are known with X-ray energies exceeding 10^{36} erg s^{-1}. As pointed out by Ostriker (1977), Jones (1977), Maraschi *et al.* (1977), van den Heuvel (1980) these strong point sources can, according to their X-ray spectra, be divided into two groups: a first class contains the sources with relatively hard X-rays, often pulsating, with early type massive stars as optical counterparts situated near the galactic plane; a second group contains the sources with softer spectra, with low mass stars as optical counterparts, generally not pulsating (exceptions, e.g., Her X-1 with a pulse period of 1.24 s and 4 U 1626–27 with a pulse period of 7.67 s (Ilovaisky and Chevalier, 1978). The second group consists of the low mass steady X-ray binaries, the galactic bulge sources and the globular cluster sources. The hard X-ray sources belong to the extreme population I with ages of 5–10 million years. The group of the soft spectrum sources belong to the old disk population, with ages of 5–13 billion years.

* Paper presented at the Lembang-Bamberg IAU Colloquium No. 80 on 'Double Stars: Physical Properties and Generic Relations', held at Bandung, Indonesia 3–7 June, 1983.

The same sources observed in our Galaxy are also found in the Magellanic clouds and in M31. The X-rays are produced by the accretion of matter, expelled by the companion, on a magnetized rotating neutron star.

Neutron stars in binary systems can be produced by the explosion of one of the components in a supernova event. About 50% of all stars are close binaries which means by definition that the components of these systems can fill their Roche lobes during the evolution. The expansion during core hydrogen burning or shell hydrogen burning can be interrupted by mass transfer and mass loss. The remnant is essentially determined by the evolutionary state of the star at the moment of onset of the mass transfer (Webbink, 1979; De Greve et al., 1978).

Several review papers on binary evolution were published, e.g., Paczynski (1971), van den Heuvel (1976), Massevich et al. (1976), Paczynski (1979), and Tutukov (1981).

The evolution of close binaries starting from homogeneous zero-age Main-Sequence (ZAMS) models can be followed, taking into account mass transfer and eventual mass loss. The existence of X-ray binaries and binary pulsars as, e.g., PSR 1913+16 are striking examples of the fact that close binaries can loss significant fractions of their mass and angular momentum during explosive stages.

2. Observational Evidence of Compact Components in Binaries

Binary X-ray sources containing compact objects can in a crude way be divided into two groups: massive sources and low mass sources.

(a) *Massive sources*: $M > 15\,M_0$.

(1) *Strong permanent sources*: the optical component, a giant or supergiant star, is nearly filling its Roche lobe.

(2) *Weak or transient sources*: the optical component is in most cases a rapidly rotating Be-star. These binary sources have large periods, and the volume of the optical companion its much smaller than its Roche volume.

(b) *Low mass sources*: $M < 2\,M_0$. For only a few sources direct evidence of binary motion exists.

(1) *Pulsating sources*, with hard spectra, example, Her X-1.

(2) *Non-pulsating sources*, with softer spectra: example, Sco X-1. They have large X-ray luminosities.

(3) *Galactic bulge X-ray sources*.

(4) *Steady sources associated* with bursters. The latter two groups have optical and X-ray spectra similar to Sco X-1. The identified optical counterparts are faint and show the spectrum of an accretion disk, comparable with the spectra of cataclysmic binaries (Cowley, 1980; Lewin and Clark, 1980).

In Cen X-4, and Aql X-1 the spectrum of a faint K dwarf can be observed (Cowley, 1980; Van Paradijs, 1980). This evidence together with the fact that the optical companion has a low luminosity suggests that bursters and bulge sources are low-mass binaries, in which the mass of the companion of the compact object is $\lesssim 1 M_0$, filling its Roche lobe (Joss and Rappaport, 1979; Lewin and Clark, 1980). Evidence that the bursters are neutron stars has been given by Van Paradijs (1978); moreover the high X-ray luminosity ($\log L_x \sim 36-38$) points to neutron stars or black holes.

The same holds for globular cluster sources in our galaxy and in M31.

Only stars with $M > 15 M_0$ or $M < 2 M_0$ provide accretion rates for long lived X-ray sources (van den Heuvel, 1981), the first group by producing a sufficiently strong stellar wind, the latter category producing a sufficiently high-mass transfer rate by Roche lobe overflow. However also in stars of intermediate mass compact objects occur.

3. Final Evolution of Helium Stars

When the primary of a close binary system is filling its Roche lobe a phase of mass transfer (and eventually mass loss) starts, which ends as a result of the ignition of He in the core, the effect being that the layer separating the $\delta R/\delta t > 0$ region and the $\delta R/\delta t < 0$ region moves outwards, causing the mass loss to stop when it reaches the surface.

Hence, after the mass exchange phase the primary has lost practically its complete H-rich envelope, and only the core remains, consisting mainly of He (and heavier elements).

The further evolution is determined completely by the helium core.

Computations of the evolution of helium stars or helium remnants have been carried out by Paczynski (1971), De Greve and de Loore (1976). Arnett (1978), Savonije (1978), Delgado and Thomas (1980), and Sugimoto and Nomoto (1980). The minimum mass of the primary leading to a neutron star after supernova explosion was estimated by De Greve and de Loore (1977), at $\sim 14\ M_0$, and between 8 and 15 M_0 by Massevitch and Tutukov (1981).

The evolution of the helium core mass (M_{He}) occurs as follows:

$$M_{He} > 2 M_0.$$

The CO-core formed by helium burning degenerates and the outer layers expand.

A second phase of mass transfer starts leaving a degenerate CO star, with $M_{remnant} < M_{ch}$, cooling off to a CO with dwarf. Figure 1 shows the evolution of a $10 M_0 + 8 M_0$ binary, with an initial period of 8 days, calculated by De Greve and de Loore (1977), showing two phases of mass transfer.

The various evolutionary stages are shown in Table I.

$$2 M_0 < M_{He} < 3 M_0$$

C-ignition occurs under not highly degenerate conditions and the star undergoes successive C-shell flashes (Miyaji et al., 1980; Nomoto, 1980; Sugimoto and Nomoto, 1980).

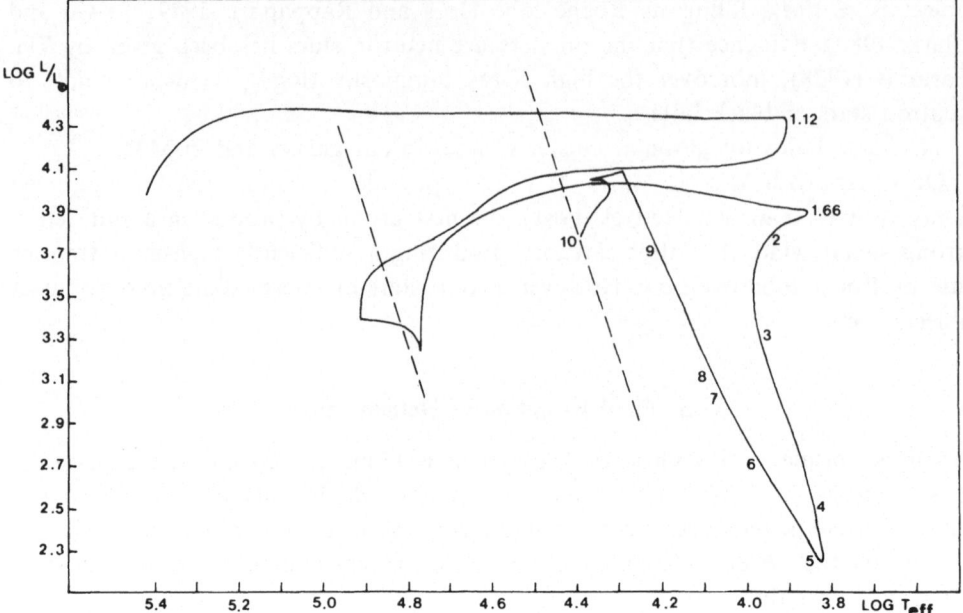

Fig. 1. The evolutionary track in the HRD of the primary of a $10\,M_0 + 8\,M_0$ binary system. The masses of the mass losing primary are indicated. ZAMS of hydrogen and helium are also given.

When the outer layers of the produced degenerate O–Ne–Mg core, which is increasing in mass, approach the He-burning shell, C burning finishes and due to He-shell burning the outer helium layers expand. Also in this case a second mass transfer stage occurs, leaving an O–Ne–Mg white dwarf with mass between $1.2\,M_0$ and $1.4\,M_0$, while

$$M_{He} > 3\,M_0.$$

The CO-core formed by He-burning is larger than the Chandrasekhar limit; Ne, O, and Si ignition occur under non degenerate condition.

Fe-photodesintegration in the core, leads to a supernova-explosion and the formation of a neutron star.

4. Mass Accretion by White Dwarfs

Accreting white dwarfs can be driven over their Chandrasekhar limit and undergo an electron capture supernova collapse (Myaji *et al.*, 1980; Sugimoto and Nomoto, 1980; Nomoto, 1980).

Computations were performed for a white dwarf of $1.2\,M_0$, with an accretion rate of $4 \times 10^{-6}\,M_0\,\mathrm{yr}^{-1}$; it was found that due to electron captures on $^{20}\mathrm{Ne}$ and $^{24}\mathrm{Mg}$ the core density increases, leading to a collapse.

TABLE I

The evolution of the 10 $M_\odot$ + 8 $M_\odot$ binary system

	$t \times 10^6$ yr	M_1	M_2	R_1	A	P	$\log L/L_\odot$	$\log T_{\text{eff}}$	M_{cc}	X_{at}	$\log T_c$	$\log \rho_c$	$\log \dot{M}$
1 Initial model	0	10	8	4.22	23.63	3.15	3.77	4.39	2.93	0.7	7.48	0.91	
2 Start 1st mass exchange	15.149 271	10	8	9.44	23.63	3.15	4.06	4.29	0.0	0.7	7.66	2.29	−5.0
3 Minimum luminosity	15.182 579	4.82	13.18	10.51	37.43	6.29	2.24	3.81	0	0.7	7.73	2.59	−3.58
4 Helium ignition	15.264 418	1.98	16.02	34.39	150.17	50.53	3.82	3.95	0	0.32	7.97	3.37	−4.86
5 End 1st mass exchange	15.310 440	1.66	16.34	44.66	205.80	81.07	3.96	3.93	0.12	0.20	8.13	3.70	−5.20
6 End core helium burning	18.107 183	1.66	16.34	0.26	205.80	81.07	3.41	4.91	0	0.0	8.37	4.07	
7 Start 2nd mass exchange	18.481 810	1.66	16.34	44.66	205.80	81.07	4.12	3.97	0	0	8.64	6.22	−5.00
8 End 2nd mass exchange	18.512 902	1.12	16.88	84.86	420.10	236.44	4.32	3.90	0	0	8.56	6.73	−5.40
9 End of the computations	18.517 200	1.12	16.88	0.045	420.10	236.44	3.95	5.43	0	0	8.49	6.94	

5. Envelope Interaction

Constraints for conservative evolution are:

(1) the mass ratio $q = M_2/M_1$ is not too low (≥ 0.3);

(2) the separation A is not too small;

(3) the envelope of the loser is in radiative equilibrium.

If these conditions are satisfied the change in the distance of the centers of gravity of the two components, given by

$$\frac{A}{A_0} = \left(\frac{M_{10}M_{20}}{M_1M_2}\right)^2,$$

is not too much reduced.

M_1 and M_2 are the masses of primary and secondary respectively, the subscripts 0 denotes the initial value. If the conditions 1 and/or 2 are not satisfied the systems evolve into contact systems with a common envelope (Webbink, 1979).

If the envelope of the loser is convective and mass transfer occurs, the effect of the mass loss will be that the envelope expands, inducing a stronger amount of mass transfer. In this way the envelope grows in a catastrophic way, engulfes its companion, and produces also here a common envelope.

In convective envelopes the specific entropy of the gas decreases outwards so that no energy supply for the restoration of the equilibrium (as is necessary in radiative envelopes) is required. These stars have a tendency to expand further when the mass loss starts.

The mass loss ratios increase and the outflowing matter can reach velocities near the velocity of sound (even for low mass stars $\dot{M} > 10^{-3} M_0 \, \mathrm{yr}^{-1}$). Only few detailed calculations have been carried out for this mode by Paczynski and Sienkiewicz (1972), Plavec et al. (1973), Webbink (1977a, b), all using the mass loss formalism developed by Jedrzejec (1969). Mass loss of this type occurs also in degenerate stars filling their Roche lobes. Classical novae and dwarf novae are associated with binaries which evolved according to this mode (Ritter, 1975, 1976; Webbink, 1975; Paczynski, 1976).

Binaries with not too long periods (< 100 to 1000 d) start mass transfer before the degenerate He-core is sufficiently massive for ignition and leave low mass He-white dwarfs of masses smaller than $\sim 0.5 \, M_0$. Systems with longer periods (1000 to 10 000 d) start mass transfer after the primary has reached the giant branch, leaving CO-white dwarfs with masses between about a half and 1.4 solar masses.

6. Evolution of Massive Close Binaries from ZAMS to the Final Stage: Primaries Leaving Neutron Stars

The evolution of ZAMS massive close binaries occurs as follows (Table II). When the primary fills its Roche lobe mass is transferred towards the secondary, and a He-remnant is left. A system in this phase represents a Wolf-Rayet binary.

TABLE II

The evolution of massive stars

Primary S_1 type	Secondary S_2 type	M_1	M_2	Age (in 10^6 yr)	
OB	OB	20	8	0	$P = 4.56$ d
OB fills Roche lobe	OB	20	8	6.17	$P = 4.56$ d
	Direct mass exchange S1 → S2				
He-star	OB	5.4	22.6	6.2	$P = 10.86$ d
	End of the first phase of mass transfer Wolf–Rayet binary Evolution to final stage of S1				
Neutron star OB Runaway	OB	2	22.6	6.78	
	1st Supernova explosion OB Runaway – Remnant of the Further Evolved Helium star is a young neutron star (Pulsar)				
Neutron star X-ray Binary	OB	2	22.6	11.186	$P = 11.7$ d
	OB star nearly fills its roche lobe enhanched stellar wind. accretion on the neutron star generates X-rays 1st X-ray phase.				
Neutron star	OB	2	22.6	11.209	$P = 11.70$
	Reverse Mass-transfer roche lobe overflow-mass leaves the system				
Neutron star	OB → He S2 on its way to WR star. accretion disk 2nd X-ray phase		Slow phase of mass transfer; optical star is 'on its way' to become a Wolf–Rayet star. (Ex. SS 433, Cir X-1).		
Neutron star	He Wolf–Rayet runaway end of the second mass loss stage. He star + neutron in common (expanding) envelope.	2	6.3	$t = 11.239 \times 10^6$ yr	
Neutron star	He Supercritical disk accretion; 3^d X-ray phase (Ex. CYG X-3) 2n supernova explosion system remains bound	2	6.3		
Neutron star	Neutron star Ex. PSR 1913 + 16 binary pulsar				$P = 7.75$ hr
Neutron star	Neutron star Coalescence of two neutron stars Ex. PSR 1937 + 214 Millisecond Pulsar. OR: system disrupted, two runaway neutron stars.				$P = 1.5$ s

The neutrino cooling shortens the C-, Ne-, Si-burning lifetimes to a few thousand years. Then the primary explodes and leaves a neutron star remnant of 1 or 2 solar masses. The system has a large probability to remain bound (de Loore *et al.*, 1975). The space velocity is 50–80 km s^{-1}, typical for OB runaways. The optical component can be removed during the optical component's lifetime to a distance of ∼ 100 pc, a typical distance for X-ray binaries.

The stellar wind matter is partially accreted by the compact companion which produces the X-rays. The X-ray luminosity remains weak until the optical component is nearly filling its Roche lobe. The Roche lobe overflow itself quenches the X-rays, and the time-scale of the X-ray phase is exactly determined by this behaviour. Thus the X-ray stage is of the order of 10^5 yr for X-ray binaries with an optical companion of ∼ 20 $M_\odot$ (Savonije, 1979). When the optical component fills its Roche lobe a common envelope is formed since the accretion by the neutron star is limited by its Eddington limit to 10^{-8} $M_\odot$ yr^{-1}, while the mass giving companion loses mass at a much higher rate. The common envelope stage was proposed by Paczynski and worked out numerically by Taam *et al.* (1978), Tutukov and Yungselson (1979).

It is suggested by Tutukov and Yungelson (1979) that η Car, P Cyg, and S Dor are common-envelope binaries. Now two possible branches exist for the further evolution: either the neutron is engulfed by its companion in the common envelope and this leads to a red supergiant with a compact core (Tutukov, 1981) or the bulk of the transferred matter will be expelled from the system, causing a rapid shrinking of the orbit as a consequence of the large specific orbital angular momentum of the expelled matter (van den Heuvel and de Loore, 1973). Remnants of such expelled envelopes can be observed during ∼ 2×10^4 yr as a bright nebula with a radius of ∼ 1 pc around a single Wolf–Rayet star (Massevich *et al.*, 1976). Mass loss rates of ≳ 10^{-5} M_0 yr^{-1} could transform the stars into IR sources with large space velocities.

In all cases neutron stars are formed with high space velocities. In this context slow pulsars ($\langle z \rangle$ ∼ 80 pc) should be the final products of wide binaries; fast pulsars ($\langle z \rangle$ > 150 pc) should be the final products of close binary evolution.

7. Evolution of Stars of Intermediate Mass

When the primary fills its Roche lobe during core hydrogen burning the possibility exists that mass exchange leads to the formation of a common envelope. Later on this envelope can partially be removed and in this way a system consisting of a He or CO dwarf and a normal companion is produced. If the low-mass star with a convective envelope undergoes Roche lobe overflow then, owing to large mass and angular momentum losses very close cataclysmic variables can be formed (Meyer and Meyer-Hofmeister, 1979; Webbink, 1979; Taam *et al.*, 1978). The life-time of the system in the semi-detached stage (He or CO core with envelope, filling its Roche lobe and a normal secondary) is of the order of the thermal time-scale of the expanding envelope. The result of the mass exchange is a system composed by a Main-Sequence star and a degenerate He or CO dwarf. The overflow stage ends when the H in the envelope is exhausted or by a central He-flash. The non-degenerate component then evolves towards a red giant (or red supergiant) not yet filling its Roche lobe, but losing matter by stellar wind. The degenerate companion grows by accretion and this can lead to a supernova explosion and the formation of a neutron star.

In this way a progenitor system for X-ray bursters can be produced. Such systems could also be formed by capture in globular clusters, or in the galactic bulge. Mass decrease by stellar wind and later on by Roche lobe overflow could possibly change the non compact object into a red dwarf. This second, inverse, mass

TABLE III

The evolution of lower mass stars

ZAMS	
The primary fills its Roche lobe.	
Mass loss leads to a common expanding envelope which disappears later.	If strong gravitational radiation occurs, the large mass and momentum losses lead to a system of a degenerate and a red dwarf – later on a SN produces a neutron star – bursters could be produced
He or CO-dwarf + normal secondary. He or CO-dwarf + normal secondary.	Degenerate dwarf grows → Roche lobe overflow.
Primary fills its Roche lobe. Semi-detached stage (ex. Algol systems).	Supernova X-ray burster
He- or CO-dwarf + red giant with stellar wind mass loss.	
Secondary fills Roche lobe. Reverse mass transfer Increase of the mass of the CO-dwarf. (Such systems could also be produced by capture in globular clusters.)	Systems with large mass loss rates evolve into double core systems with a common envelope. (Ex. UU Sge, planetary nebula). Friction causes mass loss.
SN explosion-X-rays (Example: bursters).	Gravitational energy very high → system becomes very narrow → Roche lobe overflow; accretion by CO dwarf; SN → NS. (Ex.: X-ray bursters). High gravitational radiation. Common envelope disappears. (Example: V471 Tau).

transfer producing the X-rays could represent the burster stage. The evolution of lower mass stars according to this picture is shown in Table III.

If the systems evolving according to case B have an average mass loss rate satisfying the following criterion

$$\langle \dot{M} \rangle = \frac{M}{t_{KH}} > 10^{-6} \, M_0 \, \mathrm{yr}^{-1},$$

where t_{KH} is the Kelvin–Helmholtz time-scale, the evolutionary scenario is different.

The expansion of the non-degenerate star leads also in this case to a Roche lobe overflow stage but now two degenerate dwarfs in a common envelope can be formed. This phase has been investigated by Meyer and Meyer-Hofmeister (1979). The double core, revolving into the common envelope produces a very large friction and as a result the envelope is removed, leaving two degenerate dwarfs, at least when the gravitational energy of the two compact cores is sufficiently large to disperse the envelope. UU Sge could be an example of such a common envelope system. Common-envelope cores of this kind were discovered by Miller *et al.* (1976); further evolved systems (i.e., 10^4 yr later, when the envelope is dispersed, and merely consisting of two degenerate dwarfs) are also known, e.g., V471 Tau (Nelson and Young, 1970) and PG 1413+0.1 (Green *et al.*, 1978). Another possibility is that two degenerate objects merge and evolve into one single degenerate core hidden in a large envelope.

For lower masses (initial primaries $\lesssim 3 \, M_0$) a long mass exchange stage may occur. The accreting degenerate dwarf thus can also be transformed into a neutron star after a supernova explosion. Further Roche lobe overflow and accretion by the neutron star can then produce X-rays. The less massive of the two degenerate dwarfs has the largest radius so that this star fills its Roche lobe first. Here again is a possibility for the formation of a neutron star by accretion in an explosive (or non explosive?) way, leading to an X-ray stage. In all these cases of neutron star formation the systems can remain bound, or can be disrupted, depending on the relative masses, the asymmetry of the explosion and the explosion conditions. In this way, either low mass X-ray binaries can be produced, or runaway neutron stars. If the mass ratio of the Roche lobe filling degenerate star and the companion is large (i.e., near 1) the former can be destroyed, and transformed into a disk surrounding the companion. Further evolution depends on the viscosity of the ring. If the viscosity is large enough the ring can be removed and a single star is formed.

8. Supernova Explosion – Probabilities for Disruption

Each of the two components of a massive close binary undergoes a supernova explosion. The effects of these events on the status of the system are completely different.

In the case of the first explosion the less massive star explodes, and consequently the probability that the system is disrupted is extremely low.

Hence, most systems remain bound so that many massive X-ray binaries exist.

During the second supernova explosion the more massive component explodes; in this case, the disruption probability is very large, so that two runaway neutron stars are produced, an old one, remnant of the first explosion and a young active one, just formed.

Computations of the disruption probabilities were carried out by De Cuyper (1981), de Loore *et al.* (1975). If we assume an extra kick of 100 km s^{-1}, a shell-expansion velocity of 10 000 km s^{-1} and a post-supernova remnant of 1.5 M_0 these probabilities are of the order of 70–80% for ZAMS primaries between 40 and 100 M_0 for all mass ratios. For lower mass ZAMS primaries (20–40 M_0) the systems always remain bound.

As well in the case of Roche lobe overflow with reverse mass transfer and mass loss as in the case of spiralling in, the outer layers with the original composition leave the primary star, and more and more layers with larger helium abundances show up at the surface. The O star is 'on its way' to become a He-star, hence, shows more and more WR characteristics. The object SS433 is possibly an example of a binary in this stage (van den Heuvel *et al.*, 1980), as well as Cir X-1.

According to Firmani and Bisiacchi (1980) and Shklovski (1981) and the mass loss rate is $\sim 10^{-4}$ M_0 yr^{-1}, with an expansion velocity of 1000 km s^{-1}, and the spectral type ranges between Of and WR (from the relative strengths of Balmer- and Pickering-lines). This agrees with the conclusions of van den Heuvel *et al.* (1980) about the mass loss rate.

Probably the X-ray system is surrounded by an envelope of matter expelled by the non compact companion, and not by remnants of the supernova shell as is discussed by Shklovski (1981). An argument is given by the life-time of the object according to evolutionary computations (11×10^6 yr) and the large distance (3.7–4.7 kps). Hence, the picture of van den Heuvel *et al.* (1980) of SS433 to be a second X-ray stage, with mass expelled by the companion, the larger part of which is not accreted by the neutron star, but stored in a disk, sounds very attractive.

As more and more matter is expelled by the companion, more He-enriched layers will appear at the surface. Hence, the next evolutionary stage should be a second Wolf–Rayet stage, i.e., a helium star with a neutron star companion (de Loore *et al.*, 1975) (Figure 2). Since such systems are descendants from binaries through a supernova event, they are assumed to have a large runaway velocity, and are supposed to be found at large distances of their place of birth.

Further evolution with further spiralling in could then lead to ultra-short period binaries like Cyg X-3 (van den Heuvel and de Loore, 1973). Finally, these systems undergo a new supernova explosion, and are in nearly all cases disrupted, leaving two neutron stars, an old one and a young pulsar.

In exceptional cases binary pulsars are formed. A list of binary pulsars is given in Table IV.

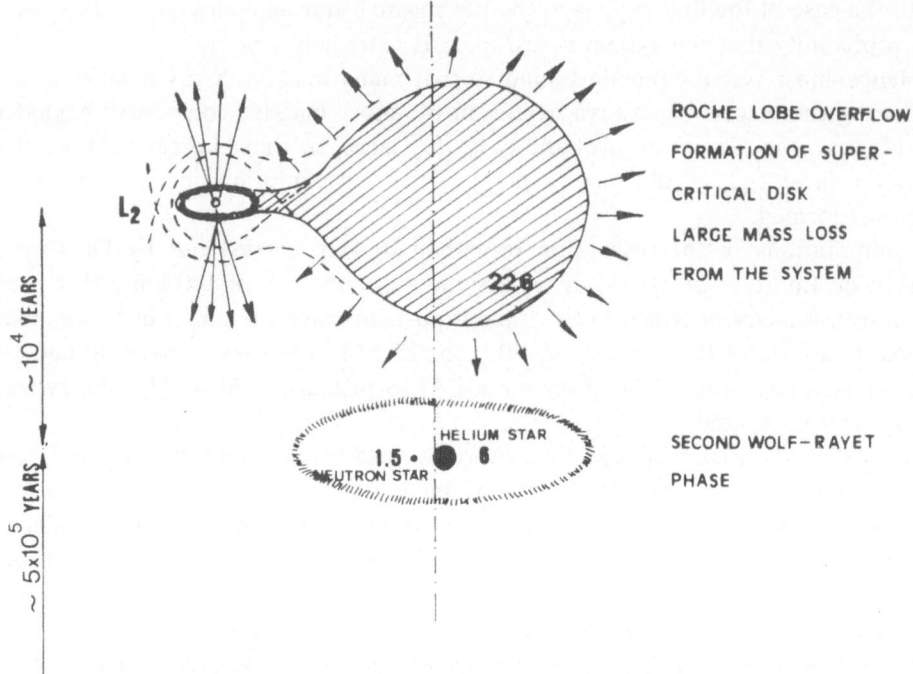

Fig. 2. The expected evolution of a massive X-ray binary after the X-ray stage. Roche lobe overflow of the non compact star transfers mass on a thermal time-scale, with mass loss rates of the order of $\sim 10^4$–10^{-3} M_0 yr^{-1}. A supercritical disk is formed, and heavy mass loss from the central disk regions occurs, probably in the form of beams.

9. The Evolutionary Status of Be-X-Ray Binaries

Massive close binaries with primaries of masses exceeding 30–40 M_0 evolve into Wolf–Rayet binaries (Conti *et al.*, 1983; Maeder, 1982).

The end products of intermediate systems after a case *B* of mass transfer could be identified with Be-systems (Vanderlinden, 1982). Further evolution of these binary systems leading to a supernova explosition of the primary, leaving a compact object should then explain the existence of the Be X-ray binaries as suggested by Rappaport and van den Heuvel (1982).

TABLE IV

Binary pulsars

Name	P_{orb}	$P_{pulse}(s)$	Eccentricity
PSR 0656+64	24^{h}41^m	0.196	0.06
PSR 0820+02	11:00 d	0.865	0
PSR 1913+16	7^{h}75^m	0.059	0.617
PSR 1937+215		0.0015	

In Figure 3 is depicted the evolution of a system of $15 M_0 + 10 M_0$, calculated simultaneously, starting with an initial period of 8 days. The results are given in Table V.

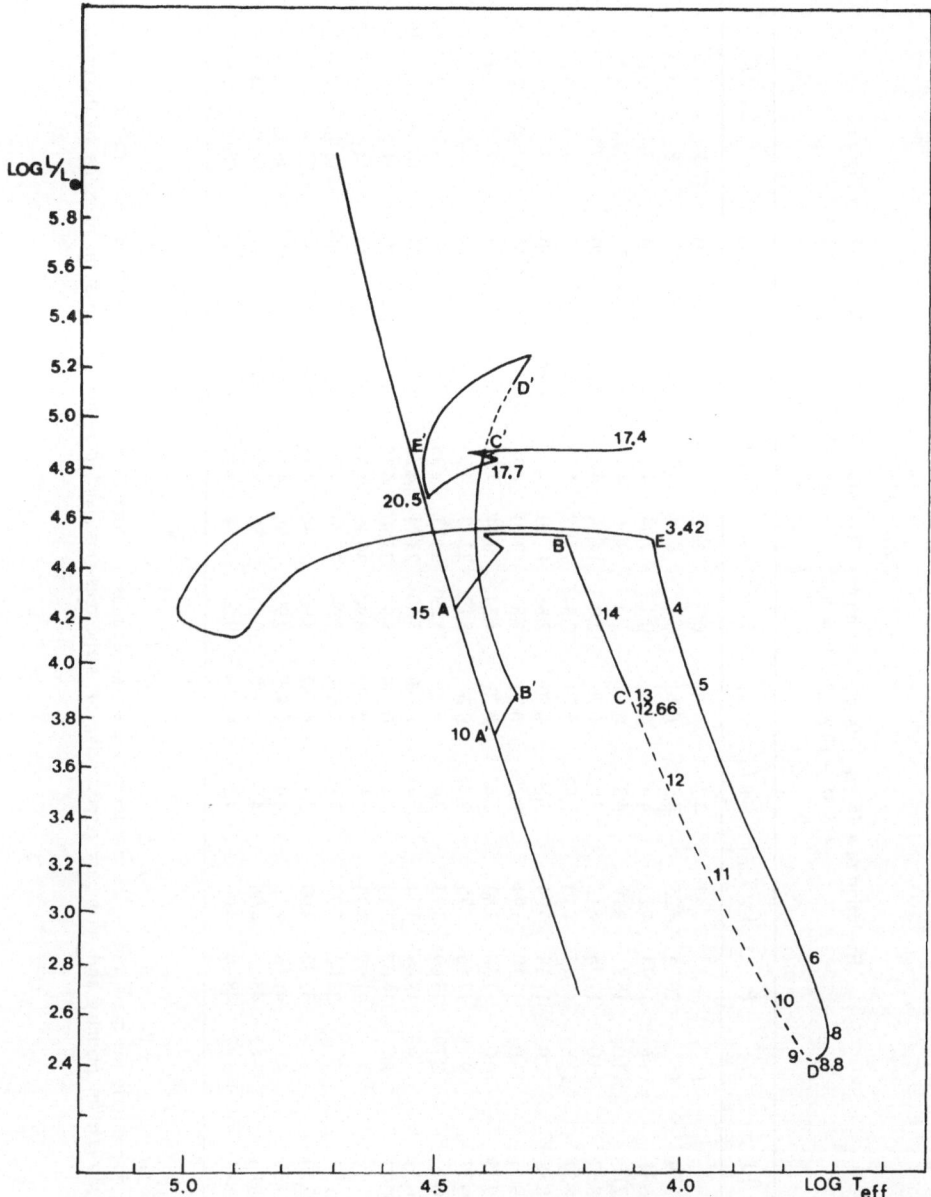

Fig. 3. The formation of a Be–X-ray binary. The simultaneous evolution of a $15 M_0 + 10 M_0$ binary with an inital period of $8 M_0$. The changed mass of primary and secondary is indicated along the evolutionary tracks. The letters refer to Table II, A to F refer to phases of the primary, A' to F' to corresponding phases of the secondary. The dotted parts of the evolutionary tracks correspond with the contact phase.

TABLE V

Evolution of a $15\,M_0 + 10\,M_0$ close binary. The letters in the first column refer to Figure 3

$(15\,M_0 + 10\,M_0, \quad P_i = 8\,\text{d}, \quad \text{contact: } 2230\,\text{yr}, \quad \Delta M = 1.10)$

	Time (10^6 yr)	M_1	R_1	$\log T_{\text{eff},1}$	$\log L_1$	$X_{\text{at}1}$	M_2	R_2	$\log T_{\text{eff},2}$	$\log L_2$	$X_{\text{at}2}$	P	$\dot{M}$	β
AA'	0	15	5.14	4.47	4.26	0.70	10	4.12	4.39	3.73	0.70	8	– –	1
BB'	13.095 59	15	20.25	4.25	4.57	0.70	10	6.04	4.35	3.91	0.70	8	$5.29\ 10^{-5}$	1
	13.096 91	14.6	19.73	4.22	4.43	0.70	10.4	7.44	4.42	3.38	0.70	7.71	$5.8\ 10^{-4}$	1
	13.097 71	14.0	18.91	4.19	4.28	0.70	11.0	9.66	4.43	4.66	0.70	7.40	$1.1\ 10^{-3}$	1
CC'	13.098 32	13.0	17.76	4.12	3.95	0.70	12.0	14.88	4.40	4.91	0.70	7.11	$1.73\ 10^{-3}$	1
	13.098 52	12.66	17.47	4.10	3.82	0.70	12.34	17.07	4.39	4.97	0.70	7.08	$1.87\ 10^{-3}$	0.998
	13.098 88	12.0	17.18	4.04	3.60	0.70	12.85	17.50	4.38	4.96	0.70	7.18	$1.96\ 10^{-3}$	0.730
	13.099 35	11.0	17.10	3.94	3.20	0.70	13.57	18.40	4.38	5.00	0.70	7.60	$2.05\ 10^{-3}$	0.69
	13.099 85	10.0	17.30	3.80	2.62	0.70	14.34	20.04	4.37	5.05	0.70	8.32	$2.15\ 10^{-3}$	0.532
	13.100 37	9	17.86	3.75	2.45	0.70	15.12	22.18	4.37	5.11	0.70	9.48	$2.21\ 10^{-3}$	0.633^a
DD'	13.100 39	8.79	18.09	3.75	2.45	0.70	15.29	22.87	4.36	5.13	0.70	9.79	$2.21\ 10^{-3}$	0.749
	13.100 55	8.43	18.49	3.74	2.45	0.699	15.60	23.94	4.36	5.15	0.70	10.39	$2.28\ 10^{-3}$	1
	13.100 73	8	19.07	3.73	2.45	0.699	15.97	25.50	4.35	5.18	0.70	11.23	$2.35\ 10^{-3}$	1
	13.101 13	7	21.07	3.73	2.51	0.699	16.90	30.45	4.33	5.24	0.70	14.03	$2.63\ 10^{-3}$	1
	13.101 52	6	25.00	3.77	2.82	0.699	17.90	35.70	4.31	5.29	0.699	18.76	$2.47\ 10^{-3}$	1
	13.102 10	5	32.32	4.00	3.98	0.582	18.90	27.44	4.36	5.26	0.697	27.5	$1.17\ 10^{-3}$	1
	13.104 31	4	41.16	4.06	4.43	0.409	19.90	10.96	4.51	5.06	0.683	46.23	$1.33\ 10^{-4}$	1^b
	13.109 96	3.62	45.56	4.06	4.51	0.330	20.28	6.86	4.54	4.78	0.673	58.77	$4.87\ 10^{-5}$	1
	13.115 41	3.42	48.40	4.06	4.56	0.284	20.50	6.64	4.54	4.75	0.666	68.46	$2.03\ 10^{-5}$	1

^{a} min L.

^{b} Start He-burning.

M, R, T_{eff}, L are respectively the mass, radius, effective temperature and luminosity. Index 1 refers to the primary, index 2, to the secondary, X_{at} is the abundance of hydrogen (in primary and secondary respectively). $\dot{M}$ is the mass loss, and β is the fraction of the mass lost by the primary accreted by the secondary.

The computations show that the system becomes semi-detached after ~ 13 million years, and evolves into a contact phase some ~ 2900 yr later. During this period, lasting about 2200 yr, matter leaves the system (about 50% of the matter expelled by the primary). The system becomes again semi-detached until 15 000 yr later, after the onset of helium burning the two components become detached again.

The final system, consisting of a primary of $3.42\ M_0$ and a secondary of $20.50\ M_0$ (hence, $1.08\ M_0$ has left the system), has a period of ~ 68 days.

The secondary component, starting from $10\ M_0$ accretes matter, and evolves parallel with the ZAMS.

When the slow phase of mass transfer begins, it moves towards the ZAMS, and later on evolves exactly like a normal ZAMS star of the same mass.

The helium remnant of the primary has meanwhile exploded. The system in this stage could represent a Be X-ray binary, with a B component of $\sim 20\ M_0$, a neutron star of $\sim 1.5\ M_0$, a period of ~ 81 d in the case of a symmetric explosion, and a runaway velocity of the system of ~ 11 km s^{-1}.

10. The Evolutionary Status of SS433

As well in the case of Roche lobe overflow with reverse mass transfer and mass loss (i.e., from the original secondary towards the compact object) as in the case of spiralling in, the outer layers, with the original composition, leave the star and deeper layers, containing more and more helium appear at the surface. The O star is 'on its way' to become a helium star; hence, shows more and more Wolf–Rayet characteristics.

The object SS433 is possibly an example of a binary in this stage (van den Heuvel *et al.*, 1980) as well as Cir X-1. According to Firmani and Bisiacchi (1980) and Shklovski (1981) the mass loss rate is $\sim 10^{-4}\ M_0$ yr^{-1}, and the spectral type, (determined from the relative strengths of Balmer- and Pickering lines) ranges between Of and WR.

Probably the X-ray system is surrounded by an envelope of matter expelled by the non-compact component. The life-time of the object, determined from evolutionary computations is $\sim 11 \times 10^6$ yr.

The compact star cannot collect more than some $10^{-8}\ M_0$ yr^{-1}, hence, the larger part of the mass expelled by the companion will be blown away by the radiation pressure produced by the compact star, when it arrives at the inner edge of an accretion disk (van den Heuvel, 1980).

The excess material will be expelled in directions perpendicular to the disk. The effect of this mass ejection will be a rapid decrease of the orbital period (van den Heuvel, 1973; van den Heuvel and de Loore, 1973). When matter is ejected at high velocities, symmetrically with respect to the center of the disk, its specific angular momentum is equal to the specific angular momentum of the compact star.

The change of the orbital period P is then (cf. van den Heuvel, 1980b) given by

$$\frac{P}{P_0} = \left(\frac{1+q_0}{1+q}\right)^2 \frac{q_0}{q} e^{3(q-q_0)};$$

where q is the ratio between the mass of the neutron star and the mass of its companion.

The subscripts zero refer to the original situation. The time scale for the shrinking of the orbital period calculated in this way is $\sim 3 \times 10^3$ yr.

The result of the evolution will be an extremely close binary system, consisting of the evolved He-core of the compact star, and a neutron star, resembling a WR star. The evolution is depicted in Table II.

11. The Millisecond Pulsar PSR 1937+214

This pulsar has an extremely short pulse period of 1.5 milliseconds and a very weak magnetic field strength of the order of 10^8 g.

The other known binary pulsars (see Table IV) have much larger magnetic fields, of the order of 10^{12} g, and longer periods.

The millisecond pulsar has a very large rotational energy and a small magnetic field, while the other binary pulsars have rotational energies three to four orders smaller, and a very large magnetic field, hence, it is not obvious that these objects evolved in a similar way. Various mechanisms for the explanation of the ms-pulsar have been presented. An analysis of these mechanisms has been carried out by Henrichs and van den Heuvel (1983) and a thorough investigation of the difficulties with the different models led them to the suggestion that the ms-pulsar is formed by coalescence of two neutrons stars.

11.1. POSSIBLE MODELS

11.1.1. *Spin-up by Accretion in a Massive Binary During a Common–Envelope Phase*
In this model an old neutron star, with an age of several 10^6 yr starts accretion, so that it spins up to an equilibrium spin period, depending mainly on the magnetic field strength and the mass loss rate. Since PSR 1913+16 and PSR 0656+64 have periods of the order of a day, a spiral-in phase with large mass loss rates and large angular momentum losses must have occurred. The high orbital eccentricity of PSR 1913+16 shows that the companion star itself exploded in a supernova event, in such a way that the system was not disrupted, and collapsed to a neutron star.

If PSR 1937+214 was descendant from a massive X-ray binary the system was disrupted in the second supernova event. Since the remnant of the exploding star is a neutron star, the initial ZAMS mass of the progenitor or the final mass after accretion by mass transfer, from its companion, must have been at least 14 M_0 (De Greve and de Loore, 1977). The maximum possible Eddington accretion rate is $\sim 1.5 \times 10^{-8}$ $M_\odot$ yr^{-1}.

The amount of required accreting matter M_0 for a spin up to the equilibrium period is $\sim 0.12\ M_\odot$ (Henrichs and van den Heuvel). Companions with masses exceeding 14 M_0 have lifetimes below $\sim 10^7$ yr, hence the ensuing average accretion rate is $\sim 1.2 \times 10^{-8}\ M_0$ yr^{-1}, this means that the average accretion rate should equal the maximum possible Eddington accretion rate: However such large transfer rates can only occur during short evolutionary stages.

The maximum possible amount of accretion is smaller than $\sim 0.01\ M_\odot$.

11.1.2. *After the Common Envelope Phase and Disappearance of the Wolf–Rayet Companion a Massive Disk > 0.1 M_0 is Present Around the Neutron Star (Alpar et al., 1982)*

The survival probability of the disk after the supernova of its companion is extremely low.

11.1.3. *Accretion From the Wind of an Intermediate Red Giant*

Spins up the neutron star to a very short period (Arons, 1983).

Consideration of the transferred mass and the possible accretion rate leads to a maximum of $\sim 0.005\ M_\odot$ for the accreted matter, which rules out also this model.

11.1.4. *Spin-up by Accretion in a Low-Mass X-ray Binary* (Alpar *et al.*, 1982; Arons, 1983).

Globular cluster sources and bulge sources contain most probably a neutron star, accreting matter from a low mass companion, in a phase of Roche lobe overflow (Lewin *et al.*, 1980, 1981), with very narrow orbits (e.g., for 4U1626–67 the orbital period is 41 min). The mass transfer in very close systems is driven by angular momentum losses produced by gravitational radiation (Verbunt and Zwaan, 1981). The companion becomes a fully convective red dwarf, a degenerate star, which can be represented by a polytrope of index $\frac{3}{2}$. In such a case the radius increases when the star is losing mass and the orbit becomes wider. The mass transfer rate decreases and vanishes (Paczynski and Sienkiewicz, 1981; Rappaport *et al.*, 1982).

According to Henrichs and van den Heuvel (1983) for a H-rich component, with an assumption on the gravitational radiation time-scale, a final period of a couple of hours is found, and a mass of $\sim 0.017\ M_\odot$ for the companion of a neutron star of 1.4 $M_\odot$. Such a system would show a periodic modulation of the pulse arrival times. The observed limits on the variation of the pulse period rule out the presence of such a companion star (Ashworth *et al.*, 1983; Backer *et al.*, 1983).

11.1.5. *Coalescence of a Close Binary System Consisting of Two Neutron Stars*

The binary radio pulsar PSR 1913 + 16 consists most probably of two neutron stars with masses of $\sim 1.4\ M_\odot$. The decay of the orbit occurs as predicted by the emission of gravitational radiation according to general relativity, leading to a coalescence of the system in $\sim 3.1 \times 10^8$ yr (Peters, 1964; Clark and Eardly, 1977). Clark and Eardley found that the two components coalesce directly if they have the same mass and the orbitis smaller than 30 km, and that a Roche lobe overflow occurs, with a spiral out and tidal disruption of the lower mass star, leaving a neutron star. The orbital period near corotation is $\sim$ 1 ms and the two magnetized neutron stars are corotating.

For rotation periods below 1.5 ms the neutron star is expected to be unstable to the radiation of gravitational waves by non-stellar modes (Papaloizou and Pringle, 1978); hence, the rotation period will not drop below 1.5 ms.

Hence, the coalescence of two orbiting neutron stars seems to be the most plausible explanation for the existence of the millisecond pulsar.

12. Global Picture of the Evolution of Massive Close Binaries

Considering typical objects such as SS 433, Cyg X-3, PSR 1913 + 16, and PSR 1937 + 214, and by comparing their characteristics with those of the standard X-ray sources a general picture for the evolution of massive stars emerges, starting from a pair of OB ZAMS components through successive stages of quiet and explosive mass loss, either leading to a binary pulsar lateron coalescing into a very short pulsar, or into two runaway neutron stars. Two massive stars with a period of $\sim$ 10 days start their mass transfer phase during shell hydrogen burning of the primary.

After the mass exchange stage a Wolf–Rayet binary is formed, later on evolving, after a SN explosion into an OB runaway, i.e., a massive star with a neutron star compagnon. As the massive star is nearly filling its Roche lobe; the enhanced stellar wind will be sufficient to produce X-rays, hence, a first X-ray stage occurs.

The secondary evolves through the shell hydrogen binary phase, expands, fills its Roche lobe. Matter has to leave the system since the neutron star is not able to accrete the material expelled by its companion. The optical star is on its way to become a Wolf–Rayet star. An accretion disk could be formed. This could represent a second X-ray stage; SS433 and Cir X-1 could be in this phase, a phase of slow mass transfer.

At the end of this second mass loss stage, the He-remnant and the orbiting neutron star represent run-away Wolf-Rayet stars.

Due to spiral in of the neutron star in the atmosphere of the WR star, a third X-ray stage could be produced, a WR-star and a neutron star with a very short period, such as Cyg X-3.

The He-star continues its subsequent nuclear burning phases, and finally explodes. If the system, remains bound a binary pulsar is produced, such as PSR 1913 + 16. The two neutron stars coalesce, hence, a very short period pulsar is formed, with a period of $\sim$ 1.5 ms.

References

Alpar, M. A., Cheng, A. R., Ruderman, M. A., and Shaham, J.: 1982, *Nature* **300**, 728.

Arnett, W. D.: 1978, in R. Giacconi and R. Ruffini (eds.), *Physics and Astrophysics of Neutron Stars and Black Holes*, North-Holland Publ. Co., Amsterdam, p. 356.

Arons, J.: 1983, *Nature* **301**, 302.

Ashworth, M., Lyne, A. G., and Smith, F. G.: 1983, *Nature* **301**, 313.

Backer, D. C., Kulkarni, S. R., and Taylor, J. H.: 1983, *Nature* **301**, 314.

Clark, J. P. A. and Eardly, D. M.: 1977, *Astrophys. J.* **125**, 311.

Conti, P. S., Garmany, C., de Loore, C., and Vanbeveren, D.: 1983, *Astrophys. J.* (in press).

Cowley, A. P.: 1980, in P. Sandford (ed.), *Compact Galactic X-Ray Sources*, Cambridge Univ. Press, Cambridge.

De Cuyper, J. P.: 1981, in W. Sieber and R. Wielebinsky (eds.), 'Pulsars', *IAU Symp.* **95**, 399.

De Greve, J. P. and de Loore, C.: 1976, *Astrophys. Space Sci.* **43**, 35.

De Greve, J. P. and de Loore, C.: 1977, *Astrophys. Space. Sci.* **50**, 75.

De Greve, J. P., de Loore, C., and van Dessel, E. L.: 1978, *Astrophys. Space. Sci.* **53**, 105.

de Loore, C., De Greve, J. P., and De Cuyper, J. P.: 1975, *Astrophys. Space Sci.* **36**, 219.

Delgado, A. and Thomas, H. C.: 1980, *Astron. Astrophys.* **36**, 142.

Firmani, C. and Bisiacchi, F.: 1980, *Proc. 5th IAU Reg. Meeting*, Liège, Belgium.

Green, R. F., Richstone, P. O., and Schmidt, M.: 1978, *Astrophys. J.* **224**, 892.

Henrichs, H. F. and van den Heuvel, F. P. J.: 1983, *Nature* **303**, 213.

Ilovaisky, S. A. and Chevalier, C.: 1978, *Astron. Astrophys.* **70**, L19.

Jedrzejec, E.: 1969, M. S. Thesis, Univ. of Warsaw.

Jones, C.: 1977, *Astrophys. J.* **214**, 956.

Joss, P. C. and Rappaport, S.: 1979, *Astron. Astrophys.* **71**, 217.

Lewin, W. H. G. and Clark, G. W.: 1980, *Ann. N.Y. Acad. Sci.* **336**, 451.

Lewin, W. H. G. and Joss, P. C.: 1981, *Space Sci. Rev.* **28**, 3.

Maraschi, L., Treves, A., and van den Heuvel, E. P. J.: 1977, *Astrophys. J.* **216**, 819.

Massevich, A. G., Tutukov, A. V., and Yungelson, L. R.: 1981, in D. Sugimoto, D. Schramm, and D. Lamb (eds.), 'Fundamental Problems in the Theory of Stellar Evolution', *IAU Symp.* **93**, 185.

Massevich, A. G., Tutukov, A. V., and Yungelson, A. R.: 1976, *Astrophys. Space Sci.* **40**, 115.

Meyer, F. and Meyer-Hofmeister, E.: 1979, *Astron. Astrophys.* **78**, 167.

Miller, J. S., Krzeminsky, W., and Priedhorsky, W.: 1976, *IAU Circ.*, No. 2974.

Miyaji, S., Nomoto, K., Yokoi, K., and Sugimoto, D.: 1980, *Publ. Astron. Soc. Japan* **32**, 303.

Nelson, B. and Young, A.: 1970, *Publ. Astron. Soc. Pacific* **82**, 7699.

Nomoto, K.: 1980, Proc. Workshop 'Type I Supernovae', Univ. of Texas.

Papaloizou, J. and Pringle, J. E.: 1978, *Monthly Notices Roy. Astron. Soc.* **184**, 501.

Paczynski, B.: 1971, *Ann. Rev. Astron. Astrophys.* **9**, 183.

Paczynski, B.: 1976, in P. Eggleton, S. Mitton, and J. Whelan (eds.), 'Structure and Evolution of Close Binary Systems', *IAU Symp.* **73**, 75.

Paczynski, B. and Sienkiewicz, R.: 1972, *Acta Astron.* **22**, 73.

Paczynski, B. and Sienkiewicz, R.: 1981, *Astrophys. J.* **248**, L27.

Peters, P. C.: 1964, *Phys. Rev.* **B136**, 1224.

Plavec, M., Ulrich, R. K., and Polidan, R. S.: 1973, *Publ. Astron. Soc. Pacific* **85**, 769.

Rappaport, S. A., Joss, P. C., and Webbink, R. F.: 1982, *Astrophys. J.* **254**, 616.

Ritter, H.: 1975, *Mitt. Astron. Ges.* **36**, 93.

Ritter, H.: 1976, *Monthly Notices Roy. Astron. Soc.* **175**, 279.

Savonije, G. J.: 1978, *Astron. Astrophys.* **62**, 317.

Savonije, G. J.: 1979, *Astron. Astrophys.* **71**, 352.

Shklovski, I. S.: 1981, *Proc. 5th IAU Regional Meeting*, Liège, Belgium.

Sugimoto, D. and Nomoto, K.: 1980, *Space Sci. Rev.* **25**, 155.

Taam, R. F., Bodenheimer, P., and Ostriker, J. P.: 1978, *Astrophys. J.* **222**, 269.

Tutukov, A. V.: 1981, in D. Sugimoto, D. Schramm, and D. Lamb (eds.), 'Fundamental Problems in the Theory of Stellar Evolution', *IAU Symp.* **93**, 137.

Tutukov, A. V. and Yungelson, L. R.: 1979, *Acta Astron.* **29**, 665.

van den Heuvel, E. P. J.: 1976, in J. Whelan and S. Mitton (eds.), 'Structure and Evolution of Close Binary Systems', *IAU Symp.* **73**, 35.

van den Heuvel, E. P. J.: 1980, in R. Giacconi and G. Setti (eds.), *X-Ray Astronomy*, D. Reidel Publ. Co., Dordrecht, Holland, p. 115.

van den Heuvel, E. P. J.: 1981, in D. Sugimoto, D. Lamb, and D. Schramm (eds.), 'Fundamental Problems in the Theory of Stellar Evolution', *IAU Symp.* **93**, 115.

van den Heuvel, E. P. J. and Heise, J.: 1972, *Nat. Sci.* **239**, 67.

van den Heuvel, E. P. J. and de Loore. C.: 1973, *Astron. Astrophys.* **25**, 387.

van den Heuvel, E. P. J., Ostriker, J. P., and Petterson, J. A.: 1980, *Astron. Astrophys.* **81**, L7.

Vanderlinden, T.: 1982, Ph.D. Thesis, Univ. Amsterdam.

Van Paradijs, J.: 1978, *Nature* **274**, 650.

Van Paradijs, J.: 1980, *IAU Circ.*, No. 3487.

Verbunt, F. and Zwaan, C.: 1981, *Astron. Astrophys.* **100**, L7.

Webbink, R. F.: 1975, Ph.D. Thesis, Univ. of Cambridge.

Webbink, R. F.: 1977a, *Astrophys. J.* **211**, 486.

Webbink, R. F.: 1977b, *Astrophys. J.* **211**, 881.

Webbink, R. F.: 1979, in F. M. Bateson, J. Smak, and I. H. Urch (eds.), 'Changing Trends in Variable Stars Research', *IAU Colloq.* **49**, 102.

MASS LOSS IN SEMI-DETACHED BINARIES*

K. D. ABHYANKAR

Centre of Advanced Study in Astronomy, Osmania University, Hyderabad, India

(Received 7 June, 1983)

Abstract. Pre-Main-Sequence contracting objects, post-Main-Sequence expanding stars and mass-losing components of semi-detached systems all occupy more or less the same region in the conventional H–R-diagram. We make a transformation to variables $\Delta (\log L)$ and $\Delta (\log T_e)$, where Δ is the difference between the observed quantity, $\log L$ or $\log T_e$, and the value of that quantity which a star of the same mass would have on the empirical Main Sequence. It is demonstrated that a plot between the new variables clearly separates the mass-losing stars from other objects which is essentially an effect of the increasing abundance of helium relative to hydrogen.

1. Introduction

H–R-diagram is a powerful tool for studying the evolution of stars mainly because of the general validity of the Russell–Vogt theorem for stars in hydrostatic and thermal equilibrium. Given the total mass and run of chemical composition within the star, one can obtain a unique configuration with specific values of effective temperature and luminosity which fix the position of the star in the H–R-diagram. The zero-age Main Sequence is characterised by homogeneous chemical composition throughout the star and energy production by hydrogen burning in the core. Since it is possible to calculate the change of chemical composition caused by thermonuclear reactions one can trace the track of the star in the H–R-diagram as it evolves away from the Main Sequence. In general, the star expands and the effective temperature drops which causes it to move towards right into the giant and subgiant region of the H–R-diagram. There is also an increase in luminosity due to the increase in mean molecular weight caused by the conversion of hydrogen into helium in the central core.

The same region of giants and subgiants is also occupied by the pre-Main-Sequence contracting stars which are not in strict hydrostatic equilibrium. They are distinguished from the post-Main-Sequence objects by other characteristics such as: surrounding nebulosities, infrared excess, presence of emission lines in spectra, flare activity, and overabundance of lithium.

There is a third group of objects which also occupy the giant and subgiant region of the H–R-diagram, viz. the evolved components of semi-detached binaries (Kopal, 1956). It is now well-established that they have lost a large amount of their original mass, particularly from their hydrogen rich envelopes. As it is desirable to distinguish them from the other two groups we give here a method of doing so by means of a transformation of variables.

* Paper presented at the Lembang-Bamberg IAU Colloquium No. 80 on 'Double Stars: Physical Properties and Generic Relations', held at Bandung, Indonesia, 3–7 June, 1983.

Astrophysics and Space Science **99** (1984) 355–361. 0004–640X/84/0992–0355$01.05.

2. The Choice of Variables

The clue to the choice of variables comes from R CMa-systems which contain secondaries of very low mass. In the case of R CMa itself the evolved secondary has a mass of $0.2 M_\odot$ (Sahade, 1963; Radhakrishnan *et al.*, 1983) which corresponds to a Main-Sequence star of spectral type M5 that would have a luminosity $3.5 \times 10^{-3} L_\odot$ and effective temperature of 2800 K. However, its actual luminosity of $0.55 L_\odot$ and effective temperature of 4500 K make it appear like a Main-Sequence star of spectral type G8. Thus the secondary of R CMa is both overluminous and hotter for its mass; this is a characteristic of increased abundance of helium relative to hydrogen (cf. Schwarzschild, 1958; p. 143). Since the evolved components of semi-detached sytems are supposed to have lost a large portion of their hydrogen envelope exposing the inner

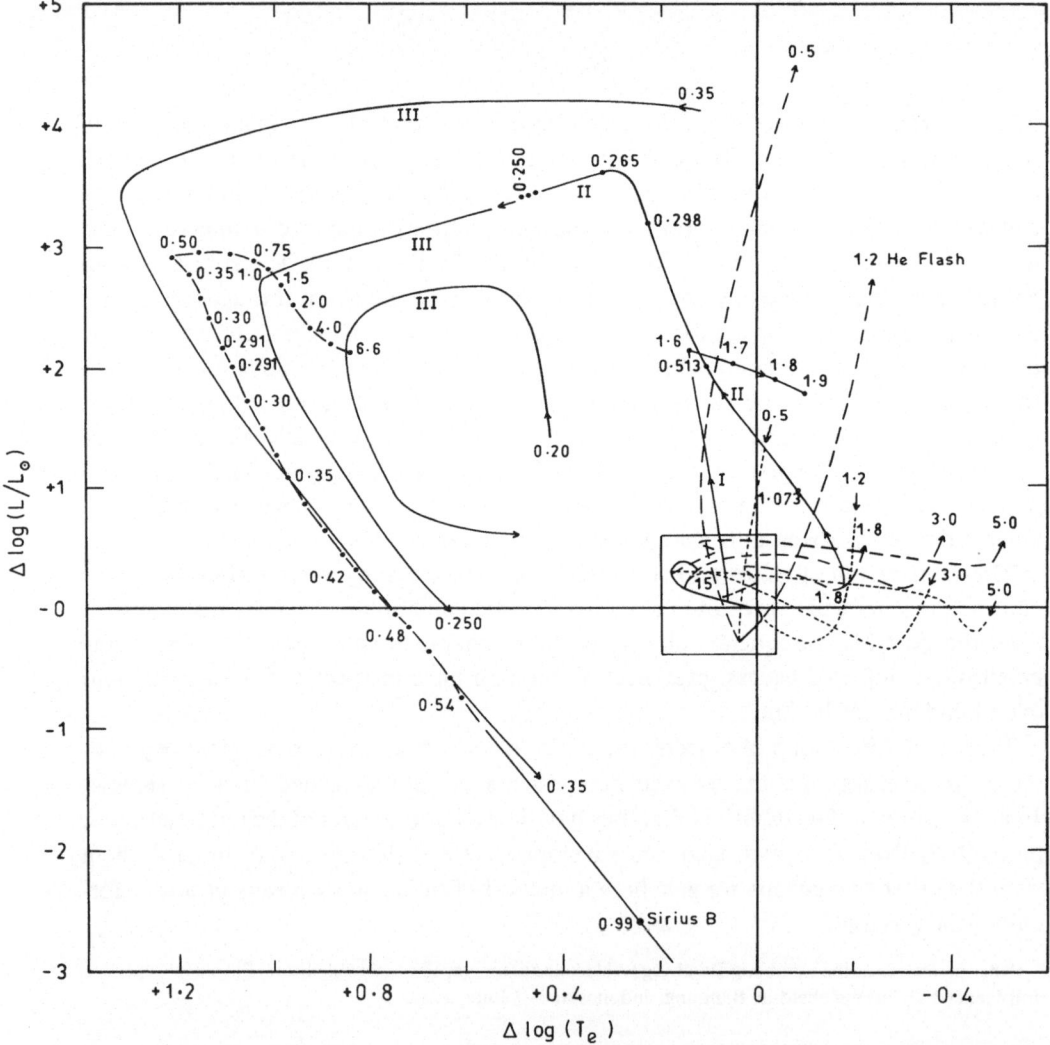

Fig. 1.   Evolutionary tracks of various objects in the $\Delta \log L - \Delta \log T_e$ plot; see text for explanation.

helium-rich core they should show both the increase in luminosity as well as effective temperature.

We choose, therefore, the following variables:

(i) $\Delta (\log L)$ = observed $\log L - \log L$ for a star of the same mass on the empirical Main Sequence.

(ii) $\Delta (\log T_e)$ = observed $\log T_e - \log T_e$ for a star of the same mass on the empirical Main Sequence.

Figure 1 shows the following groups of stars in the $\Delta (\log L) - \Delta (\log T_e)$ plane:

(a) Empirical Main Sequence is represented by just one point viz. the origin.

(b) The zero-age Main Sequence based on calculations of Iben (1965) for $X = 0.708$, $Z = 0.02$ is shown by the S-shaped curve near the origin. The difference from empirical Main Sequence arises because at the upper end it contains somewhat evolved stars while at the lowest end it contains stars which have just reached the zero-age Main Sequence. The square surrounding ZAMS is a measure of uncertainty in the estimation of our variables.

(c) Tracks (small dashed curves) of pre-Main-Sequence contracting single stars of various masses (marked on the curves) according to Iben (1965).

(d) Tracks (long-dashed curves) of post-Main-Sequence single stars of various masses (marked on the curves) due to Iben (1964).

(e) Track of a rapidly mass accreting star of mass $1.5 M_\odot$ in a binary (continuous curve marked I) according to Neo et al. (1977); the mass of the star at each stage is marked on the track.

(f) Track of the mass-losing component in a binary with original masses $M_1 = 1.8 M_\odot$, $M_2 = 0.7 M_\odot$ (continuous curve marked II) as calculated by Refsdal and Weigert (1969); here also the mass at various stages is marked on the track.

(g) Evolutionary tracks of helium white dwarfs in close binaries (continuous curves marked (III) according to Webbink (1975)); mass of the star is indicated on each curve.

(h) Helium Main Sequence (dash dot curve) due to Hansen et al. (1972); numbers of this curve give the mass of the helium star.

(i) A typical white dwarf, Sirius B, in the lower part.

It is seen that the pre-Main-Sequence contracting stars and post-Main-Sequence evolving stars both occupy a region to the right of the origin. The rapidly mass accreting component of a binary first moves slightly to the left due to the heating of the photosphere by the infalling material, but it also eventually moves to the right as the envelope grows in size. This track also shows that a star cannot accrete much of the mass lost by its companion. Most of the mass lost by the evolving component leaves the system through the outer Lagrangian points.

The track of the mass losing component is, however, quite distinct in that it moves up and to the left which corresponds to an increase in luminosity as well as surface temperature. This is obviously the effect of the decrease in the hydrogen content by mass loss from the envelope. The star eventually moves towards the helium Main Sequence and ultimately becomes a white dwarf. We thus see that the track of a mass-losing star in the $\Delta (\log L) - \Delta (\log T_e)$ plane is markedly different from those for pre-Main-

Sequence and post-Main-Sequence single stars which have not lost mass during the course of evolution.

3. Components of Semi-Detached Systems

Figure 2 is an enlarged version of Figure 1 with components of semi-detached systems plotted therein. The open symbols represent the more massive primaries while the filled

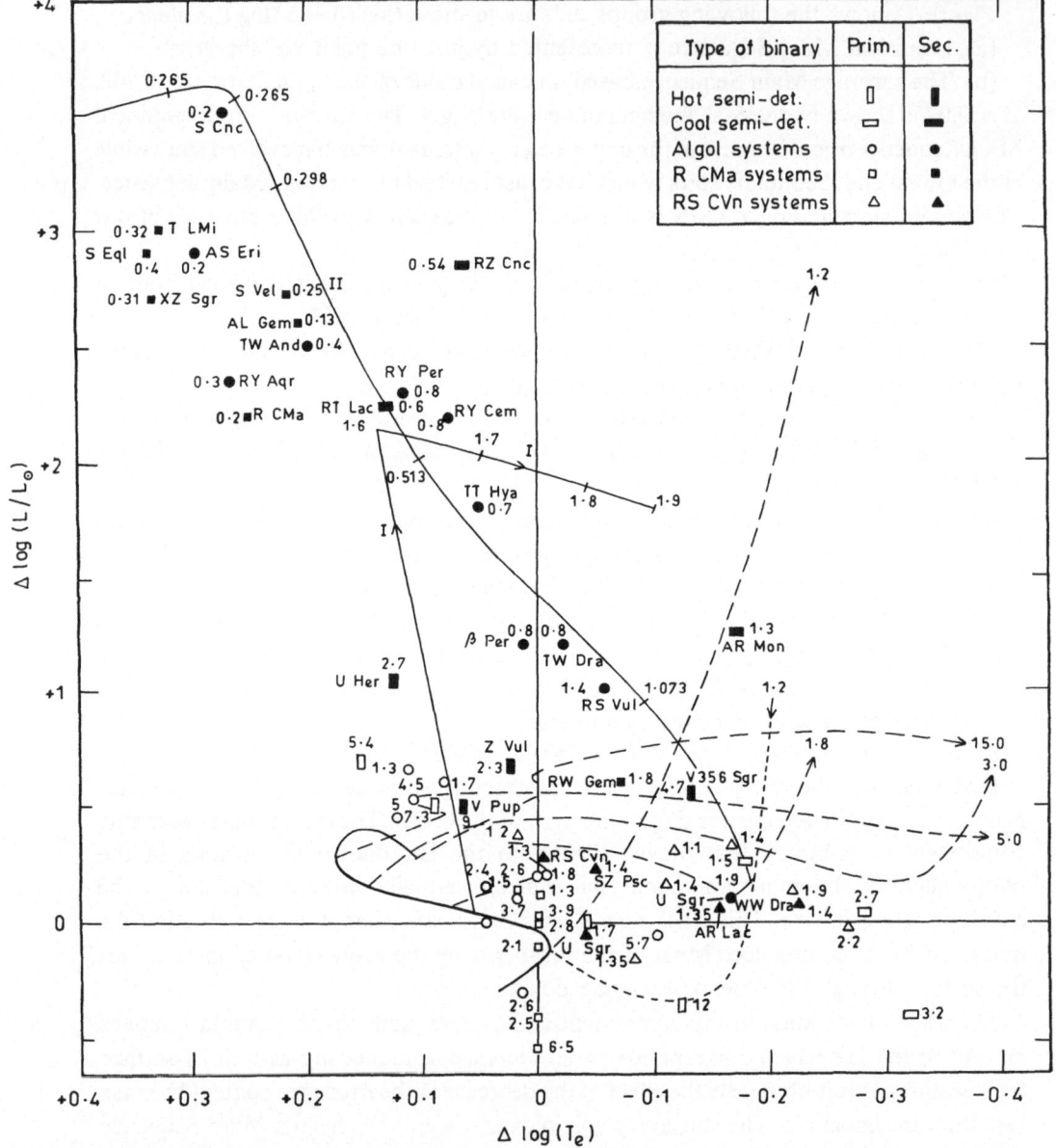

Fig. 2.   Position of the components of semi-detached systems in the $\Delta \log L - \Delta \log T_e$ plot; see text for explanation.

symbols denote the less massive secondaries; the numbers indicate masses of the two components. The following groups of binaries are shown in Figure 2:

(a) Hot semi-detached systems (vertical rectangles) listed by Popper (1980).

(b) Cool semi-detached systems (horizontal rectangles) listed by Popper (1980).

(c) Algol-systems (circles) listed by Popper (1980).

(d) R CMa-systems (squares) listed by Cester et al. (1979). The data for R CMa itself is ours (Radhakrishnan et al., 1983); in the case of other stars the more massive star was assumed to be a Main-Sequence star of observed spectral type for deriving the mass of the secondary.

(e) Six RS CVn-systems for which data are taken from Popper (1961, RS CVn), Arnold et al. (1979, SS Cam), Jakate et al. (1976, SZ Psc), Popper (1956, Z Her), Mardirossian et al. (1980; WW Dra) and Chambliss (1976; AR Lac).

It is clearly seen that the secondary components of most semi-detached systems occupy the region of mass losing stars in which the hydrogen content is considerably reduced. We would like to draw attention to TT Hya in which Kulkarni and Abhyankar (1981) had considered the secondary to be in pre-Main-Sequence contraction phase. From its position in Figure 2 it is clear that the star has definitely lost mass. Hence, its undersize nature which is confirmed by Kaul and Abhyankar (1982), has to be attributed to the second phase of contraction after considerable loss of mass from the envelope through the outer Lagrangian points by normal evolutionary process.

It may be noted that the secondaries of semi-detached systems invariably lie below the tracks of helium white dwarfs of same masses as calculated by Webbink (1975). This underluminosity indicates that they have lost mass so rapidly that the shell burning of hydrogen is also getting suppressed.

4. Conclusions

We have demonstrated that the $\Delta (\log L) - \Delta \log(T_e)$ plot separates the mass losing evolved components of semi-detached systems from the pre-Main-Sequence contracting stars and post-Main-Sequence expanding single stars. It is to be noted that only a small fraction of the mass lost by the evolved component can be accomodated on the companion. Hence, most of the mass is lost from the system through the outer Lagrangian points, and the more massive primary has essentially retained its original character with only a small increase in its mass. Since the evolved secondary must have been originally the more massive component the quantity $1 - q$, where $q = m_2/m_1$ is the mass ratio, is a measure of the minimum fraction of mass lost by it. Hence, there should be a correlation between the mass ratio and the state of evolution of the secondary.

In order to find this correlation we consider the differential evolution of the two components of the semi-detached system in the $\Delta (\log L) - \Delta (\log T_e)$ plane. Defining $\delta = \Delta$ (secondary) $- \Delta$ (primary) we have plotted $\delta (\Delta \log L)$ vs $\delta (\Delta \log T_e)$ in Figure 3 for all the systems shown in Figure 2. We find a linear correlation (line AB) between these quantities such that systems with small value of q, which lie at the upper end of the line AB, show larger differential evolution by mass loss as compared to the systems

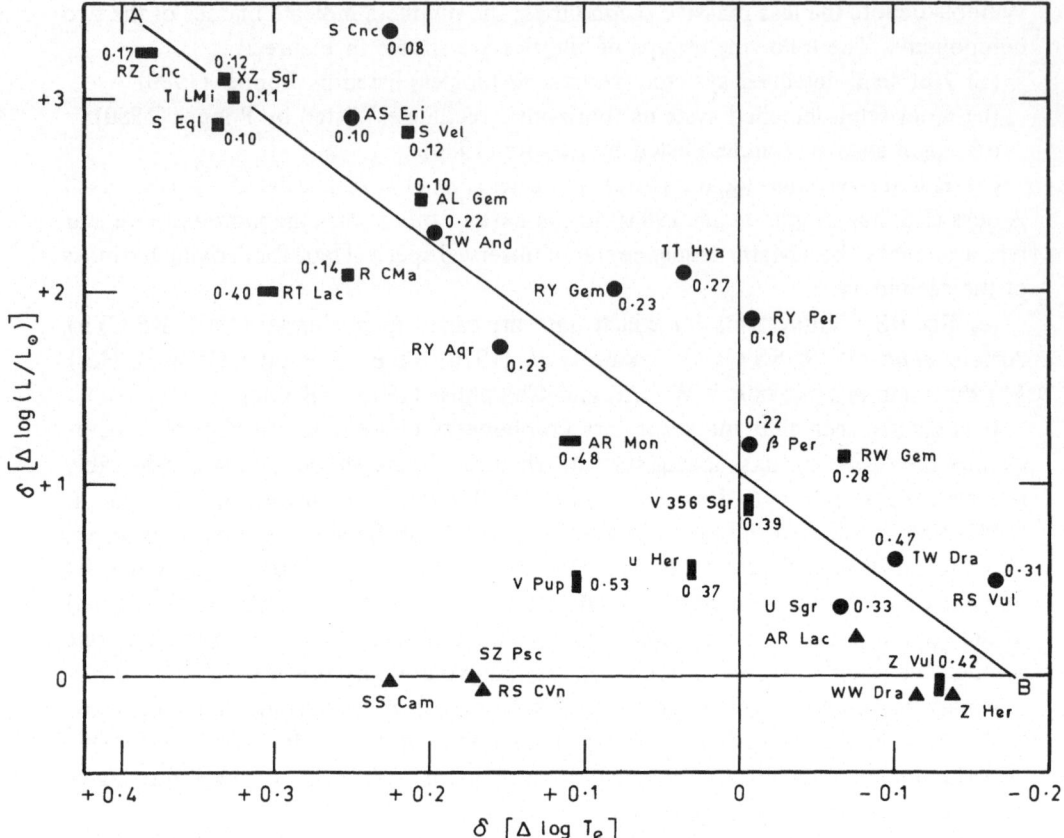

Fig. 3.   Differential evolution of the components of the semi-detached systems; $q = m_2/m_1$ values are marked for each system.

with larger values of q which lie at the lower end. The RS CVn-systems show the least differential evolution by mass loss and that also along the axis $\delta(\Delta \log L) = 0$ as is expected from their almost equally massive detached components.

The correlation of overluminosity with q was noted long ago by Struve (1954). We now find that the excess of effective temperature is also similarly correlated with q. As the evolved secondary loses mass with attendant reduction in the value of q, its hydrogen content is also reduced and the star becomes both overluminous and hotter for its residual mass.

References

Arnold, C. N., Hall, D. S., Montle, R. F., and Stuhlinger, T. W.: 1979, *Acta Astron* **29**, 243.

Cester, B., Giuricin, G., Mardirossian, F., Mizzetti, M., and Milano, L.: 1979, *Astron. Astrophys. Suppl.* **36**, 273.

Chambliss, C. R.: 1976, *Publ. Astron. Soc. Pacific* **88**, 762.

Hansen, C. J., Cox, J. P., and Herz, M. A.: 1972, *Astron. Astrophys.* **19**, 144.

Iben, I., Jr.: 1964, *Astrophys. J.* **140**, 1631.

Iben, I., Jr.: 1965, *Astrophys. J.* **142**, 993.

Jakate, S., Bakos, G. A., Fernie, J. D., and Heard, J. F.: 1976, *Astron. J.* **81**, 250.

Kaul, J. and Abhyankar, K. D.: 1982, *J. Astrophys. Astron.* **3**, 93.

Kopal, Z.: 1956, *Ann. Astrophys.* **19**, 298.

Kulkarni, A. G. and Abhyankar, K. D.: 1981, *J. Astrophys. Astron.* **2**, 119.

Mardirossian, F., Mezzetti, M., Cester, B., and Giuricin, G.: 1980, *Astron. Astrophys. Suppl.* **39**, 73.

Neo, S., Miyaji, S., and Nomoto, K.: 1977, *Publ. Astron. Soc. Japan* **29**, 249.

Popper, D. M.: 1956, *Astrophys. J.* **124**, 196.

Popper, D. M.: 1961, *Astrophys. J.* **133**, 148.

Popper, D. M.: 1980, *Ann. Rev. Astron. Astrophys.* **18**, 115.

Radhakrishnan, K. R., Sarma, M. B. K., and Abhyankar, K. D.: 1983, *Astrophys. Space Sci.* **99**, 229 (this volume).

Refsdal, S. and Weight, A.: 1969, *Astron. Astrophys.* **1**, 167.

Sahade, J.: 1963, *Ann. Astrophys.* **26**, 80.

Schwarzschild, M.: 1958, *Structure and Evolution of the Stars*, Princeton University Press, Princeton.

Struve, O.: 1954, *Mem. Soc. Roy. Liège* **14**, 236.

Webbink, R. F.: 1975, *Monthly Notices Roy. Astron. Soc.* **171**, 555.

CONSTRAINTS FOR CATACLYSMIC BINARY EVOLUTION AS DERIVED FROM SPACE DISTRIBUTIONS*

H. W. DUERBECK

Observatorium Hoher List der Universitäts-Sternwarte Bonn, Daun, F.R.G.

(Received 30 August, 1983)

Abstract. Space densities and galactic z-distributions of novae, recurrent novae, dwarf novae and symbiotic stars are newly determined and discussed in the context of earlier determinations. The data are then compared with the distributions of single and binary stars of possibly related types (late type giants, Mira variables, Algol systems, W UMa systems).

Novae and dwarf novae have similar distributions, those of fairly young stellar populations. The observed space density of potential novae (novalike objects) indicates that the mean recurrence time of novae might be as small as a few hundred years, which leads, with given nova shell masses and mass transfer rates in the minimum stage, to a secular decrease of the masses of the components undergoing nova outbursts.

Recurrent novae and symbiotic stars have distributions of older stellar populations, similar to those of late type giants and Mira variables.

On the basis of galactic distribution, novae and dwarf novae are closely related and may be final stages of W UMa systems, as well as progenitors of supernovae of type I. A small fraction of W UMa systems seems to belong to an older population. If evolutionary transitions between these types of stars can be substantiated, the presence of a minority of novae and dwarf novae in globular clusters and of supernovae I in elliptical galaxies can be explained.

Due to the lack of sufficiently well determined space distributions of Algol binaries, the suggestion that long-period Algol systems might be the progenitors of cataclysmic binaries can as yet neither be substantiated nor refuted. A very high space density of long-period Algol systems in the solar neighbourhood is derived. The observed space density of cataclysmic binaries could be explained by the transformation of a small percentage of the long-period Algol systems by common envelope evolution.

1. Introduction

With the advent of the space age and the possibility of studying celestial objects at a multitude of wavelengths, the interest in cataclysmic binaries (CBs) has increased, and the progress of understanding the physical processes in these systems has increased likewise. Numerous review papers dealing with various aspects of these systems and other types of binaries give an eloquent testimony of this development (for a review of reviews, see Trimble, 1983).

The determination of the evolutionary state of CBs receives much attention at the present time. What types of stars evolve into CBs? Where to do CBs evolve? Some outlines of binary star evolution (see, e.g., Paczynski, 1980; Vilhu, 1981; Yungelson, 1982; Eggleton, 1983) have placed the CBs into evolutionary schemes, using physical similarities of the relevant groups of stars: masses, angular momenta, periods; and plausible physical interactions of the components in the course of their evolution: Roche-lobe overflow, common envelopes, spiralling-in. Another crucial point for

* Paper presented at the Lembang-Bamberg IAU Colloquium No. 80 on 'Double Stars: Physical Properties and Generic Relations', held at Bandung, Indonesia, 3–7 June, 1983.

Astrophysics and Space Science **99** (1984) 363–385. 0004–640X/84/0992–0363$03.45.

establishing evolutionary relations between objects of different types is the study of their galactic distribution, and any proposed evolutionary scheme must not be in conflict with data from stellar statistics.

In this contribution, I shall review our knowledge of the galactic distribution and the space density of several types of CBs, and I shall try to compare them with data of possibly related types of binary and single stars. Since the data found in the literature are often fragmentary and imprecise, this contribution is based to a large extent on newly derived data of galactic distributions of various types of binary stars.

2. Current Evolutionary Schemes Including CBs

In the late 50s (Sahade, 1959) and early 60s (Kraft, 1962, 1967), the scheme

$$EW \rightarrow CB \tag{1}$$

was proposed (EW = W UMa type eclipsing binary; the abbreviation here also includes non-eclipsing systems). This scheme could be reconciled with the space densities, galactic distributions, space motions, and periods of these types of objects, but it proved difficult in the following decades to find a feasible mechanism for the production of CBS out of EWs. The scheme was revived and extended by Warner (1974a) to

$$EW \rightarrow CB \rightarrow SNI . \tag{2}$$

The second half of this scheme, the production of supernovae of type I from CBs, where the mass of the white dwarf is pushed, by accretion, beyond the Chandrasekhar limit, has received much attention in recent years (e.g., Starrfield *et al.*, 1981).

The discontinuity theory of EW structure is in favour of scheme (1) (e.g., Shu, 1980), while the thermal relaxation oscillation theory yields

$$EW \rightarrow single\ star \tag{3}$$

(e.g., Webbink, 1976), so that another process of producing CBs must be found. One possible scheme is the evolution of a long period binary system (noted here as EA) through case C or case B evolution, including a spiralling-in of the secondary star (Paczynski, 1976; Ritter, 1976; see also Law and Ritter, 1983):

$$EA\ (P:\ months \ldots years) \rightarrow giant\ stage \rightarrow CB\ (M_A \gtrsim 0.45 M_\odot),$$

$$EA\ (P:\ days \ldots months) \rightarrow CB\ (M_A \lesssim 0.45 M_\odot), \tag{4}$$

where M_A indicates the mass of the primary (white dwarf) component. We will postpone the discussion of these concurring schemes to Section 10.

3. Methods

Previous studies of the galactic distribution of various types of objects were based on different methods, and the parameters were defined in different ways. We have tried to convert the results into a set of parameters that allows intercomparison. The following quantities were collected or extracted from published data, or were newly determined by the straightforward method described by Kurkarkin (1954), using data taken from the *General Catalogue of Variable Stars* (CGVS) and its three supplements, unless mentioned otherwise:

ρ_0 = space density of the objects in the galactic plane, in the neighbourhood of the Sun (pc^{-3});

z_0 = scale height of the distribution perpendicular to the galactic plane in the solar neighbourhood, assuming a density law

$$\rho(z) = \rho_0 \exp(-|z|/z_0);$$ (5a)

or

$$\rho(z) = \rho_0 \exp(-z^2/2z_0^2).$$ (5b)

We have not attempted to derive the radial density gradient, $\partial \rho / \partial R$, and we have not corrected the samples for the slight displacement of the Sun from the galactic plane. Furthermore, space motions of the various groups of stars are not considered. Earlier material on space motions has already been discussed (e.g., Kraft, 1965), and more recent studies are not available. It is expected that the HIPPARCOS mission will provide data for a more detailed study of the kinematics of various stellar types.

In this text we generally discuss the parameters of approximations using form (5a). It should be noted, however, that distributions of the form (5b) can also be fitted to the data, yielding fits of similar quality. The results of the approximations of the form (5b) are given in Table IX which summarizes all results.

4. Galactic Distribution of Novae

The population assignment of novae has always been a matter of discussion. Baade (1958) counts them among Population II objects because of the occurrence of novae in globular clusters. Plaut (1965) emphasizes the concentration of novae towards the galactic plane and the galactic centre, stating that they form a 'disc population'. Iwanowska and Burnicki (1962) who assume that any group can be described as a mixture of Population I and Population II objects, find evidence for their assumption. Observations of novae in M31 also yield discordant results (Rosino, 1973, 'disc population'; Wenzel and Meinunger, 1978, 'halo population'). Table I summarizes the results of previous investigations.

For the rediscussion of the galactic distribution of novae, the GCVS was used as a source list. Crucial parameters for the determination of the galactic distribution and the space density of a given group of stars are:
(a) the absolute magnitude (or the magnitude range) of the stars in the group;
(b) the completeness of the search as a function of distance;
(c) the interstellar absorption.

TABLE I

Previous determinations of the galactic distribution of novae

Author	ρ_0 (pc^{-3})	z_0 (pc)	Comments
Kukarkin (1954)	–	250	(originally published 1949)
Parenago (1955)	5.5×10^{-6}	109	
Kopylov (1957)	–	182	
Schmidt-Kaler (1957)	–	434	
Oort (1958)	–	440	
Iwanowska and Burnicki (1962)	$\sim 10^{-7}$	72	(Population I objects only)
Plaut (1965)	–	290	(source: Schmidt-Kaler (1957) (?!))
Richter (1968)	–	182	(source: Kopylov (1957))
Allen (1973)	–	300	
Warner (1974)	$10^{-8}–10^{-7}$	–	

We rely on the luminosity calibration and distance determinations by Duerbeck (1981). Only novae with recorded outbursts are included, the few recurrent novae are treated separately, WZ Sge-type objects are omitted. A problem which hardly influences the z-distribution, but very much the space density, is that of completeness. Are all novae that erupt during a given time interval discovered? Novae that are faint during maximum light, southern novae, novae in the more distant past had a better chance to be overlooked. The seasonal variation of the night sky also plays a role, especially for fast novae. Novae that erupted between 1848 and 1981 have been included in our list, covering 133 yr of observation. But only since the beginning of the 20th century the sky has been searched in a more or less regular way, and, taking into account the seasonal effect, the 'effective' observing time is certainly closer to 50 yr, the value used in the subsequent calculations. 20 novae are found in a cylinder with a radius $d_{xy} = 1500$ pc and infinite height, centered in the Sun and extending into the z-direction of the Galaxy, i.e., 0.40 novae per year (Table II). Novae at the edge of this cylinder reach apparent maximum magnitudes between 4^m and 5^m. A study of the novae in a cylinder with $d_{xy} = 1000$ pc yields a similar z-distribution, however, it contains more novae than expected from the ratio of the volumes of the cylinders, indicating that the discovery of novae is not complete (or, at least, was not complete in the past) to a limiting magnitude 4^m to 5^m. This confirms the findings of Allen (1954) and Schmidt-Kaler (1957) that only novae with apparent magnitudes brighter than $3^m.0$ (distance less or equal to 1000 pc) were discovered with certainty (Figure 1). The space/time density ρ^* and the distribution were derived from novae in the cylinder with $d_{xy} = 1$ kpc, yielding

$$\rho^*(z) = (0.38 \pm 0.04) \times 10^{-9} \exp(-|z|/(125 \pm 22)) \, (\text{pc}^{-3} \, \text{yr}^{-1}) \qquad (6)$$

(see Figure 2). The space density can be obtained when an estimate of the mean recurrence time of outbursts is available. Let us first assume that the nova phenomenon is a quasi-stationary process, i.e., the mass ejected during outburst is equal to the mass accreted between subsequent outbursts. Shell masses are estimated to lie in the range $(2–10) \times 10^{-5} M_\odot$ (Duerbeck, 1980). The accretion rate can be estimated from the luminosity in the minimum state. Smak (1982) has studied the energy distributions and

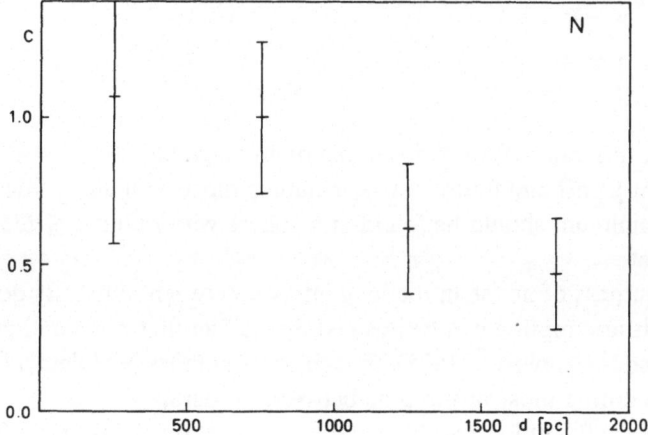

Fig. 1.   Completeness of the discovery of novae in different cylinders with radii d_{xy} (in pc). The completeness is set 1.0 for the cylinder with d_{xy} = 1000 pc. The completeness decreases for radii (or nova distances) which are larger than 1000 pc.

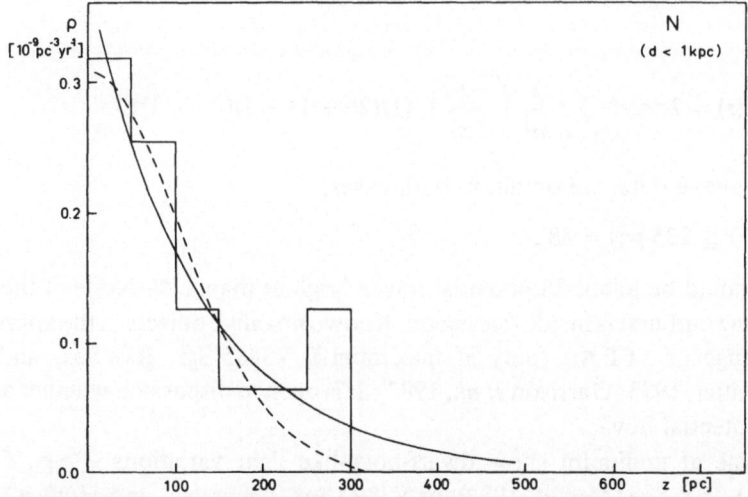

Fig. 2.   The galactic distribution of novae in the z-direction. All novae with $d_{xy} \leqq 1000$ pc are included. The approximations of the form (5a), solid line; and (5b), broken line; are also shown.

luminosities of several novae and other CBs. His data indicate that the nova accretion rates are in the range $(0.8-3) \times 10^{-8} M_\odot$ yr^{-1}. The time interval between outbursts is then

$$t = 5 \times 10^{-5}/1.5 \times 10^{-8} = 3000 \text{ yr},\qquad(7)$$

with lower and upper limits of 700–12 000 yr. The mean time interval yields a space density

$$\rho_0 = 1.27 \times 10^{-6} \, pc^{-3} \tag{8}$$

for Novae.

Let us check this value. With a mean minimum magnitude $M_V^{min} = 4\overset{m}{.}2 \pm 1\overset{m}{.}8$ (s.d.) (Duerbeck, 1980), past and future novae, including those with unrecorded outbursts, of $m_V \leq 11^m$ at minimum should be found in a sphere with radius $r \leq 235$ pc, under the assumption that:

(a) The brightness of novae in the long interval between outbursts does not change noticeably. This assumption can be justified by the fact that post- and pre-novae have similar magnitudes (Robinson, 1975), though some authors postulate a less active, less luminous state during most of the time between outbursts.

(b) Interstellar absorption can be neglected for distances up to $r = 235$ pc.

(c) The large dispersion of minimum magnitudes is not taken into account.

If we assume $M_V^{min} = +4\overset{m}{.}2$ for all novae, the number of novae in the minimum state with apparent magnitudes brighter than 11^m can be calculated, using the density distribution function (5a):

$$N(r) = 2\pi\rho_0(z_0 r^2 + 2z_0^2 r \, e^{-r/z_0} + 2z_0^3 \, (e^{-r/z_0} - 1)) \, ; \tag{9a}$$

or, with (5b),

$$N(r) = 2\pi\rho_0 r^3 \sum_{n=0}^{\infty} \frac{1}{n!} \left(\frac{-r^2}{2z_0^2} \right)^n (1/(2n+1) - 1/(2n+3)). \tag{9b}$$

Inserting the above data, we obtain in both cases,

$$N(r \leq 235 \, pc) = 38 \, .$$

Thus, there should be about 38 potential novae brighter than 11^m. None of the novae observed during outburst is inside this region. Known novalike objects in the appropriate magnitude range are TT Ari (only at maximum!), V3885 Sgr, RW Sex, and CPD $-48°1577$ (Ritter, 1983; Garrison et al., 1982). It is open to discussion whether all these objects are potential novae.

A few novae at minimum show dwarf-nova-like light variations – e.g., GK Per (Hudec, 1981), WY Sge (Moffat, 1982), or V3890 Sgr (Dinerstein and Hoffleit, 1973). Thus some systems with unrecorded outbursts might well be classified as dwarf novae, they have, however, higher luminosities than ordinary dwarf novae. Since this class seems to constitute a small percentage of the novae, potential novae among the catalogued dwarf novae will not significantly make up the deficit. In any case, the observed space density is possibly of the order of 10 times smaller than the expected one.

Let us check the parameters which enter the determination of space density for their probable errors:

(a) The 'effective' time interval of nova observations, 50 yr, can hardly be in error by more than a factor 2.

(b) The accretion rates may be variable. A secular change of the accretion rate would, however, hardly improve the situation: lowering by a factor 10 would increase the time between outbursts by a factor 10, but the minimum magnitude would also decline, by about $2\overset{m}{.}5$. Thus only novae with $r \leq 74$ pc would be brighter than 11^m. Then

$$N(r \leq 74 \text{ pc}) = 20 \,,$$

which is still to large. The minimum magnitude of novae has only a weak influence of their space density.

(c) The mean shell masses might be lower than $5 \times 10^{-5} M_{\odot}$ (contrary to observational evidence), or the nova eruptions cause a *secular decrease* of the mass of the erupting star, i.e., the eruption not only removes the accreted mass, but also genuine material of the white dwarf, which must then comprise a large fraction of the shell mass. In both cases, the mean recurrence time of nova eruptions would only be a few hundred years. It might be rewarding to review old written sources for probable earlier outbursts of bright novae.

TABLE II

Galactic novae with $d_{xy} < 1.5$ kpc

Nova	b	d_{xy} (pc)	z (pc)	Nova	b	d_{xy} (pc)	z (pc)
V603 Aql	+ 0.84	330	5	DI Lac	− 4.86	895	26
T Aur	− 1.71	600	18	DK Lac	− 5.35	1495	140
Q Cyg	− 7.55	1485	200	BT Mon	− 2.63	1000	46
V1500 Cyg	− 0.07	1350	2	V841 Oph	+ 17.79	855	275
HR Del	− 13.97	775	212	GK Per	− 10.11	515	92
DN Gem	+ 14.70	435	114	RR Pic	− 25.67	360	173
DQ Her	+ 26.44	235	116	CP Pup	− 0.84	1500	22
V446 Her	+ 4.71	785	65	FH Ser	+ 5.78	645	65
V533 Her	+ 24.27	620	280	LV Vul	+ 0.85	820	12
CP Lac	− 0.84	1000	15	NQ Vul	+ 1.28	1200	27

In the discussion concerning secular mass loss the recurrent nova T Pyx may play an important role. It is the only slow recurrent nova ($t_3 = 113^d$). Its spectral appearance at outburst, which is similar to that of normal novae, its fairly strong surrounding nebulosity (Duerbeck and Seitter, 1979), its blue colour at minimum light (Bruch *et al.*, 1981) separate it from the properties of all other recurrent novae, and link it to the group of ordinary slow novae. The only peculiarity is the outburst interval (about 19 ± 5 yr) which cannot be reconciled with any reasonable accretion rate if steady state nova evolution is assumed. A depletion of the mass of the accreting object, as indicated by our statistical study of all novae, would resolve the problem. A thorough investigation of this unique object, which is likely to show another outburst soon, might considerably improve our knowledge of evolutionary trends in nova systems.

Taking into account the above mentioned ambiguities, the space density (8) should be taken as an upper limit; there is evidence that ρ_0 may be smaller by at least a factor 10.

5. Galactic Distribution of Recurrent Novae

The group of recurrent novae is small and inhomogeneous. A recent list can be found in Duerbeck and Seitter (1982). It is still subject to discussion whether their light eruptions are thermonuclear runaways or accretion events (Webbink, 1978). Each member of the group shows characteristics that make its memberhsip uncertain:

(a) T CrB can be explained as an accretion event onto a non-degenerate star (Webbink, 1978);

(b) VY Aqr has a very blue colour; evidence for a fairly short outburst interval (Richter, 1983) make its membership in the group of dwarf novae likely;

(c) WZ Sge, showing typical dwarf nova characteristics during outburst, has already been removed from the group;

(d) T Pyx, the only 'slow' recurrent nova, has certainly no giant secondary, however, it has a noticeable shell, thus it cannot be a dwarf nova (see Section 4);

(e) U Sco possibly has also a dwarf companion (Barlow *et al.*, 1981);

(f) V1017 Sgr has been classified as a symbiotic star by Payne-Gaposchkin (1977), however, the observed 'secondary maxima' have some resemblance to the photometric behaviour of T CrB (Vidal and Rodgers, 1974).

An attempt was made to obtain at least a rough estimate of z_0. Distances were derived under the assumption that the spectroscopically detectable secondaries are giant (or dwarf) stars; the result for T CrB, $r = 1200$ pc, is in good agreement with the nebular parallax (Duerbeck, 1981). For U Sco, a dwarf secondary of spectral type $G-K$ was assumed. Details are given in Table III.

TABLE III

List of recurrent novae

Object	b	m_V (min)	A_V	Spectrum		M_V	d (pc)	z (pc)
T CrB	+ 48.2	$9\overset{m}{.}9$ (1)	$0\overset{m}{.}1$ (2)	M3 III	(3)	$-0\overset{m}{.}6$	1200	300
RS Oph	+ 10.4	11.4 (1)	2.4 (2)	M2 III/A7Q(3)		-0.6	1800	325
U Sco	+ 21.9	19.0 (2)	0.6 (2)	G0–K0V?	(4)	4.5–6.0	$3–6 \times 10^3$	1100–2200
V1017 Sgr	-9.1	13.6 (1)	1.0 (2)	G5 III p	(3)	$+0.9$	2200	350

References: (1) Bruch (1982); (2) Webbink (1978); (3) Duerbeck and Seitter (1982); (4) Barlow *et al.* (1981).

TABLE IV

Previous determinations of the galactic distribution of recurrent novae

Author	ρ_0 (pc^{-3})	z_0 (pc)	Comments
Kopylov (1957)	–	480	
Richter (1968)	–	480	(source: Kopylov (1957))
Allen (1973)	–	500	
Warner (1974)	8×10^{-10}	–	

The resulting mean z_0, 660 ± 330 pc, is not unlike that of symbiotic stars (see Section 7), and very different from that of classical novae and dwarf novae. Previously derived data are summarized in Table IV.

6. Galactic Distribution of Dwarf Novae

Previous estimates of the scale height of dwarf novae vary noticeably (Table V). We have therefore determined the distribution of dwarf novae again.

TABLE V

Previous determinations of the galactic distribution of dwarf novae

Author	ρ_0 (pc^{-3})	z_0 (pc)	Comments
Kopylov (1957)	–	2600	
Schmidt-Kaler (1957)	–	90	
Kraft (1962)	–	48*	
Popov (1964)	$\sim 10^{-5}$*	63/39*	
Kraft (1965)	–	60/84*	
Richter (1968)	5×10^{-6}	111	(also $z_0 = 2600$ quoted)
Warner (1974)	4×10^{-7}	–	

* Values marked with an asterisk were derived from the original data, using approximations (5a) and (5b).

Of the ~ 300 dwarf novae listed in the GCVS and its supplements, 72 are found in a cylinder of 400 pc radius and infinite height. The crucial parameter that determines the derived distribution is the absolute magnitude. Different statistical studies yield for the absolute magnitude at minimum, $\langle M_V^{min} \rangle$, values around $+7\overset{m}{.}5$ (Kraft and Luyten, 1965; Primkulov, 1968; Voikhanskaya, 1973). However, trigonometric parallaxes of several dwarf novae yield values of $\langle M_V^{min} \rangle = 9\overset{m}{.}5 \pm 1\overset{m}{.}2$ (s.d.) (Vasilevskis et al., 1975). A recent discussion of the parallaxes of SS Cyg, U Gem, and SS Aur by Kamper (1979) yields $\langle M_V^{min} \rangle = +8\overset{m}{.}9 \pm 0.8$ and $\langle M_V^{max} \rangle = 4\overset{m}{.}4 \pm 1\overset{m}{.}3$. A study based on K-line photometry of four objects by Bailey (1981) yields $\langle M_V^{min} \rangle = +8\overset{m}{.}5 \pm 1\overset{m}{.}9$ and $\langle M_V^{max} \rangle = +4\overset{m}{.}4 \pm 1\overset{m}{.}1$. Vogt (1981) showed that the luminosities of the discs of all dwarf novae during outburst seems to be equal, corresponding to $\langle M_V^{max} \rangle = +4\overset{m}{.}7$. The mean outburst amplitude of all dwarf novae for which reliable data exist is $3\overset{m}{.}9$ magnitudes. The old assumption, $\langle M_V^{min} \rangle = +7\overset{m}{.}5$, gives $\langle M_V^{max} \rangle = +3\overset{m}{.}6$, a brightness that is much higher than the recently obtained results.

We have adopted for the following discussing

$$\langle M_V^{max} \rangle = +4\overset{m}{.}5 ,$$

corresponding to $\langle M_V^{min} \rangle = +8\overset{m}{.}4$. It is advantageous to use the brightness at maximum, because it has probably a lower intrinsic scatter, and is better documented.

The 44 objects that are included in a cylinder of 250 pc radius are listed in Table VI, where distances and z-values are given. The absorption was estimated by using the study

TABLE VI

Dwarf novae with $d_{xy} < 250\,pc$

Object	b	m_V^{max}	Type	d_{xy} (pc)	z (pc)
RX And	− 21.5	10.3	ZC	135	53
DX And	− 16.7	11.2	UG	210	62
Z Aps	− 9.4	10.7	ZC	175	29
VZ Aqr	− 36.3	11.6	UG	195	142
UU Aql	− 18.8	11.3	UG	220	66
SS Aur	+ 13.8	10.5	UG	155	38
TT Boo	+ 60.7	12.7	UG	215	379
Z Cam	+ 32.6	10.2	ZC	120	75
SY Cnc	+ 36.4	10.9	ZC	155	113
AA Cnc	+ 33.9	11.8	UG?	240	159
BV Cen	+ 7.5	10.7	UG	160	21
WW Cet	− 71.7	9.6	ZC	31	95
WX Cet	− 79.1	10.8	WZ	35	172
Z Cha	− 22.1	11.5	UG/SU	220	90
AL Com	+ 76.4	13.3	WZ	135	559
GO Com	+ 88.7	13.4	WZ	15	605
SS Cyg	− 7.1	8.2	UG	55	6
SY For	− 62.4	12.3	UG	145	279
U Gem	+ 23.4	8.2	UG	50	22
IR Gem	+ 11.6	10.7	UG	145	30
AH Her	+ 38.2	10.5	ZC	125	96
EX Hya	+ 33.6	11.7	UG	230	152
VW Hyi	− 38.1	8.4	UG/SU	45	37
WX Hyi	− 51.6	9.6	UG/SU	65	82
T Leo	+ 60.5	10.3	UG	70	126
X Leo	+ 45.1	11.1	UG	150	149
TU Leo	+ 44.0	12.0	UG	225	219
TU Men	− 33.7	11.7	UG/SU?	230	152
CN Ori	− 15.6	11.5	ZC	225	63
RU Peg	− 34.8	9.0	UG	60	41
EZ Peg	− 32.5	9.8	UG?	95	59
FO Peg	− 0.5	11.8	ZC	250	2
KT Per	− 11.3	11.0	ZC	155	31
WZ Sge	− 7.9	7.3	WZ	35	5
V1830 Sgr	− 4.8	11.8	UG	245	21
UZ Ser	+ 1.8	12.3	UG	250	8
EK TrA	− 6.3	12.2	UG/SU	240	26
SU UMa	+ 33.1	11.0	UG/SU	170	109
SW UMa	+ 37.0	11.1	UG	170	126
BC UMa	+ 65.1	11.2	UG?	95	199
BZ UMa	+ 38.8	10.8	UG?	140	113
CH UMa	+ 42.6	.11.0	UG	145	135
CU Vel	+ 2.6	11.0	UG/SU	190	9
TW Vir	+ 54.6	12.1	UG	190	269

of the distribution of interstellar extinction by Neckel (1967), and, for stars near the galactic equator, by Neckel and Klare (1980).

It is important to find out the limiting distance for which completeness of the sample is guaranteed. The surface density Σ_{DN},

$$\Sigma_{DN} = 2\rho_{0,\,DN}\, z_{0,\,DN} \tag{10}$$

was determined for cylinders with various radii d_{xy}. If we assume $\Sigma_{DN} = 1$ for the cylinder with $d_{xy} = 200$ pc, the relative surface density drops noticeably for cylinders with $d_{xy} > 250$ pc, that is, our knowledge of these systems is incomplete beyond 250 pc, corresponding $m_V^{max} = 11^m\!.5$ (Figure 3). The z-distribution in the cylinder with

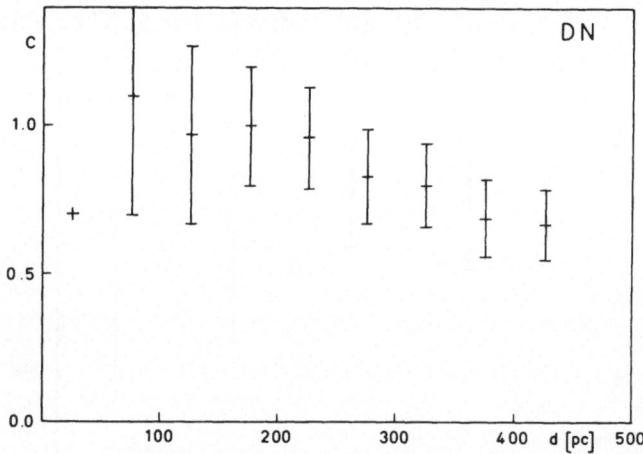

Fig. 3. Completeness of the discovery of dwarf novae in different cylinders with radii d_{xy} (in pc). The completeness is set to 1.0 for the cylinder with $d_{xy} = 200$ pc. The completeness decreases slowly for radii which are larger than 250 pc.

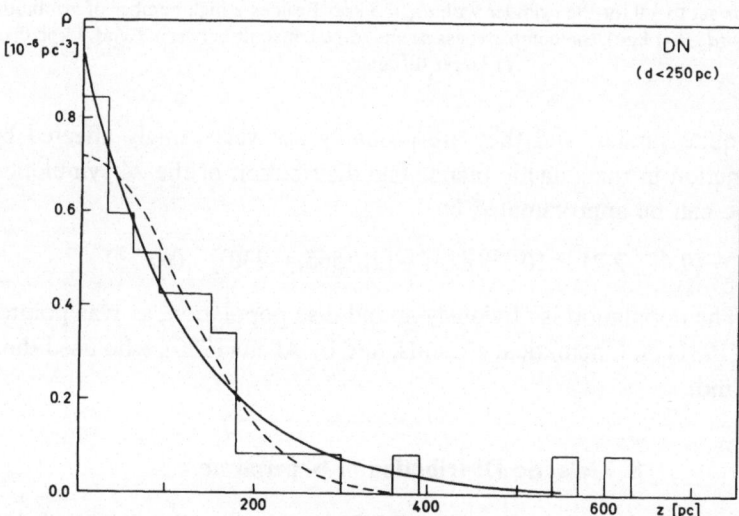

Fig. 4. The galactic distribution of dwarf novae in the z-direction. All dwarf novae with $d_{xy} \leq 250$ pc are included.

d_{xy} = 250 pc was determined. It can be approximated by

$$\rho(z) = (0.95 \pm 0.05) \times 10^{-6} \exp(-|z|/(119 \pm 9)) \quad (pc^{-3}) \tag{11}$$

(Figure 4). Since the value of z_0 is well determined by the objects with z-distances up to 250 pc, it does not have to be corrected for incompleteness.

7. Galactic Distribution of Symbiotic Stars

The catalogue of symbiotic stars by Allen (1979) wa used; the distance estimates were taken from Allen (1980). More than most other groups studied here, the group of symbiotic stars suffers from small number statistics. The completeness and distribution in cylinders with d_{xy} = 1, 2, 3, and 4 kpc was studied (Figure 5). The results for d_{xy} = 2

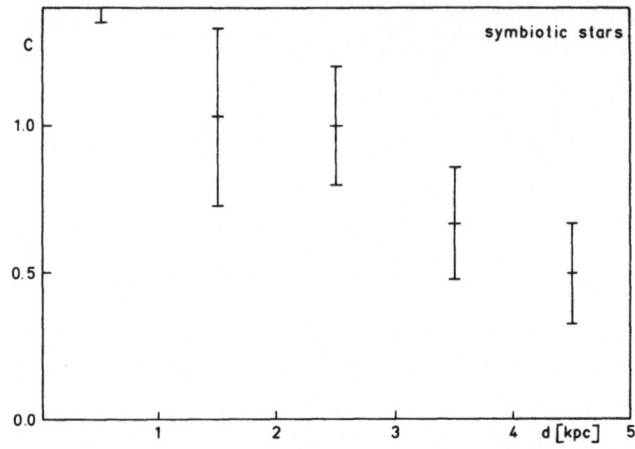

Fig. 5. Completeness of the discovery of symbiotic stars in different cylinders with radii d_{xy} (in kpc). The completeness is set to 1.0 for the cylinder with d_{xy} = 3 kpc. Besides a high number of symbiotic stars in the solar vicinity ($d_{xy} \leqq 1$ kpc), the completeness seems to be constant between 2 and 3 kpc decreases at larger distances.

and 3 kpc are quite similar, and they are probably not yet strongly affected by the interstellar extinction in the galactic plane. The distribution of the 24 symbiotic stars with $d_{xy} \leqq 3$ kpc can be approximated by

$$\rho(z) = (9.4 \pm 2.2) \times 10^{-10} \exp(-|z|/(565 \pm 230)) \quad (pc^{-3}) \tag{12}$$

(see Figure 6). The population is obviously an old disc population, as was pointed out by Wallerstein (1981) on kinematical grounds, and by Allen (1980), who used the same material as we did.

8. Galactic Distribution of Supernovae

No own investigation was carried out. Table VII summarizes the results of different authors.

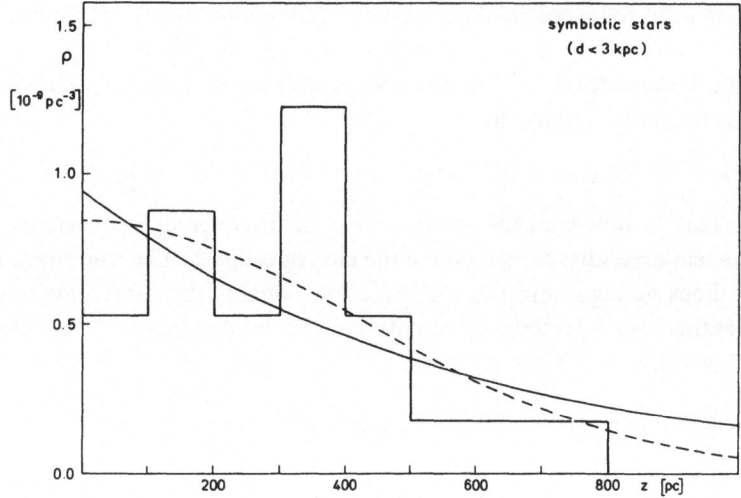

Fig. 6. The galactic distribution of symbiotic stars in the z-direction. All symbiotic stars with $d_{xy} \leqq 3$ kpc are included.

TABLE VII

Previous determinations of the galactic distribution of supernovae

Author	ρ_0 (pc^{-3})	z_0 (pc)	Comments
Kopylov (1957)	–	83	
Johnson and MacLeod (1963)	–	135	(type I)
		68	(type II)
Richter (1968)	–	83	(source: Kopylov (1957))
Tsvetkov (1981)	–	100	(type I and II)

9. Other Types of Binaries and Single Stars

9.1. W UMa SYSTEMS (EW)

In the past, CBs were thought to be descendants of EWs; more recently, this scheme is treated with more reservation because of major theoretical difficulties. We thought, however, that it might be useful to include a study of the galactic distribution of EWs, since more data are now available, as compared with the first comparative studies (Kraft, 1962, 1965; Popov, 1964).

The GCVS and the Finding List for Observers of Interacting Binaries were consulted. It was assumed that systems with known spectral types have luminosities equal to those of Main-Sequence stars of the same spectral types. The luminosity calibration of Schmidt-Kaler (1982) was used. The period is the only known parameter of systems for which no spectral type is available. A period – spectral type relation, similar to the well-known period – colour relation, was established from the data of well-investigated systems and was used for a coarse determination of the luminosities of poorly known systems.

Systems in a cylinder with $d_{xy} = 110$ pc were used for the derivation of the z-distribution, which can be approximated by

$$\rho(z) = (10.7 \pm 0.7) \times 10^{-6} \exp(-|z|/(76 \pm 6)) \quad (\text{pc}^{-3}) \tag{13}$$

(see Figure 7). This is, however, the space density of discovered EW systems. About $\frac{2}{3}$ of the systems remain undiscovered due to the lack of eclipses. The true space density is about three times as high, and the z-distribution remains the same only under the assumption that there is no preferential orientation of orbital planes (e.g., parallel to the plane of the Galaxy).

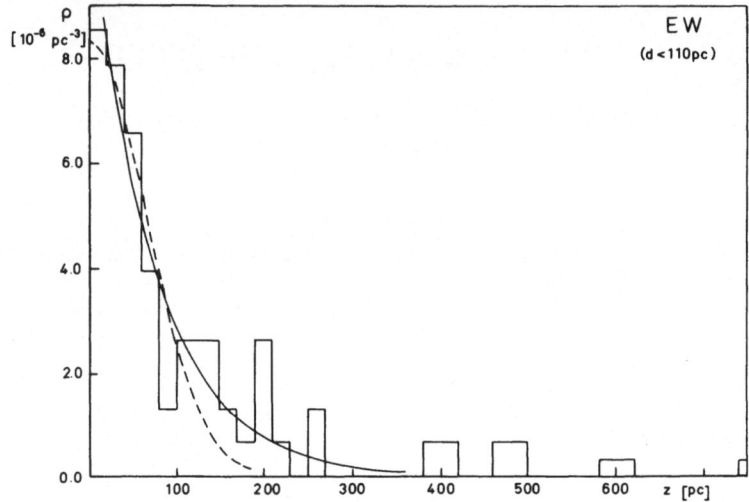

Fig. 7.   The galactic distribution of W UMa type binaries in the z-direction. All systems with $d_{xy} \leqq 110$ pc are included.

It should also be noted that the omission of EW systems with periods $P > 0.5$ days leads to a somewhat different distribution:

$$\rho(z) = (7.5 \pm 0.5) \times 10^{-6} \exp(-|z|/(96 \pm 10)) \quad (\text{pc}^{-3}). \tag{14}$$

Earlier determinations are summarized in Table VIII.

Van 't Veer's (1975) claim of a short lifetime for the EW systems is in good agreement with the small z-value determined above. However, when the distribution (Figure 7) is inspected more closely, a small percentage of old systems seems to be present at high galactic z-distances. The systems VW CVn, VY Cet, SS Com, AQ Com, DD Com, EK Com, BL Leo, AD Phe, and SZ Scl are objects with $z > 200$ pc. Not all of them have been investigated thoroughly so far, but at least half of them have published or unpublished light curves (Hoffmann, 1983), thus a confusion with RRc variables can be excluded. It would certainly be rewarding to carry out a comparative study of these

TABLE VIII

Previous determinations of the galactic distribution of W UMa systems

Author	ρ_0 (pc^{-3})	z_0 (pc)	Comments
Kraft (1962)	–	60*	Improved by Kraft (1965)
Popov (1964)	$\sim 2 \times 10^{-5}$*	144*	
Kraft (1965)	–	62/87*	
Kraft (1967)	10^{-6}	–	0.2% of G stars
van 't Veer (1975)[a]	3.6×10^{-5}	–	Spectral type F; 0.8% of F stars
	7.5×10^{-5}	–	Spectral type G; 1.2% of G stars
	11×10^{-5}	–	All types

* Values marked with an asterisk were derived from the original data, using approximations (5a) and (5b).

[a] van 't Veer (1975) gives percentages of W UMa systems among the F- and G-type Main-Sequence stars; these were converted to absolute space densities, using Upgren's (1963) space densities of Main-Sequence stars.

systems with those in the solar neighbourhood, to find out whether ageing effects are present or whether these systems evolved into contact only a short time ago.

9.2. ALGOL SYSTEMS AND OTHER BINARIES OF LONG PERIOD

In recent theoretical studies, long period Algol systems are regarded as the most probable progenitors of CBs. Systems listed in the GCVS and the Finding List were examined. Data for systems with periods $P < 1\overset{d}{.}5$, $1\overset{d}{.}5 < P < 5^d$, and $P > 5^d$ and spectral types of the primaries earlier than F5 were analyzed separately. Distances were calculated under the assumption that the primaries are Main-Sequence stars (unless other information was supplied by the catalogues). The galactic z-distributions of the

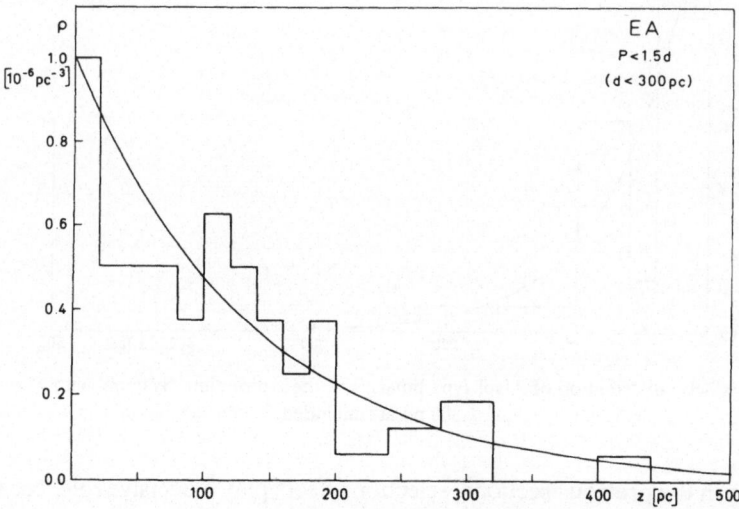

Fig. 8. The galactic distribution of Algol type binaries in the z-direction. Systems with $P < 1.5$ days and $d_{xy} \leqq 300$ pc are included.

systems in cylinders with $d_{xy} \leq 100$, 200, and 300 pc are poorly defined, and the space densities decrease noticeably when the larger volumes are considered. Rough estimates, which should be treated with some reservation, are given in Table IX (see also Figures 8, 9, and 10).

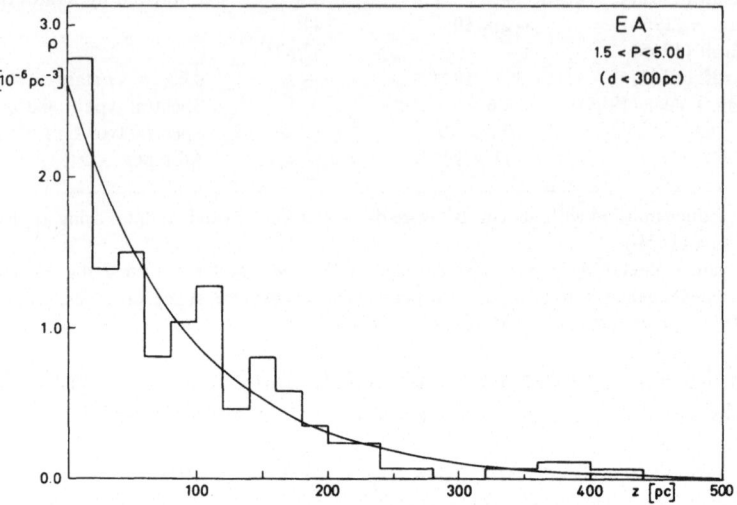

Fig. 9.   The galactic distribution of Algol type binaries in the z-direction. Systems with 1.5 days $< P < 5$ days and $d_{xy} \leq 300$ pc are included.

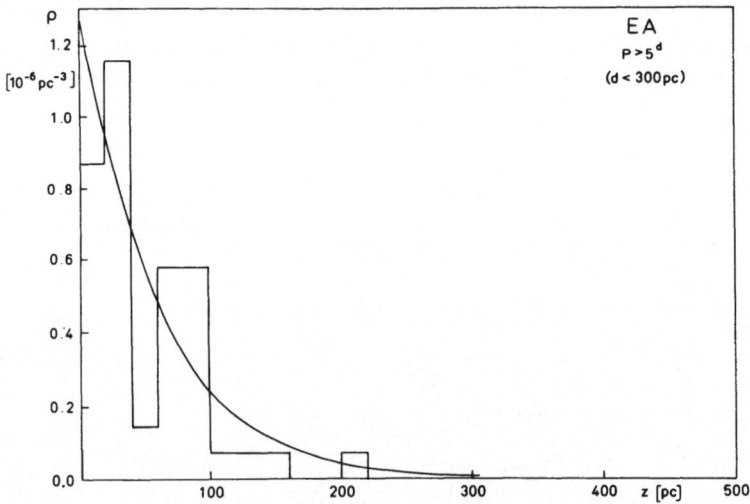

Fig. 10.   The galactic distribution of Algol type binaries in the z-direction. Systems with $P > 5$ days and $d \leq 300$ pc are included.

More than in the previous sections, selection effects play a decisive role, because only systems showing eclipse effects are catalogued. The space densities listed in Table IX have not been corrected for the probability of discovery, and are crude lower limits of

TABLE IX

Galactic distribution of cataclysmic binaries and possibly related types

Group	ρ_0 (pc^{-3})	z_0 (pc)	ρ_0' (pc^{-3})	z_0' (pc)	Comments/reference
N	$(0.38 \pm 0.04) \times 10^{-9}$	125 ± 22	$(0.31 \pm 0.04) \times 10^{-9}$	107 ± 15	novae per year
RN	$(0.8) \times 10^{-9}$	660 ± 330			Warner (1974)
DN	$(0.95 \pm 0.05) \times 10^{-6}$	119 ± 9	$(0.72 \pm 0.03) \times 10^{-6}$	114 ± 6	
SN		135			Johnson and MacLeod (1963)
Symb.	$(0.94 \pm 0.22) \times 10^{-9}$	565 ± 230	$(0.84 \pm 0.14) \times 10^{-9}$	425 ± 85	
EW (all)	$(10.7 \pm 0.7) \times 10^{-6}$	76 ± 6	$(8.3 \pm 0.5) \times 10^{-6}$	65 ± 5	uncorrected
EW ($P < 0^{\mathrm{d}}5$)	$(7.5 \pm 0.5) \times 10^{-6}$	96 ± 10	$(5.5 \pm 0.3) \times 10^{-6}$	93 ± 7	uncorrected
EA ($P < 1^{\mathrm{d}}5$)	$(1.0 \pm 0.2) \times 10^{-6}$	135 ± 20			uncorrected
EA ($P < 5^{\mathrm{d}}$)	$(2.2 \pm 0.3) \times 10^{-6}$	95 ± 15			uncorrected
EA ($P > 5^{\mathrm{d}}$)	$(1.3 \pm 0.3) \times 10^{-6}$	60 ± 15			uncorrected
M I			$(54 \pm 3) \times 10^{-9}$	213 ± 35	Mikami and Ishida (1981)
K0–K4 III			$(40 \pm 3) \times 10^{-6}$	170 ± 17	Mikami and Ishida (1981)
K5–M1 III			$(3.5 \pm 0.4) \times 10^{-6}$	202 ± 5	Mikami and Ishida (1981)
M2–M4 III			$(3.1 \pm 0.3) \times 10^{-6}$	313 ± 11	Mikami and Ishida (1981)
M5–M6 III			$(0.9 \pm 0.1) \times 10^{-6}$	392 ± 15	Mikami and Ishida (1981)
M7–M10 III			$(0.6 \pm 0.2) \times 10^{-6}$	382 ± 14	Mikami and Ishida (1981)
Mira (S, C)	$(25) \times 10^{-9}$	150 ± 25			Richter (1968)
Mira ($P > 400^{\mathrm{d}}$)	$(12.5 \pm 7.5) \times 10^{-9}$	300 ± 70			Richter (1968)
Mira ($P > 350^{\mathrm{d}}$)	$(20 \pm 10) \times 10^{-9}$	300 ± 30			Richter (1968)
Mira ($P > 300^{\mathrm{d}}$)	$(6.3 \pm 3.0) \times 10^{-9}$	320 ± 30			Richter (1968)
Mira ($P > 250^{\mathrm{d}}$)	$(6.3 \pm 4.7) \times 10^{-9}$	640 ± 140			Richter (1968)
Mira ($P > 200^{\mathrm{d}}$)	$(2.5 \pm 1.5) \times 10^{-9}$	430 ± 70			Richter (1968)
Mira ($P > 150^{\mathrm{d}}$)	$(2.5 \pm 0.5) \times 10^{-9}$	690 ± 150			Richter (1968)
Mira ($P > 90^{\mathrm{d}}$)		450 ± 100			Richter (1968)
Plan. Neb.	$(0.05) \times 10^{-9}$	120 ± 20			Cahn and Wyatt (1978)

Comments: ρ_0, z_0 are derived by assuming a density law of the form (5a), ρ_0', z_0' are derived by assuming a density law of the form (5b).

the true space densities. The decreasing z_0-values for systems of longer period can be explained either by different mean ages of the considered groups, or, more likely, as an indication of a preferential orientation parallel to the galactic plane.

It would certainly be rewarding to carry out a more detailed study, which should also take into account the evolutionary status of the EA systems (e.g., Budding's (1983) EA1/EA2, Kholopov's (1981) DM/DS, or Svechnikov *et al.*'s (1980) RGP/PR/RS groups).

The GCVS becomes a completely insufficient source list when systems with periods of the order of months or years are evaluated: the probability of eclipses is low, and the true space density is much higher than the observed one. A study based on the *Catalogue of Spectroscopic Orbits* (Batten *et al.*, 1978) was carried out. It comprised systems with $200^{\mathrm{d}} \leq P \leq 1000^{\mathrm{d}}$ and evolved (giant) primaries. It was noted that the incompleteness of data for systems with distances > 80 pc makes it impossible to derive the z-distribution. The observed ρ_0 of these systems in the solar neighbourhood is of the order 2.5×10^{-5} pc^{-3}. Because of effects of inclination, the true density is somewhat larger.

The space density of this special group of binaries is of the same order of magnitude as that of the EW systems, only the probability of eclipses is very much reduced. This should be kept in mind when the estimate of Shapley (1948) concerning the frequency of EW systems is quoted. The argument that only EWs and CBs have similar, high space densities (Kraft, 1965) is not valid; the space density of EA systems with giant primaries and periods between 200 and 1000 days is several times higher than that of the CBs.

9.3. OTHER TYPES OF STARS

No new studies of the galactic distribution and space density of other types of objects were carried out. Data for Mira variables were taken from Richter (1968); data for late type giants and supergiants were taken from Mikami and Ishida (1981). Results are shown in Table IX.

10. Discussion of the Results

10.1. STATISTICAL TESTS FOR EVOLUTIONARY DEPENDENCES

When a group of stars evolves from state i into state k, the following relations hold for a Galaxy with constant rates of star formation:

$$z_{0,i} \lesseqgtr z_{0,k} \tag{15}$$

$$t_i/\Sigma_i = t_k/\Sigma_k \tag{16}$$

where t_i is the lifetime in state i and Σ_i the surface density of objects of type i, as defined in (10).

10.2. RELATIONS BETWEEN NOVAE, DWARF NOVAE, AND SUPERNOVAE OF TYPE I

Let us first examine the possible relationships between different groups of CBs and their connection with SN I.

Novae and dwarf novae have similar z-distributions, the space density of N is about $\frac{1}{10}$ that of DN. The galactic distribution of these objects indicates that, at least in the solar neighbourhood, the majority of these systems belongs to the older Population I.

In addition to the majority of young objects, a few objects of old age seem to exist. Detached evolved binary systems like V471 Tau and AA Dor (LB 3459) might represent potential CBs that will enter their active state only after billions of years when gravitational radiation or evolutionary expansion of the secondary will transform the systems into semi-detached ones.

Let us review the observational evidence for a nonvanishing fraction of Population II CBs (see also Webbink, 1980):

(a) Nova T Sco in the globular cluster M80 (NGC 6093). It is a fast nova ($t_3 = 22^d$) of light curve type A (smooth), and has an absolute magnitude in good agreement with that of other novae of type A, when it is placed at the cluster distance (Sawyer, 1938; Duerbeck, 1981).

(b) Nova Oph (1938) in the globular cluster M14 (NGC 6402). Its light curve is poorly covered, but it is possibly a slow nova. With an observed maximum brightness of 16^m, it could have been as bright as 10^m at true maximum, corresponding to $M = -6^m$. Membership is likely (Sawyer-Hogg and Wehlau, 1964).

(c) V1148 Sgr near the globular cluster NGC 6553. It is not close to the cluster, it is poorly observed, and membership is unlikely (Mayall, 1949).

Several candidates for dwarf novae in globular clusters are listed by Webbink (1980). A recently found object is V101 in the globular cluster M5 (NGC 5904) (Margon *et al.*, 1981). Assuming cluster membership, their absolute magnitudes are, in general, unusually bright ($M^{max} = +2^m.7$ for V101), and it is not unlikely that they are foreground objects. Deeper searches, which should record stars of at least $m = 19^m$ are necessary to look successfully for DN candidates at maximum light in those globular clusters which are relatively close to the Sun.

A similar dichotomy between young and old members of the class exists in the case of planetary nebulae: formerly assigned to Population II (Oort, 1958), they show a strong concentration towards the galactic plane, with $z_0 = 120$ pc (Cahn and Wyatt, 1978), and a much rarer occurrence in Population II aggregates, e.g., the object Küstner 648 in the globular cluster M15 (NGC 7078).

Concerning the scheme CB → SN I, we note that

$$z_{0, \text{DN}} < z_{0, \text{SN I}},$$

using the z-distribution of galactic SN I, as determined by Johnson and MacLeod (1963). Most DN are Population I objects, thus most SN I will also be Population I objects. The existence of SN I in elliptical galaxies can be explained by the fact that N and probably also DN occur in Population II stellar aggregates.

While relation (15) is thus fulfilled, (16) cannot be checked because of lack of data. However, some conclusions can be drawn from the death rate of cataclysmic binaries.

An estimate of the lifetime of a CB can be derived from the following arguments: dwarf novae show mean mass transfer rates of $\sim 1.5 \times 10^{-9} M_\odot \text{ yr}^{-1}$ (Smak, 1982), and mean secondary masses of $\sim 0.5 M_\odot$. Assuming that all DN start with $M_{\text{sec}} = 1 M_\odot$, and that they use up the total mass of the secondary during the DN state, the maximum length of the active state is 6×10^8 yr. It can, however, not be excluded that DN switch, during the course of their evolution, to the nova state and vice versa (Duerbeck, 1980). Novae have mean mass transfer rates that are higher by a factor 10, leading under the above assumptions to a lifetime of 6×10^7 yr. If both types of variability occur in identical systems, it can be concluded from the observed space densities that half of the mass is transferred in the low (DN) state, half during the high (N) state.

The lifetimes of CBs are thus at most between 6×10^7 and 6×10^8 yr. For the following discussion, a mean lifetime of 2×10^8 yr is adopted.

The total number of DN in the Galaxy can be calculated by Parenago's formula (Richter, 1968). Using the space density ρ_0 and the scale-height z_0 as determined in (11) and the radial density gradient $\partial \log \rho / \partial R = -0.24$ (Richter, 1968), Parenago's formula yields 1.1×10^6 DN. With a lifetime of 2×10^8 yr, the death rate, one DN in 180 yr, is not very far from the estimate of the SN I frequency in the Galaxy.

10.3. CATACLYSMIC BINARIES AS DESCENDANTS OF UNEVOLVED BINARIES

So far, it is not possible to prove, by statistical arguments, evolutionary schemes including normal binaries and CBs. We will, however, collect the statistical evidence.

Case 1: EW → CB

$z_{0, EW} < z_{0, CB}$ is valid; in addition, an old population of EWs seems to exist, corresponding to Population II novae (see Sections 4, 6, 9.1, and 10.2).

For the application of Equation (16), estimates of the surface densities and lifetimes must be available. With novae $\frac{1}{10}$ as frequent as dwarf novae, the surface densities are:

$$\Sigma_{EW} = 49 \times 10^{-4}\, pc^{-2}; \qquad \Sigma_{CB} = 2.5 \times 10^{-4}\, pc^{-2}.$$

With $t_{CB} = 2 \times 10^8$ yr, the lifetime in the EW state is

$$t_{EW} = 3.3 \times 10^9\, yr.$$

The strong concentration of EW systems towards the galactic plane suggests a slightly lower value for the mean age of the (young) W UMa population, about 10^9 yr (Scheffler, 1983). In view of the uncertainties in the derivation, the agreement is surprisingly good, and the result is also not in conflict with most theoretical predictions (see, e.g., Eggleton *et al.*, 1976). It seems worthwhile to further revive and substantiate the scheme EW → CB.

Concerning the older W UMa generation and the occurrence of novae in globular clusters, the following should also be considered:

EW binaries are tightly bound and should survive in globular clusters where gravitational interaction between the cluster members is noticeable. These EW systems would be good candidates for progenitors of CBs in globular clusters. Systems with larger separation going through case B or C of binary evolution are more likely subject to disruption; if these systems are the progenitors of CBs, the percentage of CBs in globular clusters should be noticeably smaller than in the field (of similar age).

A careful search for CBs and eclipsing binaries of all types in globular clusters is a necessary prerequisite for the establishment of evolutionary sequences.

Case 2: EA → CB

$z_{0, EA} < z_{0, CB}$ is valid, but because of a possible preferential orientation of orbital planes, this relation is not conclusive (see Section 9.2).

Equation (16) is difficult to apply because of the unknown value of z_0 in the case of spectroscopic binaries or the uncertain value of ρ_0 in the case of eclipsing binaries. Assuming $z_{0, CB} = z_{0, EA}$, the surface densities can be replaced by the space densities, and we obtain, using the space density of spectroscopic binaries:

$$t_{EA} = 2 \times 10^8 \times 2.5 \times 10^{-5}/1.2 \times 10^{-6}. \tag{17}$$

If we introduce a factor f, the fraction of Algol systems which evolve into CBs is

$$t_{EA} = f \times 4 \times 10^9 \, \text{yr} . \tag{18}$$

It is assumed that the lifetime of the giant components, for which our statistical study has been carried out, is of the order of 3×10^7 yr (Meyer-Hofmeister, 1982), the factor f can be determined:

$$f = 7 \times 10^{-3},$$

i.e., less than 1% of all Algol systems with giant components and periods between 200 and 1000 days must undergo a spiralling-in in the course of their evolution in order to account for the observed space density of CBs. This very attractive result needs, however, still much more confirmation both from the theoretical and the stellar-statistical side before it can be used as an argument in favour of the scheme EA → CB.

10.4. RECURRENT NOVAE AND SYMBIOTIC STARS

Finally, a few words on recurrent novae and symbiotic stars are appropriate. These objects seem to have similar z-distributions, those of an older disc population. The (not unequivocally accepted) models of both groups, binary systems consisting of a giant and an accreting (white?) dwarf, are also very similar. The red giants in both groups are sometimes long-period variables (RS Oph, RX Pup, V1016 Cyg). The galactic distribution of both groups is similar to that of M giant stars (Mikami and Ishida, 1981) and Mira variables of comparable period (Richter, 1968).

The evolutionary scheme (e.g., Rudak, 1982), that a close, long-period binary having undergone case C evolution without the formation of a common envelope and thus having avoided a spiralling-in, will become a symbiotic star when the secondary evolves through the giant, or, sometimes, the Mira stage, is in agreement with the galactic distribution.

11. Conclusion

From the study of the galactic distribution it is found that

> novae, dwarf novae, perhaps supernovae of type I

and

> recurrent novae and symbiotic stars

form two distinct groups.

Both schemes EW → CB and EA → CB are not in disagreement with the galactic distributions of these systems. The importance of observations of these types of stars in globular clusters is stressed. A summary of the newly determined galactic distributions of cataclysmic binaries and related binary and single stars is given in Table IX.

Acknowledgements

I thank Messrs F. Th. Lentes and B. Nelles for their untiring computational help, and the colleagues of the Universitätssternwarte München for their interest in an earlier presentation of this topic as a colloquium. The participation at IAU Colloquium No. 80 was made possible by a travel grant of the Deutsche Forschungsgemeinschaft, Bonn-Bad Godesberg (477/832/83(2)).

References

Allen, C. W.: 1954, *Monthly Notices Roy. Astron. Soc.* **114**, 387.
Allen, C. W.: 1973, *Astrophysical Quantities*, 3rd edition (and previous editions), Athlone Press, London.
Allen, D. A.: 1979, in F. M. Bateson *et al.* (eds.), 'Changing Trends in Variable Star Research', *IAU Colloq.* **46**, 125.
Allen, D. A.: 1980, *Monthly Notices Roy. Astron. Soc.* **192**, 521.
Baade, W.: 1958, in D. J. K. O'Connell (ed.), 'Stellar Populations', *Ric. Astron. Spec. Vatican* **5**, 165.
Bailey, J.: 1981, *Monthly Notices Roy. Astron. Soc.* **197**, 31.
Barlow, M. J., Brodie, J. P., Brunt, C. C., Hanes, D. A., Hill, P. W., Mayo, S. K., Pringle, J. E., Ward, M. J., Watson, M. G., Whelan, J. A. J., and Willis, A. J.: 1981, *Monthly Notices Roy. Astron. Soc.* **195**, 61.
Batten, A. H., Fletcher, J. M., and Mann, P. J.: 1978, *Publ. Dominion Astrophys. Obs. Victoria* **15**, 121.
Bruch, A.: 1982, Ph.D. thesis, Münster University.
Bruch, A., Duerbeck, H. W., and Seitter, W. C.: 1981, *Mitt. Astron. Ges.* **52**, 34.
Budding, E.: 1983, *Astrophys. Space Sci.* **99**, (this issue).
Cahn, J. H. and Wyatt, S. P.: 1978, in Y. Terzian (ed.), 'Planetary Nebulae', *IAU Symp.* **76**, 3.
Dinerstein, H. and Hoffleit, D.: 1973, *Inf. Bull. Var. Stars*, No. 845.
Duerbeck, H. W.: 1980, Habilitation thesis, Bonn.
Duerbeck, H. W.: 1981, *Publ. Astron. Soc. Pacific* **93**, 165.
Duerbeck, H. W. and Seitter, W. C.: 1979, *ESO Messenger* **17**, 1; see also *Mitt. Astron. Ges.* **50**, 90 (1980).
Duerbeck, H. W. and Seitter, W. C.: 1982, in K. Schaifers and H. H. Voigt (eds.), *Landolt-Börnstein N.S.* **VI/2**, 197.
Eggleton, P. P.: 1983, in M. Livio and G. Shaviv (eds.), 'Cataclysmic Variables and Related Objects', *IAU Colloq.* **72**, 239.
Eggleton, P. P., Mitton, S., and Whelan, J. A. J. (eds.): 1976, in 'Structure and Evolution of Close Binary Stars', *IAU Symp.* **73**.
Garrison, R. F., Hiltner, W. A., and Schild, R. E.: 1982, *IAU Circ.*, No. 3730.
Hoffmann, M.: 1983, private communication.
Hudec, R.: 1981, *Bull. Astron. Inst. Czech.* **32**, 93.
Iwanowska, W. and Burnicki, A.: 1962, *Bull. Acad. Pol. Sci. Ser. Sci. Math. Astron. Phys.* **10** (10), 537.
Johnson, H. M. and MacLeod, J. M.: 1963, *Publ. Astron. Soc. Pacific* **75**, 123.
Kamper, K. W.: 1979, in H. M. van Horn and V. Weidemann (eds.), 'White Dwarfs and Variable Degenerate Stars', *IAU Colloq.* **53**, 494.
Kholopov, P. N.: 1981, *Perem. Zvezdy* **21**, 465.
Kopylov, I. M.: 1957, in G. H. Herbig (ed.), 'Non-Stable Stars', *IAU Symp.* **3**, 71.
Kraft, R. P.: 1962, *Astrophys. J.* **135**, 408.
Kraft, R. P.: 1965, *Astrophys. J.* **142**, 1588.
Kraft, R. P.: 1967, *Publ. Astron. Soc. Pacific* **79**, 395.
Kraft, R. P. and Luyten, W.: 1965, *Astrophys. J.* **142**, 1041.
Kukarkin, B. W.: 1954, *Untersuchung der Struktur und Entwicklung der Sternsysteme auf der Grundlage des Studiums veränderlicher Sterne*, Akademie-Verlag, Berlin.
Law, W. Y. and Ritter, H.: 1983, *Astron. Astrophys.* **123**, 33.
Margon, B., Downes, R. A., and Gunn, J. E.: 1981, *Astrophys. J.* **247**, L89.
Mayall, M. W.: 1949, *Publ. Astron. Soc. Pacific* **54**, 191.
Meyer-Hofmeister, E.: 1982, in K. Schaifers and H. H. Voigt (eds.), *Landolt-Börnstein N.S.* **VI/2b**, 152.

Mikami, T. and Ishida, K.: 1981, *Publ. Astron. Soc. Japan* **33**, 135.

Moffat, A. F.: 1982, private communication.

Neckel, Th.: 1967, *Veröff. Landssternwarte Heidelberg* **19**.

Neckel, Th. and Klare, G.: 1980, *Astron. Astrophys. Suppl.* **42**, 251.

Oort, J. H.: 1958, in D. J. K. O'Connell (ed.), 'Stellar Populations', *Ric. Astron. Spec. Vatican.* **5**, 415.

Paczynski, B.: 1976, in P. Eggleton, S. Mitton, and J. A. J. Whelan (eds.), 'Structure and Evolution of Close Binary Systems', *IAU Symp.* **73**, 27.

Paczynski, B.: 1980, in P. A. Weyman (ed.), *Highlights of Astronomy 5*, D. Reidel Publ. Co., Dordrecht, Holland, p. 27.

Parenago, P. P.: 1955, in 'Principes fondamentaux de classification stellaire', *Colloque International du CNRS*, CNRS, Paris, p. 13.

Payne-Gaposchkin, C. H.: 1977, in M. Friedjung (ed.), *Novae and Related Stars*, D. Reidel Publ. Co., Dordrecht, Holland, p. 3.

Plaut, L.: 1965, in A. Blaauw and M. Schmidt (eds.), *Stars and Stellar Systems V*, University of Chicago Press, Chicago, p. 311.

Popov, M. V.: 1964, *Perem. Zvezdy* **15**, 115.

Primkulov, Sh.: 1968, *Tsirk. Astron. Tashkent* **8**, No. 255.

Richter, G.: 1968, *Veröffentl. Sternw. Sonneberg* **7**, 229.

Richter, G.: 1983, *Inf. Bull. Var. Stars*, No. 2267.

Ritter, H.: 1976, *Monthly Notices Roy. Astron. Soc.* **175**, 279.

Ritter, H.: 1983, *Catalogue of Cataclysmic Binaries*, MPI Astrophysik Garching, (MPA-51).

Robinson, E. L.: 1975, *Astron. J.* **80**, 515.

Rosino, L.: 1973, *Astron. Astrophys. Suppl.* **9**, 347.

Rudak, B.: 1982, in M. Friedjung and R. Viotti (eds.), 'The Nature of Symbiotic Stars', *IAU Colloq.* **70**, 275.

Sahade, J.: 1959, Liège Symp. No. 9, 'Modèles d'étoiles et évolution stellaire', *Mém. Soc. R. Liège Sér.* **5**, t. 3, p. 76.

Sawyer, H.: 1938, *J. Roy. Astron. Soc. Canada* **32**, 69.

Sawyer-Hogg, H. and Wehlau, A.: 1964, *J. Roy. Astron. Soc. Can.* **58**, 163.

Scheffler, H.: 1983, in K. Schaifers and H. H. Voigt (eds.), *Landolt-Börnstein N.S.* **VI/2c**, 175.

Schmidt-Kaler, Th.: 1957, *Z. Astrophys.* **41**, 182.

Schmidt-Kaler, Th.: 1962, *Kl. Veröffentl. Remeis-Sternw. Bamberg* **34**. 109.

Schmidt-Kaler, Th.: 1982, in K. Schaifers and H. H. Voigt (eds.), *Landolt-Börnstein N. S.* **VI/2b**, 1.

Shapley, H.: 1948, *Harvard Obs. Monograph* **7**, 249.

Shu, F. H.: 1980, in M. J. Plavec, D. M. Popper, and R. K. Ulrich (eds.), 'Close Binary Stars: Observations and Interpretation', *IAU Symp.* **88**, 477.

Smak, J.: 1982, *Acta Astron.* **32**, 213.

Starrfield, S., Truran, J. W., and Sparks, W. M.: 1981, *Astrophys. J.* **243**, L27.

Svetchnikov, M. A., Istomin, L. F., and Grehova, O. A.: 1980, *Perem. Zvezdy* **21**, 399.

Trimble, V.: 1983, *Nature* **303**, 137.

Tsvetkov, D. Yu.: 1981, *Soviet Astron. Letters* **7**, 254.

Upgren, A. R.: 1963, *Astron. J.* **68**, 475.

Van 't Veer, F.: 1975, *Astron. Astrophys.* **40**, 167.

Vasilevskis, S., Harlan, E. A., Klemola, A. R., and Wirtanen, C. A.: 1975, *Publ. Lick Obs.* **22**, Pt. 5.

Vidal, N. V. and Rodgers, A. W.: 1974, *Publ. Astron. Soc. Pacific* **86**, 26.

Vilhu, O.: 1981, *Astrophys. Space Sci.* **78**, 401.

Voikhanskaya, N. F.: 1973, *Astrofiz. Issled* **5**, 89.

Vogt, N.: 1981, Habilitation thesis, Bochum University.

Wallerstein, G.: 1981, *Observatory* **101**, 172.

Warner, B.: 1974a, *Monthly Notices Roy. Astron. Soc.* **167**, 61p.

Warner, B.: 1974b, *Monthly Notices Astron. Soc. Southern Africa* **33**, 21.

Webbink, R. F.: 1976, *Astrophys. J.* **209**, 829.

Webbink, R. F.: 1978, *Publ. Astron. Soc. Pacific* **90**, 57.

Webbink, R. F.: 1980, in M. J. Plavec, D. M. Popper, and R. K. Ulrich (eds.), 'Close Binary Stars: Observations and Interpretation', *IAU Symp.* **88**, 561.

Wenzel, W. and Meinunger, I.: 1978, *Astron. Nachr.* **299**, 237.

Yungelson, L. R.: 1982, in J. Tremko (ed.), *Ejection and Accretion of Matter in Binary Systems*, Veda, Bratislava, p. 11.

A COMPARISON BETWEEN BINARY STAR LIGHT CURVES
AND THOSE OF POSSIBLE BINARY ASTEROIDS*

H. J. SCHOBER

Institut für Astronomie, Graz, Austria

(Received 1 August, 1983)

Abstract. Since about ten years coordinated programs of photoelectric observations of asteroids are carried out to derive rotation rates and light curves. Quite a number of those asteroids exhibit features in their light curves, with similar characteristics as variable stars and especially eclipsing binaries. This would allow also an interpretation that there might be an evidence for the binary nature of some asteroids, based on observational hints. A few examples are given and a list of indications for the possible binary nature of asteroids, based on their light curve features, is presented.

1. Introduction

Right from the beginning, I should state that if we observe asteroids, variable stars, or eclipsing stars, using photoelectric photometry, there is no difference in the technical procedure to do that – or in reducing the photometric data. And as we shall see, asteroid light curves easily can be interpreted as binary light curves in a number of cases.

In the last *Ephemerides of Minor Planets for 1983* the number of asteroids with known orbits was given with about 2500. Roughly the major part of orbits is found to be in between Mars and Jupiter (semi-major axis $a = 1.52$ AU and $a = 5.20$ AU), but meanwhile we do know a few asteroids with exceptional orbits – e.g., closely approaching the Earth as 1566 Icarus with $a = 1.08$ AU, 1620 Geographos, $a = 1.24$ AU or 433 Eros with $a = 1.46$ AU. A group of new discovered objects are called Atens and they have orbits completely inside of that of the Earth (2062 Aten, $a = 0.97$ AU or 2100 Ra-Shalom with $a = 0.83$ AU). Further out of the general system of asteroids there is 944 Hidalgo at $a = 5.79$ AU, but with a very eccentric orbit, and far from that a new type of asteroid (or planet?) 2060 Chiron with a semi-major axis of $a = 13.70$ AU was discovered; this either proves that we are starting to see another 'asteroid belt' between Jupiter and Saturn, or that there are more planets than we do know; this would be a matter of where to put the lower limit for the diameter of a solar-system planetary object, of course.

From the physical observations of asteroids since about 1970, when we started really to consider physics of asteroids to be more important, photometry and spectrophotometry was applied and a new picture was developed; compositional types could be defined as C (carbanaceous, dark), S (stony, metallic, high albedo), M (metallic), E (enstatic), R (reddish), and U (unknown, unusual) as given by Bowell *et al.* (1978). Also the bimodality due to the albedo for C- and S-types was established. The distribution

---

* Paper presented at the Lembang-Bamberg IAU Colloquium No. 80 on 'Double Stars: Physical Properties and Generic Relations', held at Bandung, Indonesia, 3–7 June, 1983.

Astrophysics and Space Science **99** (1984) 387–392. 0004–640X/84/0992–0387$00.90.

in different orbital zones, their correlation with diameters, rotational periods, light curve amplitudes as an indication for the irregularities of the geometric forms might be a basic clue to the understanding of the origin and formation of the asteroid belt as well as possibly for the solar system in future.

From the combination of radiometry in the infrared, polarimetry, and *UBV* photometry in addition we know diameters and albedos of about 800 asteroids, now (1500 with good *UBV*-data), among them approximately 250 with well determined diameters of 10% accuracy, as listed by Bowell *et al.* (1979). Spectrophotometric data in a 11-filter system we do have for about 300 asteroids.

From photoelectric light-curve observations, usually done in *UBV*, we have now a statistical sample of 350 (300 well-) known rotation periods of asteroids, ranging between 2 hr up to 145 hr (six days) with an asteroid, which probably has 48 days of rotation period; a histogram of rotation rates was given by Schober *et al.* (1982). The light variations are mostly due to irregular shape, albedo differences on the surface for nearly spherical object like 1 Ceres, but there are also very elongated (or binary?) asteroids like 1620 Geographos with a cigar-shape form of 1 : 6 axial ratio. Also the spin axis orientation in space was measured, but only for a very small number of asteroids.

Usually the light curves of asteroids show a double wave characteristic, like eclipsing binaries, especially close ones, with two maxima M1, M2, and two minima m1, m2, sometimes with additional features in the light curves. A typical example is shown in

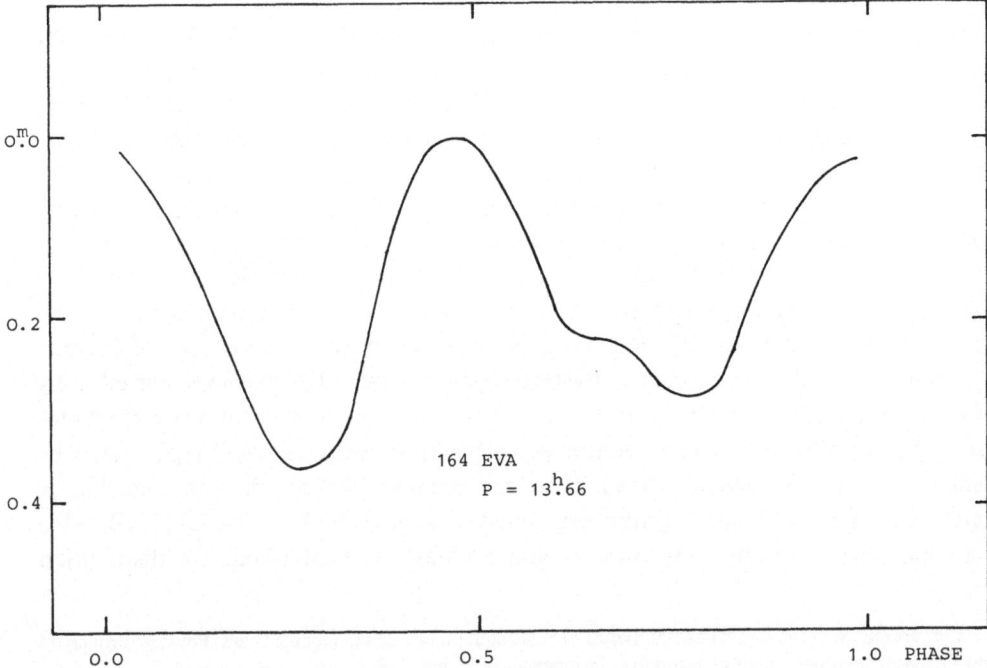

Fig. 1.   Typical light curve of a rotating asteroid (164 Eva), which also can be interpreted as a binary-object light curve.

Figure 1 with the light curves of the asteroid 164 Eva, obtained by Schober (1982), which easily can be interpreted as a binary-star light curve as well.

2. Existence of Binary Asteroids or Satellites of Asteroids?

Up to the present date there never was discovered a binary asteroid or a satellite of an asteroid on photographic plates. But, of course, in the past there was no need to look systematically for such a phenomenon; and searches initiated in the last years also were not successful.

The first indications for a possible satellite of an asteroid were given in an occultation report by Bowell *et al.* (1978). Before and after the occultation of the bright SAO star 120774 on 7 June, 1978 by the asteroid 532 Herculina with 20.6 s duration, secondary events within two minutes of each 0.5–4.0 s duration were reported, and observed photoelectrically – but only two degrees above the horizon. A model was given, that the major-secondary occultation was caused by a satellite: 532 Herculina with a mean diameter of 220 km should have a satellite with 50 km diameter in a distance of about 1000 km from the asteroid center. This distance corresponds to about 0.7 arc sec if observed from the Earth at 2 AU.

Evidences for satellites were found from occultations also for other asteroids, e.g., 18 Melpomene, 2 Pallas, or 65 Cybele (?), but the occultation measurements are difficult to obtain, to organize and to predict for a certain observatory. Observatories, where precise measurements can be made are not distributed well enough, so a real confirmation of a binary asteroid or a satellite by direct observations never was possible.

3. Evidences for the Binary Nature of Asteroids from Photoelectric Light Curves

In the last years we have given high priority to observe light curves of asteroids to derive rotation rates for the statistical purposes. But it turns out that this kind of observation still delivers the best indication to detect possible binary objects. On the other hand, also the number of asteroids already observed is rather large now, a few of them were observed even during several oppositions.

Binary asteroids, or satellites of asteroids with a relative large size compared with the major body, should show some details in the light curves, if observed in a suitable geometric configuration between Earth-Sun-asteroid-satellite. Striking similarities of light curves were stated by van Flandern *et al.* (1979); the light curve of 433 Eros is similar to that of β Lyrae type variables, that of 44 Nysa is strikingly similar to the contact binary light curve of W Ursae Majoris stars. Algol like light curves of 171 Ophelia, 49 Pales, or 46 Hestia are remarked, with flat constant parts in the extrema and sometimes with very deep and sharp minima in the light curve. This has drawn attention to the fact that asteroids also can be interpreted as binary systems.

Van Flandern *et al.* (1979) have summarized the problem of satellite of asteroids and gave a list of six points for asteroid light curves (1)–(6), where the possibility is high to have observed two bodies instead of a single one during a rotational cycle. Based on my own observing experience I have added four more indications (7)–(10) as a hint for possible binary nature:

(1) light curve maxima sharper than minima: e.g., 129 Antigone;

(2) complex light curves: e.g., 24 Themis, 29 Amphitrite, 51 Nemausa;

(3) increase in light curve amplitude with increasing solar phase angle: e.g., 349 Dembowska, 944 Hidalgo;

(4) light curves with two maxima and minima per rotational cycle at one opposition, but only one of each at another: e.g., 532 Herculina;

(5) triple maxima and minima in the light curve per rotation cycle: e.g., 1580 Betulia, 337 Devosa, 37 Fides (quadruple ?), 51 Nemausa; assuming bound rotation;

(6) contact binary light curves: e.g., 44 Nysa;

(7) nonperiodic irregular features in the light curves, or showing up with periods different from the rotation rate: e.g., 37 Fides, 423 Diotima;

(8) color variation during rotation, if not interpreted as spotted surface: e.g., 48 Doris;

(9) slowly spinning asteroids with rotation periods of 2–6 days instead of the usual range of hours: e.g., 1689 Floris-Jan (possible tidal evolution of binary systems);

(10) large light curve amplitudes (1–2 mag.): e.g., 433 Eros, 216 Kleopatra.

In Figure 2 there is shown the example of the triple light curve of 51 Nemausa, a very important object for which an international campaign is running, as obtained by Gammelgaard and Kristensen (1983). In Figure 3 the most recent example for a possible

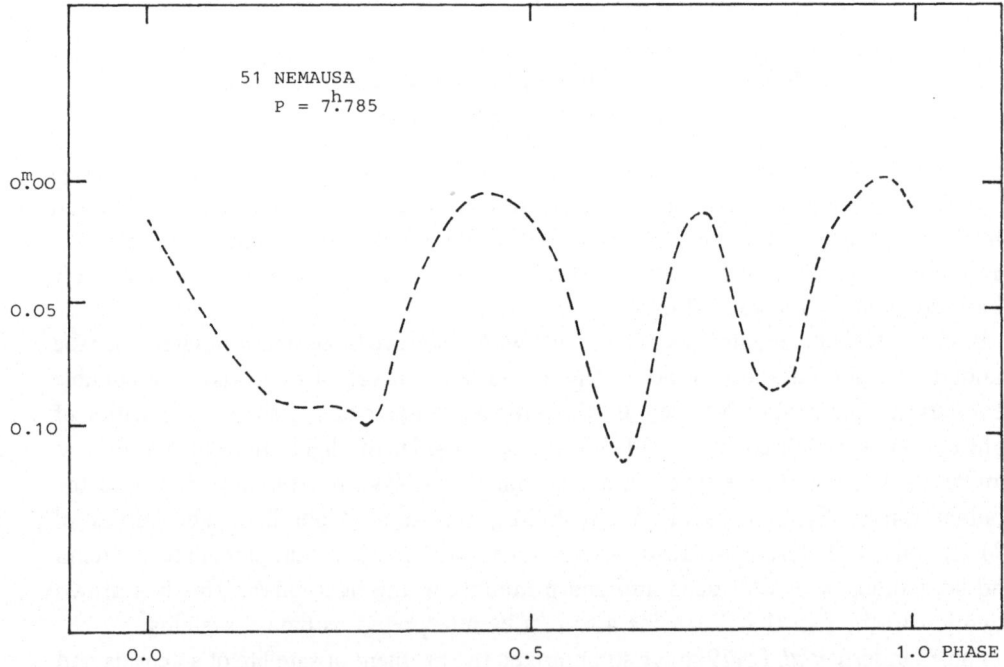

Fig. 2.   The complex light curve of the asteroid 51 Nemausa with triple extrema.

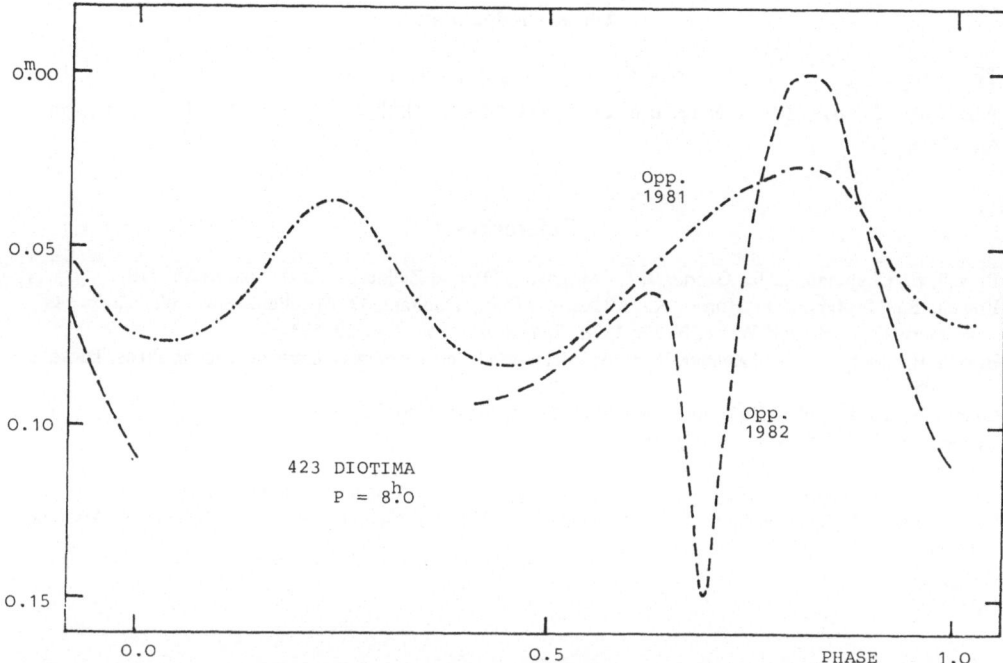

Fig. 3. Light curves of the asteroid 423 Diotima obtained during two different oppositions; the additional 'absorption' feature might give an indication for the possible detection of a satellite of 423 Diotima.

satellite is shown for 423 Diotima, as obtained by Schober (1983), where an additional 'absorption' is remarked in a different opposition, which looks like an occultation of the major body of 423 Diotima by a satellite.

4. Concluding Remarks

Of course, the presented list of hints does supply only indications for a possible binary nature of asteroids; multiple asteroids cannot be observed directly with convential techniques, but might be detected in future using the space telescope or the astrometric satellite Hipparcos; but also more sophisticated methods can be used, such as speckle interferometry in all available wavelengths (the best would be in infrared), radar scatter measurements – or excellently organized occultation-measurements (weather conditions and participation must be granted!), if the worldwide interest would be high enough.

Due to van Flandern *et al.* (1979) the gravitational boundary sphere before perturbations begin to dominate, should be roughly hundred times the mean diameter of the asteroid; this would give for 532 Herculina a distance of about 25 000–30 000 km, e.g.

In the future high preference should be given to observe large asteroids in detail for satellites, and small asteroids for binary nature. They could be very elongated solid bodies or 'ribble-piles', but also 'semi-detached binaries' as it might be concluded on the base of their light curves.

Acknowledgement

The project of searching possible binary asteroids is supported financially by the Austrian 'Fonds zur Förderung der Wissenschaftlichen Forschung' under project No. P4852.

References

Bowell, E., Chapman, C. R., Gradie, J. C., Morrison, D., and Zellner, B.: 1978, *Icarus* **35**, 313.

Bowell, E., McMahon, J., Horne, K., A'Hearn, M. F., Dunham, D. W., Penhallow, W., Taylor, G., Wasserman, L. H., and White, N. M.: 1978, *Bull. Am. Astron. Soc.* **10**, 594.

Bowell, E., Gehrels, T., and Zellner, B.: 1979, in T. Gehrels (ed.), *Asteroids*, Univ. of Arizona Press, Tucson, p. 1108.

Gammelgaard, P. and Kristensen, L. K.: 1983, *The ESO-Messenger* **32**, 29.

Schober, H. J.: 1982, *Astron. Astrophys. Suppl.* **48**, 57.

Schober, H. J.: 1983, *Astron. Astrophys.* (in press).

Schober, H. J., Surdej, J., Harris, A. W., and Young, J. W.: 1982, *Astron. Astrophys.* **115**, 257.

Van Flandern, T. C., Tedesco, E. F., and Binzel, R. P.: 1979, in T. Gehrels (ed.), *Asteroids*, Univ. of Arizona Press, Tucson, p. 443.

ASTRONOMY IN JAPAN*

SATIO HAYAKAWA

Department of Astrophysics, Nagoya University, Nagoya, Japan

and

MAMORU SAITO

Department of Astronomy, Kyoto University, Kyoto, Japan

(Received 5 August, 1983)

Abstract. The research activity in Japanese astronomy is described, taking into account the social and historical background. A trend in the last two decades is shown by the numbers of papers in 13 branches of astronomy. Major research facilities and international collaboration programs are summerized. Future programs under consideration are briefly discussed.

1. Introduction

The history of astronomy in Japan is relatively young compared with that in other Asian countries which pioneered in the advancement of world civilization. Although some astronomical records were found in diaries of intellectuals, few professional astronomers existed before Japan was exposed to modern civilization in 1870's. Even in 1950's there were only three universities which had departments of astronomy and accepted only a little more than ten students each year.

In contrast to a small community of astronomical science, astronomy is very popular among Japanese people. A number of planetaria in many big cities are crowded by visitors, particularly by school children, many articles of astronomy appear in popular scientific journals, and astronomy is one of the most popular TV programs. Among about 2000 members of the Astronomical Society of Japan nearly $\frac{3}{4}$ are amateurs who often contribute to astronomy by the discoveries of comets and novae, as frequently reported in the IAU Circular.

In comparison with the great enthusiasm in astronomy, the progress in astronomy as science is not so sound as it ought to be. In schools astronomy is taught as minor part of earth science. As a consequence, most teachers are not well educated in astronomy, and schools absorb few astronomy graduates as teachers. Funding for astronomical research has been poor in comparison with that for other branches of science. Hence, the number of telescopes available for astronomical research is extremely short compared with the number of astronomers, although Japan is producing many good telescopes for amateurs.

---

* Paper presented at the Lembang-Bamberg IAU Colloquium No. 80 on 'Double Stars: Physical Properties and Generic Relations', held at Bandung, Indonesia, 3–7 June, 1983.

Astrophysics and Space Science **99** (1984) 393–402. 0004–640X/84/0992–0393$01.50.
© 1984 *by D. Reidel Publishing Company.*

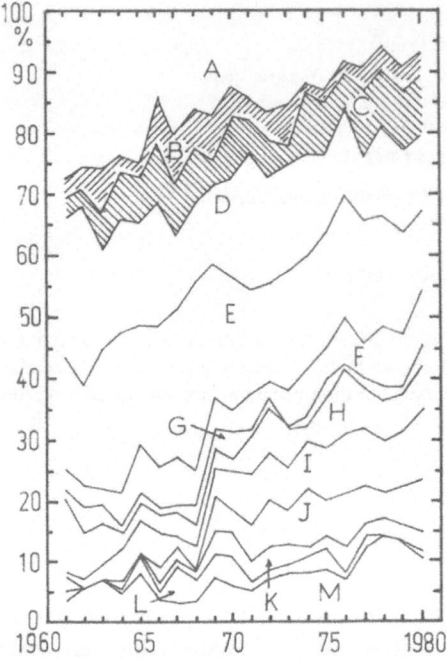

Fig. 1. Relative numbers of papers in 13 branches presented at Astronomical Society meetings. Branches to which the papers belong are indicated by alphabets as explained in Table I.

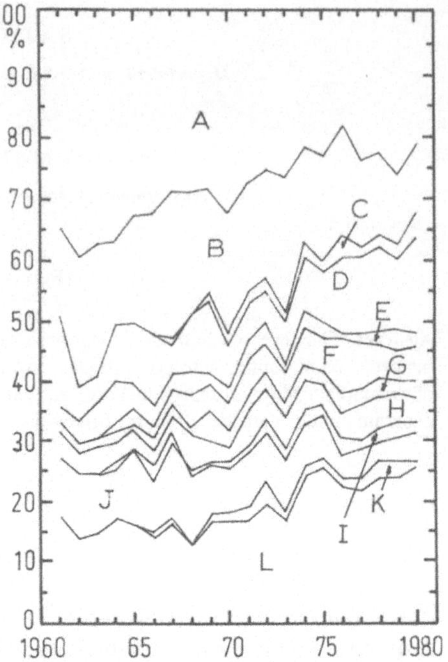

Fig. 2. Relative numbers of speakers of various institutions at Astronomical Society meetings. Affiliated institutions are indicated by the following alphabets. A: TAO, B: University of Tokyo excluding TAO, C: ISAS, D: Kyoto University excluding HAO (KAO), E: HAO (KAO), F: Nagoya University excluding RIA, G: RIA, H: Tohoku University, I: Hokkaido University, J: ILO, K: Radio Research Laboratory, and L: others.

TABLE I

Trend of the papers in 13 branches[a] of astronomy

Fractional number	Real number		
	Increased	Unchanged	Decreased
Increased	I, J, K, M		
Unchanged	C, F	G, H, L	
Decreased	D, E	B	A

[a] A: Postion astronomy, B: Celestial mechanics, C: Solar system, D: Solar physics, E: Stellar physics, F: Nebulae and interstellar matter, G: Stellar systems, H: The Galaxy, I: Galaxies, J: High energy astronomy, K: Cosmology, L: Physical processes and theory, M: Instrumentation and data processing.

Recently, however, the research in astronomy has developed with a wide step and will, we hope, develop further, if the future programs under planing come into being. We will briefly describe a change in our astronomical community in the last quarter century and present our future prospect.

2. Recent Trend in Astronomy

We begin with introducing an interesting analysis of papers presented at bi-annual meetings of the Astronomical Society of Japan (Tsubaki and Nakano, 1982). A total of 4338 papers presented in 1961–1980 are classified into 13 branches. The number of papers per year has increased as

$$N = 9.88(t - 1960) + 113.2 ,$$

according to the least square fitting. The percentages of papers in 13 branches have changed as shown in Figure 1. The result is qualitatively summarized in Table I.

Before 1961 position astronomy and solar physics are most popular branches. In Tokyo Astronomical Observatory, to which about 40% of speakers are affiliated, the largest majority of astronomers worked in position astronomy and the next in solar physics. Radioastronomy which started in 1950's at Tokyo Astronomical Observational (TAO), University of Tokyo and at Research Institute of Atmospherics (RIA), Nagoya University had been dedicated to solar radio astronomy, until a 6 m telescope for mm waves was constructed at TAO in early 1970's. In order to encourage astrophysics, 188 cm and 91 cm reflectors were constructed at Okayama Astrophysical Observatory of TAO in 1960 and have been available for nation-wide astronomers. The relative numbers of papers in Figure 1 reflect that such a change occurred in 1960's.

A jump in the trend occurred when space astronomy started in 1960's, owing to the availability of sounding rockets and balloons operated by Institute of Space and Aeronautical Science (ISAS), University of Tokyo, which is now named as Institute of Space and Astronautical Science separated from University of Tokyo. Observations of X-rays, ultraviolet radiation and infrared radiation were first performed by a group at Nagoya University in mid-1960's. This explains a sudden increase in the numbers of papers in high-energy astronomy. ISAS has launched 9 satellites, of which three have been dedicated to astronomy.

The studies of galaxies have been encouraged by the availability of a Schmidt telescope at Kiso Observatory of TAO in 1970's. Non-solar radio astronomy is rapidly expanding since Nobeyama Radio Observatory started to operate a 45 m dish in 1982. A number of astronomical facilities of high quality have been available since late 1970's. Among them we mention a solar telescope at Hida Astronomical Observatory (HAO) of Kyoto University, an expanded institution of Kasan Astronomical Observatory (KAO), and new facilities for position astronomy at International Latitude Observatory at Mizusawa (ILO).

It is important to note the role played by Research Institute of Fundamental Physics (RIFP) of Kyoto University. Director, H. Yukawa, who was once an adjoint professor of astrophysics and encouraged C. Hayashi to work in astrophysics and cosmology right after the World War II, M. Taketani and one of the authors (S.H.) planned to organize a Workshop to discuss astrophysical problems which were supposed to be closely related to fundamental physics. The first meeting in 1956 was held by active participation by J. Hitotsuyanagi, T. Hatanaka and a number of enthusiastic physicists. Workshops continued by the leadership of M. Taketani and created activities in theoretical astrophysics. Some of observational and experimental activities were also born out of a series of workshops, see Hayakawa (1981).

One of the outcomes of the Workshops was the establishment of astrophysics groups in physics departments of several universities such as Hokkaido University, Nagoya University and Kyoto University. These groups have not only conducted theoretical research but also observational and experimental works. The contributions of the physics community to astronomical research may be seen by a trend of the relative numbers of speakers at Astronomical Society meetings, as shown in Figure 2. In 1961 speakers were affiliated to 19 institutions, mostly to astronomical observatories and departments of astronomy and astrophysics, while in 1970 they were affiliated to 80 institutions. Space astronomy is dominated by physics graduates, and the leaders moved mainly from cosmic-ray research.

3. Background of Future Programs

Despite the rapid progress in recent years, there are several serious problems to overcome. Planning of future astronomy programs has to take these problems into account.

(1) Shortage of research and teaching positions. There are 10 graduate courses for astronomy and astrophysics, and about 20 are graduated therefrom each year. Only a few of them are lucky enough to get research and/or teaching jobs, about one-third get fellowships of short (1–2 yr) terms, and the rest of them (about 5% at present) support themselves for continuing research. There are at least three reasons for the accumulation of about 70 jobless astronomers of reasonable capability. First teaching jobs of the undergraduate and high-school levels are limited because of the educational system as mentioned in Section 2. Second, industry is reluctant to take astronomy graduates. Third, the government keeps policy to reduce the number of public servants which include the staff of national universities and research institutions. For example, the number of staff (including administration staff) of Nobeyama Radio Observatory is only about 20, compared with about 200 at Max-Planck Institut für Radiophysik. The shortage of positions is very serious, since many of scientists who will be responsible for future development would be lost.

(2) Shortage of optical telescopes. The nation's largest optical telescope is the 188 cm reflector at Okayama Astrophysical Observatory of TAO. The observation time is chopped into 4–7 night units, and are allocated for 60–70 groups in 45 weeks a year. The 91 cm telescope is as crowded as the 188 cm telescope, and the solar telescope is used at night for stellar spectroscopy by several groups. If the observation time allocated

is compared with about 150 clear nights, one can understand how many astronomers miss the chance of observation.

The future programs are discussed at Committee of Astronomy, Science Council of Japan which represents the voices of astronomers in various branches. The space astronomy is planned by ISAS, the national institute for space science, which has a solid program in 1980's. Other programs are not yet matured, so that only our hope in the future can be described.

4. Space Astronomy

Space science in Japan started with the participation in IGY by launching sounding rockets for geophysical research. Balloon launching developed by cosmic-ray physicists in 1950's was also adopted as a part of activities of ISAS in early 1960's. Satellite launching began in 1970's and now is the central activity of ISAS. Astronomical satellites which were launched and will be launched are listed in Table II.

TABLE II

List of astronomical satellites

Year of launch	Name	Objectives
Feb. 1979	Hakucho	Galactic X-rays
Feb. 1981	Hinotori	Solar X-rays and γ-rays
Feb. 1983	Tenma	Galactic and extragalactic X-rays
1987	ASTRO-C	Galactic and extragalactic X-rays, γ-ray bursts

X-ray astronomy is strongly emphasized because of the following reasons. First, this is an important field to be pushed forward. Second, our space astronomy began with X-ray astronomy, and there had been quite a few experienced scientists in this field. Third, Japanese space astronomers have considered themselves to be responsible for the observation of X-ray sources during the shortage of other X-ray astronomy satellites. The third point has been found important, since Hakucho detected several X-ray burst sources which were rarely active and recorded pulsation periods of several X-ray pulsars, for which both periods and the rates of period change varied.

Several satellites are planned to be launched in 1990's, though not approved yet. They include satellites for ultraviolet observation (UVSAT), for high-energy solar physics (HESP) at the next solar maximum, for cosmic X-rays and γ-rays (CXGT) and so forth. Their details will have to be defined. A shuttle infrared telescope (IRTS) was planned but has been given up because of funding difficulty. However, infrared astronomers are looking for a chance of having an orbiting telescope of much lower cost.

Because of the busy satellite schedule, neither balloons nor rockets have been used by X-ray astronomers in recent years, although they were active users in 1970's. Instead, infrared astronomers are actively using balloons and rockets, and attitude controlled rockets are mostly used for ultraviolet observations.

It should be emphasized that the international collaboration has been important for rockets and balloons. Rocket launching at Thumba, India, in collaboration with Indian X-ray astronomers, was conducted while the Japanese rocket range was closed in 1967 and 1968. A series of soft X-ray observations were performed by rockets launched from Hawaii in collaboration with a Dutch group. Another collaboration with Indian X-ray astronomers were successfully carried out with balloons launched from Hyderabad in 1969–1975. The X-ray structure of the Crab nebula was observed jointly with the UCSD group by balloons launched from Texas. Several balloons for near-infrared observations were launched from Mildura and Alice Springs, Australia in collaboration with Melbourne astronomers. Since the recovery of balloons is not easy in Japan, balloong flights in foreign countries are beneficial. A balloon infrared telescope (BIRT) is looking for launch sites of easy recovery.

Space astronomy stimulates ground-based astronomy in several ways. The first successful achievement was the optical identification of Sco X-1, the brightest galactic X-ray source, with a variable star of about 13 magnitude in June 1966 at Okayama Astrophysical Observatory. Simultaneous optical and X-ray observations were attempted. Having noticed a small probability of success, ISAS constructed a 40 cm telescope in the rocket range for this specific purpose, and this yielded a continuous optical record of an X-ray/optical nova A0620–00 in 1975. Successful simultaneous observations of Sco X-1 were performed by a rocket and a telescope at Okayama in 1971 and by balloons launched from Hyderabad and a telescope at Nizamiah Observatory, Hyderabad in 1971 to support the Comptonization model. Simultaneous observations of X-ray bursts from XB 1636–536 by Hakucho and South American telescopes in 1979 and 1980 resulted in the understanding of properties of this low-mass binary system. During a high-activity period of the Rapid Burster in August 1979, the X-ray source was observed in infrared ranges in Hawaii, but no simultaneous infrared burst was detected, although prominent infrared bursts from the direction of the globular cluster containing the Rapid Burster were discovered in March 1979 by Indian astronomers.

In the course of near-infrared survey of the Galaxy, a hump was found at $l = 355°$, $b = -1°$ by the Kyoto and Nagoya groups. This region was surveyed at 2.2 μm at Bosscha Observatory and was found to be populated by red supergiants. In order to correlate the near-infrared surface brightness distribution of the Galaxy observed by balloons with the distribution of stars and H II regions, systematic surveys from the ground have been made at observatories in Indonesia, Hawaii, and Australia.

Intensive effort of coordination has been successfully carried out in connection with the Hinotori observation of solar X-rays. Several X-ray and γ-ray detectors on board Hinotori show various faces of solar flares. Ground-based facilities have devoted themselves to simultaneous observations of flares in the optical band at HAO, Mitaka of TAO, and Norikura of TAO as well as in the radio band at Toyokawa, Nobeyama, and Nagoya.

Hinotori stopped working in the middle of 1982 because of malfunctioning of the tape recorder control. Hakucho is still active but is not fully operated because of man power shortage, since X-ray astronomers who are also responsible for attitude control, data acquisition and data analysis are busy in the operation of Tenma.

5. Radio Astronomy

Solar radio emission is observed by radiometers and interferometers over a wide-frequency range. Facilities active at present are listed in Table III.

TABLE III

Solar radio telescopes

Location/Admin.	Frequency (GHz)	Description
Hiraiso/RRL[a]	0.1, 0.2, 0.6	cir. pol. radiometers
Toyokawa/RIA	1, 2, 3.75, 9.4	cir. pol. radiometers
	3.75, 9.4	EW and 2d interferometers
Nobeyama/TAO	17, 35[b], 80[b]	spectrometer, radiometers
	0.16, 17	EW + NS and EW grating interfero-meters
Nagoya/Nagoya Univ.	35	EW interferometer

[a] Radio Research Lab.
[b] Under construction.

Most of the telescopes are rather old and will have to be renewed in coming years. The construction of a solar imaging facility with resolution of 10″ is under investigation.

For the solar system RIA operates three dipole arrays at Toyokawa, Fujigane, and Sugadaira to observe the solar wind structure by quasar scintillation. Tohuku University operates antennae at Zaō near Sendai to observe decametric emission from Jupiter.

Non-solar radio astronomy is gradually coming up. A 45 m dish at Nobeyama is now available for all users. An interferometer which consists of the 45 m dish and five 10 m dishes is almost ready for operation. The 45 m dish is the world largest radio telescope in the mm range and has the spatial resolution of about $1500''/f$ (GHz). Effort is being made to improve the system temperatures and the pointing accuracy. In early observations a great number of molecular lines have been observed, and line profiles have been obtained with a resolution of 0.3 km s^{-1}.

TABLE IV

Non-solar radio telescopes

Location/Admin.	Diameter	Description
Kisarazu/KTC[a]	1.5 m	115 GHz CO survey
Nagoya/Nagoya Univ.	1.5 m	115 GHz CO survey
	4 m	80–120, 200–300 GHz
Mitaka/TAO	6 m	90, 115 GHz. Finding telescope for 45 m dish
Nobeyama/TAO	45 m	1.4, 5, 10, 22, 40, 120 GHz
	5 × 10 m	Interferometry

[a] Kisarazu Tech. College.

Complementary to large telescopes, small telescopes are useful for mapping extended sky regions. There are several small telescopes working in the mm range, as listed in Table IV. The most recent one is the 4 m dish at Nagoya University, which started operation a few months ago.

Much effort is being made to keep NRO facilities running. It will take some more time until the whole facilities are completely ready for visitors of short stay at NRO. As in the case of satellite observations, scientists have to run around between the control panel and the computer, and rush between the telescope and the main building across a snowy field in winter.

6. Optical and Infrared Astronomy

Major telescopes available for solar, stellar, and galactic astronomy are listed in Table V. Among them three telescopes at Okayama and the Schmidt telescope at Kiso are open for all astronomers in Japan. Others may also be available for visitors of good will.

The 188 cm reflector was the largest in Asia when this was constructed in 1960 but is one of tens of medium-size telescopes at present; the world record is maintained only in its crowdedness. It is therefore natural to plan the construction of new, larger telescopes. This is the central issue of the future astronomy programs in Japan. However, the program is not yet materialized despite hot discussions in recent years. There are three ways of the future plan.

(1) A reflector with diameter larger than 3 m inside the country is favored on account of advantages of easy administration and access. Because of poor seeing attainable, stellar astronomy will be emphasized by some sacrifice of the observation of galaxies and infrared astronomy.

(2) A reflector with a little smaller size, say 2 m in diameter, constructed at a foreign site of good seeing is pushed forward by those who are interested in galaxies and infrared observations. However, a high barrier is expected because of the isolationistic attitude of Japanese government and difficult access.

TABLE V

Optical and infrared telescopes

Telescope	Location	Administration	Description
65 cm Coudé	Okayama	TAO	mainly solar
25 cm coronagraph	Norikura	TAO	
60 cm Coudé	Hida	HAO	solar
65 cm refractor	Hida	HAO	planetary
188 cm reflector	Okayama	TAO	
91 cm reflector	Okayama	TAO	
91 cm reflector	Dodaira	TAO	
50 cm Schmidt	Dodaira	TAO	comets
40 cm Schmidt	Ouda	Kyoto Univ.	
105 cm Schmidt	Kiso	TAO	
100 cm reflector	Agematsu	Kyoto Univ.	IR

(3) The participation in an international program NTT is anticipated in the near future, and technological studies for this purpose shall be encouraged.

These ideas are not mutually conflicting, but they cannot go on simultaneously because of man power and funding problems. It will not be too difficult to coordinate these ideas and to find out a pragmatic way.

7. Theory and Numerical Simulation

Theoretical activities have received a high reputation in the fields of celestial mechanics, stellar structure, and high-energy astrophysics since early days. These activities have recently been developed further by the use of fast computers. The numerical simulation has yielded interesting results in the stellar evolution in the very early and advanced stages, supernova explosions, gravitational collapses with relativistic effects, star formation, the formation of galactic arms and so forth. Computers at Mitaka, Nobeyama, ISAS, and Institute of Plasma Physics, Nagoya University as well as at respective universities are available also for visitors and are intensively used for numerical simulation.

The computer disease which is apt to spread out does not seem to become serious yet, owing to the clever guidance of theoretical astrophysicists who understand physics taking part in simulated phenomena.

It may be worth pointing out that theoretical astrophysicsts play active roles to make bridges to other fields of science. A group of solar system study involves geophysics, geology, and cosmochemistry. Close relations are maintained with plasma physics. Cosmology and particle physics are developed by mutual stimulation. Solar physics has had a long history of collaboration with solar-terrestrial physics. Celestial mechanis is an indispensable part of geodesy and planetology. Such active cooperation with neighboring fields by theory and observation increases the appreciation of astronomy and astrophysics in the whole community of science.

8. International Collaboration

Japanese science has faced a high barrier against international collaboration. This is rooted on our history that the nation was closed for about 300 years, our geography that our land is located at an edge of the world separated by the sea, and our language which is so different from other languages that most Japanese are poor speakers of foreign languages. Effort has been made in recent years to overcome these barriers.

In astronomy the international collaboration is indispensable, as already mentioned in connection with the site of a future telescope and simultaneous observations from the ground and space. One of the most successful international programs may be the collaboration between Indonesia and Japan. The collaboration consists not only of a temporal program for the present solar eclipse but also of a long-term program in 1979–1984 supported by the Directorate General for Higher Education, Indonesia and the Japan Society for Promotion of Science. The details of the long-term program shall be described by Prof. Hidayat in this session. One of the authors (M.S.) is here under this collaboration program.

References

Hayakawa, S.: 1981, *Prog. Theor. Phys. Suppl.* **70**, 1.
Tsubaki, T. and Nakano, N.: 1982, *Bull. Astron. Soc. Japan* **75**, 259.

GIANT EQUATORIAL RADIO TELESCOPE*

G. SWARUP, B. HIDAYAT**, and S. SUKUMAR

Tata Institute For Fundamental Research, Bombay, India

(Received 2 November, 1983)

Abstract. A giant radio telescope for observing galactic and extragalactic radio sources at metre wavelengths is proposed. By locating a parabolic cylindrical antenna at a site close to the Equator such that its axis lies parallel to earth's axis, it is possible to construct a large collecting area economically. The proposed instrument will be very powerful for studying compact and diffuse features of radio sources, monitoring their variability, recombination and deuterium line work, studies of interplanetary medium and pulsar search.

1. Introduction

High-resolution studies of radio sources at meter and longer wavelengths of the electromagnetic spectrum provide vital information on the energetics of the most powerful radio emitters. Many existing radio telescopes provide high frequency (i.e., cm wavelengths and above) data on these sources. The need for a sensitive and high resolution radio telescope operating at meter wavelengths has long been felt which could throw more light thus covering a wide range in the radio frequency.

The nature of radio galaxies and quasars are still intriguing to understand their energy mechanism. The high resolution studies on radio galaxies indicate very intense emitting regions in them which may play an important role in the evolution of radio sources. To study these intense hot spots we need resolutions of 0.1 to 1 arc sec. These hot spots reside in broader lobes of a few arcsecs in angular size. One of the ways of achieving high resolution in radio astronomy is to use the method of lunar occultation. The Giant Equatorial Radio Telescope efficiently exploits the essence of this method to achieve high resolution.

We here present the salient features of the Giant Equatorial Radio Telescope which is proposed to be constructed as a colloborative effort among developing countries.

2. Design Considerations of the Giant Equatorial Radio Telescope

In a year the Moon covers about 1% of the sky and by restoring the diffraction pattern which is observed when a radio source gets occulted by the Moon's limb, a strip scan across the source is obtained. The resolution obtained depends upon the signal to noise

* Paper presented at the Lembang–Bamberg IAU Colloquium No. 80 on 'Double Stars: Physical Properties and Generic Relations', held at Bandung, Indonesia, 3–7 June, 1983.
** Bosscha Observatory, Institute of Technology, Bandung, Indonesia.

ratio of the observations which is a function of the size of the antenna and receiver bandwidth. Within a few years, occultation observations of several hundred extragalactic radio sources with resolution of about 0.1 to 0.5 arc sec at meter wavelengths could be achieved. The radio telescope should have an effective area of about 60 000 m^2 and be steerable $\pm 30°$ in declination and about ± 6 hr in hour angle to track the Moon continuously every day. For a successful occultation observation the antenna beamwidth be made larger in the east–west direction and much smaller in the north–south direction. Apart from these, the antenna should have phase-switched outputs for minimising the changing baseline due to the Moon's brightness temperature.

So the telescope will be a long parabolic cylinder oriented in the north–south direction with the axis of the telescope made parallel to the axis of the Earth. This will accomplish an equatorial mount and a source can be followed for ± 6 hr by a simple mechanical rotation. The cost of such a telescope will be much lower than a parabolic dish of equivalent collecting area.

The antenna part of the proposed GERT consists of a 2 km long and 50 m wide parabolic cylinder. Since the north–south size of GERT is 2 km, it should preferably be located at the equator or very close to it, so that the slope required to facilitate equatorial mounting is reasonably small.

The reflecting surface of the parabolic cylinder will be a mesh of about 3 cm × 3 cm size made out of stainless steel wires of about 0.4 mm in diameter. The mesh will be supported by a grid of parallel stainless steel wires or ropes each 2 km long and placed every 1 or 1.5 m apart along the curved surface of the cylinder. These ropes will be supported by about 86 equidistant parabolic frames about 24 m apart.

Each wire will be given a suitable tension so that the maximum sag of the mesh for zero wind speed is only about 30 mm thus allowing operation up to about 50 cm wavelength. The bolted galvanised structural steel will have a total tonnage of about 2500 tonnes. The 86 parabolic frames would be connected together to a 2-km long drive shaft through suitable gear reduction boxes driven by about 10 electric motors of about 25 HP each for slewing and by one servo motor of about 10 HP for tracking.

The telescope will be operating at two frequencies, 325 MHz and 38 MHz. A rotatable feed with orthogonal dipole arrays with two independent receiver systems will facilitate observations at these frequencies including polarization. At 325 MHz there will be about 4000 dipoles along the focal line and the beam steering in declination will be done electronically by means of a 4 bit-diode phase shifter. The diode phase shifters will provide very rapid steerability in the range of $\pm 45°$ in declination.

The electronics systems for GERT will use the latest technology in VHF, UHF communication engineering. It will fully utilise the facilities provided by the large collecting area of GERT. At 325 MHz, the 4000 dipoles along the focal line will be divided into 84 blocks of 48 dipoles in each block. Each block will be followed by low noise RF amplifier with a bandwidth of about 12 MHz, a mixer and an IF amplifier at 38 MHz. The system temperature will be about 175 K at 325 MHz. But this could be reduced to 140 K if each of the 4000 dipoles is followed by a low noise RF amplifier and a phase shifter.

The IF signals from every 2 adjacent blocks will be combined and brought to a central receiver room, further amplified and equalised in amplitude and phase across the band. A multichannel digital delay line system and a digital correlator (1722 channels) system will provide outputs of all possible interferometer pairs made out of 42 sections of GERT. A computer processing of these data will produce about 80 independent beams at 325 MHz displaced by about 0.45 arc min apart in the north–south direction. Certain interferometer pairs will be suppressed by suitable weighting during computer processing to minimize the effect of the Moon's drift.

The bandwidth of the 80 independent beams will be restricted to about 7 MHz. In addition, the 42 outputs of GERT, each of 12 MHz bandwidth will be combined in an analogue manner to provide a few phase-switched beams with a multichannel receiver, each channel with a bandwidth of about 50 kHz, and a similar system at 38 MHz too. A de-dispersion receiver along with the above mentioned multiband receivers will be built exclusively for pulsar work. The proposed receiver will be easily adaptable to a line receiver system or a synthesis interferometer. The development effort in modern electronics will provide excellent experience and training for a large number of young engineers and scientists from the participating countries.

3. The Synthesis Radio Telescope around GERT

The second phase of development involves construction of a synthesis radio telescope around GERT which will be spread over an area of 14 × 12 km. Fouteen low-cost antennas of size 50 × 15 m will form the array. Such an array will provide a resolution of about 10 × 10 arc sec for sources of declination beyond 30°. The array configuration will be optimised in such a way we get a good resolution of 10 × 20 arc sec even for zero degree declination sources.

The spatial frequency coverage of a synthesis instrument is an indication of the quality of instrument. As we see from the figures the percentage of UV cells covered over a 12 hr observation is quite high. The rms sensitivity of the synthesis telescope will be about 50μ Jansky (1 Jy $= 10^{-26}$ W m^{-2} sr^{-1}) which means the array is one of the most sensitive one at meter wavelengths and comparable to its counterparts operating at cm wavelengths. However the limitation to the dynamic range of the instrument may be due to the rms phase and amplitude errors caused by the varying ionosphere. But the recent results on closure phase method and self calibration procedures give encouraging indications that ionospheric problems could be alleviated and the attainable dynamic range will be greater than 25 dB. Another way of reducing the ionospheric problems is to have frequent calibration. This will be feasible because of the fast declination suitching that can be accomplished by the diode phase shifters.

4. Scientific Programs with GERT

As mentioned earlier, the lunar occultation observations on several hundred weak radio sources with less than arc sec resolution will provide enormous amount of data with

which we can study in detail;
  (a) the energetics of radio galaxies,
  (b) the evolution of radio galaxies,
  (c) the evolutionary trend of hot spots.

Apart from extragalactic radio sources, lunar occultation observations can give fine-structure studies on galactic radio sources and also of nearby galaxies.

The lunar occultation strip scans will be supplemented by the full synthesis maps from the synthesis radio telescope around GERT.

The interferrometer system will be extremely useful to map extended sources with low surface brightness such as halo around spiral galaxies etc.

5. The Fast Survey

The versatality of GERT includes its great ability to survey 40 000 radio sources within 24 hr! Such a fast survey when periodically performed will give a wealth of data on variable radio sources and may lead to discoveries of novae and supernovae in distant parts of our Galaxy.

6. Pulsars

The mechanism of pulsars is not yet clearly understood. Their distribution in the Galaxy is not uniquely established. The high sensitivity of GERT will enable to add atleast 400 new pulsars to the existing list of about 300 pulsars. And the time required to do the pulsar search requires approximately 1000 hr of telescope time. It may be possible to make such a search fully ON-LINE by using an array processor along with a powerful minicomputer.

7. Interplanetary Scintillations

The Interplanetary scintillation observations will provide valuable data on the inter-planetary medium especially for regions which are close to the Sun or at high ecliptic latitudes and therefore not readily accesible to satellite probes. The high sensitivity and rapid steerability of GERT will allow monitoring of several hundred radio sources around Sun on a daily basis. The information will throw much light about the structure and characteristics of the solar wind. Also detection of travelling disturbances in the interplanetary medium is possible which provides early warning systems for geo-magnetic disturbances.

The high sensitivity of GERT is ideally suited to study radio emission from several classes of stars and recombination lines at low frequencies from cold dense clouds. Another important line-work at meter wavelengths is the detection of the hyperfine transition of deuterium at 327.4 MHz. Besides, occultation of weak radio sources by planets could also be studied simultaneously at 327 and 38 MHz which would enable us to study their ionosphere and magnetosphere. Studies on cometary occultations are also viable programs with GERT.

Above all, the very large baseline interferometry at meter wavelengths with GERT and other antennas located in AFRO–ASIAN countries will enable us to study very compact components in radio sources with milli arc second resolution and extreme sensitivity. With the improvement in the closure phase techniques and self calibration procedure a high quality mapping using VLBI network is quite feasible.

8. Conclusions

GERT is proposed as a collaborative effort among developing countries and construction of such a telescope should train many engineers and scientists in the sophisticated art of constructing antenna systems and extremely sensitive electronic hardware. GERT has flexibility in its design in order to obtain optimum performance. The initial feasibility study indicated the project can be completed within five years with the material and technology available in developing countries. Preliminary surveys were conducted for a possible location of GERT in West Sumatra, Indonesia. Two sites were identified closer to equator with the site near Piobang closer to the town Bukittinggi being a promising candidate. The preliminary seismitivity study by Indonesian experts reveal that a suspected seismitivity fault running down this proposed site is not active.

A scientific committee has been formed in Indonesia for project GERT. More discussions and formal approaches are planned by the participating countries in the near future.

The IAU and UNESCO and the Governments of the Republic of Indonesia and India are extremely interested in this astronomical project by developing nations. When completed, GERT will be one of the foremost experimental facilities available to the astronomical community.

References

1. INISSE and GERT, a proposal, November 1979, by G. Swarup, T. R. Odhiambo, and S. E. Okoye.
2. Giant Equatorial Radio Telescope (GERT) in West Sumatra, Indonesia, A report on the field study in September 1982 by the UNESCO Team, October 1982.

A PLAN FOR A NEW GENERATION 2M-CLASS TELESCOPE IN INDONESIA*

KAREL A. VAN DER HUCHT

Space Research Laboratory, The Astronomical Institute at Utrecht, Holland

(Received 29 July, 1983)

Abstract. One of the prime astrophysical interests of the Observatorium Bosscha is, and has always been, double star research: visual double star research with the double-60 cm Zeiss telescope (dedicated in 1928), and theoretical research of evolved massive spectro- scopic binaries (since 1972). For one thing, this is the very reason that this IAU Colloquium No. 80, celebrating the 60th anniversary of the Observatorium Bosscha in Lembang, is devoted to binary astrophysics.

Up to now, visual, photographic, and photometric tools have been used for binary research at the Observatorium Bosscha. An important, essential additional tool for binary research is spectrographic equipment, in order to measure radial velocities of binary components.

Therefore, we suggest to make a plan for a new modern telescope, a reflector with a primary mirror of about 2 m in diameter and with a modern spectrograph/detector combination for radial velocity measurements.

At a number of major astronomical observatories scientists have been considerating to erect new telescopes devoted primarily to radial velocity measurements. The reason for this is that at the end of this decade the parallax and proper motion measurements to be made by the ESA astrometric satellite Hipparcos will become available of more than 100000 single stars and double stars. At that time there will be a compelling need for radial velocity measurements of all these stars to complement the parallax and proper motion measurements. With the combination of this data enormous progress will be made in double star research, and in the study of galactic dynamics, another topic of interest of the Observatorium Bosscha. If it could be realized to build such a dedicated radial velocity telescope in Indonesia, Indonesian astronomers could take a leading role in this field of research.

Without going into technical details, we would like to emphasize here that such a new instrument should be a true *New Generation Telescope*, and that the Institut Teknologi Bandung should participate from the very beginning in its design, construction and assembling, and the subsequent servicing; ITB could participate in the field of optics, mechanics, and electronics. Modern astronomy offers tremendous challenges to techno- logy, which are of great interest to technological institutes. The new telescope should be computer controlled, and the spectrograph should have a modern digital read-out

---

* Paper presented at the Lembang-Bamberg IAU Colloquium No. 80 on 'Double Stars: Physical Properties and Generic Relations', held at Bandung, Indonesia 3–7 June, 1983.

(Reticon, IPCS, or CCD). The telescope should have one of those recently becoming available *thin mirrors*, allowing more mechanical freedom. It could be a telescope with a siderostat which feeds the light into a fixed telescope, thus improving both the stability of the telescope and that of the spectrograph. In this way the staff and students of ITB, as well as the technical staff of the Observatorium Bosscha will be drawn into modern techniques of many varieties. And for ITB such an enterprise may even have a spin-off into other fields than astronomy.

One aspect which is of great importance for the new telescope is the selection of its site. The present site of the Observatorium Bosscha in Lembang is a good one, but for a new modern telescope one wants to make sure that it is going to be located at the most ideal site.

Therefore an Indonesian site-survey should be initiated promptly. Site survey equipment is available at many big observatories and could be borrowed. The site survey should extend over at least 4–5 years to monitor the meteorological and environmental situation at many sites.

In the meantime the design and fund rainsing can be considered. Modern day astronomy depends on financial support from governments and inter-governmental organizations. Therefore it is urged that a proposal for a new telescope as indicated above clearly describes the advantages of such a new telescope both for astrophysical research in Indonesia, and for the introduction of new technologies in Indonesian technological institutes.

The recently formed Steering Committee for Indonesian-Netherlands Astrophysics (INA) is willing to explore the possibilities for this plan. We hope that after investigating the interest of ITB in this matter, a proposal could be made before the end of this year.

ANALYSIS OF 11 LATE-TYPE CLOSE BINARY SYSTEMS*

K. C. LEUNG [a], D. S. ZHAI [a,b], R. X. ZHANG [b], Q. Y. LIU [c], J. T. ZHANG [b],
and Y. L. YANG [c]

(Received 18 July, 1983)

Abstract. This is a preliminary report on the joint research project between 3 observatories: Beijing, Yunnan, and Behlen Observatories from China and the United States. The systems we have been dealing with are primary of late spectral types and with short periods. Most of the observations were secured from the observatories in China. The computational analysis is carried out in University of Nebraska, Lincoln. The photometric solutions are based on the Wilson and Devinney method. Out of 11 systems analyzed 6 of them: AO Cam, ER Ori, BX Peg, BB Peg, U Peg, and SW Lac are found to be contact systems. All of them are having their primary eclipses at occultation. Therefore they can be classified as W-type W UMa systems. They also show other W-type characteristics.

Three systems: ZZ Aur, RZ Dra, and AX Vir are found to be semi-detached systems, with low mass components filling their Roche surfaces. The massive components are having their radii fairly close to Roche surfaces and are larger than their contact companions.

The last two systems: AT Cam and AZ Cam are found to be detached systems, but with their low mass components almost filling their Roche surfaces. This type of configuration is of great interest to our understanding of close binary evolution. We believe that these systems are at phase immediately after the normal Algol phase, where the mass losing components detached from the Roche surfaces at the conclusion active mass lost phase.

Acknowledgements

We would like to acknowledge the support of the NSF INT 8 120 404 grant, and the Academia Sinica.

---

* Paper presented at the Lembang-Bamberg IAU Colloquium No. 80 on 'Double Stars: Physical Properties and Generic Relations', held at Bandung, Indonesia, 3–7 June, 1983.
[a] Behlen Observatory, University of Nebraska, Lincoln, Nebr., U.S.A.
[b] Beijing Observatory, Academia Sinica, Beijing, P.R.C.
[c] Yunnan Observatory, Academia Sinica, Kunming, P.R.C.

ANNOUNCEMENT

Double Stars, Physical Properties and Generic Relations

Proceedings of IAU Colloquium No. 80 held at Lembang, Java, June 1983.

Editors: Bambang Hidayat, Zdeněk Kopal, Jürgen Rahe.

Please note that a hardback edition of this special double issue of *Astrophysics and Space Science*, Vol. 99, Nos. 1-2 (February 1984), is available from the publishers.
ISBN 90-277-1748-6
Price: Dfl. 230,– / $ 90.00 / £ 58.50

Astrophysics and Space Science 99 (1984) 412.

TABLE OF CONTENTS

ARTICLES

TABLE OF CONTENTS

TABLE OF CONTENTS

NEW REIDEL TITLES

Galactic Astrophysics and Gamma-Ray Astronomy

Proceedings of a Meeting organized in the Context of the XVIII IAU General Assembly, held in Patras, Greece, August 19, 1982

Edited by
G. E. MORFILL and R. BUCCHERI

1983, v + 335 pp.
Cloth Dfl. 150,–/US $ 60.00   ISBN 90–277–1645–5
Reprinted from the journal, SPACE SCIENCE REVIEWS, *Vol. 36, Nos. 1, 2, and 3*

The aim of the Meeting was to determine the role of galactic gamma-ray astronomy within the general concept of galactic astrophysics. The timing, at the end of the COS-B mission, was regarded as ideal because it gives interested astrophysicists the opportunity for interdisciplinary studies using the existing gamma-ray data base (e.g. comparison with infrared, radio, X-ray, etc. astronomies), as well as offering the possibility of theoretical studies. The next generation of gamma-ray detectors will probably not be in operation for another five to 10 years, and it is therefore hoped that the proceedings in this book can be used (in the intermediate period) as a basis for further studies, as a stimulation for more theoretical work and as an important contribution for defining the aims and operation of future gamma-ray missions. The interrelationship with other branches of astronomy, the astrophysical implications, and the study of relevant physical processes using available measurements in the near-Earth environment are among the important results discussed.

D. Reidel Publishing Company

P.O. Box 17, 3300 AA Dordrecht, the Netherlands
190 Old Derby St., Hingham, MA 02043, U.S.A.

Astrophysical Jets

Proceedings of an International Workshop held in Torino, Italy, October 7–9, 1982

Edited by
ATTILIO FERRARI
University of Torino, Italy

A. G. PACHOLCZYK
University of Arizona, Tucson, U.S.A.

1983, xvi + 327 pp.
Cloth Dfl. 110,– / US $ 48.00   ISBN 90–277–1627–7
ASTROPHYSICS AND SPACE SCIENCE LIBRARY 103

Recent high-resolution observations at various frequencies (radio, optical, X-ray) have revealed that, in many cases, active astrophysical objects, from stellar size sources to galactic nuclei, can eject supersonic (eventual relativistic) flows. In particular, these flows involve a substantial fraction of the global energetics of their sources. The study of the physical processes producing 'jets', and supporting them in their rich morphological forms over extended regions for long lifetimes, is still at an early stage and many proposals and experimental tests are being actively discussed in an attempt at reaching a complete understanding of the phenomenon. The Torino Workshop on Astrophysical Jets was organized to provide a specific opportunity for these discussions, involving observers and theoreticaians. The Workshop was well attended by about 90 scientists gathered from Europe and the United States. The important contributions collected in the Proceedings give an updated picture of the main topics of interest in the field as well as a state-of-the-art survey of the astrophysics of jets to the end of 1982.

 **D. Reidel Publishing Company**

P.O. Box 17, 3300 AA Dordrecht, the Netherlands
190 Old Derby St., Hingham, MA 02043, U.S.A.